D0207401

Essentials of Ecology

Seventh Edition

ABOUT THE COVER PHOTO

The beautiful monarch butterfly is one of North America's best-known butterfly species. Because monarchs cannot survive cold winters, they migrate south to a warmer climate during the fall to hibernate in dense colonies in certain types of trees (as in the photo, right). Monarchs that live west of the Rocky Mountains migrate to the area around Pacific Grove, California where they cluster in groves of eucalyptus trees. Monarchs that live east of the Rockies migrate as far as 4,000 kilometers (2,500 miles) to live in oyamel fir trees in Mexico.

Monarchs go through four generations in a year. Each generation has a life cycle of four stages: egg, larva, pupa, and adult. In the spring, the first generation's eggs hatch as *larvae* (or baby caterpillars) that eat their egg cases and then feed only on milkweed. After about 2 weeks, the full-grown caterpillars build a hard protective case around themselves (called a *pupa*, or *cocoon*), attached to a twig or leaf. About 10 days later, they emerge as *adult* butterflies, who feed on nectar from a variety of flowers for 2 to 6 weeks until they die.

This cycle takes place four times during the summer, but the fourth-generation monarchs live for 6 to 8 months. During that time, they make their long journey south in the fall, hibernate for much of the winter, and make the return trip in the spring to lay their eggs and set the stage for the next annual cycle. Curiously, the fourth-generation monarchs somehow return to the same areas where their first-generation ancestors began their lives.

Because monarch caterpillars feed only on milkweed, the butterflies have a compound in their tissue that makes them poisonous or foul-tasting to predators such as birds, mice, frogs, and lizards. Some of these predators learn to avoid monarchs by recognizing their bright colors.

Monarch butterfly populations in North America have declined sharply as more of the trees they depend on in their winter habitats are being cleared each year. In the north, many of the milkweed plants that caterpillars feed on are also being cleared away as more land is developed. In addition, climate change will likely disrupt the monarch's annual migration pattern by altering the long-term weather conditions in their summer and winter habitats. Efforts are underway to classify the monarch as a protected species.

JOEL SARTORE/National Geographic Creative

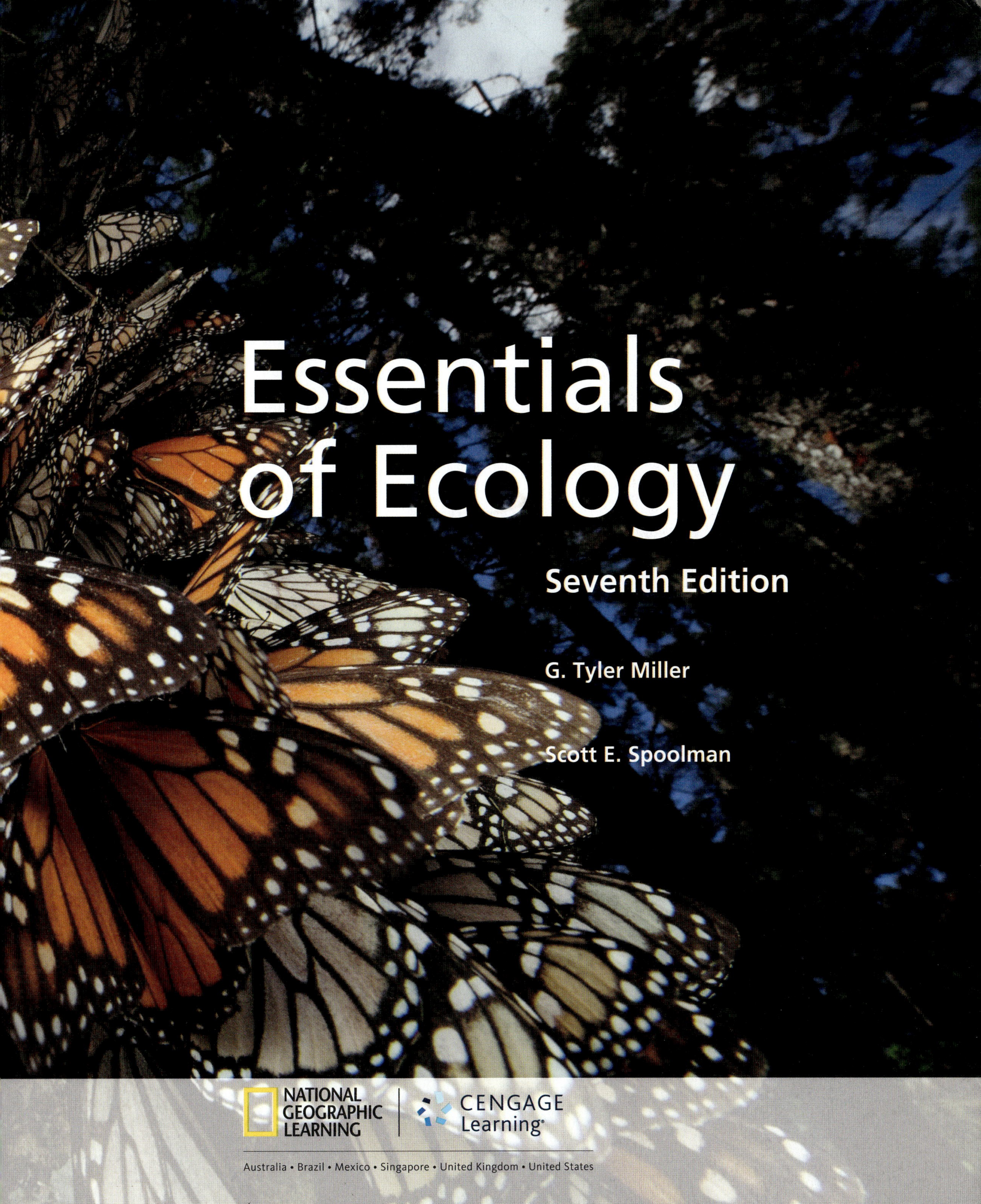

Essentials of Ecology

Seventh Edition

G. Tyler Miller

Scott E. Spoolman

NATIONAL GEOGRAPHIC LEARNING | CENGAGE Learning

Australia • Brazil • Mexico • Singapore • United Kingdom • United States

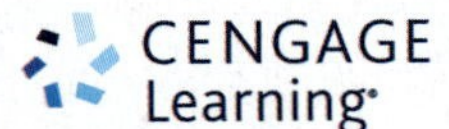

***Essentials of Ecology*, Seventh Edition**
G. Tyler Miller, Scott E. Spoolman

Senior Product Team Manager: Yolanda Cossio
Content Developer: Jake Warde
Content Coordinator: Kellie Petruzzelli
Product Development Manager: Alexandria Brady
Executive Brand Manager: Nicole Hamm
Senior Market Development Manager: Tom Ziolkowski
Content Project Manager: Harold P. Humphrey
Senior Art Director: Pamela Galbreath
Manufacturing Planner: Karen Hunt
Senior Rights Acquisitions Specialist: Dean Dauphinais
Production Service: Dan Fitzgerald, Graphic World Inc.
Photo Researcher: Christina Ciaramella, PreMedia Global
Text Researcher: Melissa Tomaselli, PreMedia Global
Copy Editor: Graphic World Inc.
Illustrator: Patrick Lane, ScEYEnce Studios
Text and Cover Designer: Jeanne Calabrese
Cover and Title Page Image: JOEL SARTORE/ National Geographic Creative
Compositor: Graphic World Inc.

ISBN-13: 978-1-285-19726-5

ISBN-10: 1-285-19726-7

Cengage Learning
200 First Stamford Place, 4th Floor
Stamford, CT 06902
USA

Cengage Learning is a leading provider of customized learning solutions with office locations around the globe, including Singapore, the United Kingdom, Australia, Mexico, Brazil, and Japan. Locate your local office at **www.cengage.com/global**.

Cengage Learning products are represented in Canada by Nelson Education, Ltd.

To learn more about Cengage Learning Solutions, visit **www.cengage.com**.

Purchase any of our products at your local college store or at our preferred online store **www.cengagebrain.com**.

Printed in Canada
1 2 3 4 5 6 7 17 16 15 14 13

BRIEF CONTENTS

CONTENTS

SUSTAINING BIODIVERSITY

PREFACE

For Instructors

With this new edition, we are happy to announce our new partnership with *The National Geographic Society,* which shares our goals, as reflected in its statement of purpose: *Inspiring people to care about the planet.* One result of this new collaboration is the addition of many stunning and informative photographs, numerous maps, and several new stories of National Geographic Explorers—people who are making a positive difference in the world. With these new tools, we continue to tell of the good news from various fields of environmental science, hoping to inspire young people to commit themselves to making our world a more sustainable place to live for their own and future generations.

What's New in This Edition?

- *Our new partnership with National Geographic* has given us access to hundreds of amazing photographs, numerous maps, and inspiring stories of *National Geographic Explorers*—people who are leading the way in environmental science, education, or entrepreneurial enterprises.
- A *stunning new design* with a National Geographic look that enhances visual learning.
- *Campus Sustainability boxes:* short descriptions about what selected U.S. colleges and universities are doing to make their institutions more sustainable.
- *Three social science principles of sustainability.* These complement the three scientific principles of sustainability that we have long used to explain how life on Earth has sustained itself for billions of years, and they act as guidelines for making a possible transition to more sustainable economies and societies.
- *New Core Case Studies* for 6 of the book's 11 chapters that serve as an integrating theme throughout each chapter. They bring important real-world stories to the forefront for use in applying those chapters' concepts and principles.
- *Two new end-of-chapter exercises: Doing Environmental Science* and *Global Environment Watch* research projects give students challenging new ways to apply the material.

Sustainability Is the Integrating Theme of This Book

Sustainability, a watchword of the 21st century for those concerned about the environment, is the overarching theme of this textbook. You can see the sustainability emphasis by looking at the Brief Contents (p. vii).

Six principles of sustainability play a major role in carrying out this book's sustainability theme. These principles are introduced in Chapter 1. They are depicted in Figure 1-2 (p. 6), in Figure 1-5 (p. 9), and on the back cover of the student edition and are used throughout the book, with each reference marked in the margin by the symbol shown here at the right (see pp. 62 and 218).

SUSTAINABILITY

We use the following five major subthemes to integrate material throughout this book.

- ***Natural Capital.*** Sustainability depends on the natural resources and ecosystem services that support all life and economies. See Figures 1-3, p. 7, and 10-4, p. 220.
- ***Natural Capital Degradation.*** We describe how human activities can degrade natural capital. See Figures 1-7, p. 11, and 7-17, p. 160.
- ***Solutions.*** We present existing and proposed solutions to environmental problems in a balanced manner and challenge students to use critical thinking to evaluate them.
- ***Trade-offs.*** The search for solutions involves trade-offs, because any solution requires weighing advantages against disadvantages. Our Trade-offs diagrams located in several chapters present the benefits and drawbacks of various environmental technologies and solutions to environmental problems.
- ***Individuals Matter.*** Throughout the book, Individuals Matter boxes and some of the Case Studies describe what various scientists and concerned citizens (including several National Geographic Explorers) have done to help us work toward sustainability (see pp. 82 and 240). Also, a number of What Can You Do? diagrams describe how readers can deal with the problems we face (see Figures 9-12, p. 202, and 10-30, p. 242).

Other Key Features of This Textbook

- ***Up-to-Date Coverage.*** Our textbooks have been widely praised for keeping users up to date in the rapidly changing field of environmental science. We have used thousands of articles and reports published in 2010–2013 to update the information and concepts in this book. Major new or updated topics include planetary boundaries that indicate ecological tipping points (Science Focus 3.3, p. 72) and the rising threat of ocean acidification (Science Focus 11.2, p. 252), along with other important topics.
- ***Concept-Centered Approach.*** To help students focus on the main ideas, we built each major chapter section around a key question and one or two key concepts, which state the section's most important take-away messages. In each chapter, all key questions are listed at the front of the chapter, and each chapter section

begins with its key question and concepts (see pp. 29 and 31). Also, the concept applications are highlighted and referenced throughout each chapter.

- ***Science-Based Coverage.*** Chapters 2–8 cover scientific principles important to the course and discuss how scientists work (see Brief Contents, p. vii). Important environmental science topics are explored in depth in Science Focus boxes distributed among the chapters throughout the book (see pp. 94 and 203) and integrated throughout the book in various Case Studies (see pp. 238 and 256) and in numerous figures.
- ***Global Perspective.*** This book also provides a global perspective, first on the ecological level, revealing how all the world's life is connected and sustained within the biosphere, and second, through the use of information and images from around the world. This includes dozens of maps in the basic text and in Supplement 6. Half of these maps are new and more than half of the new maps are from National Geographic. At the end of each chapter is a Global Environment Watch exercise that applies this global perspective (see p. 245).
- ***Core Case Studies.*** Each chapter opens with a Core Case Study (see pp. 78 and 190), which is applied throughout the chapter. These applications are indicated by the notation (**Core Case Study**) wherever they occur (see pp. 88 and 155). Each chapter ends with a *Tying It All Together* box (see pp. 96 and 213), which connects the Core Case Study and other material in the chapter to some or all of the principles of sustainability.
- ***Case Studies.*** In addition to the 11 Core Case Studies, many additional Case Studies (see pp. 92, 200, and 238) appear throughout the book (and are listed in the Detailed Contents, pp. viii–xvi). Each of these provides an in-depth look at specific environmental problems and their possible solutions. We also have included very brief descriptions of efforts on several college campuses to study or apply principles of sustainability in our new *Campus Sustainability* stories that appear in several of the book's chapters (see pp. 210 and 270).
- ***Critical Thinking.*** The Preface for Students (p. xxiii) describes critical thinking skills, and specific critical thinking exercises are used throughout the book in several ways:
    - As more than 100 *Thinking About* exercises that ask students to analyze material immediately after it is presented (see pp. 35 and 264).
    - In all *Science Focus* boxes.
    - In dozens of *Connections* boxes that stimulate critical thinking by exploring the often surprising connections related to environmental problems (see pp. 18 and 195).
    - In the captions of many of the book's figures (see Figures 3-15, p. 63, and 9-8, p. 198).
    - In end-of-chapter questions (see pp. 97 and 214).
- ***Visual Learning.*** With a new design heavily influenced by material from National Geographic and hundreds of photographs, this is the most visually appealing environmental science textbook available (see Figures 3-21, p. 71; 7-16, p. 159; and 10-18, p. 229). Also included are many diagrams designed to present complex ideas in understandable ways relating to the real world (see Figures 3-3, p. 54; 3-17, p. 66; and 4-2, p. 79).
- ***In-Text Study Aids.*** Each chapter begins with a list of *Key Questions* showing how the chapter is organized (see p. 217). When a new *key term* is introduced and defined, it is printed in boldface type, and all such terms are summarized in the glossary at the end of the book. More than 100 *Thinking About* exercises reinforce learning by asking students to think critically about the implications of various environmental issues and solutions immediately after they are discussed in the text (see p. 226). The captions of many figures contain similar questions that get students to think about the figure content (see Figure 10-28, p. 238). In their reading, students also encounter *Connections* boxes, which briefly describe connections between human activities and environmental consequences, environmental and social issues, and environmental issues and solutions (see p. 226). Finally, the text of each chapter wraps up with three *Big Ideas* (see p. 242), which summarize and reinforce three of the major take-away messages from each chapter, and a *Tying It All Together* section that relates the Core Case Study and other chapter content to the principles of sustainability (see p. 243). Again, this reinforces the main messages of the chapter along with the themes of sustainability to give students a stronger understanding of how it all ties together.

Each chapter ends with a *Chapter Review* section containing a detailed set of review questions that include all the chapter's key terms in bold type; *Critical Thinking* questions that encourage students to think about and apply what they have learned to their lives; *Doing Environmental Science*—an exercise that will help students to experience the work of various environmental scientists; a *Global Environment Watch* exercise taking students to Cengage's GREENR site, where they can use this tool for interesting research related to chapter content; and a *Data Analysis* or *Ecological Footprint Analysis* problem built around ecological footprint data or some other environmental data set. (See pp. 243–245.) And at the end of the book, we have included a comprehensive glossary that includes definitions of all key terms as well as many other terms that are important to environmental science.

Supplements for Instructors

- ***Environmental Science MindTap.*** MindTap is a new personal learning experience that combines all your digital assets—readings, multimedia, activities, and assessments—into a singular learning path to improve student outcomes.
- ***Instructor Companion Site.*** Everything you need for your course in one place! This collection of book-specific lecture and class tools is available online via www.cengage.com/login. Access and download PowerPoint presentations, images, instructor's manual, videos, and more.
- ***Cognero.*** Cengage Learning Testing Powered by Cognero is a flexible, online system that allows you to do the following:
  - author, edit, and manage test bank content from multiple Cengage Learning solutions
  - create multiple test versions in an instant
  - deliver tests from your LMS, your classroom, or wherever you want
- ***Transparencies.*** Online Transparency Correlation Guide. This guide correlates the transparency set created for *Living in the Environment 17e, Environmental Science 13e, Sustaining the Earth 10e,* and *Essentials of Ecology 6e* to the new editions of these texts: *Living in the Environment 18e, Environmental Science 14e, Sustaining the Earth 11e,* and *Essentials of Ecology 7e.* To acquire the set of 250 printed transparencies and 250 electronic masters, please ask your local Cengage Learning Sales Representative or call 1-800-423-0563.
- ***Aplia.*** Aplia™ is a Cengage Learning online homework system dedicated to improving learning by increasing student effort and engagement. Aplia makes it easy for instructors to assign frequent online homework assignments. Aplia provides students with prompt and detailed feedback to help them learn as they work through the questions, and features interactive tutorials to fully engage them in learning course concepts. Automatic grading and powerful assessment tools give instructors real-time reports of student progress, participation, and performance, and Aplia's easy-to-use course management features let instructors flexibly administer course announcements and materials online. With Aplia, students will show up to class fully engaged and prepared, and instructors will have more time to do what they do best. . . teach.
- ***BBC Videos for Environmental Science.*** This large library of BBC clips are informative, short clips of current news stories on environmental issues from around the world. These clips are a great way to start a lecture or spark a discussion. Available on the Instructor Companion Site and within MindTap.
- ***Global Environment Watch.*** Updated several times a day, the Global Environment Watch is a focused portal into GREENR—the Global Reference on the Environment, Energy, and Natural Resources—an ideal one-stop site for classroom discussion and research projects. This resource center keeps courses up to date with the most current news on the environment. Users get access to information from trusted academic journals, news outlets, and magazines, as well as statistics, an interactive world map, videos, primary sources, case studies, podcasts, and much more.
- ***Virtual Field Trips in Environmental Issues.*** This supplement brings the field to you, with dynamic panoramas, videos, photographs, maps, and quizzes covering important topics within environmental science. A case study approach covers the issues of keystone species, the role of climate change in extinctions, invasive species, the evolution of a species in relation to its environment, and an ecosystem approach to sustaining biodiversity. Students are engaged, interacting with real issues to help them think critically about the world around them.

Help Us Improve This Book or Its Supplements

Let us know how you think this book can be improved. If you find any errors, bias, or confusing explanations, please e-mail us about them at:

- mtg89@hotmail.com
- spoolman@tds.net

Most errors can be corrected in subsequent printings of this edition, as well as in future editions.

Acknowledgments

We wish to thank the many students and teachers who have responded so favorably to the 17 previous editions of *Living in the Environment,* the 14 editions of *Environmental Science,* the 10 editions of *Sustaining the Earth,* and the 6 editions of *Essentials of Ecology,* and who have corrected errors and offered many helpful suggestions for improvement. We are also deeply indebted to the more than 300 reviewers, who pointed out errors and suggested many important improvements in the various editions of these three books.

It takes a village to produce a textbook, and the members of the talented production team, listed on the copyright page, have made vital contributions. Our special thanks go to development editor Jake Warde, production editors Hal Humphrey and Dan Fitzgerald, designer Pam Galbreath, copy editor Chris DeVito, compositor Craig Beffa, photo researcher Christina Ciaramella, artist Patrick Lane, media developer Alexandria Brady, assistant editor Alexis Glubka, product assistant Kellie Petruzzelli, and Cengage Learning's hardworking sales staff. Finally, we are very fortunate to have the guidance, inspiration, and unfailing support of Life Sciences Senior Product Team Manager Yolanda Cossio and her dedicated team of highly talented people who have made this and other book projects such a pleasure to work on.

G. Tyler Miller

Scott E. Spoolman

Guest Essayists

Guest essays by the following authors are available online: **M. Kat Anderson**, ethnoecologist with the National Plant Center of the USDA's Natural Resource Conservation Center; **Lester R. Brown**, president, Earth Policy Institute; **Alberto Ruz Buenfil**, environmental activist, writer, and performer; **Robert D. Bullard**, professor of sociology and director of the Environmental Justice Resource Center at Clark Atlanta University; **Michael Cain**, ecologist and adjunct professor at Bowdoin College; **Herman E. Daly**, senior research scholar at the School of Public Affairs, University of Maryland; **Lois Marie Gibbs**, director, Center for Health, Environment, and Justice; **Garrett Hardin**, professor emeritus (now deceased) of human ecology, University of California, Santa Barbara; **John Harte**, professor of energy and resources, University of California, Berkeley; **Paul G. Hawken**, environmental author and business leader; **Jane Heinze-Fry**, environmental educator; **Paul F. Kamitsuja**, infectious disease expert and physician; **Amory B. Lovins**, energy policy consultant and director of research, Rocky Mountain Institute; **Bobbi S. Low**, professor of resource ecology, University of Michigan; **John J. Magnuson**, Director Emeritus of the Center for Limnology, University of Wisconsin, Madison; **Lester W. Milbrath**, director of the research program in environment and society, State University of New York, Buffalo; **Peter Montague**, director, Environmental Research Foundation; **Norman Myers**, tropical ecologist and consultant in environment and development; **David W. Orr**, professor of environmental studies, Oberlin College; **Noel Perrin**, adjunct professor of environmental studies, Dartmouth College; **David Pimentel**, professor of insect ecology and agricultural sciences, Cornell University; **John Pichtel**, Ball State University; **Andrew C. Revkin**, environmental author and environmental reporter for the New York Times; **Vandana Shiva**, physicist, educator, environmental consultant; **Nancy Wicks**, ecopioneer and director of Round Mountain Organics; and **Donald Worster**, environmental historian and professor of American history, University of Kansas.

Dr. Dean Goodwin and his colleagues Berry Cobb, Deborah Stevens, Jeannette Adkins, Jim Lehner, Judy Treharne, Lonnie Miller, and Tom Mowbray provided excellent contributions to the Data Analysis and Ecological Footprint Analysis exercises. Mary Jo Burchart of Oakland Community College wrote the in-text Global Environment Watch exercises.

Cumulative List of Reviewers

Barbara J. Abraham, Hampton College; Donald D. Adams, State University of New York at Plattsburgh; Larry G. Allen, California State University, Northridge; Susan Allen-Gil, Ithaca College; James R. Anderson, U.S. Geological Survey; Mark W. Anderson, University of Maine; Kenneth B. Armitage, University of Kansas; Samuel Arthur, Bowling Green State University; Gary J. Atchison, Iowa State University; Thomas W. H. Backman, Lewis-Clark State College; Marvin W. Baker, Jr., University of Oklahoma; Virgil R. Baker, Arizona State University; Stephen W. Banks, Louisiana State University in Shreveport; Ian G. Barbour, Carleton College; Albert J. Beck, California State University, Chico; Eugene C. Beckham, Northwood University; Diane B. Beechinor, Northeast Lakeview College; W. Behan, Northern Arizona University; David Belt, Johnson County Community College; Keith L. Bildstein, Winthrop College; Andrea Bixler, Clarke College; Jeff Bland, University of Puget Sound; Roger G. Bland, Central Michigan University; Grady Blount II, Texas A&M University, Corpus Christi; Lisa K. Bonneau, University of Missouri–Kansas City; Georg Borgstrom, Michigan State University; Arthur C. Borror, University of New Hampshire; John H. Bounds, Sam Houston State University; Leon F. Bouvier, Population Reference Bureau; Daniel J. Bovin, Université Laval; Jan Boyle, University of Great Falls; James A. Brenneman, University of Evansville; Michael F. Brewer, Resources for the Future, Inc.; Mark M. Brinson, East Carolina University; Dale Brown, University of Hartford; Patrick E. Brunelle, Contra Costa College; Terrence J. Burgess, Saddleback College North; David Byman, Pennsylvania State University, Worthington–Scranton; Michael L. Cain, Bowdoin College; Lynton K. Caldwell, Indiana University; Faith Thompson Campbell, Natural Resources Defense Council, Inc.; John S. Campbell, Northwest College; Ray Canterbery, Florida State University; Ted J. Case, University of San Diego; Ann Causey, Auburn University; Richard A. Cellarius, Evergreen State University; William U. Chandler, Worldwatch Institute; F. Christman, University of North Carolina, Chapel Hill; Lu Anne Clark, Lansing Community College; Preston Cloud, University of California, Santa Barbara; Bernard C. Cohen, University of Pittsburgh; Richard A. Cooley, University of California, Santa Cruz; Dennis J. Corrigan; George Cox, San Diego State University; John D. Cunningham, Keene State College; Herman E. Daly, University of Maryland; Raymond F. Dasmann, University of California, Santa Cruz; Kingsley Davis, Hoover Institution; Edward E. DeMartini, University of California, Santa Barbara; James Demastes, University of Northern Iowa; Charles E. DePoe, Northeast Louisiana University; Thomas R. Detwyler, University of Wisconsin; Bruce DeVantier, Southern Illinois University at Carbondale; Peter H. Diage, University of California, Riverside; Stephanie Dockstader, Monroe Community College; Lon D. Drake, University of Iowa; Michael

Draney, University of Wisconsin–Green Bay; David DuBose, Shasta College; Dietrich Earnhart, University of Kansas; Robert East, Washington & Jefferson College; T. Edmonson, University of Washington; Thomas Eisner, Cornell University; Michael Esler, Southern Illinois University; David E. Fairbrothers, Rutgers University; Paul P. Feeny, Cornell University; Richard S. Feldman, Marist College; Vicki Fella-Pleier, La Salle University; Nancy Field, Bellevue Community College; Allan Fitzsimmons, University of Kentucky; Andrew J. Friedland, Dartmouth College; Kenneth O. Fulgham, Humboldt State University; Lowell L. Getz, University of Illinois at Urbana–Champaign; Frederick F. Gilbert, Washington State University; Jay Glassman, Los Angeles Valley College; Harold Goetz, North Dakota State University; Srikanth Gogineni, Axia College of University of Phoenix; Jeffery J. Gordon, Bowling Green State University; Eville Gorham, University of Minnesota; Michael Gough, Resources for the Future; Ernest M. Gould, Jr., Harvard University; Peter Green, Golden West College; Katharine B. Gregg, West Virginia Wesleyan College; Paul K. Grogger, University of Colorado at Colorado Springs; L. Guernsey, Indiana State University; Ralph Guzman, University of California, Santa Cruz; Raymond Hames, University of Nebraska, Lincoln; Robert Hamilton IV, Kent State University, Stark Campus; Raymond E. Hampton, Central Michigan University; Ted L. Hanes, California State University, Fullerton; William S. Hardenbergh, Southern Illinois University at Carbondale; John P. Harley, Eastern Kentucky University; Neil A. Harriman, University of Wisconsin, Oshkosh; Grant A. Harris, Washington State University; Harry S. Hass, San Jose City College; Arthur N. Haupt, Population Reference Bureau; Denis A. Hayes, environmental consultant; Stephen Heard, University of Iowa; Gene Heinze-Fry, Department of Utilities, Commonwealth of Massachusetts; Jane Heinze-Fry, environmental educator; John G. Hewston, Humboldt State University; David L. Hicks, Whitworth College; Kenneth M. Hinkel, University of Cincinnati; Eric Hirst, Oak Ridge National Laboratory; Doug Hix, University of Hartford; S. Holling, University of British Columbia; Sue Holt, Cabrillo College; Donald Holtgrieve, California State University, Hayward; Michelle Homan, Gannon University; Michael H. Horn, California State University, Fullerton; Mark A. Hornberger, Bloomsberg University; Marilyn Houck, Pennsylvania State University; Richard D. Houk, Winthrop College; Robert J. Huggett, College of William and Mary; Donald Huisingh, North Carolina State University; Catherine Hurlbut, Florida Community College at Jacksonville; Marlene K. Hutt, IBM; David R. Inglis, University of Massachusetts; Robert Janiskee, University of South Carolina; Hugo H. John, University of Connecticut; Brian A. Johnson, University of Pennsylvania, Bloomsburg; David I. Johnson, Michigan State University; Mark Jonasson, Crafton Hills College; Zoghlul Kabir, Rutgers, New Brunswick; Agnes Kadar, Nassau Community College; Thomas L. Keefe, Eastern Kentucky University; David Kelley, University of St. Thomas; William E. Kelso, Louisiana State University; Nathan Keyfitz, Harvard University; David Kidd, University of New Mexico; Pamela S. Kimbrough; Jesse Klingebiel, Kent School; Edward J. Kormondy, University of Hawaii–Hilo/West Oahu College; John V. Krutilla, Resources for the Future, Inc.; Judith Kunofsky, Sierra Club; E. Kurtz; Theodore Kury, State University of New York at Buffalo; Troy A. Ladine, East Texas Baptist University; Steve Ladochy, University of Winnipeg; Anna J. Lang, Weber State University; Mark B. Lapping, Kansas State University; Michael L. Larsen, Campbell University; Linda Lee, University of Connecticut; Tom Leege, Idaho Department of Fish and Game; Maureen Leupold, Genesee Community College; William S. Lindsay, Monterey Peninsula College; E. S. Lindstrom, Pennsylvania State University; M. Lippiman, New York University Medical Center; Valerie A. Liston, University of Minnesota; Dennis Livingston, Rensselaer Polytechnic Institute; James P. Lodge, air pollution consultant; Raymond C. Loehr, University of Texas at Austin; Ruth Logan, Santa Monica City College; Robert D. Loring, DePauw University; Paul F. Love, Angelo State University; Thomas Lovering, University of California, Santa Barbara; Amory B. Lovins, Rocky Mountain Institute; Hunter Lovins, Rocky Mountain Institute; Gene A. Lucas, Drake University; Claudia Luke, University of California, Berkeley; David Lynn; Timothy F. Lyon, Ball State University; Stephen Malcolm, Western Michigan University; Melvin G. Marcus, Arizona State University; Gordon E. Matzke, Oregon State University; Parker Mauldin, Rockefeller Foundation; Marie McClune, The Agnes Irwin School (Rosemont, Pennsylvania); Theodore R. McDowell, California State University; Vincent E. McKelvey, U.S. Geological Survey; Robert T. McMaster, Smith College; John G. Merriam, Bowling Green State University; A. Steven Messenger, Northern Illinois University; John Meyers, Middlesex Community College; Raymond W. Miller, Utah State University; Arthur B. Millman, University of Massachusetts, Boston; Sheila Miracle, Southeast Kentucky Community & Technical College; Fred Montague, University of Utah; Rolf Monteen, California Polytechnic State University; Debbie Moore, Troy University Dothan Campus; Michael K. Moore, Mercer University; Ralph Morris, Brock University, St. Catherine's, Ontario, Canada; Angela Morrow, Auburn University; William W. Murdoch, University of California, Santa Barbara; Norman Myers, environmental consultant; Brian C. Myres, Cypress College; A. Neale, Illinois State University; Duane Nellis, Kansas State University; Jan Newhouse, University of Hawaii, Manoa; Jim Norwine, Texas A&M University, Kingsville; John E. Oliver, Indiana State University; Mark Olsen, University of Notre Dame; Carol Page, copy editor; Bill Paletski, Penn State University; Eric Pallant, Allegheny College; Charles F. Park, Stanford University; Richard J. Pedersen, U.S. Department of Agriculture, Forest Service; David

Pelliam, Bureau of Land Management, U.S. Department of the Interior; Murray Paton Pendarvis, Southeastern Louisiana University; Dave Perault, Lynchburg College; Rodney Peterson, Colorado State University; Julie Phillips, De Anza College; John Pichtel, Ball State University; William S. Pierce, Case Western Reserve University; David Pimentel, Cornell University; Peter Pizor, Northwest Community College; Mark D. Plunkett, Bellevue Community College; Grace L. Powell, University of Akron; James H. Price, Oklahoma College; Marian E. Reeve, Merritt College; Carl H. Reidel, University of Vermont; Charles C. Reith, Tulane University; Erin C. Rempala, San Diego City College; Roger Revelle, California State University, San Diego; L. Reynolds, University of Central Arkansas; Ronald R. Rhein, Kutztown University of Pennsylvania; Charles Rhyne, Jackson State University; Robert A. Richardson, University of Wisconsin; Benjamin F. Richason III, St. Cloud State University; Jennifer Rivers, Northeastern University; Ronald Robberecht, University of Idaho; William Van B. Robertson, School of Medicine, Stanford University; C. Lee Rockett, Bowling Green State University; Terry D. Roelofs, Humboldt State University; Daniel Ropek, Columbia George Community College; Christopher Rose, California Polytechnic State University; Richard G. Rose, West Valley College; Stephen T. Ross, University of Southern Mississippi; Robert E. Roth, Ohio State University; Dorna Sakurai, Santa Monica College; Arthur N. Samel, Bowling Green State University; Shamili Sandiford, College of DuPage; Floyd Sanford, Coe College; David Satterthwaite, I.E.E.D., London; Stephen W. Sawyer, University of Maryland; Arnold Schecter, State University of New York; Frank Schiavo, San Jose State University; William H. Schlesinger, Ecological Society of America; Stephen H. Schneider, National Center for Atmospheric Research; Clarence A. Schoenfeld, University of Wisconsin, Madison; Madeline Schreiber, Virginia Polytechnic Institute; Henry A. Schroeder, Dartmouth Medical School; Lauren A. Schroeder, Youngstown State University; Norman B. Schwartz, University of Delaware; George Sessions, Sierra College; David J. Severn, Clement Associates; Don Sheets, Gardner-Webb University; Paul Shepard, Pitzer College and Claremont Graduate School; Michael P. Shields, Southern Illinois University at Carbondale; Kenneth Shiovitz; F. Siewert, Ball State University; E. K. Silbergold, Environmental Defense Fund; Joseph L. Simon, University of South Florida; William E. Sloey, University of Wisconsin, Oshkosh; Robert L. Smith, West Virginia University; Val Smith, University of Kansas; Howard M. Smolkin, U.S. Environmental Protection Agency; Patricia M. Sparks, Glassboro State College; John E. Stanley, University of Virginia; Mel Stanley, California State Polytechnic University, Pomona; Richard Stevens, Monroe Community College; Norman R. Stewart, University of Wisconsin, Milwaukee; Frank E. Studnicka, University of Wisconsin, Platteville; Chris Tarp, Contra Costa College; Roger E. Thibault, Bowling Green State University; Nathan E. Thomas, University of South Dakota; William L. Thomas, California State University, Hayward; Shari Turney, copy editor; John D. Usis, Youngstown State University; Tinco E. A. van Hylckama, Texas Tech University; Robert R. Van Kirk, Humboldt State University; Donald E. Van Meter, Ball State University; Rick Van Schoik, San Diego State University; Gary Varner, Texas A&M University; John D. Vitek, Oklahoma State University; Harry A. Wagner, Victoria College; Lee B. Waian, Saddleback College; Warren C. Walker, Stephen F. Austin State University; Thomas D. Warner, South Dakota State University; Kenneth E. F. Watt, University of California, Davis; Alvin M. Weinberg, Institute of Energy Analysis, Oak Ridge Associated Universities; Brian Weiss; Margery Weitkamp, James Monroe High School (Granada Hills, California); Anthony Weston, State University of New York at Stony Brook; Raymond White, San Francisco City College; Douglas Wickum, University of Wisconsin, Stout; Charles G. Wilber, Colorado State University; Nancy Lee Wilkinson, San Francisco State University; John C. Williams, College of San Mateo; Ray Williams, Rio Hondo College; Roberta Williams, University of Nevada, Las Vegas; Samuel J. Williamson, New York University; Dwina Willis, Freed-Hardeman University; Ted L. Willrich, Oregon State University; James Winsor, Pennsylvania State University; Fred Witzig, University of Minnesota at Duluth; Martha Wolfe, Elizabethtown Community and Technical College; George M. Woodwell, Woods Hole Research Center; Todd Yetter, University of the Cumberlands; Robert Yoerg, Belmont Hills Hospital; Hideo Yonenaka, San Francisco State University; Brenda Young, Daemen College; Anita Závodská, Barry University; Malcolm J. Zwolinski, University of Arizona.

For Students

Students who can begin early in their lives to think of things as connected, even if they revise their views every year, have begun the life of learning.

Mark Van Doren

Why Is It Important to Study Environmental Science?

Welcome to **environmental science**—an *interdisciplinary* study of how the earth works, how we interact with the earth, and how we can deal with the environmental problems we face. Because environmental issues affect every part of your life, the concepts, information, and issues discussed in this book and the course you are taking will be useful to you now and throughout your life.

Understandably, we are biased, but *we strongly believe that environmental science is the single most important course that you could take.* What could be more important than learning about the earth's life-support system, how our choices and activities affect it, and how we can reduce our growing environmental impact? Evidence indicates strongly that we will have to learn to live more sustainably by reducing our degradation of the planet's life-support system. We hope this book and the learning opportunities available to you online will inspire you to become involved in this change in the way we view and treat the earth, which sustains us, our economies, and all other living things.

You Can Improve Your Study and Learning Skills

Maximizing your ability to learn involves trying to *improve your study and learning skills.* Here are some suggestions for doing so:

- ***Develop a passion for learning.***
- ***Get organized.***
- ***Make daily to-do lists.*** Put items in order of importance, focus on the most important tasks, and assign a time to work on these items. Shift your schedule as needed to accomplish the most important items.
- ***Set up a study routine in a distraction-free environment.*** Study in a quiet, well-lit space. Take breaks every hour or so. During each break, take several deep breaths and move around; this will help you to stay more alert and focused.
- ***Avoid procrastination.*** Do not fall behind on your reading and other assignments. Set aside a particular time for studying each day and make it a part of your daily routine.
- ***Make hills out of mountains.*** It is psychologically difficult to read an entire book, read a chapter in a book, write a paper, or cram to study for a test. Instead, break these large tasks (mountains) down into a series of small tasks (hills). Each day, read a few pages of a book or chapter, write a few paragraphs of a paper, and review what you have studied and learned.
- ***Ask and answer questions as you read.*** For example, "What is the main point of a particular subsection or paragraph?" "How does it relate to the key question and key concepts addressed in each major chapter section?"
- ***Focus on key terms.*** Use the glossary in your textbook to look up the meaning of terms or words you do not understand. This book shows all key terms in **bold** type and lesser, but still important, terms in *italicized* type. The MindTap online edition of this text provides direct links to definitions for all bold-type terms. The *Chapter Review* questions at the end of each chapter also include the chapter's key terms in bold. Flash cards for testing your mastery of key terms for each chapter are available on the website for this book, or you can make your own.
- ***Interact with what you read.*** You could highlight key sentences and paragraphs and make notes in the margins. You might also mark important pages that you want to return to. The MindTap edition supports extensive note-taking features.
- ***Review to reinforce learning.*** Before each class session, review the material you learned in the previous session and read the assigned material.
- ***Become a good note taker.*** Learn to write down the main points and key information from any lecture using your own shorthand system. Review, fill in, and organize your notes as soon as possible after each class.
- ***Check what you have learned.*** At the end of each chapter, you will find review questions that cover all of the key material in each chapter section. We suggest that you try to answer each of these questions after studying each chapter section.
- ***Write out answers to questions to focus and reinforce learning.*** Write down your answers to the critical thinking questions found in the *Thinking About* boxes throughout the chapters, in many figure captions, and at the end of each chapter. These questions are designed to inspire you to think critically about key ideas and connect them to other ideas and to your own life. Also, write down your answers to all chapter-ending review questions. Additional quizzes can be found online as well. Save your answers for review and test preparation.
- ***Use the buddy system.*** Study with a friend or become a member of a study group to compare notes, review material, and prepare for tests. Explaining

something to someone else is a great way to focus your thoughts and reinforce your learning. Attend any review sessions offered by instructors or teaching assistants.

- ***Learn your instructor's test style.*** Does your instructor emphasize multiple-choice, fill-in-the-blank, true-or-false, factual, or essay questions? How much of the test will come from the textbook and how much from lecture material? Adapt your learning and studying methods to this style.
- ***Become a good test taker.*** Avoid cramming. Eat well and get plenty of sleep before a test. Arrive on time or early. Calm yourself and increase your oxygen intake by taking several deep breaths. (Do this also about every 10–15 minutes while taking the test.) Look over the test and answer the questions you know well first. Then work on the harder ones. Use the process of elimination to narrow down the choices for multiple-choice questions. For essay questions, organize your thoughts before you start writing. If you have no idea what a question means, make an educated guess. You might earn some partial credit and avoid getting a zero. Another strategy for getting some credit is to show your knowledge and reasoning by writing something like this: "If this question means so and so, then my answer is ________."
- ***Take time to enjoy life.*** Every day, take time to laugh and enjoy nature, beauty, and friendship.

You Can Improve Your Critical Thinking Skills

Critical thinking involves developing skills to analyze information and ideas, judge their validity, and make decisions. Critical thinking helps you to distinguish between facts and opinions, evaluate evidence and arguments, and take and defend informed positions on issues. It also helps you to integrate information and see relationships and to apply your knowledge to dealing with new and different problems, as well as to your own lifestyle choices. Here are some basic skills for learning how to think more critically.

- ***Question everything and everybody.*** Be skeptical, as any good scientist is. Do not believe everything you hear and read, including the content of this textbook, without evaluating the information you receive. Seek other sources and opinions.
- ***Identify and evaluate your personal biases and beliefs.*** Each of us has biases and beliefs taught to us by our parents, teachers, friends, role models, and our own experience. What are your basic beliefs, values, and biases? Where did they come from? What assumptions are they based on? How sure are you that your beliefs, values, and assumptions are right and why? According to the American psychologist and philosopher William James, "A great many people think they are thinking when they are merely rearranging their prejudices."
- ***Be open-minded and flexible.*** Be open to considering different points of view. Suspend judgment until you gather more evidence, and be willing to change your mind. Recognize that there may be a number of useful and acceptable solutions to a problem and that very few issues are either black or white. Try to take the viewpoints of those you disagree with. Understand that there are trade-offs involved in dealing with any environmental issue, as you will learn in reading this book.
- ***Be humble about what you know.*** Some people are so confident in what they know that they stop thinking and questioning. To paraphrase American writer Mark Twain, "It's what we know is true, but just ain't so, that hurts us."
- ***Find out how the information related to an issue was obtained.*** Are the statements you heard or read based on firsthand knowledge and research or on hearsay? Are unnamed sources used? Is the information based on reproducible and widely accepted scientific studies or on preliminary scientific results that may be valid but need further testing? Is the information based on a few isolated stories or experiences or on carefully controlled studies that have been reviewed by experts in the field involved? Is it based on unsubstantiated and dubious scientific information or beliefs?
- ***Question the evidence and conclusions presented.*** What are the conclusions or claims based on the information you're considering? What evidence is presented to support them? Does the evidence support them? Is there a need to gather more evidence to test the conclusions? Are there other, more reasonable conclusions?
- ***Try to uncover differences in basic beliefs and assumptions.*** On the surface, most arguments or disagreements involve differences of opinion about the validity or meaning of certain facts or conclusions. Scratch a little deeper and you will find that many disagreements are based on different (and often hidden) basic assumptions concerning how we look at and interpret the world around us. Uncovering these basic differences can allow the parties involved to understand one another's viewpoints and to agree to disagree about their basic assumptions, beliefs, or principles.
- ***Try to identify and assess any motives on the part of those presenting evidence and drawing conclusions.*** What is their expertise in this area? Do they have any unstated assumptions, beliefs, biases, or values? Do they have a personal agenda? Can they

benefit financially or politically from acceptance of their evidence and conclusions? Would investigators with different basic assumptions or beliefs take the same data and come to different conclusions?

- ***Expect and tolerate uncertainty.*** Recognize that scientists cannot establish absolute proof or certainty about anything. However, the reliable results of science have a high degree of certainty.
- ***Check the arguments you hear and read for logical fallacies and debating tricks.*** Here are six of many examples of such debating tricks: *First,* attack the presenter of an argument rather than the argument itself. *Second,* appeal to emotion rather than facts and logic. *Third,* claim that if one piece of evidence or one conclusion is false, then all other related pieces of evidence and conclusions are false. *Fourth,* say that a conclusion is false because it has not been scientifically proven (scientists never prove anything absolutely, but they can often establish high degrees of certainty). *Fifth,* inject irrelevant or misleading information to divert attention from important points. *Sixth,* present only either/or alternatives when there may be a number of options.
- ***Do not believe everything you read on the Internet.*** The Internet is a wonderful and easily accessible source of information that includes alternative explanations and opinions on almost any subject or issue—much of it not available in the mainstream media and scholarly articles. Blogs of all sorts have become a major source of information, even more important than standard news media for some people. However, because the Internet is so open, anyone can post anything they want to some blogs and other websites with no editorial control or review by experts. As a result, evaluating information on the Internet is one of the best ways to put into practice the principles of critical thinking discussed here. Use and enjoy the Internet, but think critically and proceed with caution.
- ***Develop principles or rules for evaluating evidence.*** Develop a written list of principles to serve as guidelines for evaluating evidence and claims. Continually evaluate and modify this list on the basis of your experience.
- ***Become a seeker of wisdom, not a vessel of information.*** Many people believe that the main goal of their education is to learn as much as they can by gathering more and more information. We believe that the primary goal is to learn how to sift through mountains of facts and ideas to find the few *nuggets of wisdom* that are especially useful for understanding the world and for making decisions. This book is full of facts and numbers, but they are useful only to the extent that they lead to an understanding of key ideas, scientific laws, theories, concepts, and connections. The major goals of the study of environmental science are to find out how nature works and sustains itself *(environmental wisdom)* and to use *principles of environmental wisdom* to help make human societies and economies more sustainable, more just, and more beneficial and enjoyable for all. As writer Sandra Carey observed, "Never mistake knowledge for wisdom. One helps you make a living; the other helps you make a life."

To help you practice critical thinking, we have supplied questions throughout this book, found within each chapter in brief boxes labeled *Thinking About,* in the captions of many figures, and at the end of each chapter. There are no right or wrong answers to many of these questions. A good way to improve your critical thinking skills is to compare your answers with those of your classmates and to discuss how you arrived at your answers.

Use the Learning Tools We Offer in This Book

We have included a number of tools throughout this textbook that are intended to help you improve your learning skills and apply them. First, consider the *Key Questions* list at the beginning of each chapter section. You can use these to preview a chapter and to review the material after you've read it.

Next, note that we use three different special notations throughout the text. Each chapter opens with a **Core Case Study**, and each time we tie material within the chapter back to this core case, we note it in bold, colored type as we did in this sentence. You will also see two icons appearing regularly in the text margins. When you see the *sustainability* icon, you will know that you have just read something that relates directly to the overarching theme of this text, summarized by our six **principles of sustainability**, which are introduced in Figures 1-2, p. 6, and 1-5, p. 9, and which appear on the back cover of the student edition. The *Good News* icon appears near each of many examples of successes that people have had in dealing with the environmental challenges we face.

SUSTAINABILITY

GOOD NEWS

We also include several brief *Connections* boxes to show you some of the often surprising connections between environmental problems or processes and some of the products and services we use every day or some of the activities we partake in. These, along with the *Thinking About* boxes scattered throughout the text (both designated by the *Consider This. . .* heading), are intended to get you to think carefully about activities and choices we take for granted and how they might be affecting the environment.

At the end of each chapter, we list what we consider to be the *three big ideas* that you should take away from the chapter. Following that list in each chapter is a *Tying It All Together* box. This feature quickly reviews the Core Case Study and how chapter material relates to it, and it explains how the principles of sustainability can be applied to deal with challenges discussed in the **Core Case Study** and throughout the chapter.

We have also included a *Chapter Review* section at the end of each chapter with questions listed for each chapter section. These questions cover all of the key material and key terms in each chapter. A variety of other exercises and projects follow this review section at the end of each chapter.

Finally, at the back of the book, we have included a comprehensive glossary. It includes definitions of all the book's key terms, as well as definitions of many other important terms.

Know Your Own Learning Style

People have different ways of learning and it can be helpful to know your own learning style. *Visual learners* learn best from reading and viewing illustrations and diagrams. *Auditory learners* learn best by listening and discussing. They might benefit from reading aloud while studying and using a tape recorder in lectures for study and review. *Logical learners* learn best by using concepts and logic to uncover and understand a subject rather than relying mostly on memory.

This book and its supporting website material contain plenty of tools for all types of learners. Visual learners can benefit from using flash cards (available online) to memorize key terms and ideas. This is a highly visual book with many carefully selected photographs and diagrams designed to illustrate important ideas, concepts, and processes. Auditory learners can make use of our ReadSpeaker app in MindTap, which can read the chapter aloud in different speeds and voices. For logical learners, the book is organized by key concepts that are revisited throughout any chapter and related carefully to other concepts, major principles, and case studies and other examples. We urge you to become aware of your own learning style and make the most of these various tools.

This Book Presents a Positive, Realistic Environmental Vision of the Future

Our goal is to present a positive vision of our environmental future based on realistic optimism. To do so, we strive not only to present the facts about environmental issues, but also to give a balanced presentation of different viewpoints. We consider the advantages and disadvantages of various technologies and proposed solutions to environmental problems. We argue that environmental solutions usually require *trade-offs* among opposing parties, and that the best solutions are *win-win* solutions. Such solutions are achieved when people with different viewpoints work together to come up with a solution that both sides can live with. And we present the good news as well as the bad news about efforts to deal with environmental problems.

One cannot study a subject as important and complex as environmental science without forming conclusions, opinions, and beliefs. However, we argue that any such results should be based on use of critical thinking to evaluate conflicting positions and to understand the trade-offs involved in most environmental solutions. To that end, we emphasize critical thinking throughout this textbook, and we encourage you to develop a practice of applying critical thinking to everything you read and hear, both in school and throughout your life.

Help Us Improve This Book

Researching and writing a book that covers and connects the numerous major concepts from the wide variety of environmental science disciplines is a challenging and exciting task. Almost every day, we learn about some new connection in nature. However, in a book this complex, there are bound to be some errors—some typographical mistakes that slip through and some statements that you might question, based on your knowledge and research. We invite you to contact us to correct any errors you find, point out any bias you see, and suggest ways to improve this book. Please e-mail your suggestions to Tyler Miller at mtg89@hotmail.com or Scott Spoolman at spoolman@tds.net.

Now start your journey into this fascinating and important study of how the earth's life-support system works and how we can leave our planet in a condition at least as good as what we now enjoy. Have fun.

Supplements for Students

You have a large variety of electronic and other supplemental materials available to you to help you take your learning experience beyond this textbook:

- ***Environmental Science MindTap.*** MindTap is a new approach to highly personalized online learning. Beyond an eBook, homework solution, digital supplement, or premium website, MindTap is a digital learning platform that works alongside your campus LMS to deliver course curriculum across the range of electronic devices in your life. MindTap is built on an "app" model, allowing enhanced digital collaboration and delivery of engaging content across a spectrum of Cengage and non-Cengage resources.
- ***Global Environment Watch.*** Updated several times a day, the Global Environment Watch is a focused portal into GREENR—the Global Reference on the Environment, Energy, and Natural Resources—an ideal one-stop site for classroom discussion and research projects. This resource center keeps courses up-to-date with the most current news on the environment.

Users get access to information from trusted academic journals, news outlets, and magazines, as well as statistics, an interactive world map, videos, primary sources, case studies, podcasts, and much more. Log in or purchase access at www.cengagebrain.com/shop/isbn/9781423929444 to complete the exercises found at the end of each chapter.

- **New! *Virtual Field Trips in Environmental Issues.*** *Virtual Field Trips in Environmental Issues* brings the field to you, with dynamic panoramas, videos, photographs, maps, and quizzes covering important topics within environmental science. A case study approach covers the issues of *keystone species, climate change's role in extinctions, invasive species, the evolution of a species due to its environment,* and *an ecosystem approach to sustaining biodiversity.* Students are engaged, interacting with real issues to help them think critically about the world around them.

Visit www.cengagebrain.com for additional materials, including free resources, at www.cengagebrain.com/shop/isbn/9781133940135.

Other student learning tools include the following:

- *Essential Study Skills for Science Students* by Daniel D. Chiras. This book includes chapters on developing good study habits, sharpening memory, getting the most out of lectures, labs, and reading assignments, improving test-taking abilities, and becoming a critical thinker. Available for students on instructor's request.
- *Lab Manual.* Edited by Edward Wells, this lab manual includes both hands-on and data analysis labs to help your students develop a range of skills. Create a custom version of this Lab Manual by adding labs you have written or ones from our collection with Cengage Custom Publishing. An Instructor's Manual for the labs will be available to adopters.
- *What Can You Do?* This guide presents students with a variety of ways that they can affect the environment, and shows them how to track the effect their actions have on their carbon footprint. Available for students on instructor's request.

G. TYLER MILLER

G. Tyler Miller has written 62 textbooks for introductory courses in environmental science, basic ecology, energy, and environmental chemistry. Since 1975, Miller's books have been the most widely used textbooks for environmental science in the United States and throughout the world. They have been used by almost 3 million students and have been translated into eight languages.

Miller has a professional background in chemistry, physics, and ecology. He has PhD from the University of Virginia and has received two honorary doctoral degrees for his contributions to environmental education. He taught college for 20 years, developed one of the nation's first environmental studies programs, and developed an innovative interdisciplinary undergraduate science program before deciding to write environmental science textbooks full time in 1975. Currently, he is the president of Earth Education and Research, devoted to improving environmental education.

He describes his hopes for the future as follows:

If I had to pick a time to be alive, it would be the next 75 years. Why? First, there is overwhelming scientific evidence that we are in the process of seriously degrading our own life-support system. In other words, we are living unsustainably. Second, within your lifetime we have the opportunity to learn how to live more sustainably by working with the rest of nature, as described in this book.

I am fortunate to have three smart, talented, and wonderful sons—Greg, David, and Bill. I am especially privileged to have Kathleen as my wife, best friend, and research associate. It is inspiring to have a brilliant, beautiful (inside and out), and strong woman who cares deeply about nature as a lifemate. She is my hero. I dedicate this book to her and to the earth.

SCOTT E. SPOOLMAN

Scott Spoolman is a writer and textbook editor with more than 30 years of experience in educational publishing. He has worked with Tyler Miller since 2003 as a contributing editor on earlier editions of *Living in the Environment*, *Environmental Science*, and *Sustaining the Earth*. With Norman Myers, he also coauthored *Environmental Issues and Solutions: A Modular Approach*.

Spoolman holds a master's degree in science journalism from the University of Minnesota. He has authored numerous articles in the fields of science, environmental engineering, politics, and business. He worked as an acquisitions editor on a series of college forestry textbooks. He has also worked as a consulting editor in the development of over 70 college and high school textbooks in fields of the natural and social sciences.

In his free time, he enjoys exploring the forests and waters of his native Wisconsin along with his family—his wife, environmental educator Gail Martinelli, and his children, Will and Katie.

Spoolman has the following to say about his collaboration with Tyler Miller.

I am honored to be working with Tyler Miller as a coauthor to continue the Miller tradition of thorough, clear, and engaging writing about the vast and complex field of environmental science. I share Tyler Miller's passion for ensuring that these textbooks and their multimedia supplements will be valuable tools for students and instructors. To that end, we strive to introduce this interdisciplinary field in ways that will be informative and sobering, but also tantalizing and motivational.

If the flip side of any problem is indeed an opportunity, then this truly is one of the most exciting times in history for students to start an environmental career. Environmental problems are numerous, serious, and daunting, but their possible solutions generate exciting new career opportunities. We place high priorities on inspiring students with these possibilities, challenging them to maintain a scientific focus, pointing them toward rewarding and fulfilling careers, and in doing so, working to help sustain life on the earth.

My Environmental Journey — *G. Tyler Miller*

My environmental journey began in 1966 when I heard a lecture on population and pollution problems by Dean Cowie, a biophysicist with the U.S. Geological Survey. It changed my life. I told him that if even half of what he said was valid, I would feel ethically obligated to spend the rest of my career teaching and writing to help students learn about the basics of environmental science. After spending six months studying the environmental literature, I concluded that he had greatly underestimated the seriousness of these problems.

I developed an undergraduate environmental studies program and in 1971 published my first introductory environmental science book, an interdisciplinary study of the connections between energy laws (thermodynamics), chemistry, and ecology. In 1975, I published the first edition of *Living in the Environment*. Since then, I have completed multiple editions of this textbook, and of three others derived from it, along with other books.

Beginning in 1985, I spent ten years in the deep woods living in an adapted school bus that I used as an environmental science laboratory and writing environmental science textbooks. I evaluated the use of passive solar energy design to heat the structure; buried earth tubes to bring in air cooled by the earth (geothermal cooling) at a cost of about $1 per summer; set up active and passive systems to provide hot water; installed an energy-efficient instant hot water heater powered by LPG; installed energy-efficient windows and appliances and a composting (waterless) toilet; employed biological pest control; composted food wastes; used natural planting (no grass or lawnmowers); gardened organically; and experimented with a host of other potential solutions to major environmental problems that we face.

I also used this time to learn and think about how nature works by studying the plants and animals around me. My experience from living in nature is reflected in much of the material in this book. It also helped me to develop the six simple principles of sustainability that serve as the integrating theme for this textbook and to apply these principles to living my life more sustainably.

I came out of the woods in 1995 to learn about how to live more sustainably in an urban setting where most people live. Since then, I have lived in two urban villages, one in a small town and one within a large metropolitan area.

Since 1970, my goal has been to use a car as little as possible. Since I work at home, I have a "low-pollute commute" from my bedroom to a chair and a laptop computer. I usually take one airplane trip a year to visit my sister and my publisher.

As you will learn in this book, life involves a series of environmental trade-offs. Like most people, I still have a large environmental impact, but I continue to struggle to reduce it. I hope you will join me in striving to live more sustainably and sharing what you learn with others. It is not always easy, but it sure is fun.

Cengage Learning's Commitment to Sustainable Practices

We the authors of this textbook and Cengage Learning, the publisher, are committed to making the publishing process as sustainable as possible. This involves four basic strategies:

- *Using sustainably produced paper.* The book publishing industry is committed to increasing the use of recycled fibers, and Cengage Learning is always looking for ways to increase this content. Cengage Learning works with paper suppliers to maximize the use of paper that contains only wood fibers that are certified as sustainably produced, from the growing and cutting of trees all the way through paper production.
- *Reducing resources used per book.* The publisher has an ongoing program to reduce the amount of wood pulp, virgin fibers, and other materials that go into each sheet of paper used. New, specially designed printing presses also reduce the amount of scrap paper produced per book.
- *Recycling.* Printers recycle the scrap paper that is produced as part of the printing process. Cengage Learning also recycles waste cardboard from shipping cartons, along with other materials used in the publishing process.
- *Process improvements.* In years past, publishing has involved using a great deal of paper and ink for the writing and editing of manuscripts, copyediting, reviewing page proofs, and creating illustrations. Almost all of these materials are now saved through use of electronic files. Very little paper and ink were used in the preparation of this textbook.

1 Environmental Problems, Their Causes, and Sustainability

No civilization has survived the ongoing destruction of its natural support system. Nor will ours.

LESTER R. BROWN

Key Questions

1-1 What are some principles of sustainability?

1-2 How are our ecological footprints affecting the earth?

1-3 Why do we have environmental problems?

1-4 What is an environmentally sustainable society?

Forests help sustain all life and economies.

CORE CASE STUDY A Vision of a More Sustainable World in 2065

Mostovyi Sergii Igorevich/Shutterstock.com

Figure 1-1 These parents—like Emily and Michael in our fictional vision of a possible world in 2065—are teaching their children about some of the world's environmental problems (left) and helping them to enjoy the wonders of nature (right).

©JOEL SARTORE/National Geographic Creative

Michael Rodriguez and Emily Briggs graduated from college in 2018. Michael earned a master's degree in environmental education, became a high-school teacher, and loved teaching environmental science. Emily established a thriving practice as an environmental lawyer.

Michael and Emily met while doing volunteer work for an environmental organization, got married, and had a child. They taught her about some of the world's environmental problems (Figure 1-1, left) and about the joys of nature that they had experienced as children (Figure 1-1, right). This led their daughter to become involved in working toward a more sustainable world.

When Michael and Emily were growing up, there had been increasing signs of stress on the land, air, water, and wildlife from the harmful environmental impacts of a growing population consuming more resources. However, by 2018, a small but growing number of people had begun shifting to more environmentally sustainable lifestyles.

In 2065, Emily and Michael celebrated the birth of their grandchild. He was born into a world where the loss of species and the degradation of land had slowed to a trickle. The atmosphere, oceans, lakes, and rivers were cleaner, 80% of the solid wastes were reused or recycled, and energy waste had been cut in half.

By 2050, significant atmospheric warming and the resulting climate change had occurred as many climate scientists had projected in the 1990s. However, the threat of further climate change and air and water pollution had begun to decrease because of greatly reduced energy waste and the gradual shift in human use of energy resources from oil and coal to cleaner energy from the sun, wind, flowing water, and other renewable resources.

By 2065, farmers producing most of the world's food had shifted to more sustainable farming practices that helped to conserve water and protect and renew much of the planet's vital topsoil. In addition, the human population had peaked at about 8 billion in 2050 and then had begun a slow decline, lessening human pressure on the earth's life-support systems. In 2065, Emily and Michael felt a great sense of pride, knowing that they and their child and countless others had helped to bring about these improvements so that current and future generations could live more sustainably on this marvelous planet that is our only home.

Sustainability is the capacity of the earth's natural systems and human cultural systems to survive, flourish, and adapt to changing environmental conditions into the very long-term future. It is the overarching theme of this textbook, as well as a focal point for understanding the environmental problems we face and for exploring possible solutions. Our goal is to present to you a realistic and hopeful vision of what could be.

1-1 What Are Some Principles of Sustainability?

CONCEPT 1-1A
Nature has been sustained for billions of years by relying on solar energy, biodiversity, and chemical cycling.

CONCEPT 1-1B
Our lives and economies depend on energy from the sun and on natural resources and ecosystem services *(natural capital)* provided by the earth.

CONCEPT 1-1C
We could shift toward living more sustainably by applying full-cost pricing, searching for win-win solutions, and committing to preserving the earth's life-support system for future generations.

Environmental Science Is a Study of Connections in Nature

The **environment** is everything around us. It includes the living and the nonliving things (air, water, and energy) with which we interact in a complex web of relationships that connect us to one another and to the world we live in. Despite our many scientific and technological advances, we are utterly dependent on the earth for clean air and water, food, shelter, energy, fertile soil, and everything else in the planet's *life-support system.*

This textbook is an introduction to **environmental science**, an *interdisciplinary* study of how humans interact with the living and nonliving parts of their environment. It integrates information and ideas from the *natural sciences* such as biology, chemistry, and geology; the *social sciences* such as geography, economics, and political science; and the *humanities* such as ethics. The three goals of environmental science are **(1)** to learn how life on the earth has survived and thrived, **(2)** to understand how we interact with the environment, and **(3)** to find ways to deal with environmental problems and live more sustainably.

A key component of environmental science is **ecology**, the biological science that studies how living things interact with one another and with their environment. These living things are called **organisms.** Each organism belongs to a **species**, a group of organisms that has a unique set of characteristics that distinguish it from other groups of organisms.

A major focus of ecology is the study of ecosystems. An **ecosystem** is a set of organisms within a defined area or volume that interact with one another and with their environment of nonliving matter and energy. For example, a forest ecosystem consists of plants (especially trees; see chapter-opening photo), animals, and various other organisms that decompose organic materials, all interacting with one another, with solar energy, and with the chemicals in the forest's air, water, and soil.

We should not confuse environmental science and ecology with **environmentalism**, a social movement dedicated to trying to protect the earth's life-support systems for all forms of life. Environmentalism is practiced more in the political and ethical arenas than in the realm of science.

Three Scientific Principles of Sustainability

How has an incredible variety of life on the earth been sustained for at least 3.5 billion years in the face of catastrophic changes in environmental conditions? Such changes included gigantic meteorites impacting the earth, ice ages lasting for hundreds of millions of years, and long warming periods during which melting ice raised sea levels by hundreds of feet.

The latest version of our species has been around for only about 200,000 years—less than the blink of an eye relative to the 3.5 billion years that life has existed on the earth. Yet, there is mounting scientific evidence that, as we have expanded into and dominated almost all of the earth's ecosystems during that short time, we have seriously degraded these natural systems that support our species and all other life forms, as well as our economies. Thus, the newest challenge for the human species is to learn how to live more sustainably (**Core Case Study**), and some scientists argue that we have no time to waste in doing so.

Many scientists contend that the earth is the only real example of a sustainable system. Our science-based research leads us to believe that three major natural factors have played the key roles in the long-term sustainability of life on this planet, as summarized below and in Figure 1-2 (**Concept 1-1A**). We use these three **scientific principles of sustainability**, or *lessons from nature,* throughout the book to suggest how we might move toward a more sustainable future. SUSTAINABILITY

- **Dependence on solar energy:** The sun warms the planet and provides energy that plants use to produce **nutrients**, or the chemicals necessary for their own life processes along with those of most other animals, including humans. The sun also powers indirect forms of **solar energy** such as wind and flowing water, which we use to produce electricity.
- **Biodiversity** (short for *biological diversity*): **Biodiversity** is the variety of genes, organisms, species, and ecosystems in which organisms exist and interact. The interactions among species, especially the feeding relationships, provide vital ecosystem services and keep any population from growing too large. Biodiversity also provides countless ways for life to adapt to changing environmental conditions, even catastrophic changes that wipe out large numbers of species.
- **Chemical cycling: Chemical cycling**, or **nutrient cycling**, is the circulation of chemicals necessary for life from the environment (mostly from soil and

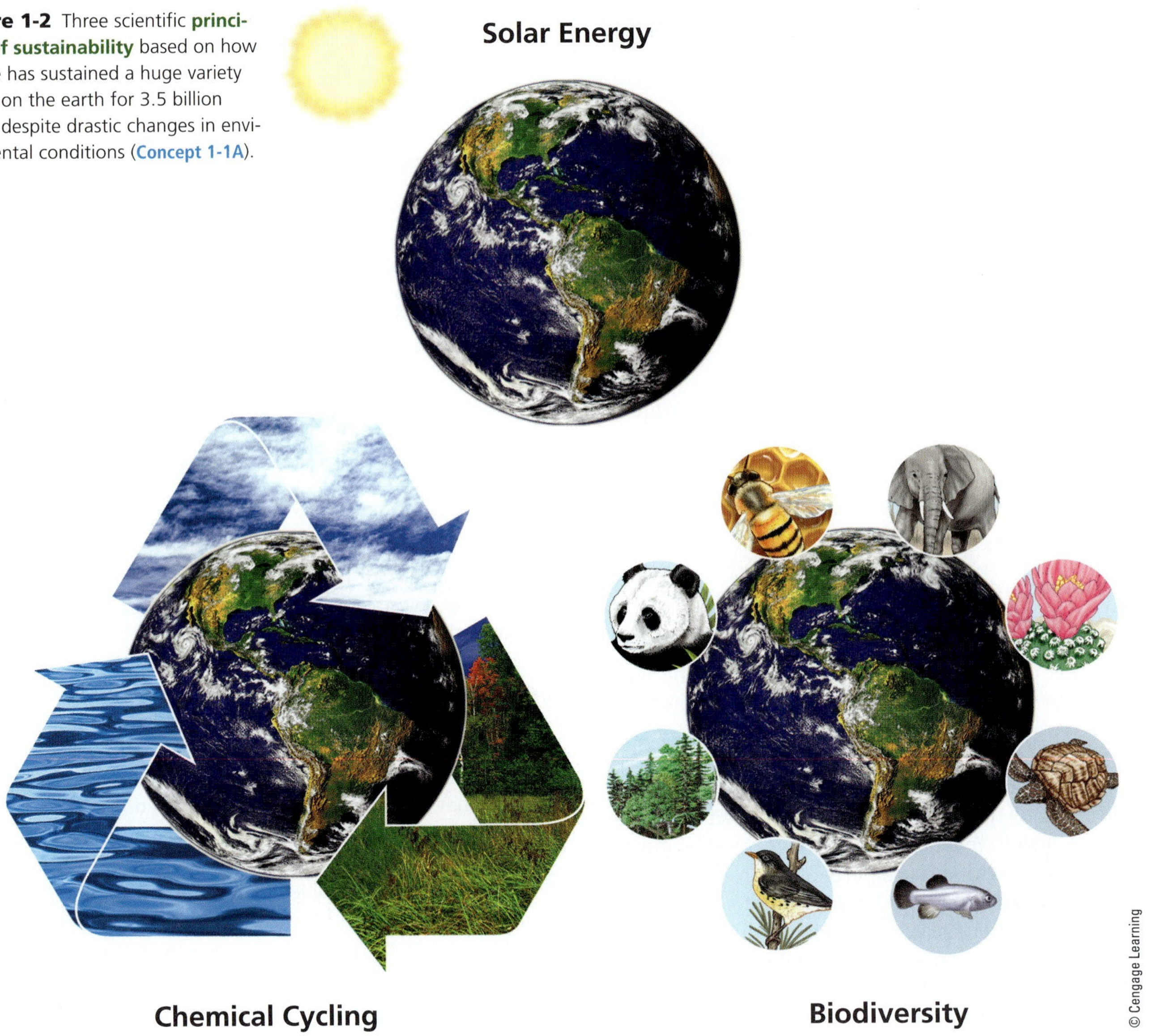

Figure 1-2 Three scientific **principles of sustainability** based on how nature has sustained a huge variety of life on the earth for 3.5 billion years, despite drastic changes in environmental conditions (**Concept 1-1A**).

Solar Energy

Chemical Cycling

Biodiversity

water) through organisms and back to the environment. Because the earth receives no new supplies of these chemicals, organisms must recycle them continuously in order to survive. This means that there is little waste in nature, other than in the human world, because the wastes of any organism become nutrients or raw materials for other organisms.

Sustainability Has Certain Key Components

Sustainability, the central integrating theme of this book, has several critical components that we use as subthemes. One such component is **natural capital**—the natural resources and natural services that keep us and other species alive and support human economies (Figure 1-3).

Natural resources are materials and energy in nature that are essential or useful to humans. They are often classified as *inexhaustible resources* (such as energy from the sun and wind), *renewable resources* (such as air, water, topsoil, plants, and animals), or *nonrenewable* or *depletable resources* (such as copper, oil, and coal). **Natural services**, or **ecosystem services**, are processes provided by healthy ecosystems. Examples include purification of air and water, renewal of topsoil, and pollination, which support life and human economies at no monetary cost to us. For example, forests help to purify air and water, regulate climate, reduce soil erosion, and provide countless species with a place to live.

One vital natural service is chemical, or nutrient, cycling—one of the three scientific **principles of sustainability**. An important component of nutrient cycling is *topsoil*—a vital natural resource that provides us and most other land-dwelling species with food. Without nutrient cycling in topsoil, life as we know it could not exist on the earth's land.

Natural capital is also supported by energy from the sun—another of the scientific **principles of sustainability** (Figure 1-2). Thus, our lives and economies depend on

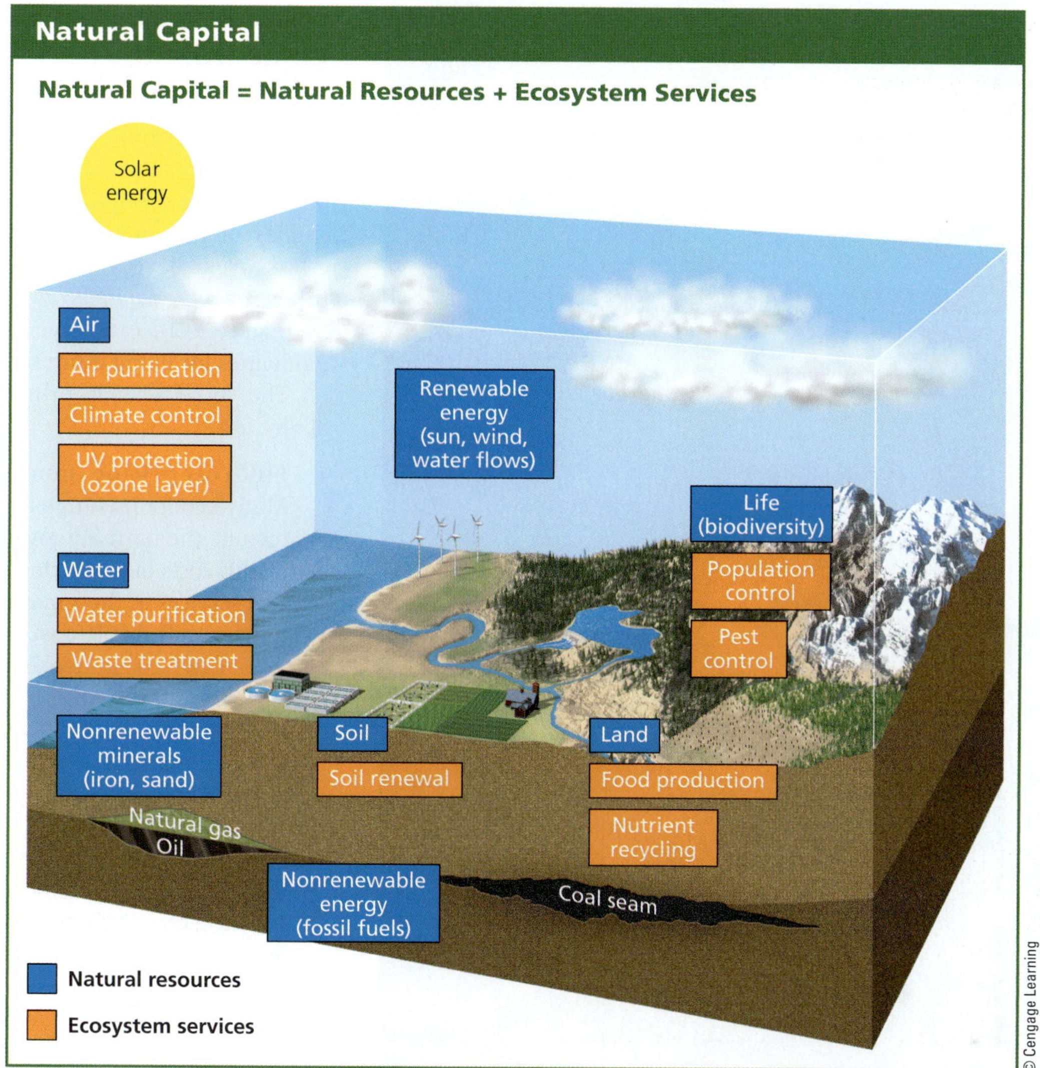

Figure 1-3 Natural capital consists of natural resources (blue) and natural or ecosystem services (orange) that support and sustain the earth's life and human economies (**Concept 1-1A**).

energy from the sun, and on natural resources and natural services *(natural capital)* provided by the earth (**Concept 1-1B**).

A second component of sustainability—and another subtheme of this text—is to recognize that many human activities can *degrade natural capital* by using normally renewable resources such as trees and topsoil faster than nature can restore them and by overloading the earth's normally renewable air and water systems with pollution and wastes. For example, in some parts of the world, we are replacing diverse and naturally sustainable forests (Figure 1-4) with crop plantations that can be sustained only with large inputs of water, fertilizer, and pesticides. We are also adding harmful chemicals and wastes to some rivers, lakes, and oceans faster than these bodies of water can cleanse themselves through natural processes.

This leads us to a third component of sustainability: *solutions.* While environmental scientists search for scientific solutions to problems such as the unsustainable degradation of forests and other forms of natural capital, social scientists are looking for economic and political solutions. For example, a scientific solution to the problems of depletion of forests is to stop burning or cutting down biologically diverse, mature forests (Figure 1-4). A scientific solution to the problem of pollution of rivers is to prevent the excessive dumping of harmful chemicals and wastes into streams and to allow them to recover naturally. However, to implement such solutions, governments might have to enact and enforce environmental laws and regulations.

The search for solutions often involves conflicts. For example, when a scientist argues for protecting a natural forest on government-owned land to help preserve its important diversity of plants and animals, the timber company that had planned to harvest the trees in that forest might protest. Dealing with such conflicts often involves making *trade-offs,* or compromises—another component of sustainability. For example, the timber company might be persuaded to plant and harvest trees in an area that it had

John Lee/Aurora Photos

Figure 1-4 Small remaining area of once diverse Amazon rain forest surrounded by vast simplified soybean fields in the Brazilian state of Motto Grosso.

already cleared or degraded, instead of clearing the trees in an undisturbed diverse natural forest. In return, the government might give the company a *subsidy,* or financial support, to meet some of the costs for planting the trees.

In making a shift toward sustainability, the daily actions of each and every individual are important. In other words, *individuals matter*—another subtheme of this book. History shows that almost all of the significant changes in human systems have come from the bottom up, through the collective actions of individuals and from individuals inventing more sustainable ways of doing things. Thus, *sustainability begins with actions at personal and local levels.*

Other Principles of Sustainability Come from the Social Sciences

Our search for solutions and trade-offs to environmental problems has led us to propose three **social science principles of sustainability** (Figure 1-5), derived from studies of economics, political science, and ethics. We believe that these, along with our three *scientific principles of sustainability* (Figure 1-2), can serve as general guidelines for living more sustainably.

The social science principles of sustainability are

- **Full-cost pricing** (from economics): Many economists urge us to find ways to include the harmful environmental and health costs of producing and using goods and services in their market prices—a practice called **full-cost pricing**. This would give consumers better information about the environmental impacts of their lifestyles, and it would allow them to make more informed choices about the goods and services they use.
- **Win-win solutions** (from political science): We can learn to work together in dealing with environmental problems by focusing on solutions that will benefit the largest possible number of people, as well as the environment. This means shifting from an *I win, you lose* approach to a *we both win* approach (*win-win* solutions), and to an *I win, you win, and the earth wins* approach (*win-win-win* solutions).
- **A responsibility to future generations** (from ethics): We should accept our responsibility to leave the planet's life-support systems in at least as good a shape as what we now enjoy, for future generations.

We will explore these principles further in this chapter and apply them throughout this textbook. For quick reference, you can find all six principles of sustainability on the back cover of this book.

Some Resources Are Renewable and Some Are Not

From a human standpoint, a **resource** is anything that we can obtain from the environment to meet our needs and wants. Some resources, such as solar energy, wind, surface water, and edible wild plants, are directly available for use. Other resources, such as petroleum, minerals, underground water, and cultivated plants, become useful to us only with some effort and technological ingenuity. For example, petroleum was merely a mysterious, oily fluid until we learned how to convert it into gasoline, heating oil, and other products.

Solar energy is called an **inexhaustible resource** because its continuous supply is expected to last for at least 6 billion years, until the sun dies. A **renewable resource** is one that can be replenished by natural processes within hours to centuries, as long as we do not use it up faster than natural processes can renew it. Examples include forests, grasslands, fishes, fertile topsoil, clean air, and freshwater.

The highest rate at which we can use a renewable resource indefinitely without reducing its available supply is called its **sustainable yield**. During most of the 10,000 years since we invented agriculture, civilization has lived on the sustainable yield of the earth's natural systems. But in recent decades we have been living unsustainably by degrading and depleting the earth's natural capital at an accelerating rate to fuel a growing population and ever-increasing resource consumption. As a result, in parts of the world, we are overharvesting forests, overgrazing grasslands, overfishing oceans, overdrawing underground water deposits (aquifers), and overloading the air and water with harmful wastes and pollutants.

Nonrenewable or **exhaustible resources** exist in a fixed quantity, or *stock*, in the earth's crust. On a time scale of millions to billions of years, geologic processes can renew such resources. However, on the much shorter human time scale of hundreds to thousands of years, we can deplete these resources much faster than nature can form them. Such exhaustible stocks include *energy resources* such as oil and coal (Figure 1-6), *metallic mineral resources* such as copper and aluminum, and *nonmetallic mineral resources* such as salt and sand.

As we deplete such resources, human ingenuity can often find substitutes. However, sometimes there is no acceptable or affordable substitute for a resource.

From an environmental and sustainability viewpoint, the priorities for more sustainable use of nonrenewable resources should be, in order: **R**efuse (don't use), **R**educe (use less), **R**euse, and **R**ecycle. Each of these steps helps to extend supplies and to reduce the environmental impacts of using these resources. According to a number of environmental scientists, we already know how to reuse or recycle at least 80% of the metal, glass, and other

GOOD NEWS

©PETE MCBRIDE/National Geographic Creative

Figure 1-6 It would take more than a million years for natural processes to replace the coal that was removed within a couple of decades from this strip mine in Wyoming (USA).

Left: ©Minerva Studio/Shutterstock.com. Center: mikeledray/Shutterstock.com. Right: ©Yuri Arcurs/Shutterstock.com.

Figure 1-5 Three social science **principles of sustainability** can help us make a transition to a more environmentally and economically sustainable future.

nonrenewable resources that we use. However, we cannot recycle or reuse nonrenewable energy resources such as oil, natural gas, and coal. Once burned, the concentrated energy they contain is dispersed as heat into the environment and is no longer available to us.

Countries Differ in Their Resource Use and Environmental Impact

The United Nations (UN) classifies the world's countries as economically more developed or less developed, based primarily on their average income per person. In using these classifications in this textbook, we do not mean to imply that either type of country is superior to the other. We are simply distinguishing between countries in terms of their economic activity because this measure is highly useful for studying the resource use and environmental impacts of nations and regions.

More-developed countries—industrialized nations with high average income—have 17% of the world's population and include the United States, Canada, Japan, Australia, and most European countries. All other nations, in which 83% of the world's people live, are classified as **less-developed countries**, most of them in Africa, Asia, and Latin America. Some are *middle-income, moderately developed countries* such as China, India, Brazil, Thailand, and Mexico. Others are *low-income, least-developed countries,* including Congo, Haiti, Nigeria, and Nicaragua. (Figure 3, p. S28, in Supplement 6 is a map of high-, upper-middle-, lower-middle-, and low-income countries.)

1-2 How Are Our Ecological Footprints Affecting the Earth?

CONCEPT 1-2
As our ecological footprints grow, we are depleting and degrading more of the earth's natural capital.

We Are Living Unsustainably

The earth is a beautiful, largely blue and white island of water, land, and life that moves around the sun at about 107,000 kilometers per hour (67,000 miles per hour). The life on this planet has been sustained for billions of years by solar energy, chemical cycling, and biodiversity (the scientific **principles of sustainability**). SUSTAINABILITY

However, a large and growing body of scientific evidence indicates that the human species, a relative newcomer on the planet, is violating these sustainability principles. In fact, we are living unsustainably by wasting, depleting, and degrading the earth's natural capital (Figure 1-3) at an accelerating rate—a process known as **environmental degradation**, summarized in Figure 1-7. Scientists also refer to this as **natural capital degradation**.

In many parts of the world, renewable forests are shrinking (Figure 1-4), deserts are expanding, topsoil is eroding, and suburbs are replacing croplands. In addition, the lower atmosphere is warming, floating ice and glaciers are melting at unexpected rates, sea levels are rising, ocean acidity is increasing, and floods, droughts, severe weather, and forest fires are more frequent in some areas. In a number of regions, rivers are running dry, harvests of many species of fish are dropping sharply, and coral reefs are dying. Species are becoming extinct at least 100 times faster than in prehuman times, and extinction rates are projected to increase by at least another 100-fold during this century.

In 2005, the UN released its *Millennium Ecosystem Assessment,* a 4-year study by 1,360 experts from 95 countries. According to this study, human activities have degraded or overused about 60% of the earth's natural or ecosystem services (Figure 1-3, orange boxes), mostly since 1950. The report's summary statement warned that "human activity is putting such a strain on the natural functions of Earth that the ability of the planet's ecosystems to sustain future generations can no longer be taken for granted." The report also concluded that we have scientific, economic, and political solutions to these problems that we could implement within a few decades in order to make the transition to a more sustainable future within your lifetime (**Core Case Study**), as you will learn in reading this book. GOOD NEWS

Pollution Comes from a Number of Sources

A major environmental problem is **pollution**, which is contamination of the environment by a chemical or other agent such as noise or heat to a level that is harmful to the health, survival, or activities of humans or other organisms. Polluting substances, or *pollutants,* can enter the environment naturally, such as from volcanic eruptions, or through human activities, such as the burning of coal and gasoline, and the dumping of chemicals into rivers, lakes, and oceans. At a high enough concentration in the air, in water, or in our bodies, almost any chemical can cause harm and be classified as a pollutant.

The pollutants we produce come from two types of sources. **Point sources** are single, identifiable sources. Examples are the smokestack of a coal-burning power or industrial plant (Figure 1-8), the drainpipe of a factory, and the exhaust pipe of an automobile. **Nonpoint sources** are dispersed and often difficult to identify. Examples are pesticides blown from the land into the air and the runoff of fertilizers, pesticides, and trash from the land into streams and lakes (Figure 1-9). It is much easier and cheaper to identify and control or prevent pollution from point sources than from widely dispersed nonpoint sources.

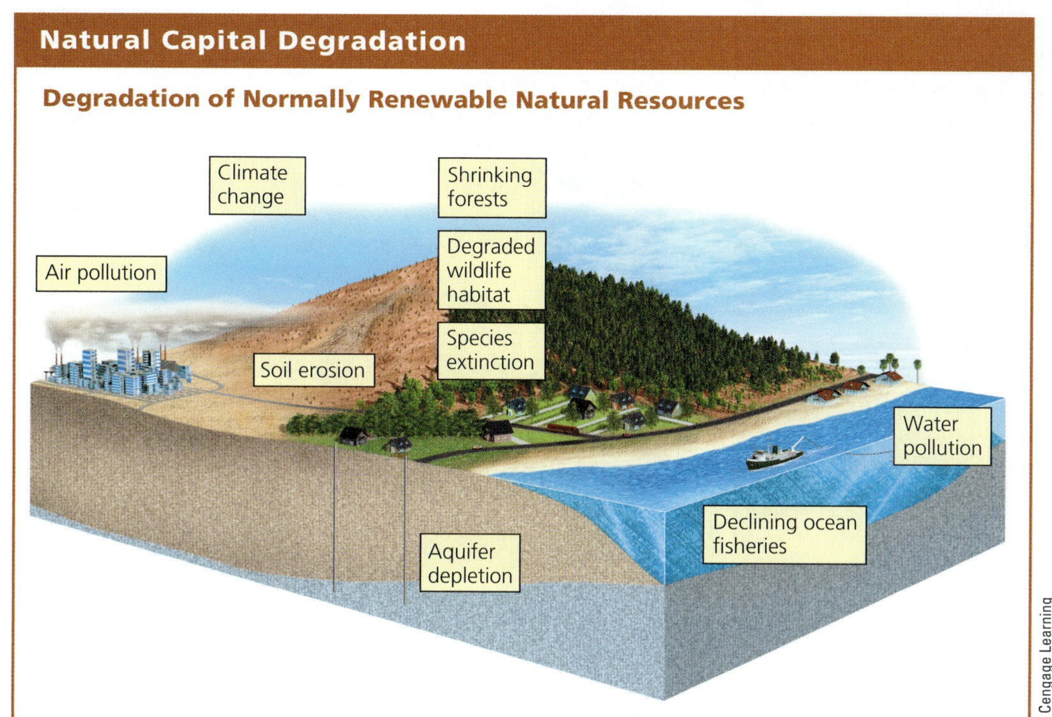

Figure 1-7 **Natural capital degradation:** Examples of the degradation of normally renewable natural resources and natural services (Figure 1-3) in parts of the world, mostly as a result of growing populations and rising rates of resource use per person.

We have tried to deal with pollution in two very different ways. One approach is **pollution cleanup**, which involves cleaning up or diluting pollutants after we have produced them. For example, we can dilute and reduce the local effects of air pollutants from smokestacks (Figure 1-8) by releasing them high into the atmosphere where winds can carry them to downwind areas. However, while tall smokestacks can reduce local air pollution they can increase air pollution in downwind areas.

The other approach is **pollution prevention**, efforts focused on greatly reducing or eliminating the production of pollutants. For example, we can enact pollution-control laws that ban, or set low levels for, the emission of various pollutants into the atmosphere or into bodies of water. Many environmental scientists say that pollution prevention is a key to a more sustainable future because it works better and in the long run is cheaper than cleanup.

Figure 1-8 *Point-source air pollution* from smokestacks in a coal-burning industrial plant.

Figure 1-9 The trash in this river came from a large area of land and is an example of *nonpoint-source water pollution*.

© Chris Fredriksson/Alamy

© Jim Pruitt/Photos.com

Figure 1-10 *Patterns of natural resource consumption:* The photo on the left shows a makeshift home for a poor family in a Mumbai, India, slum with few possessions. The photo on the right shows a typical suburban home for a U.S. family in Maryland that is filled with their possessions.

The Tragedy of the Commons: Degrading Commonly Shared Renewable Resources

Some renewable resources, known as *open-access renewable resources,* can be used by almost anyone. Examples are the atmosphere, the open ocean, and its fishes. Other examples of less open, but often *shared resources,* are grasslands and forests. Many of these commonly held resources have been environmentally degraded. In 1968, biologist Garrett Hardin (1915–2003) called such degradation the *tragedy of the commons.*

Degradation of a shared or open-access resource occurs because each user of the resource reasons, "If I do not use this resource, someone else will. The little bit that I use or pollute is not enough to matter, and anyway, it's a renewable resource." When the number of users is small, this logic works. Eventually, however, the cumulative effect of many people trying to exploit a shared resource can degrade it and eventually exhaust or ruin it, and then no one can benefit from it. That is the tragedy.

There are two major ways to deal with this difficult problem. One is to use a shared or open-access renewable resource at a rate well below its estimated sustainable yield by using less of the resource, regulating access to the resource, or doing both. For example, governments can establish laws and regulations limiting the annual harvests of various types of ocean fishes or regulating the amount of pollutants that we add to the atmosphere or the oceans.

The other way is to convert shared renewable resources to private ownership. The reasoning is that if you own something, you are more likely to protect your investment. However, history shows that shared renewable resources such as forests and soils can be quickly degraded, once they are privately owned. Also, this approach is not practical for open-access resources such as the atmosphere, the oceans, and our global life-support system, which cannot be divided up and sold as private property.

Ecological Footprints: Our Environmental Impacts

Many poor people in less-developed countries struggle to survive. Their individual use of resources and the resulting environmental impact is low and is devoted mostly to meeting their basic needs and living in makeshift housing (Figure 1-10, left). However, a large number of such people in a given area can have a large total environmental impact. By contrast, many individuals in more-developed nations enjoy **affluence**, or wealth, which allows them to consume large amounts of resources far beyond their basic needs and live in large homes filled with possessions (Figure 1-10, right). Their environmental impact is more a product of their rates of resource use per person.

When people use renewable resources, it can result in natural capital degradation (Figure 1-7), pollution, and wastes. We can think of this harmful environmental impact as an **ecological footprint**—the amount of land and water needed to supply a person or an area with renewable resources such as food and water, and that are needed to absorb and recycle the wastes and pollution produced by such resource use. (The developers of this tool for measuring environmental impacts chose to focus on renewable resources, although the use of nonrenewable resources also contributes to our ecological footprints.)

The **per capita ecological footprint** is the average ecological footprint of an individual in a given country or

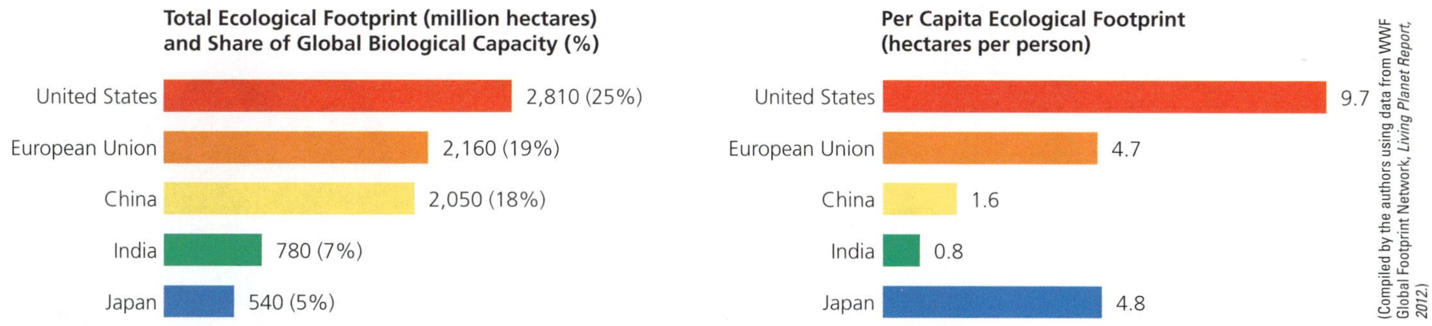

(Compiled by the authors using data from WWF Global Footprint Network, *Living Planet Report, 2012.*)

Figure 1-11 **Natural capital use and degradation:** These graphs show the total and per capita ecological footprints of selected countries.

area. We can use various simple online tools to estimate our ecological footprints. Figure 1-11 compares the total and per capita ecological footprints for selected countries. Figure 1-12 shows the human ecological impact in different parts of the world. In Supplement 6, see a map of the ecological footprint for North America (Figure 8, p. S33).

If the total ecological footprint for a city, a country, or the world is larger than its *biological capacity* to replenish its renewable resources and absorb the resulting wastes and pollution, it is said to have an *ecological deficit.* In other words, its people are living unsustainably by depleting their natural capital instead of living off the renewable

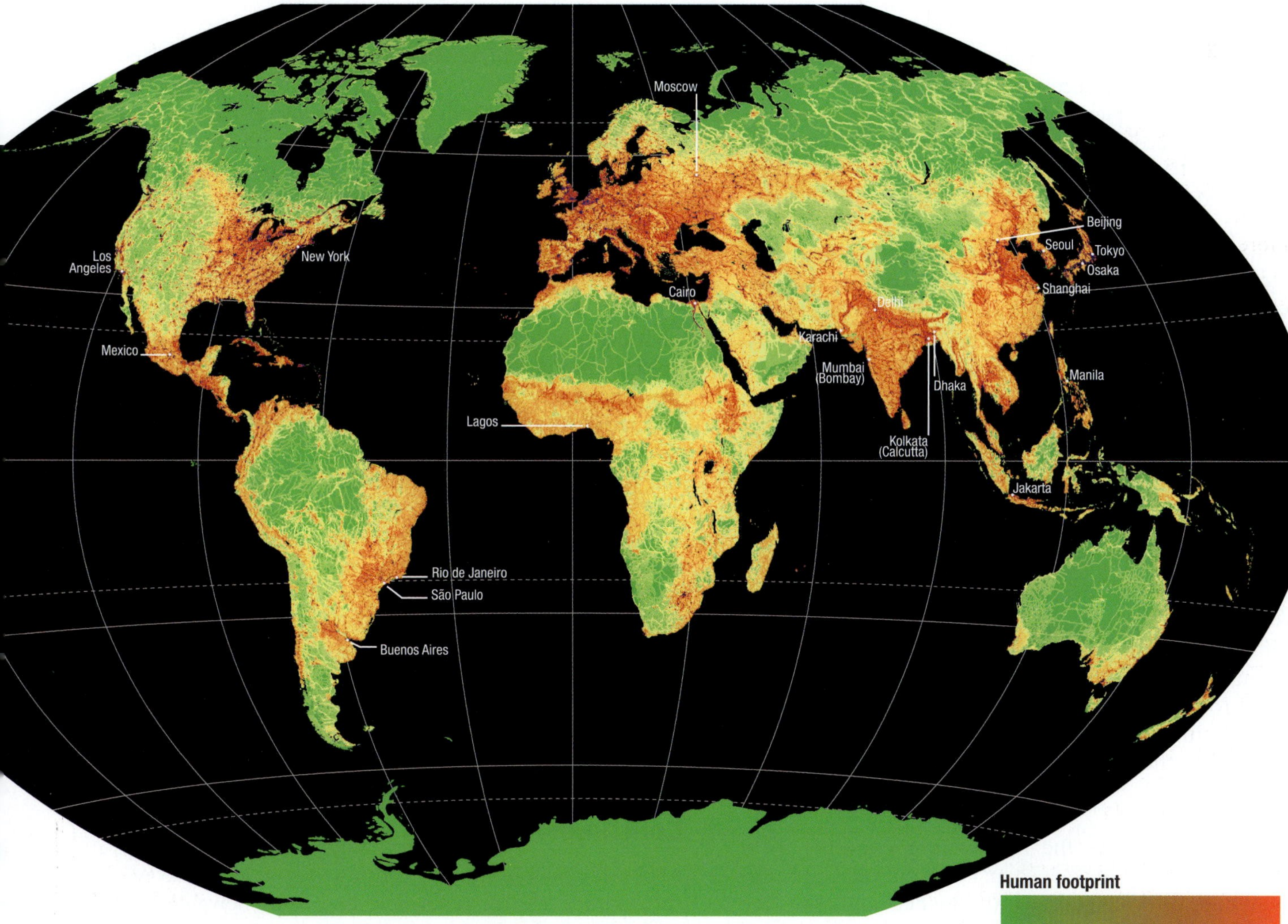

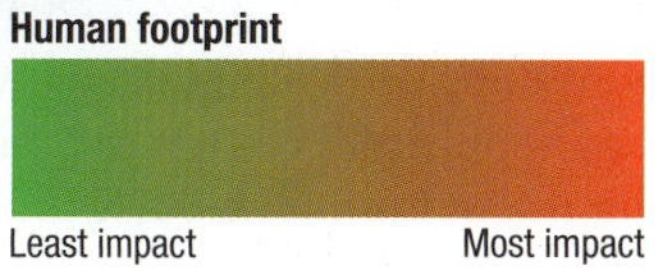

(National Geographic Earth Pulse; data from Mark Levy at Columbia University's Center for International Earth Science Information, 2002, and WWF, *Living Planet Report 2012.*)

Figure 1-12 **Natural capital use and degradation:** The human ecological footprint has an impact on about 83% of the earth's total land surface.

(Compiled by the authors using data from WWF, *Living Planet Report 2012*.)

Figure 1-13 **Natural capital degradation:** Estimated number of planet Earths needed to indefinitely support our ecological footprints today (left), in 2030 (middle), and today if everyone were to have the same ecological footprint as the average American now has. ***Question:*** If we are living beyond the earth's renewable biological capacity, why do you think the human population and per capita resource consumption are still growing rapidly?

supply of resources provided by such capital. Globally we are running up a huge ecological deficit (Figure 1-13). (In Supplement 6, see Figure 4, p. S29 for a map of countries that are either ecological debtors or ecological creditors.)

Ecological footprint data and models, though imperfect, provide useful rough estimates of our individual, national, and global environmental impacts. We can use existing and emerging technologies and economic tools to reduce the size of our ecological footprints over the next few decades to the point where we need only one earth to support the human population (**Core Case Study**). We discuss ways to make this shift throughout this book.

IPAT Is Another Environmental Impact Model

In the early 1970s, scientists Paul Ehrlich and John Holdren developed a simple model showing how population size (P), affluence, or resource consumption per person (A), and the beneficial and harmful environmental effects of technologies (T) help to determine the environmental impact (I) of human activities. This provides another rough estimate of how much humanity is degrading the natural capital it depends on. We can summarize this model by the simple equation:

$$\text{Impact (I)} = \text{Population (P)} \times \text{Affluence (A)} \times \text{Technology (T)}$$

Figure 1-14 shows the relative importance of these three factors in less-developed and more-developed countries. While the ecological footprint model emphasizes the use of renewable resources, this model includes the per capita use of both renewable and nonrenewable resources.

Some forms of technology such as polluting factories, gas-guzzling motor vehicles, and coal-burning power plants increase environmental impact by raising the harmful T factor (Figure 1-14, red arrows). However, other technologies reduce environmental impact by decreasing the T factor (Figure 1-14, green arrows). Examples are pollution control and prevention technologies, fuel-efficient cars, and wind turbines and solar cells that generate electricity with a low environmental impact.

In most less-developed countries, the key factor in total environmental impact (Figure 1-14, top) is usually population size (P) as a growing number of poor people struggle to stay alive. For example, while each individual in a rural village needs only a small amount of wood for heating and cooking, the total population of the village can eventually clear most or all of the trees in the nearby area for firewood. In more-developed countries, the key factor in overall environmental impact is affluence (A), which can result in high rates of per capita resource use, pollution, and resource depletion (Figure 1-14, bottom). Affluence is also a key factor in less-developed countries whose economies and populations are growing (see the following Case Study).

CASE STUDY

China's Growing Number of Affluent Consumers

About 1.4 billion affluent consumers put immense pressure on the earth's renewable and nonrenewable natural capital. Most of these middle-class consumers live in more-developed countries, but an estimated 300 million of

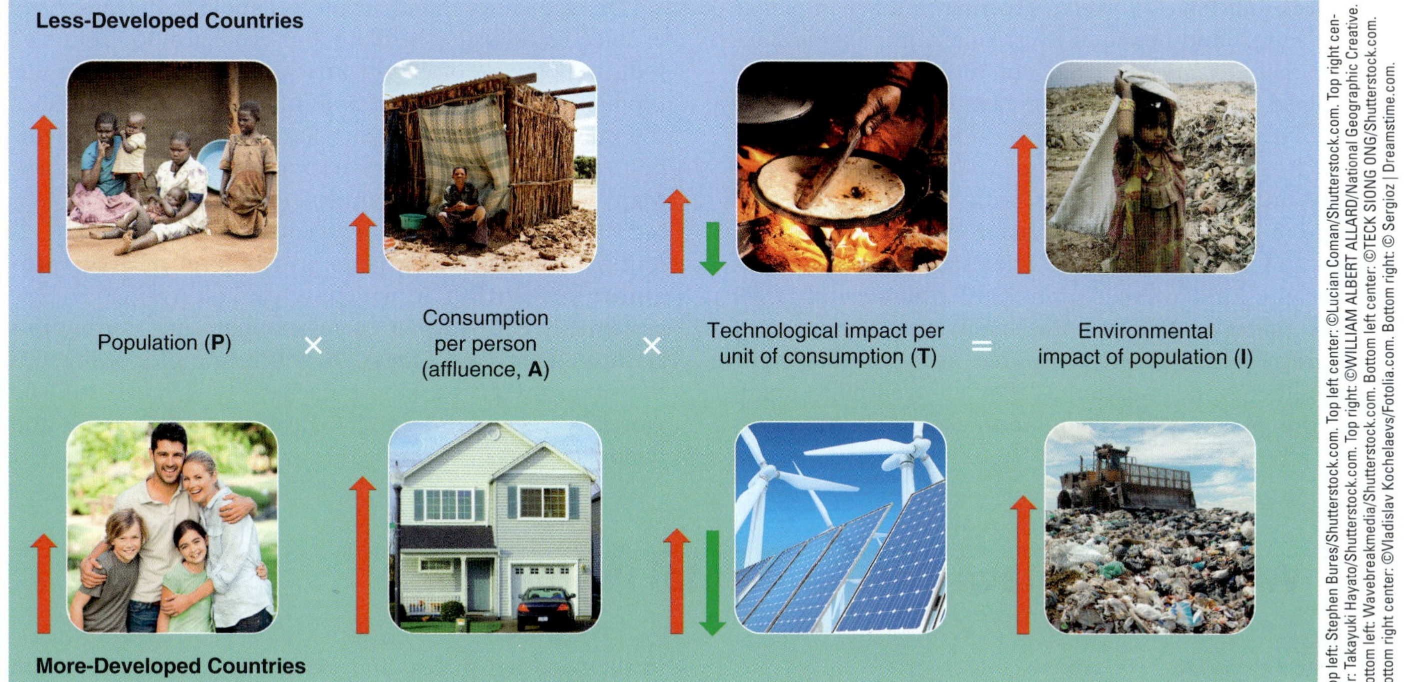

Figure 1-14 This simple model demonstrates how three factors—population size, affluence (resource use per person), and technology—help to determine the environmental impacts of populations in less-developed countries (top) and more-developed countries (bottom). Red arrows show generalized harmful impacts and green arrows show generalized beneficial impacts.

them—a number almost the size of the U.S. population—live in China, and this number may double by 2020.

China has the world's largest population and second-largest economy. It is the world's leading consumer of wheat, rice, meat, coal, fertilizer, steel, cement, and oil. China also leads the world in the production of goods such as televisions, cell phones, and refrigerators. It has also produced more wind turbines than any other country and will soon become the world's largest producer of solar cells and fuel-efficient cars. Between 2010 and 2025, China expects to build 10 cities the size of New York City.

Now, after 20 years of industrialization, China contains two-thirds of the world's most polluted cities. Some of its major rivers are choked with waste and pollution, and some areas of its coastline are basically devoid of fishes and other ocean life. A massive cloud of air pollution, largely generated in China, affects China and other Asian countries, areas of the Pacific Ocean, and even parts of the West Coast of North America.

Suppose that China's economy continues to grow at an overall rapid rate and its population size reaches 1.5 billion by around 2025, as projected by UN population experts. Environmental policy expert Lester R. Brown estimates that if such projections are accurate, China will need two-thirds of the world's current grain harvest, twice the amount of paper now consumed in the world, and more than all the oil currently produced in the world.

According to Brown:

> The western economic model—the fossil fuel–based, automobile-centered, throwaway economy—is not going to work for China. . . or for the other 3 billion people in developing countries who are also dreaming the "American dream."

For more details on China's growing ecological footprint, see the online Guest Essay by Norman Myers on this topic.

Cultural Changes Can Grow or Shrink Our Ecological Footprints

Until about 10,000 to 12,000 years ago, we were mostly *hunter–gatherers* who obtained food by hunting wild animals or scavenging their remains, and gathering wild plants. Early hunter–gatherers lived in small groups, consumed few resources, had few possessions, and moved as needed to find enough food for their survival.

Since then, three major cultural changes have occurred. *First* was the *agricultural revolution,* which began 10,000–12,000 years ago when humans learned how to grow and breed plants and animals for food, clothing, and other purposes. *Second* was the *industrial–medical revolution,* beginning about 275 years ago when people invented machines for the large-scale production of goods in factories. This involved learning how to get energy from fossil fuels (such as coal and oil) and how to grow large quantities of food in an efficient manner. It also included medical advances that have allowed a growing number of people to live longer and healthier lives. Finally, the *information–globalization revolution* began about 50 years ago, when we developed new technologies for gaining rapid access to all kinds of information and resources on a global scale.

Each of these three cultural changes gave us more energy and new technologies with which to alter and control more of the planet's resources to meet our basic needs and increasing wants. They also allowed expansion of the human population, mostly because of larger food supplies and longer life spans. In addition, they each resulted in greater resource use, pollution, and environmental degradation as they allowed us to dominate the planet and expand our ecological footprints (Figure 1-13).

On the other hand, some technological leaps have enabled us to begin shrinking our ecological footprints by reducing our use of energy and matter resources and our production of wastes and pollution. For example, use of the energy-efficient compact fluorescent and LED lightbulbs and energy-efficient cars and buildings is on the rise. Many environmental scientists and other analysts see such developments as evidence of an emerging fourth major cultural change in the form of a **sustainability revolution**, in which we could learn to live more sustainably with smaller ecological footprints, during this century (**Core Case Study**). A key factor in making this transition involves asking ourselves: How much stuff do we really need in order to live healthful, comfortable, and happy lives?

GOOD NEWS

1-3 Why Do We Have Environmental Problems?

CONCEPT 1-3A
Major causes of environmental problems are population growth, unsustainable resource use, poverty, avoidance of full-cost pricing, and increasing isolation from nature.

CONCEPT 1-3B
Our environmental worldviews play a key role in determining whether we live unsustainably or more sustainably.

Experts Have Identified Several Causes of Environmental Problems

According to a number of environmental and social scientists, the major causes of the environmental problems we face are **(1)** population growth, **(2)** wasteful and unsustainable resource use, **(3)** poverty, **(4)** failure to include the harmful environmental costs of goods and services in their market prices, and **(5)** increasing isolation from nature (Figure 1-15) (Concept 1-3A).

We discuss each of these causes in detail in later chapters of this book. Let us begin with a brief overview of them.

The Human Population Is Growing at a Rapid Rate

Exponential growth occurs when a quantity such as the human population increases at a fixed percentage per unit of time, such as 1% or 2% per year. Exponential growth starts off slowly, but after only a few doublings, it grows to enormous numbers because each doubling is twice the total of all earlier growth.

Here is an example of the immense power of exponential growth. Fold a piece of paper in half to double its thickness. If you could continue doubling the thickness of

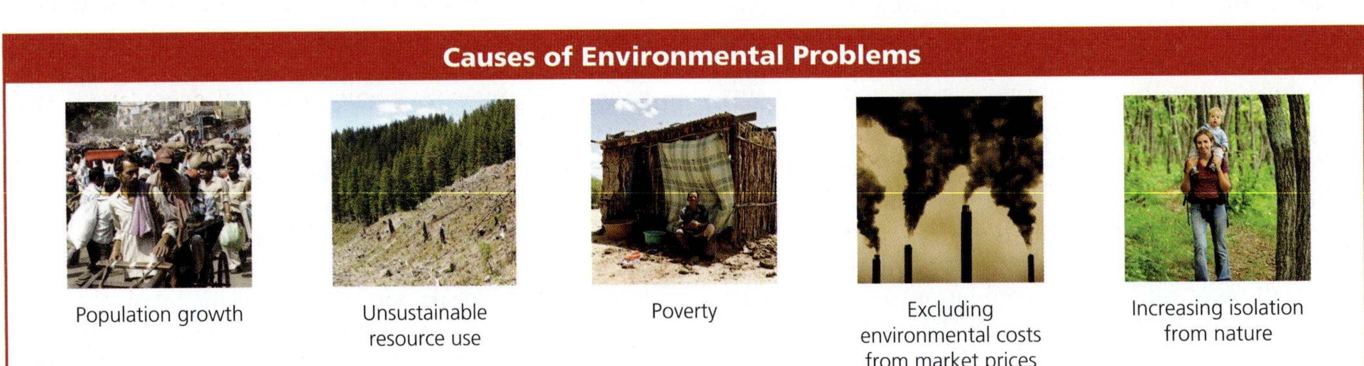

Figure 1-15 Environmental and social scientists have identified five basic causes of the environmental problems we face (Concept 1-3A). ***Question:*** For each of these causes, what are two environmental problems that result?

Left: Jeremy Richards/Shutterstock.com. Left center: steve estvanik/Shutterstock.com. Center: ©Lucian Coman/Shutterstock.com. Right center: El Greco/Shutterstock.com. Right: ©Maxim Tupikov/Shutterstock.com.

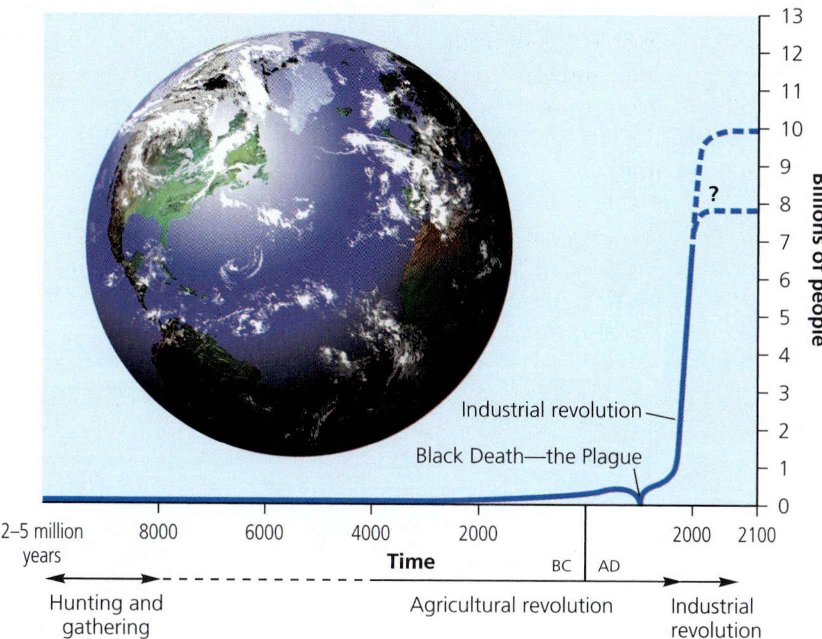

Figure 1-16 *Exponential growth:* The J-shaped curve represents past exponential world population growth, with projections to 2100 showing possible population stabilization as the J-shaped curve of growth changes to an S-shaped curve. (This figure is not to scale.)

(Compiled by the authors using data from the World Bank, United Nations, and Population Reference Bureau.)

the paper 50 times, it would be thick enough to almost reach the sun—150 million kilometers (93 million miles) away! Hard to believe, isn't it?

The human population has been growing exponentially (Figure 1-16). There are now about 7.1 billion people on the earth with about 84 million more people added each year. There could be 9.6 billion of us by 2050. This projected addition of 2.5 billion more people within your lifetime is more than 8 times the current U.S. population and 2 times that of China, the world's most populous nation.

CONSIDER THIS. . .

CONNECTIONS Exponential Growth and Doubling Time: The Rule of 70

The doubling time of the human population or any factor can be calculated by using the rule of 70: doubling time (years) = 70/annual growth rate (%). The world's population is growing at about 1.2% per year. At this rate how long will it take to double its size? If you have an investment that grows by 10% a year, how long will it take to double your money?

No one knows how many people the earth can support indefinitely, and at what level of average resource consumption per person, without seriously degrading the ability of the earth to support us, our economies, and the millions of other forms of life. However, our expanding total and per capita ecological footprints (Figures 1-12 and 1-13) are disturbing warning signs.

We could slow population growth with the goal of having it level off at around 8 billion by 2050, as suggested in the **Core Case Study**, instead of growing to 9.6 billion or more. Some ways to do this include reducing poverty through economic development, promoting family planning, and elevating the status of women, as discussed in Chapter 6.

Affluence Has Harmful and Beneficial Environmental Effects

The lifestyles of many consumers in more-developed countries and in less-developed countries such as China and India are built on growing affluence, which results in high levels of resource consumption and the resulting environmental degradation, wastes, and pollution.

These harmful environmental effects of affluence can be dramatic. In its 2012 *Living Planet Report,* the World Wildlife Fund (WWF) estimated that the United States is responsible for almost half of the global ecological footprint. The average American consumes about 30 times as much as the average Indian and 100 times as much as the average person in the world's poorest countries. According to some ecological footprint estimates, it takes about 27 large tractor-trailer loads of resources per year to support one typical American, and in 2012, there were 315 million Americans.

The problem is that providing each of these tractor-trailer loads represents environmental degradation, such as air pollution and water pollution from factories and motor vehicles and land degradation from the mining of raw materials used to make the products we consume. This is why higher levels of consumption expand a person's ecological footprint. Another downside to wealth is that it allows affluent consumers to obtain their resources from almost anywhere in the world without seeing the harmful environmental and health impacts of their high-consumption lifestyles.

On the other hand, affluence can allow for better education, which can lead people to become more concerned about environmental quality. It also provides money for developing technologies to reduce pollution, environ-

Figure 1-17 Poor settlers in Peru have cleared and burned this small plot of tropical rain forest in the Amazon and planted it with maize seedlings.

mental degradation, and resource waste. As a result, in the United States and most other affluent countries, the air is clearer, drinking water is purer, and most rivers and lakes are cleaner than they were in the 1970s. In addition, the food supply is more abundant and safer, the incidence of life-threatening infectious diseases has been greatly reduced, life spans are longer, and some endangered species are being rescued from extinction hastened by human activities. GOOD NEWS

These improvements in environmental quality were achieved because of greatly increased scientific research and technological advances financed by affluence. Education also spurred many citizens to insist that businesses and governments work toward improving environmental quality.

Poverty Has Harmful Environmental and Health Effects

Poverty is a condition in which people are unable to fulfill their basic needs for adequate food, water, shelter, health care, and education. According to the World Bank, about 900 million people—almost three times the U.S. population—live in *extreme poverty* (Figure 1-10, left), struggling to live on the equivalent of less than \$1.25 a day. This is less than what many people spend for a bottle of water or a cup of coffee. About one of every three, or 2.6 billion, of the world's people struggles to live on less than \$2.25 a day. Could you do this?

Poverty causes a number of harmful environmental and health effects. The daily lives of the world's poorest people are focused on getting enough food, water, and cooking and heating fuel to survive. Desperate for short-term survival, these individuals collectively degrade forests (Figure 1-17), topsoil, grasslands, fisheries, and wildlife. They do not have the luxury of worrying about long-term environmental quality or sustainability. Even though the poor in less-developed countries use very few resources per person, the large population size in some countries leads to a high overall environmental impact (Figure 1-14, top).

However, poverty does not necessarily lead to environmental degradation. Some of the world's poor people have learned how to take care of their environment as a part of their long-term survival strategy. For example, many small-scale farmers in African countries plant and nurture trees and work to conserve the soils that they depend on.

CONSIDER THIS. . .

CONNECTIONS Poverty and Population Growth

To many poor people, having more children is a matter of survival. Their children help them gather fuel (mostly wood and animal dung), haul drinking water, and tend crops and livestock. The children also help to care for their parents in their old age (their 40s or 50s in the poorest countries) because they do not have social security, health care, and retirement funds. This is largely why populations in some of the poorest less-developed countries continue to grow at high rates.

While poverty can increase some types of environmental degradation, the reverse is also true. Pollution and environmental degradation have a severe impact on the poor and can worsen their poverty. Consequently, many of the world's poor people die prematurely from several preventable health problems. One such problem is *malnutrition* caused by a lack of protein and other nutrients needed for good health (Figure 1-18). The resulting weakened condition can increase an individual's chances of death from starvation and normally nonfatal ailments such as diarrhea and measles.

A second health problem is limited access to adequate sanitation facilities and clean drinking water. More than one-third of the world's people have no real bathroom facilities and are forced to use backyards, alleys, ditches, and streams. As a result, about one of every eight of the world's people get water for drinking, washing, and cooking from sources polluted by human and animal feces. A third health problem is severe respiratory disease that people get from breathing the smoke of open fires or poorly vented stoves used for heating and cooking inside their dwellings. This indoor air pollution kills about 2.4 million people a year, according to the World Health Organization.

In 2010, the World Health Organization estimated that one or more of these factors, mostly related to poverty, cause premature death for about 7 million children under the age of 5 each year. Some hopeful news is that this number of annual deaths is down from about 10 million in 1990. Even so, every day an average of at least 19,000 young children die prematurely from these causes. This is equivalent to *95 fully loaded 200-passenger airliners crashing every day with no survivors.* The news media rarely cover this ongoing human tragedy.

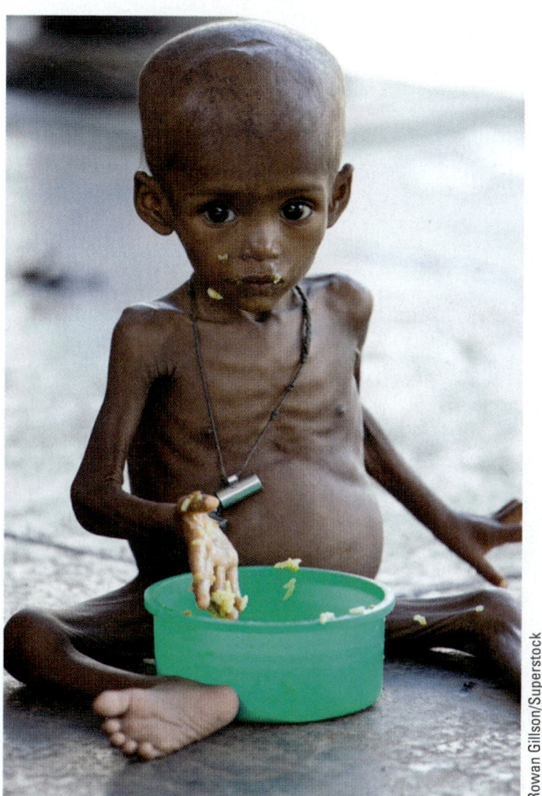
Rowan Gillson/Superstock

Figure 1-18 One of every three children younger than age 5 in less-developed countries, including this starving child in Bangladesh, suffers from severe malnutrition caused by a lack of calories and protein.

CONSIDER THIS. . .

THINKING ABOUT The Poor, the Affluent, and Rapidly Increasing Population Growth

Some see the rapid population growth of the poor in less-developed countries as the primary cause of our environmental problems (Figure 1-14, top). Others say that the much higher resource use per person in more-developed countries is the more important factor (Figure 1-14, bottom). Which factor do you think is more important? Why?

Prices of Goods and Services Do Not Include Harmful Environmental Costs

Another basic cause of environmental problems has to do with how goods and services are priced in the marketplace. Companies using resources to provide goods for consumers generally are not required to pay for most of the harmful environmental and health costs of supplying such goods. For example, timber companies pay the cost of clear-cutting forests but do not pay for the resulting environmental degradation and loss of wildlife habitat. The primary goal of a company is to maximize profits for its owners or stockholders. Indeed, it would be economic suicide for a company to add these costs to its prices unless government regulations were to create a level economic playing field by using taxes or regulations to require all businesses to pay for the environmental and health costs of producing their products.

Because the prices of goods and services do not include most of their harmful environmental and human health costs, consumers have no effective way to evaluate the harmful effects, on their own health and on the earth's life-support systems, of producing and using these goods and services. For example, producing and using gasoline result in air pollution and other problems that damage the environment and people's health. Scientists and economists have estimated that the real cost of gasoline to U.S. consumers would be about $3.18 per liter ($12 per gallon) if the estimated short- and long-term harmful environmental and health costs were included in its pump price.

Partly because of this inadequate pricing system, many people do not understand the importance and truly high value of the natural resources and natural services that make up the earth's natural capital (Figure 1-3). This lack of information is a major reason for why we are degrading these key components of our life-support system (Figure 1-7).

Another problem arises when governments (taxpayers) give companies *subsidies* such as tax breaks and pay-

Figure 1-19 These ecotourists atop endangered Asian elephants in India's Kaziranga National Park are learning about threatened barasingha deer and numerous other species.

ments to assist them with using resources to run their businesses. This helps to create jobs and stimulate economies, but environmentally harmful subsidies encourage the depletion and degradation of natural capital. (See the online Guest Essay on this topic by Norman Myers.)

We could live more sustainably by finding ways to include in market prices the harmful environmental and health costs of the goods and services that we use. Implementing such full-cost pricing would be a way to apply one of the three social science **principles of sustainability** (Figure 1-5). Two ways to do this over the next two decades would be to shift from environmentally harmful government subsidies to environmentally beneficial subsidies, and to tax pollution and waste heavily while reducing taxes on income and wealth. We discuss such *subsidy shifts* and *tax shifts* in Chapter 23.

We Are Increasingly Isolated from Nature

Today, three out of four people in the more-developed countries and one of every two people in the world live in urban areas, and this shift from rural to urban living is continuing at a rapid pace. Our artificial urban environments and our increasing use of cell phones, computers, and other electronic devices isolate us from the natural world that provides the food, water, and most of the raw materials used to produce the consumer goods that we depend on.

Thus, it is not surprising that many people do not know the full story of where their food, water, and other goods come from. Similarly, many people are unaware of the amounts of wastes and pollutants they produce, of where they go, and of how these wastes and pollutants affect the environment. Many do not understand that life on earth has been sustained largely by the recycling of wastes—one of the scientific **principles of sustainability** (Figure 1-2).

According to some analysts, many of us are increasingly suffering from *nature deficit disorder.* They suggest that by not having enough contacts with the natural world (chapter-opening photo and Figure 1-19) a person can be more likely to suffer from stress, have health problems, show unwarranted irritability or aggression, and be less adaptable to changes in life. These analysts also argue that this disorder helps to explain why we are rapidly degrading our life-support system. They ask: How will we shrink our ecological footprints and live more sustainably if we do not experience and understand our utter dependence on the earth's natural systems and the natural capital they provide for us?

People Have Different Views about Environmental Problems and Their Solutions

Another challenge we face is that people differ over the seriousness of the world's environmental problems and what we should do to help solve them. Differing opinions about environmental problems and solutions arise mostly out of differing environmental worldviews. Your **environmental worldview** is your set of assumptions and values reflecting how you think the world works and what you think your role in the world should be. **Environmental ethics**, which is the study of our various beliefs about what is right and wrong with how we treat the environment, provides important ways to examine our worldviews. Here are some important *ethical questions* relating to the environment:

- Why should we care about the environment?
- Are we the most important beings on the planet or are we just one of the earth's millions of different forms of life?
- Do we have an obligation to see that our activities do not cause the extinction of other species? Should we try to protect all species or only some? How do we decide which to protect?
- Do we have an ethical obligation to pass on to future generations the extraordinary natural world in a condition that is at least as good as what we inherited?
- Should every person be entitled to equal protection from environmental hazards regardless of race, gender, age, national origin, income, social class, or any other factor? This is the central ethical and political issue for what is known as the *environmental justice* movement. (See the online Guest Essay on this topic by Robert D. Bullard.)
- How do we promote sustainability?

CONSIDER THIS. . .

THINKING ABOUT Our Responsibilities

How would you answer each of the questions above? Compare your answers with those of your classmates. Record your answers and, at the end of this course, return to these questions to see if your answers have changed.

People with widely differing environmental worldviews can take the same data, be logically consistent with it, and arrive at quite different answers to such questions because they start with different assumptions and moral, ethical, or religious beliefs. Environmental worldviews are discussed in detail in Chapter 25, but here is a brief introduction.

The **planetary management worldview** holds that we are separate from and in charge of nature, that nature exists mainly to meet our needs and increasing wants, and that we can use our ingenuity and technology to manage the earth's life-support systems, mostly for our benefit, into the distant future.

The **stewardship worldview** holds that we can and should manage the earth for our benefit, but that we have an ethical responsibility to be caring and responsible managers, or *stewards,* of the earth. It says we should encourage environmentally beneficial forms of economic growth and development and discourage environmentally harmful forms.

The **environmental wisdom worldview** holds that we are part of, and dependent on, nature and that the earth's life-support system exists for all species, not just for us. According to this view, our success depends on learning how the earth sustains itself (Figure 1-2 or back cover of this book) and integrating such *environmental wisdom* into the ways we think and act. SUSTAINABILITY

1-4 What Is an Environmentally Sustainable Society?

CONCEPT 1-4

Living sustainably means living off the earth's natural income without depleting or degrading the natural capital that supplies it.

The More Environmentally Sustainable Societies Protect Natural Capital and Live Off Its Income

According to most environmental scientists, our ultimate goal should be to achieve an **environmentally sustainable society**—one that meets the current and future basic resource needs of its people in a just and equitable manner without compromising the ability of future generations to meet their basic needs (**Core Case Study**).

Imagine that you win $1 million in a lottery. Suppose you invest this money (your capital) and earn 10% interest per year. If you live on just the interest, or the income made by your capital, you will have a sustainable annual income of $100,000 that you can spend each year indefinitely without depleting your capital. However, if you spend $200,000 per year, while still allowing interest to accumulate, all of your money will be gone early in the seventh year. Even if you spend only $110,000 per year and allow the interest to accumulate, you will be bankrupt within 25 years.

individuals matter 1.1

Tuy Sereivathana: Elephant Protector

Courtesy of Tom Dusenbery

As a young child in rural Cambodia, Tuy Sereivathana developed a respect for nature, became fascinated with elephants, and decided that he wanted to be an elephant protector.

Since 1970, Cambodia's rain forest cover has dropped from over 70% of the country's land area to 3% primarily because of population growth, rapid development, illegal logging, and warfare. This severe loss of forests forced elephants to search for food and water on farmlands, and it set up a conflict between elephants and poor farmers who killed the elephants to protect their food supply. Elephants have also been killed illegally for their valuable ivory tusks.

Since 1995, Sereivathana, with a master's degree in forestry, has been on a mission to accomplish two goals. One is to raise the population of Cambodia's endangered Asian elephants from less than 400 in 2010 to 1,000 by 2030. The other is to show poor farmers that protecting elephants and other forms of wildlife can help them escape poverty.

Sereivathana directs the Cambodian Elephant Conservation Group, devoted to reducing poaching and helping farmers work together to use low-cost and innovative ways to protect their crops without having to kill elephants. Affectionately known as Uncle Elephant, he helped farmers set up nighttime lookouts for elephants. He taught villagers to scare raiding elephants away by using foghorns and fireworks and using solar-powered electric fences to mildly shock them. He also encouraged farmers to stop growing watermelons and bananas, which elephants love, and to grow crops such as eggplant and chile peppers that elephants shun. Measures such as these have lowered crop losses, and farmers have benefited economically.

Since 2005, mostly because of Sereivathana's efforts, no elephants have been killed in Cambodia as a result of conflicts with humans. His community-based model for elephant and wildlife conservation is being used in neighboring communities and in other countries such as Indonesia and Vietnam. His model is an outstanding example of applying the win-win **principle of sustainability** (Figure 1-5).

SUSTAINABILITY

In 2010, Sereivathana was one of the six recipients of the Goldman Environmental Prize (often dubbed the "Nobel Prize for the environment"). In 2011 he was named a National Geographic Emerging Explorer.

Background photo: Ekkachai/Shutterstock.com

The lesson here is an old one: *Protect your capital and live on the income it provides.* Deplete or waste your capital and you will move from a sustainable to an unsustainable lifestyle.

The same lesson applies to our use of the earth's natural capital (Figure 1-3)—the global trust fund that nature has provided for us, for future generations, and for the earth's other species. *Living sustainably* means living on **natural income**, the renewable resources such as plants, animals, soil, clean air, and clean water, provided by the earth's natural capital. It also means not depleting or degrading the earth's natural capital, which supplies this income, and providing the human population with adequate and equitable access to this natural income for the foreseeable future (**Concept 1-4**).

CONSIDER THIS. . .

CAMPUS SUSTAINABILITY* Arizona State University's School of Sustainability

Wrigley Hall houses Arizona State University's School of Sustainability, the first comprehensive degree-granting program in sustainability in the United States for both undergrads and graduate students. Established in 2007, it is an interdisciplinary program focused on finding practical solutions to environmental problems while considering important economic and social factors. The courses emphasize learning by doing, researching with faculty members, interacting with the business world and with K–12 students, committing to community service, and leadership training.

A More Sustainable Future Is Possible

Moving toward environmental sustainability will require an overall attitude that combines environmental wisdom with compassion for all forms of life, including humans. Some people seem to be born with this sort of attitude (see Individuals Matter 1.1). Others learn it as they experience nature in various ways.

Making a shift toward a more sustainable future will involve some tough challenges. However, here are two pieces of good news: *First,* research by social scientists suggests that it takes only 5–10% of the population of a community, a country, or the world to bring about major social change. *Second,* such research also shows that significant social change can occur in a much shorter time than most people think. Anthropologist Margaret Mead summarized our potential for social change: "Never doubt that a small group of thoughtful, committed citizens can change the world. Indeed, it is the only thing that ever has."

GOOD NEWS

Evidence from the physical sciences and the social sciences indicates that we have perhaps 50 years and no more than 100 years to make a new cultural shift from unsustainable living to more sustainable living, if we start now. One of the goals of this book is to provide a realistic vision of a more environmentally sustainable future (**Core Case Study**) based on energizing, realistic hope rather than on immobilizing fear, gloom, and doom.

Three strategies for reducing our ecological footprints, helping to sustain the earth's natural capital, and making a transition to more sustainable lifestyles and economies are summarized in the *three big ideas* of this chapter:

Big Ideas

- **A more sustainable future will require that we rely more on energy from the sun and other renewable energy sources, protect biodiversity through the preservation of natural capital, and avoid disrupting the earth's vitally important chemical cycles.**
- **A major goal for becoming more sustainable is full-cost pricing—the inclusion of harmful environmental and health costs in the market prices of goods and services.**
- **We will benefit ourselves and future generations if we commit ourselves to finding win-win-win solutions to our problems and to leaving the planet's life-support system in at least as good a shape as what we now enjoy.**

*In several chapters of this book, through this Campus Sustainability feature, we are highlighting the efforts of certain colleges and universities in various areas of sustainability.

TYING IT ALL TOGETHER A Vision of a More Sustainable Earth

Pecold/Shutterstock.com

We face an array of serious environmental problems. This book is about understanding these problems and finding *solutions* to them. A key to most solutions is to apply the three scientific **principles of sustainability** (Figure 1-2) and the three social science **principles of sustainability** (Figure 1-5) to the design of our economic and social systems, as well as to our individual lifestyles. We can use such strategies to try to slow the rapidly increasing losses of biodiversity, to sharply reduce production of wastes and pollution, to switch to more sustainable sources of energy, and to promote more sustainable forms of agriculture and other uses of land and water. We can also use these principles to sharply reduce poverty and slow human population growth.

SUSTAINABILITY

Suppose that we begin to make environmental choices during this century that promote sustainability, as the fictional characters Emily and Michael and people like them did in the **Core Case Study** that opens this chapter. Then, chances are that we will help to create an extraordinary and more sustainable future for ourselves, for future generations, and for most other forms of life on our planetary home. If we get it wrong, we face ecological disruption that could set humanity back for centuries and wipe out as many as half of the world's species, as well as much of the human population.

You have the good fortune to be a member of the 21st century's *transition generation* that will decide which path humanity takes. This means confronting the urgent challenges presented by the major environmental problems discussed in this book. It is an incredibly exciting and challenging time to be alive as we struggle to develop a more sustainable relationship with this planet that is our only home.

GOOD NEWS

Chapter Review

Core Case Study

1. Summarize the authors' vision of a more sustainable world, which could be attainable by 2065.

Section 1-1

2. What are the three key concepts for this section? Define **sustainability**. Define **environment**. Distinguish among **environmental science**, **ecology**, and **environmentalism**. Distinguish between an **organism** and a **species**. What is an **ecosystem**? What are three **scientific principles of sustainability** derived from how the natural world works? What is **solar energy** and why is it important to life on the earth? What is **biodiversity** and why is it important to life on the earth? Define **nutrients**. Define **chemical** or **nutrient cycling** and explain why it is important to life on the earth.

3. Define **natural capital**. Define **natural resources** and **natural services,** or **ecosystem services,** and give two examples of each. Give three examples of how we are degrading natural capital. Explain how finding solutions to environmental problems involves making trade-offs. Explain why individuals matter in dealing with the environmental problems we face. What are three **social science principles of sustainability**? What is **full-cost pricing** and why is it important?

4. What is a **resource**? Distinguish between an **inexhaustible resource** and a **renewable resource** and give an example of each. What is the **sustainable yield** of a renewable resource? Define and give two examples of a **nonrenewable** or **exhaustible resource**. Explain why the suggested priorities for more sustainable use of nonrenewable resources are, in order: **R**efuse, **R**educe, **R**euse, and **R**ecycle. What percentage of the metals and other nonrenewable materials that we use could be reused or recycled? Distinguish between **more-developed countries** and **less-developed countries** and give an example of a high-income, middle-income, and low-income country.

Section 1-2

5. What is the key concept for this section? Define and give three examples of **environmental degradation (natural capital degradation)**. About what percentage of the earth's natural or ecosystem services has been degraded by human activities? Define **pollution**. Distinguish between **point sources** and **nonpoint sources** of pollution and give an example of each. Distinguish between **pollution cleanup** and **pollution prevention** and give an example of each. What is the *tragedy of the commons*? What are two ways to deal with this effect? Explain why they don't work for some systems.

6. What is **affluence**? What is an **ecological footprint**? What is a **per capita ecological footprint**? Compare the total and per capita ecological footprints of the United States and China. Use the ecological footprint concept to explain how we are living unsustainably in terms of the estimated number of planet Earths that we need to sustain ourselves now and in the future.

7. What is the IPAT model for estimating our environmental impact? Explain how we can use this model to estimate the impacts of the human populations in less-developed and more-developed countries. Describe the environmental impacts of China's new affluent consumers. Describe three major cultural changes that have occurred since humans were hunter–gatherers and how they have increased our overall environmental impact. What would a **sustainability revolution** involve?

Section 1-3

8. What are the two key concepts for this section? Identify five basic causes of the environmental problems that we face. What is **exponential growth**? What is the rule of 70? What is the current size of the human population? How many people are added each year? How many people might be here by 2050? How do Americans, Indians, and the average people in the poorest countries compare in terms of average consumption per person? What are two types of environmental damage resulting from growing affluence? How can affluence help us to solve environmental problems? What is **poverty** and what are three of its harmful environmental and health effects? About how many of the world's people struggle to live on the equivalent of $1.25 a day? How many try to live on $2.25 a day? Explain the connection between poverty and population growth. List three major health problems suffered by many of the world's poor.

9. Explain how excluding the harmful environmental and health costs of production from the prices of goods and services affects the environmental problems we face. What is the connection between government subsidies, resource use, and environmental degradation? What are two ways to include the harmful environmental and health costs of the goods and services that we use in their market prices? Explain how lack of knowledge of the nature and importance of natural capital and our increasing isolation from nature can intensify the environmental problems that we face. What is an **environmental worldview**? What are **environmental ethics**? Distinguish among the **planetary management**, **stewardship**, and **environmental wisdom worldviews**.

Section 1-4

10. What is the key concept for this section? What is an **environmentally sustainable society**? What is **natural income** and what does it mean to live off of natural income? Describe Tuy Sereivathana's efforts to prevent elephants from becoming extinct in Cambodia and to reduce the country's poverty. What are two pieces of good news about making the transition to a more sustainable society? Based on the three scientific **principles of sustainability** and the three social science **principles of sustainability**, what are three important ways to make a transition to sustainability as summarized in this chapter's *three big ideas*? Explain how we can use the six principles of sustainability to move us closer to the vision of a more sustainable world described in the **Core Case Study** that opens this chapter.

Note: Key terms are in bold type. Knowing the meanings of these terms will help you in the course you are taking.

Critical Thinking

1. Do you think you are living unsustainably? Explain. If so, what are the three most environmentally unsustainable components of your lifestyle? List two ways in which you could apply each of the three scientific **principles of sustainability** (Figure 1-2) and each of the three social science **principles of sustainability** (Figure 1-5) to making your lifestyle more environmentally sustainable.

2. Do you believe that a vision such as the one described in the **Core Case Study** that opens this chapter is possible? Why or why not? What, if anything, do you believe will be different from that vision of the future? Explain. If your vision of what it will be like in 2065 is sharply different from that in the **Core Case Study**, write a description of your vision. Compare your answers to this question with those of your classmates.

3. For each of the following actions, state one or more of the three scientific **principles of sustainability** that are involved: **(a)** recycling aluminum cans; **(b)** using a rake instead of a leaf blower; **(c)** walking or bicycling to class instead of driving; **(d)** taking your own reusable bags to the grocery store to carry your purchases home; **(e)** volunteering to help restore a prairie; and **(f)** lobbying elected officials to require that at least 20% of your country's electricity be produced with renewable wind and solar power by 2020.

4. Explain why you agree or disagree with the following propositions:
   a. Stabilizing population is not desirable because, without more consumers, economic growth would stop.
   b. The world will never run out of resources because we can use technology to find substitutes and to help us reduce resource waste.

5. What do you think when you read that the average American consumes 30 times more resources than the average citizen of India? Are you skeptical, indifferent, sad, helpless, guilty, concerned, or outraged by this fact? Do you think that these differences in consumption have led to problems? If so, describe them and propose some possible solutions.

6. When you read that at least 19,000 children age 5 and younger die each day (13 per minute) from preventable malnutrition and infectious disease, how does it make you feel? Can you think of something that you and others could do to address this problem? What might that be?

7. Explain why you agree or disagree with each of the following statements: **(a)** humans are superior to other forms of life; **(b)** humans are in charge of the earth; **(c)** the value of other forms of life depends only on whether or not they are useful to humans; **(d)** based on records of past extinctions and the history of life on the earth over the last 3.5 billion years, biologists hypothesize that all forms of life eventually become extinct, and we should not worry about whether our activities hasten their extinction; **(e)** all forms of life have an inherent right to exist; **(f)** all economic growth is good; **(g)** nature has an almost unlimited storehouse of resources for human use; **(h)** technology can solve our environmental problems; **(i)** I do not believe I have any obligation to future generations; and **(j)** I do not believe I have any obligation to other forms of life.

8. What are the basic beliefs within your environmental worldview? Record your answer. Then, at the end of this course, return to your answer to see if your environmental worldview has changed. Are the beliefs included in your environmental worldview consistent with the answers you gave to Question 7 above? Are your actions that affect the environment consistent with your environmental worldview? Explain.

Doing Environmental Science

Estimate your own ecological footprint by using one of the many estimator tools available on the Internet. Is your ecological footprint larger or smaller than you thought it would be, according to this estimate? Why do you think this is so? List three ways in which you could reduce your ecological footprint. Try one of them for a week, and write a report on this change.

Global Environment Watch Exercise

Using the world maps in Figure 1, p. S24 and Figure 3, p. S28 in Supplement 6, choose one more-developed country and one less-developed country to compare their ecological footprints (found under Quick Facts on the country portal). Click on the ecological footprint number to view a graph of both the ecological footprint and biocapacity of each country. Using those graphs, determine which country appears to be living more sustainably. What would be some reasons for the differences in their levels of sustainability?

Ecological Footprint Analysis

If the *ecological footprint per person* of a country or the world is larger than its *biological capacity per person* to replenish its renewable resources and absorb the resulting waste products and pollution, the country or the world is said to have an *ecological deficit.* If the reverse is true, the country or the world has an *ecological credit* or *reserve.* Use the data below to calculate the ecological deficit or credit for the countries listed and for the world. (For a map of ecological creditors and debtors see Figure 4, p. S29, in Supplement 6.)

Place	Per Capita Ecological Footprint (hectares per person)	Per Capita Biological Capacity (hectares per person)	Ecological Credit (+) or Debit (−) (hectares per person)
World	2.2	1.8	−0.4
United States	9.8	4.7	
China	1.6	0.8	
India	0.8	0.4	
Russia	4.4	0.9	
Japan	4.4	0.7	
Brazil	2.1	9.9	
Germany	4.5	1.7	
United Kingdom	5.6	1.6	
Mexico	2.6	1.7	
Canada	7.6	14.5	

Compiled by the authors using data from WWF *Living Planet Report 2006.*

1. Which two countries have the largest ecological deficits? Why do you think they have such large deficits?
2. Which two countries have an ecological credit? Why do you think each of these countries has an ecological credit?
3. Rank the countries in order from the largest to the smallest per capita ecological footprint.

2 Science, Matter, Energy, and Systems

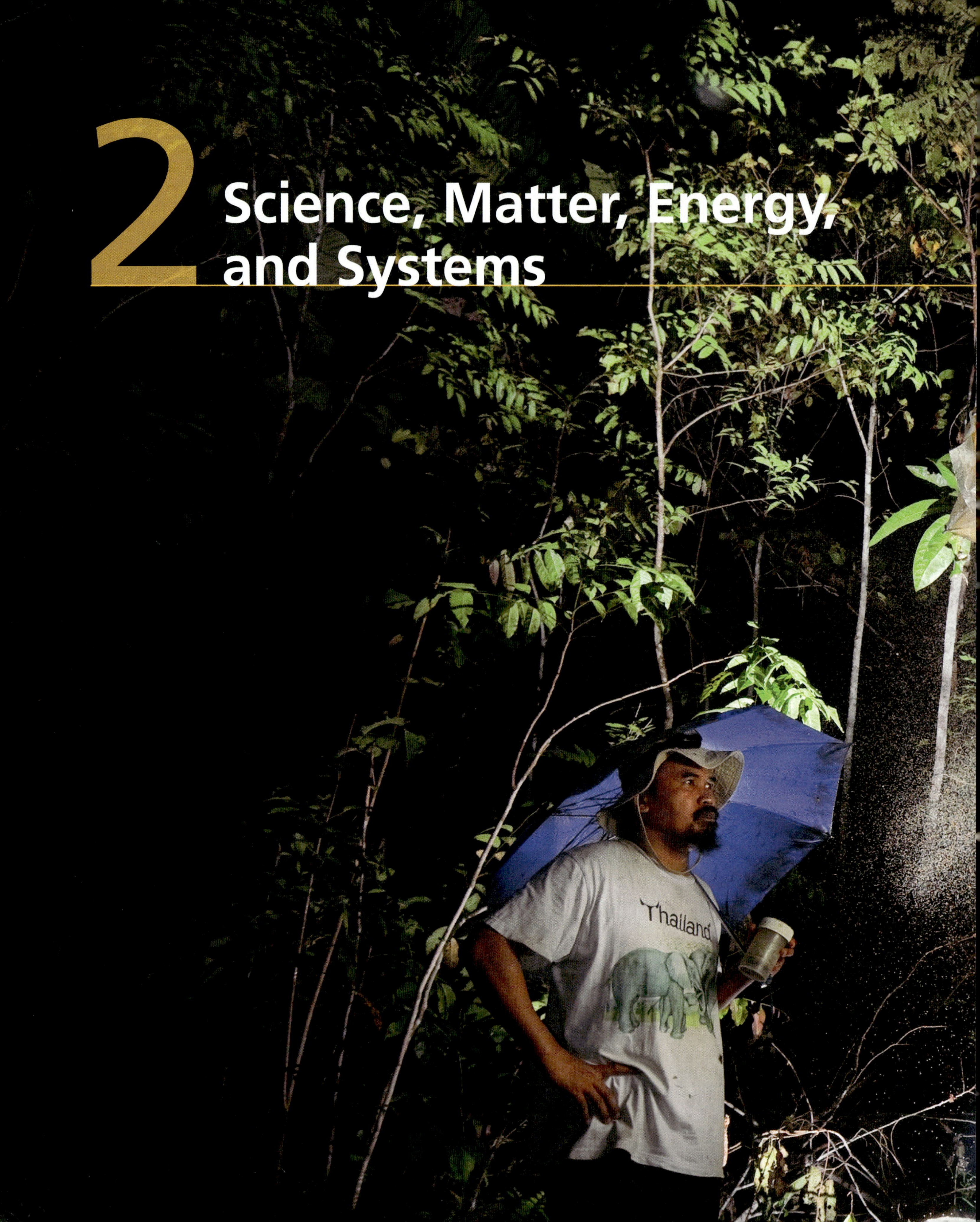

Science is built up of facts, as a house is built of stones; but an accumulation of facts is no more a science than a heap of stones is a house.

HENRI POINCARÉ

Key Questions

2-1 What do scientists do?

2-2 What is matter and what happens when it undergoes change?

2-3 What is energy and what happens when it undergoes change?

2-4 What are systems and how do they respond to change?

Scientist studying biodiversity in a tropical rain forest in Indonesia by using a light to trap moths.

Tim Laman/National Geographic Creative

CORE CASE STUDY How Do Scientists Learn about Nature? Experimenting with a Forest

Suppose a logging company plans to cut down all of the trees on a hillside behind your house. You are very concerned and want to know about the possible harmful environmental effects of this action.

One way to learn about such effects is to conduct a *controlled experiment,* just as environmental scientists do. They begin by identifying key *variables,* such as water loss and soil nutrient content, that might change after the trees are cut down. Then, they set up two groups. One is the *experimental group,* in which a chosen variable is changed in a known way. The other is the *control group,* in which the chosen variable is not changed. They then compare the results from the two groups.

In 1963, botanist F. Herbert Bormann, forest ecologist Gene Likens, and their colleagues began carrying out such a controlled experiment. Their goal was to compare the loss of water and soil nutrients from an area of uncut forest (the *control site*) with one that had been stripped of its trees (the *experimental site*).

They built V-shaped concrete dams across the creeks at the bottoms of several forested valleys in the Hubbard Brook Experimental Forest in New Hampshire (Figure 2-1). The dams were designed so that all surface water leaving each forested valley had to flow across a dam, where scientists could measure its volume and dissolved nutrient content.

First, the researchers measured the amounts of water and dissolved soil nutrients flowing from an undisturbed forested area in one of the valleys (the control site, Figure 2-1, left). These measurements showed that an undisturbed mature forest is very efficient at storing water and retaining chemical nutrients in its soils.

Next, they set up an experimental forest area in a nearby valley (Figure 2-1, right). One winter, they cut down all the trees and shrubs in that valley, left them where they fell, and sprayed the area with herbicides to prevent the regrowth of vegetation. Then, for 3 years, they compared outflow of water and nutrients in this experimental site with those in the control site.

The scientists found that, with no plants to help absorb and retain water, the amount of water flowing out of the deforested valley increased by 30–40%. As this excess water ran rapidly over the ground, it eroded soil and carried dissolved nutrients out of the topsoil in the deforested site. Overall, the loss of key soil nutrients from the experimental forest was 6–8 times that in the nearby uncut control forest.

In this chapter, you will learn more about how scientists study nature and about the matter and energy that make up the world within and around us. You will also learn about three *scientific laws,* or rules of nature, that govern the changes that matter and energy undergo. And you will learn the important difference between a scientific hypothesis and a scientific theory.

Figure 2-1 This controlled field experiment measured the loss of water and soil nutrients from a forest due to deforestation. The forested valley (left) was the control site; the cutover valley (right) was the experimental site.

2-1 What Do Scientists Do?

CONCEPT 2-1
Scientists collect data and develop hypotheses, theories, models, and laws about how nature works.

Science Is a Search for Order in Nature

Science is an attempt to discover how nature works and to use that knowledge to describe what is likely to happen in nature. It is based on the assumption that events in the physical world follow orderly cause-and-effect patterns that can be understood through careful observation, measurements, experimentation, and modeling. Figure 2-2 summarizes the scientific process.

There is nothing mysterious about this scientific process. You use it all the time in making decisions. As the famous physicist Albert Einstein put it, "The whole of science is nothing more than a refinement of everyday thinking."

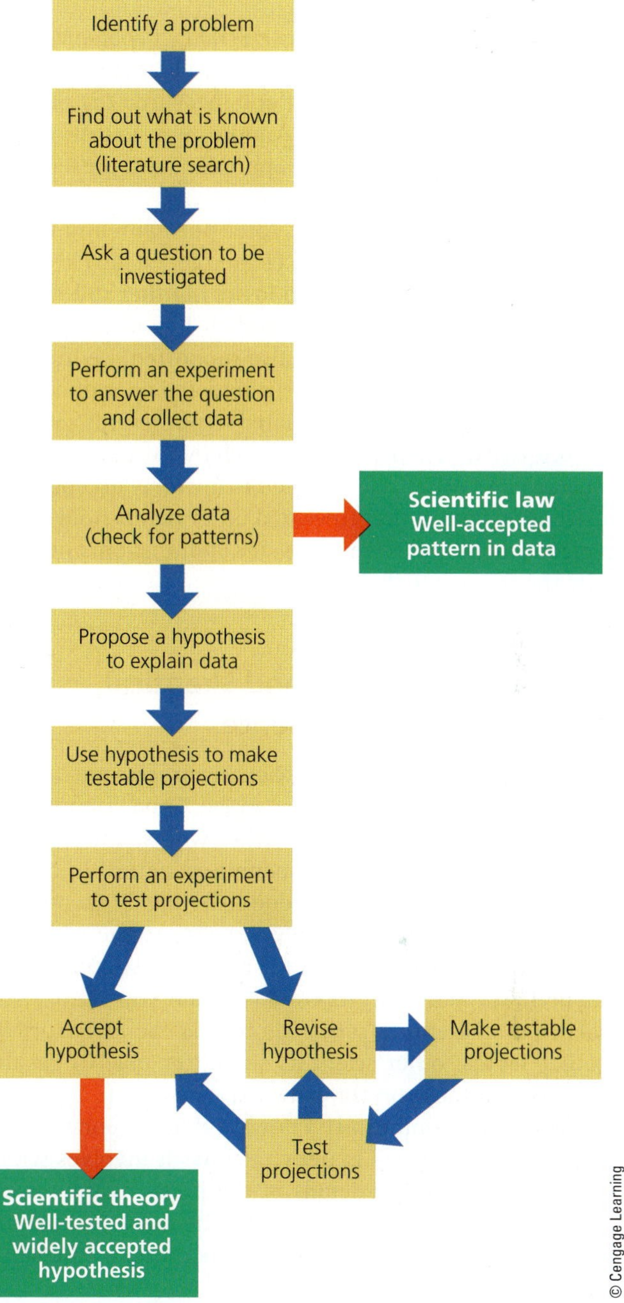

Figure 2-2 This diagram illustrates the general process that scientists use for discovering and testing ideas about how the natural world works.

Scientists Use Observations, Experiments, and Models to Answer Questions about How Nature Works

Here is a more formal outline of the steps scientists often take in trying to understand the natural world, although they do not always follow the steps in the order listed. The outline is based on the scientific experiment carried out by Bormann and Likens (**Core Case Study**), which illustrates the nature of the scientific process shown in Figure 2-2.

- *Identify a problem.* Bormann and Likens identified the loss of water and soil nutrients from cutover forests as a problem worth studying.
- *Find out what is known about the problem.* They searched the scientific literature to find out what scientists knew about both the retention and the loss of water and soil nutrients in forests.
- *Ask a question to investigate.* The scientists asked, "How does clearing forested land affect its ability to store water and retain soil nutrients?"
- *Perform an experiment and collect and analyze data to answer the question.* To collect **data**—information needed to answer their questions—scientists often perform experiments and make observations and measurements, as Bormann and Likens did (Figure 2-1). (Sometimes, they simply observe and measure natural phenomena to collect data without doing an experiment.)
- *Propose a hypothesis to explain the data.* Scientists suggest a **scientific hypothesis**—a possible and testable answer to a scientific question or explanation of what scientists observe in nature. Bormann and Likens came up with the following hypothesis to explain their data: When a forest is cleared of its vegetation and exposed to rain and melting snow, it retains less water and loses large quantities of soil nutrients.
- *Use the hypothesis to make projections that can be tested.* Scientists make projections about what should happen if their hypothesis is correct and then run experiments to test the projections. Bormann and Likens projected

that if their hypothesis was valid for nitrogen, then a cleared forest should also lose other soil nutrients such as phosphorus over a similar time period and under similar weather conditions.

- *Test the projections with further experiments or observations.* To test their projection, Bormann and Likens repeated their controlled experiment and measured the phosphorus content of the soil. Another way to test projections is to use a **model**, an approximate representation or simulation of a system.
- *Accept or revise the hypothesis.* After Bormann and Likens confirmed that the soil in a cleared forest also loses phosphorus, they measured losses of other soil nutrients, which further supported their hypothesis. The research of other scientists also supported the hypothesis. A well-tested and widely accepted scientific hypothesis or a group of related hypotheses is called a **scientific theory**. The research by Bormann and Likens and other scientists led to a widely accepted scientific theory that trees and other plants hold soil in place and help it to retain water and nutrients needed by the plants for their growth.

Scientists Are Curious and Skeptical, and They Demand Evidence

Four important features of the scientific process are *curiosity, skepticism, reproducibility,* and *peer review.* Good scientists are extremely curious about how nature works, and they are keen observers of what is happening in nature (see Individuals Matter 2.1). Scientists tend to be highly skeptical of new data and hypotheses. They say, "Show me your evidence and explain the reasoning behind the scientific ideas or hypotheses that you propose to explain your data." Any evidence that scientists gather should also be reproducible. In other words, other scientists should be able to get the same results if they run the same experiments.

Science is a community effort, and an important part of the scientific process is **peer review**. It involves scientists openly publishing details of the methods they used, the results of their experiments, and the reasoning behind their hypotheses for other scientists working in the same field (their peers) to evaluate.

For example, Bormann and Likens (**Core Case Study**) submitted the results of their forest experiments to a respected scientific journal. Before publishing this report, the journal's editors asked other soil and forest experts to review it. Other scientists have repeated the measurements of soil content in undisturbed and cleared forests of the same type and also in different types of forests, and their results have also been subjected to peer review. In addition, computer models of forest systems have been used to evaluate this problem, with the results also subjected to peer review.

Scientific knowledge advances in this self-correcting way, with scientists continually questioning the measurements and data produced by their peers. They also collect new data and sometimes come up with new and better hypotheses (Science Focus 2.1).

Critical Thinking and Creativity Are Important in Science

Scientists use logical reasoning and critical thinking skills (p. xxiv) to learn about the natural world. Thinking critically involves four important steps:

1. Be skeptical about everything you read or hear.
2. Look at the evidence and evaluate it and any related information, along with inputs and opinions from a variety of reliable sources.
3. Be open to many viewpoints and evaluate each one before coming to a conclusion.
4. Identify and evaluate your personal assumptions, biases, and beliefs. As the American psychologist and philosopher William James observed, "A great many people think they are thinking when they are merely rearranging their prejudices."

Logic and critical thinking are very important tools in science, but imagination, creativity, and intuition are just as vital. According to physicist Albert Einstein, "There is no completely logical way to a new scientific idea."

Scientific Theories and Laws Are the Most Important and Certain Results of Science

The real goal of scientists is to develop theories and laws, based on facts and data that explain how the natural world works, as illustrated in the quotation that opens this chapter. *We should never take a scientific theory lightly.* It has been tested widely, is supported by extensive evidence, and is accepted as being a useful explanation of some phenomenon by most scientists in a particular field or related fields of study.

Because of this rigorous testing process, scientific theories are rarely overturned unless new evidence discredits them or scientists come up with better explanations. So when you hear someone say, "Oh, that's just a theory," you will know that he or she does not have a clear understanding of what a scientific theory is. In sports terms, developing a widely accepted scientific theory is roughly equivalent to winning a gold medal in the Olympics.

Another important and reliable outcome of science is a **scientific law**, or **law of nature**—a well-tested and widely accepted description of what we find happening repeatedly and in the same way in nature. An example is the *law of gravity.* After making many thousands of observations and measurements of objects falling from different heights, scientists developed the following scientific law: all objects fall to the earth's surface at predictable speeds.

We can break a society's law, for example, by driving faster than the speed limit. But *we cannot break a*

individuals matter 2.1

JENS SCHLUETER/AFP/Getty Images

Jane Goodall: Chimpanzee Researcher and Protector

Jane Goodall is a primatologist and environmental educator with a PhD from Cambridge University. She is also a National Geographic Explorer-in-Residence Emeritus. At age 26, she began a 50-year career of studying chimpanzee social and family life in the Gombe Stream Game Reserve in the African country of Tanzania. She is shown above with one of the chimpanzees she studied. By carefully observing the chimps while living with them, she discovered that chimps have more complex social interactions than had previously been known. For example, she found that they can use tools, eat meat as well as vegetation, and carry on extended fights with one another.

One of her major scientific discoveries was that chimpanzees have tool-making skills. She observed some chimpanzees modifying twigs or blades of grass and then poking them into termite mounds. When the termites latched on to these primitive tools, the chimpanzees pulled them out and ate the termites. Goodall has also observed that chimps can learn simple sign language, do simple arithmetic, play computer games, develop relationships, and worry about and protect one another.

In 1977, she established the Jane Goodall Institute, a nonprofit organization that works to preserve great ape populations and their habitats. Her research encouraged the Tanzanian government to convert the game preserve into the Gombe Stream National Park, which her institute now supports.

Goodall has been a strong advocate for animal rights and has led campaigns against sport hunting, keeping animals in zoos, and using them for medical research. In 1991 she started Roots and Shoots, an environmental education program for youth that is active in more than 100 countries. She has received many awards and prizes for her scientific contributions and conservation efforts. She has also written 23 books for adults and children and produced 14 films about the lives and importance of chimpanzees.

In 2011, the movie *Jane's Journey* was made about her life's work. Now in her late 70s, Goodall still spends nearly 300 days a year traveling and educating people throughout the world about chimpanzees and the need to protect the environment. She says, "I can't slow down. . . . If we're not raising new generations to be better stewards of the environment, what's the point?"

Background photo: namatae/Shutterstock.com

SCIENCE FOCUS 2.1

SOME REVISIONS IN A POPULAR SCIENTIFIC HYPOTHESIS

For years, the story of Easter Island has been used in textbooks as an example of how humans can seriously degrade their own life-support system and as a warning about what we are doing to our life-support system.

What happened on this small island in the South Pacific is a story about environmental degradation and the demise of an ancient civilization of Polynesians living there. Years ago, scientists studied the island and its remains, including more than 300 huge statues (Figure 2-A). They hypothesized that over time, the Polynesians began living unsustainably as their population grew, and they used the island's forest and soil resources faster than they could be renewed. They further hypothesized that when the forests were depleted, there was no firewood for cooking or keeping warm and no wood for building large canoes in order to leave the island. They also hypothesized that, with the forest cover gone, soils eroded, crop yields plummeted, famine struck, the population dwindled, and the civilization collapsed.

In 2006, anthropologist Terry L. Hunt evaluated the accuracy of past measurements and other evidence and carried out new research to reevaluate the hypothesis about what happened on Easter Island. He used his data to formulate an alternative hypothesis to try to explain the human tragedy on Easter Island, and he came to some new conclusions. *First,* the Polynesians arrived on the island about 800 years ago, not 2,900 years ago, as had been thought. *Second,* their population size probably never exceeded 3,000, contrary to the earlier estimate of up to 15,000.

Third, the Polynesians did use the island's trees and other vegetation in an unsustainable manner, and visitors reported that by 1722, most of the island's trees were gone. However, one question not answered by this earlier hypothesis was, why did the trees never grow back? Recent evidence and Hunt's new hypothesis suggest that rats (which either came along with the original settlers as stowaways or were brought along as a source of protein for the long voyage) played a key role in the island's permanent deforestation. Over the years, the rats multiplied rapidly into the millions and devoured the seeds that would have regenerated the forests. According to this new hypothesis, the rats played a key role in the fall of the civilization on Easter Island.

This story is an excellent example of how science works. The gathering of new scientific data and the reevaluation of older data led to a revised hypothesis that challenged earlier thinking about the decline of civilization on Easter Island. As a result, the tragedy may not be as clear an example of human-caused ecological collapse as was once thought.

Note that the original Easter Island story was a scientific hypothesis, not a widely tested and accepted scientific theory. And Hunt's research presents another scientific hypothesis based on new data. Further research may convert Hunt's hypothesis to the status of a scientific theory, or more research may lead to other insights into what happened on this island. This is how science works.

Figure 2-A These and several thousand other statues were created by an ancient civilization of Polynesians on Easter Island. Some of them are as tall as a five-story building and weigh as much as 89 metric tons (98 tons).

modestlife/Shutterstock.com

Critical Thinking

Does the new doubt about the original Easter Island hypothesis mean that we should not be concerned about using resources unsustainably on the island in space that we call Earth? Explain.

scientific law, unless we discover new evidence that leads to changes in the law.

The Results of Science Can Be Tentative, Reliable, or Unreliable

Sometimes, preliminary scientific results that capture news headlines have not been widely tested and accepted by peer review. They are not yet considered reliable, and can be thought of as **tentative science** or **frontier science**. Some of these results and hypotheses will be validated and classified as reliable and some will be discredited and classified as unreliable. At the frontier stage, it is normal for scientists to disagree about the meaning and accuracy of data and the validity of hypotheses and results. This is how scientific knowledge advances.

By contrast, **reliable science** consists of data, hypotheses, models, theories, and laws that are widely accepted by all or most of the scientists who are considered experts in the field under study. The results of reliable science are based on the self-correcting process of testing, open peer review, and debate. New evidence and better hypotheses may discredit or alter widely accepted scientific theories, although this is rare. But until that happens, those theories are considered to be the results of reliable science.

Scientific hypotheses and results that are presented as reliable without having undergone the rigors of widespread peer review, or that have been discarded as a result of peer review, are considered to be **unreliable science**. Here are some critical thinking questions you can use to uncover unreliable science:

- Was the experiment well designed? Did it involve a control group? (**Core Case Study**)
- Does the proposed hypothesis explain the data?
- Are there no other, more reasonable explanations of the data?
- Are the investigators unbiased in their interpretations of the results? Did their funding come from unbiased sources?
- Have the data and conclusions been subjected to peer review?
- Are the conclusions of the research widely accepted by other experts in this field?

If "yes" is the answer to each of these questions, then you can classify the results as reliable science. Otherwise, the results may represent tentative science that needs further testing and evaluation, or you can classify them as unreliable science.

Science Has Some Limitations

Environmental science and science in general have four important limitations. *First,* scientists cannot prove or disprove anything absolutely, because there is always some degree of uncertainty in scientific measurements, observations, and models. Instead, scientists try to establish that a particular scientific theory or law has a very high *probability* or *certainty* (at least 90%) of being useful for understanding some aspect of the natural world.

Many scientists do not use the word *proof* because it implies "absolute proof" to people who don't understand how science works. For example, most scientists will rarely say something like, "Cigarettes cause lung cancer." Rather, they might say, "Overwhelming evidence from thousands of studies indicates that people who smoke regularly for many years have a greatly increased chance of developing lung cancer."

CONSIDER THIS. . .

THINKING ABOUT Scientific Proof

Does the fact that science can never prove anything absolutely mean that its results are not valid or useful? Explain.

A *second* limitation of science is that scientists are human and thus are not totally free of bias about their own results and hypotheses. However, the high standard of evidence required through peer review helps to uncover or greatly reduce personal bias and expose occasional cheating by scientists who falsify their results.

A *third* limitation—especially important to environmental science—is that many systems in the natural world involve a huge number of variables with complex interactions. This makes it difficult, too costly, and too time consuming to test one variable at a time in controlled experiments such as the one described in the **Core Case Study** that opens this chapter. To try to deal with this problem, scientists develop *mathematical models* that can take into account the interactions of many variables. Running such models on high-speed computers can sometimes overcome the limitations of testing each variable individually, saving both time and money. In addition, scientists can use computer models to simulate global experiments on phenomena such as climate change that cannot be done in a controlled physical experiment.

A *fourth* limitation of science involves the use of statistical tools. For example, there is no way to measure accurately how many metric tons of soil are eroded annually worldwide. Instead, scientists use statistical sampling and other mathematical methods to estimate such numbers. However, such results should not be dismissed as "only estimates" because they can indicate important trends.

Despite these limitations, science is the most useful way that we have of learning about how nature works and projecting how it might behave in the future. But we still know too little about how the earth works, about its present state of environmental health, and about the current and future environmental impacts of our activities.

GOOD NEWS

2-2 What Is Matter and What Happens When It Undergoes Change?

CONCEPT 2-2A

Matter consists of elements and compounds, which in turn are made up of atoms, ions, or molecules.

CONCEPT 2-2B

Whenever matter undergoes a physical or chemical change, no atoms are created or destroyed (the law of conservation of matter).

Matter Consists of Elements and Compounds

To begin our study of environmental science, we look at matter—the stuff that makes up life and its environment. **Matter** is anything that has mass and takes up space. It can exist in three *physical states*—solid, liquid, and gas—and two *chemical forms*—elements and compounds.

Figure 2-3 Mercury (left) and gold (right) are chemical elements. Each has a unique set of properties and cannot be broken down into simpler substances.

Andraž Cerar/Shutterstock.com

Morgan Lane Photography/Shutterstock.com

An **element** is a fundamental type of matter that has a unique set of properties and cannot be broken down into simpler substances by chemical means. For example, the elements gold (Figure 2-3, left) and mercury (Figure 2-3, right) cannot be broken down chemically into any other substance. Chemists arrange the known elements on the basis of their chemical behavior in what is called the Periodic Table of Elements (see Figure 1, p. S12, Supplement 4).

Some matter is composed of one element, such as mercury or gold (Figure 2-3). And some elements such as hydrogen (H), nitrogen (N), oxygen (O), and chlorine (Cl) are found in nature as combinations of two of their atoms represented by the chemical formulas H_2, N_2, O_2, and Cl_2. However, most matter consists of **compounds**, combinations of two or more different elements held together in fixed proportions. For example, water is a compound made of the elements hydrogen and oxygen that have chemically combined with one another. (See Supplement 4, p. S12, for an expanded discussion of basic chemistry.)

To simplify things, chemists represent each element by a one- or two-letter symbol. Table 2-1 lists the elements and their symbols that you need to know to understand the material in this book.

Table 2-1 Chemical Elements Used in This Book

Element	Symbol	Element	Symbol
arsenic	As	lead	Pb
bromine	Br	lithium	Li
calcium	Ca	mercury	Hg
carbon	C	nitrogen	N
copper	Cu	phosphorus	P
chlorine	Cl	sodium	Na
fluorine	F	sulfur	S
gold	Au	uranium	U

Atoms, Molecules, and Ions Are the Building Blocks of Matter

The most basic building block of matter is an **atom**, the smallest unit of matter into which an element can be divided and still have its distinctive chemical properties. The idea that all elements are made up of atoms is called the **atomic theory** and it is the most widely accepted scientific theory in chemistry.

Atoms are incredibly small. For example, more than 3 million hydrogen atoms could sit side by side on the period at the end of this sentence. If you could view atoms with a supermicroscope, you would find that each different type of atom contains a certain number of three types of *subatomic particles:* **neutrons (n)** with no electrical charge; **protons (p)**, each with a positive electrical charge (+); and **electrons (e)**, each with a negative electrical charge (−).

Each atom has an extremely small center called the **nucleus**, which contains one or more protons and, in most cases, one or more neutrons. Outside of the nucleus we find one or more electrons in rapid motion (Figure 2-4). We cannot determine the exact locations of the electrons. Instead, scientists can estimate the *probability* that they will be found at various locations outside the nucleus in certain spatial patterns that are called *electron probability clouds.* This is somewhat like saying that there are a number of bees flying around inside a cloud. We do not know their exact locations, but the cloud represents an area in which there is a high probability of finding them.

Each atom in its basic form has equal numbers of positively charged protons and negatively charged electrons. Because these electrical charges cancel one another, *an atom in its basic form has no net electrical charge.*

Each element has a unique **atomic number** equal to the number of protons in the nucleus of its atom. Carbon (C), with 6 protons in its nucleus, has an atomic number of 6, whereas uranium (U), a much larger atom, has 92 protons in its nucleus and thus an atomic number of 92.

Because electrons have so little mass compared to protons and neutrons, *most of an atom's mass is concentrated in its nucleus.* The mass of an atom is described by its **mass number**, the total number of neutrons and protons in its

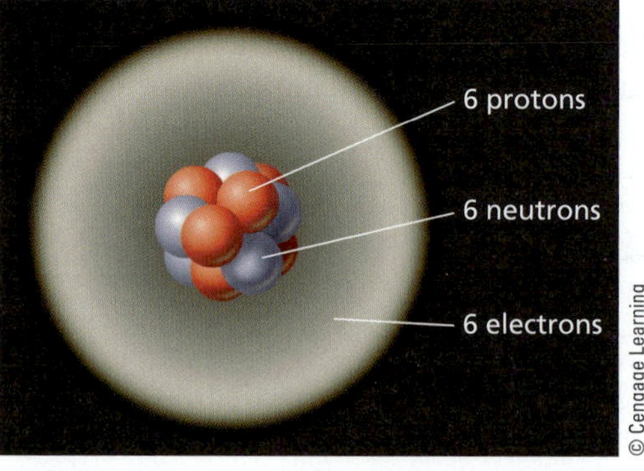

Figure 2-4 This is a greatly simplified model of a carbon-12 atom. It consists of a nucleus containing six protons, each with a positive electrical charge, and six neutrons with no electrical charge. Six negatively charged electrons are found outside its nucleus.

Table 2-2 Chemical Ions Used in This Book

Positive Ion	Symbol	Components
hydrogen ion	H^{+}	One hydrogen atom, one positive charge
sodium ion	Na^{+}	One sodium atom, one positive charge
calcium ion	Ca^{2+}	One calcium atom, two positive charges
aluminum ion	Al^{3+}	One aluminum atom, three positive charges
ammonium ion	NH_4^{+}	One nitrogen atom, four hydrogen atoms, one positive charge
Negative Ion	**Symbol**	**Components**
chloride ion	Cl^{-}	One chlorine atom, one negative charge
hydroxide ion	OH^{-}	One oxygen atom, one hydrogen atom, one negative charge
nitrate ion	NO_3^{-}	One nitrogen atom, three oxygen atoms, one negative charge
carbonate ion	CO_3^{2-}	One carbon atom, three oxygen atoms, two negative charges
sulfate ion	SO_4^{2-}	One sulfur atom, four oxygen atoms, two negative charges
phosphate ion	PO_4^{3-}	One phosphorus atom, four oxygen atoms, three negative charges

nucleus. For example, a carbon atom with 6 protons and 6 neutrons in its nucleus has a mass number of 12, and a uranium atom with 92 protons and 143 neutrons in its nucleus has a mass number of 235 (92 + 143 = 235).

Each atom of a particular element has the same number of protons in its nucleus. But the nuclei of atoms of a particular element can vary in the number of neutrons they contain and, therefore, in their mass numbers. The forms of an element having the same atomic number but different mass numbers are called **isotopes** of that element. Scientists identify isotopes by attaching their mass numbers to the name or symbol of the element. For example, the three most common isotopes of carbon are carbon-12 (with six protons and six neutrons), carbon-13 (with six protons and seven neutrons), and carbon-14 (with six protons and eight neutrons). Carbon-12 makes up about 98.9% of all naturally occurring carbon.

A second building block of matter is a **molecule**, a combination of two or more atoms of the same or different elements held together by forces called *chemical bonds.* Molecules are the basic building blocks of many compounds (see Figure 3, p. S14, in Supplement 4 for examples). An example of a molecule is that of water, or H_2O, which consists of two atoms of hydrogen and one atom of oxygen held together by chemical bonds. Another example is methane, or CH_4 (the major component of natural gas), which consists of four atoms of hydrogen and one atom of carbon.

A third building block of some types of matter is an **ion**—an atom or a group of atoms with one or more net positive or negative electrical charges. Like atoms, ions are made up of protons, neutrons, and electrons. (See p. S13 in Supplement 4 for details on how ions form.) Chemists use a superscript after the symbol of an ion to indicate how many positive or negative electrical charges it has, as shown in Table 2-2.

The nitrate ion (NO_3^{-}) is a nutrient essential for plant growth. Figure 2-5 shows measurements of the loss of nitrate ions from the deforested area (Figure 2-1, right) in the controlled experiment run by Bormann and Likens (**Core Case Study**). Numerous chemical analyses of the water flowing through the dam at the cleared forest site showed an average 60-fold rise in the concentration of NO_3^{-} compared to water running off the forested site. After a few years, however, vegetation began growing back in the cleared valley and nitrate levels in its runoff returned to normal levels.

Ions are also important for measuring a substance's **acidity** in a water solution, a chemical characteristic that helps determine how a substance dissolved in water will interact with and affect its environment. The acidity of a water solution is based on the comparative amounts of hydrogen ions (H^{+}) and hydroxide ions (OH^{-}) contained in a particular volume of the solution. Scientists use **pH** as

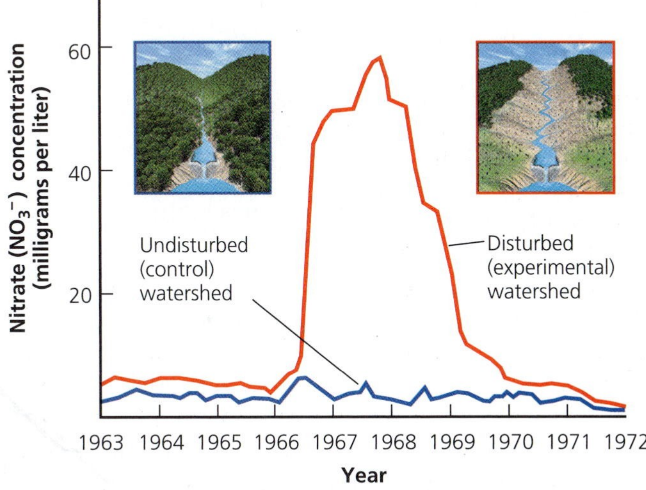

Figure 2-5 This graph shows the loss of nitrate ions (NO_3^{-}) from a deforested watershed in the Hubbard Brook Experimental Forest (**Core Case Study**).

(Based on data from F. H. Bormann and Gene Likens.)

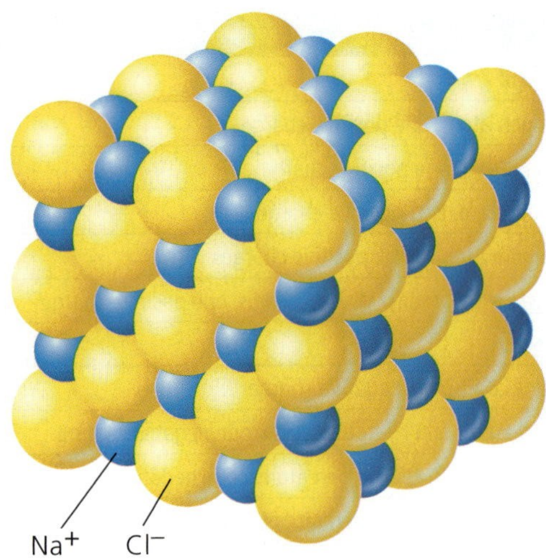

Figure 2-6 A solid crystal of an ionic compound such as sodium chloride (NaCl) consists of a three-dimensional array of oppositely charged ions held together by the strong forces of attraction between oppositely charged ions.

a measure of acidity. Pure water (not tap water or rainwater) has an equal number of H^+ and OH^- ions. It is called a *neutral solution* and has a pH of 7. An *acidic solution* has more hydrogen ions than hydroxide ions and has a pH less than 7. A *basic solution* has more hydroxide ions than hydrogen ions and has a pH greater than 7. (See Figure 4, p. S15, in Supplement 4 for more details.)

Chemists use a **chemical formula** to show the number of each type of atom or ion in a compound. This shorthand contains the symbol for each element present (Table 2-1) and uses subscripts to show the number of atoms or ions of each element in the compound's basic structural unit. Examples of compounds and their formulas encountered in this book are sodium chloride (NaCl) and water (H_2O, read as "H-two-O"). Sodium chloride is an ionic compound with a three-dimensional array of sodium ions (Na^+) and chloride ions (Cl^-) (Figure 2-6). These and other compounds important to our study of environmental science are listed in Table 2-3.

You might want to mark these pages containing Tables 2-1, 2-2, and 2-3, because they show the key elements, ions, and compounds used in this book. Think of them as lists of some main chemical characters in the story of matter that makes up the natural world.

Organic Compounds Are the Chemicals of Life

Plastics (Figure 2-7), table sugar, vitamins, aspirin, penicillin, and most of the chemicals in your body are called **organic compounds**, because they contain at least two carbon atoms combined with atoms of one or more other elements. All other compounds are called **inorganic compounds**. One exception, methane (CH_4), has only one carbon atom but is considered an organic compound.

Table 2-3 Compounds Used in This Book

Compound	Formula	Compound	Formula
sodium chloride	NaCl	methane	CH_4
sodium hydroxide	NaOH	glucose	$C_6H_{12}O_6$
carbon monoxide	CO	water	H_2O
carbon dioxide	CO_2	hydrogen sulfide	H_2S
nitric oxide	NO	sulfur dioxide	SO_2
nitrogen dioxide	NO_2	sulfuric acid	H_2SO_4
nitrous oxide	N_2O	ammonia	NH_3
nitric acid	HNO_3	calcium carbonate	$CaCO_3$

The millions of known organic (carbon-based) compounds include the following:

- *Hydrocarbons:* compounds of carbon and hydrogen atoms. One example is methane (CH_4), the main component of natural gas and the simplest organic compound. Another is octane (C_8H_{18}), a major component of gasoline.
- *Chlorinated hydrocarbons:* compounds of carbon, hydrogen, and chlorine atoms. An example is the insecticide DDT ($C_{14}H_9Cl_5$).
- *Simple carbohydrates (simple sugars):* certain types of compounds of carbon, hydrogen, and oxygen atoms. An example is glucose ($C_6H_{12}O_6$), which most plants and animals break down in their cells to obtain energy. (For more details, see Figure 5, p. S16, in Supplement 4.)

Larger and more complex organic compounds, essential to life, are composed of *macromolecules.* Some of these molecules are called *polymers,* formed when a number of simple organic molecules *(monomers)* are linked together by chemical bonds—somewhat like rail cars linked in a freight train. The three major types of organic polymers are

- *complex carbohydrates* such as cellulose and starch, which consist of two or more monomers of simple sugars such as glucose (see Figure 5, p. S16, in Supplement 4), important sources of energy in the food we eat;
- *proteins* formed by monomers called *amino acids* (see Figure 6, p. S16, in Supplement 4), important for building certain tissues in our bodies; and
- *nucleic acids* (DNA and RNA) formed by monomers called *nucleotides* (see Figures 7 and 8, pp. S16 and S17, in Supplement 4), and key chemicals in the reproductive processes of many organisms.

Lipids, which include fats and waxes, are not made of monomers but are a fourth type of macromolecule essential for life (see Figure 9, p. S17, in Supplement 4).

Figure 2-7 Plastics, which are found in an amazing variety of widely used and useful products, consist of organic compounds.

Matter Comes to Life through Cells, Genes, and Chromosomes

All organisms are composed of one or more **cells**—the fundamental structural and functional units of life. They are minute compartments covered with a thin membrane, and within them, the processes of life occur. The idea that all living things are composed of cells is called the *cell theory* and it is the most widely accepted scientific theory in biology.

Above, we mentioned nucleotides in DNA (see Figures 7 and 8, pp. S16 and S17, in Supplement 4). Within some DNA molecules are certain sequences of nucleotides called **genes**. Each of these distinct pieces of DNA contains instructions, or codes, called *genetic information,* for making specific proteins. Each of these coded units of genetic information leads to a specific **trait**, or characteristic, passed on from parents to offspring during reproduction in an animal or plant.

In turn, thousands of genes make up a single **chromosome**, a double helix DNA molecule (see Figure 8, p. S17, in Supplement 4) wrapped around some proteins. Genetic information coded in your chromosomal DNA is what makes you different from an oak leaf, an alligator, or a mosquito, and from your parents. The relationships of genetic material to cells are depicted in Figure 2-8.

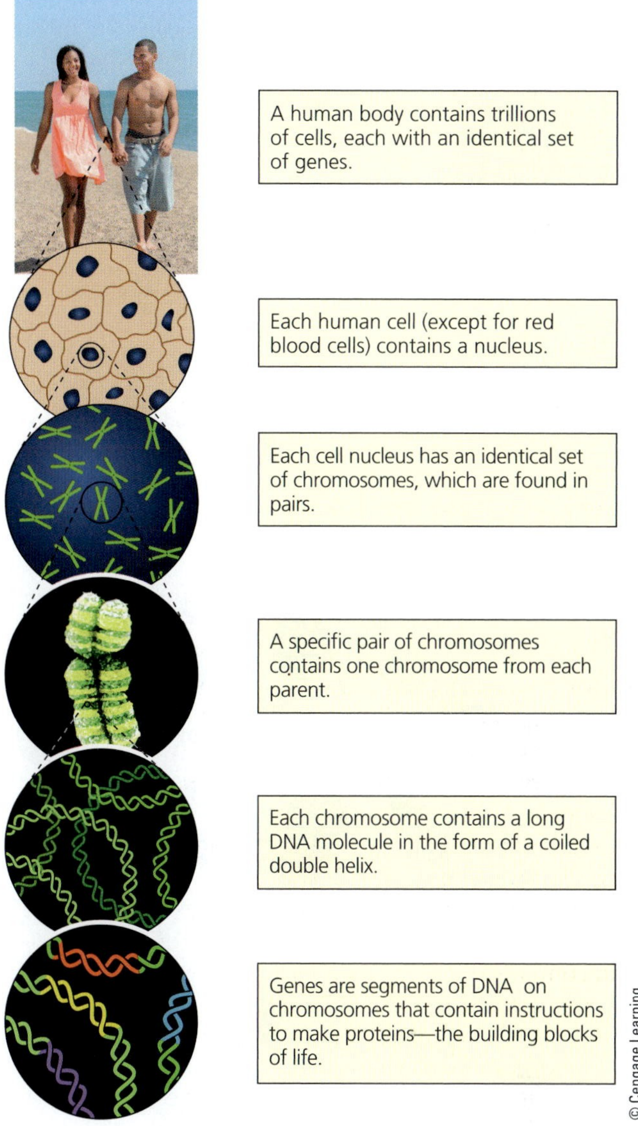

Figure 2-8 This diagram shows the relationships among cells, nuclei, chromosomes, DNA, and genes.

Photo: ©Flashon Studio/Shutterstock.com

Matter Undergoes Physical, Chemical, and Nuclear Changes

When a sample of matter undergoes a **physical change**, there is no change in its *chemical composition.* A piece of aluminum foil cut into small pieces is still aluminum foil. When solid water (ice) melts and when liquid water boils, the resulting liquid water and water vapor are still made up of H_2O molecules.

When a **chemical change**, or **chemical reaction**, takes place, there is a change in the chemical composition of the substances involved. Chemists use a *chemical equation* (and a process called *balancing the equation,* see p. S17 in Supplement 4) to show how chemicals are rearranged in a chemical reaction. For example, coal is made up almost entirely of the element carbon (C). When coal is burned completely in a power plant, the solid carbon in the coal combines with oxygen gas (O_2) from the atmosphere to form the gaseous compound carbon dioxide (CO_2). Chemists use the following shorthand chemical equation to represent this chemical reaction:

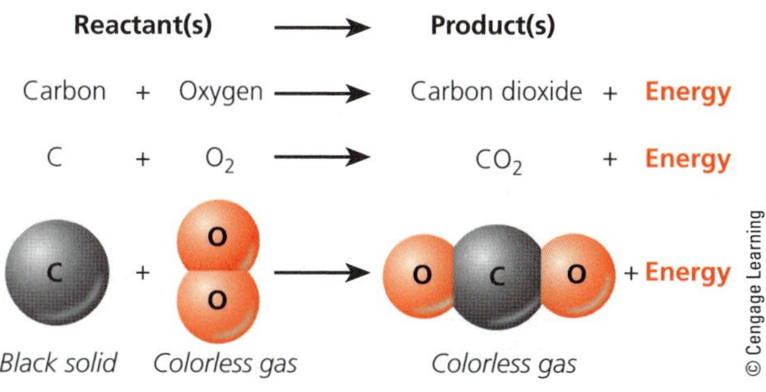

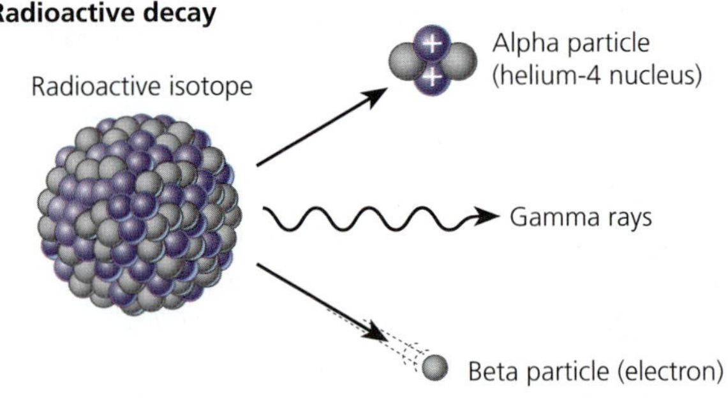

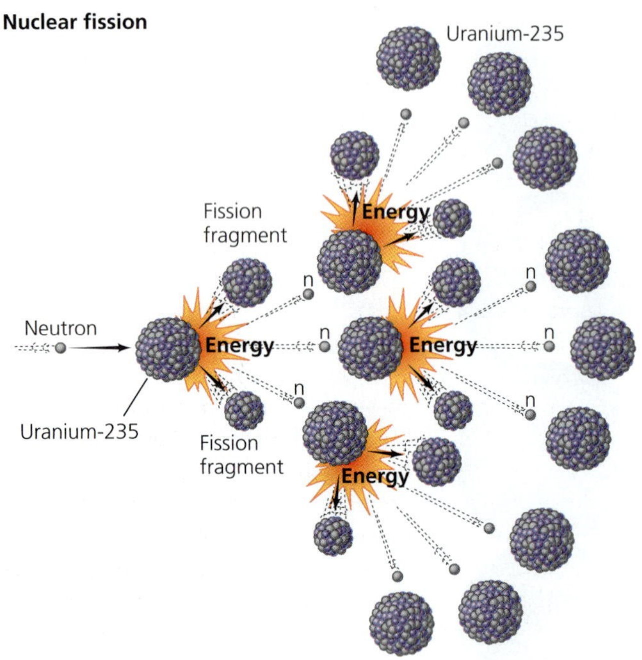

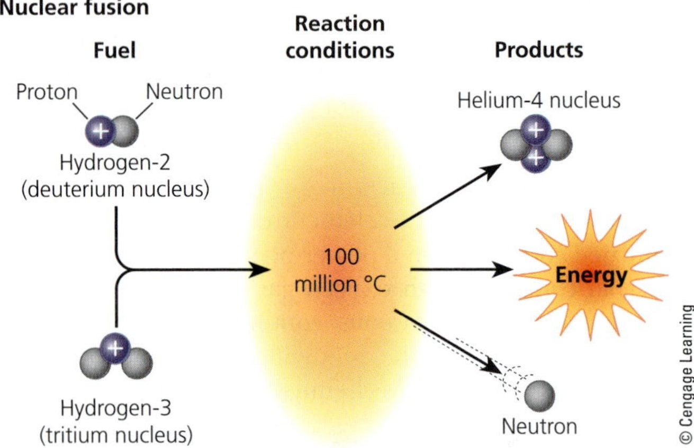

Figure 2-9 There are three types of nuclear changes: natural radioactive decay (top), nuclear fission (middle), and nuclear fusion (bottom).

In addition to physical and chemical changes, matter can undergo three types of **nuclear change**, or change in the nuclei of its atoms (Figure 2-9). **Radioactive decay** occurs when the nuclei of unstable isotopes spontaneously emit fast-moving chunks of matter (alpha particles or beta particles), high-energy radiation (gamma rays), or both at a fixed rate (Figure 2-9, top). **Nuclear fission** occurs when the nuclei of certain isotopes with large mass numbers (such as uranium-235) are split apart into lighter nuclei when struck by a neutron and release energy. Each fission releases neutrons, which can cause more nuclei to fission. This cascade of fissions can result in a chain reaction that releases an enormous amount of energy in a short time (Figure 2-9, middle). **Nuclear fusion** occurs when two nuclei of lighter atoms, such as hydrogen, are forced together at extremely high temperatures until they fuse to form a heavier nucleus and release a tremendous amount of energy (Figure 2-9, bottom).

We Cannot Create or Destroy Atoms: The Law of Conservation of Matter

We can change elements and compounds from one physical or chemical form to another, but we cannot create or destroy any of the atoms involved in any physical or chemical change. All we can do is rearrange the atoms, ions, or molecules into different spatial patterns (physical changes) or chemical combinations (chemical changes). These facts, based on many thousands of measurements, describe a scientific law known as the **law of conservation of matter**: Whenever matter undergoes a physical or chemical change, no atoms are created or destroyed (Concept 2-2B).

CONSIDER THIS. . .

CONNECTIONS Waste and the Law of Conservation of Matter

The law of conservation of matter means we can never really throw anything away because the atoms in any form of matter cannot be destroyed as it undergoes physical or chemical changes. Stuff that we put out in the trash may be buried in a sanitary landfill, but we have not really thrown it away because the atoms in this waste material will always be around in one form or another. We can burn trash, but we then end up with ash that must be put somewhere, and with gases emitted by the burning that can pollute the air. We can reuse or recycle some materials and chemicals, but the law of conservation of matter means we will always face the problem of what to do with some quantity of the wastes and pollutants we produce because their atoms cannot be destroyed.

2-3 What Is Energy and What Happens When It Undergoes Change?

CONCEPT 2-3A
Whenever energy is converted from one form to another in a physical or chemical change, no energy is created or destroyed (first law of thermodynamics).

CONCEPT 2-3B
Whenever energy is converted from one form to another in a physical or chemical change, we end up with lower-quality or less-usable energy than we started with (second law of thermodynamics).

Figure 2-10 Kinetic energy, created by the gaseous molecules in a mass of moving air, turns the blades of this wind turbine. The turbine then converts this kinetic energy to electrical energy, which is another form of kinetic energy.

Energy Comes in Many Forms

Suppose you find this book on the floor and you pick it up and put it on your desktop. To do this you have to use a certain amount of muscular force or work to move the book from one place to another. In scientific terms, work is done when any object is moved a certain distance (work = force × distance). Also, whenever you touch a hot object such as a stove, heat flows from the stove to your finger. Both of these examples involve **energy:** the capacity to do work or to transfer heat.

There are two major types of energy: *moving energy* (called kinetic energy) and *stored energy* (called potential energy). Matter in motion has **kinetic energy**, or energy associated with motion. Examples are flowing water, a car speeding down the highway, electricity (electrons flowing through a wire or other conducting material), and wind (a mass of moving air that we can use to produce electricity, as shown in Figure 2-10).

Another form of kinetic energy is **heat**, or **thermal energy**, the total kinetic energy of all moving atoms, ions, or molecules in an object, a body of water, or the atmosphere. If the atoms, ions, or molecules in a sample of matter move faster, it will become warmer. When two objects at different temperatures come in contact with one another, heat flows from the warmer object to the cooler object. You learned this the first time you touched a hot stove.

In another form of kinetic energy, called **electromagnetic radiation**, energy travels in the form of a *wave* as a result of changes in electrical and magnetic fields. There are many different forms of electromagnetic radiation (Figure 2-11), each having a different *wavelength* (the distance between successive peaks or troughs in the wave) and *energy content.* Forms of electromagnetic radiation with short wavelengths, such as gamma rays, X-rays, and ultraviolet (UV) radiation, have more energy than do forms with longer wavelengths, such as visible light and

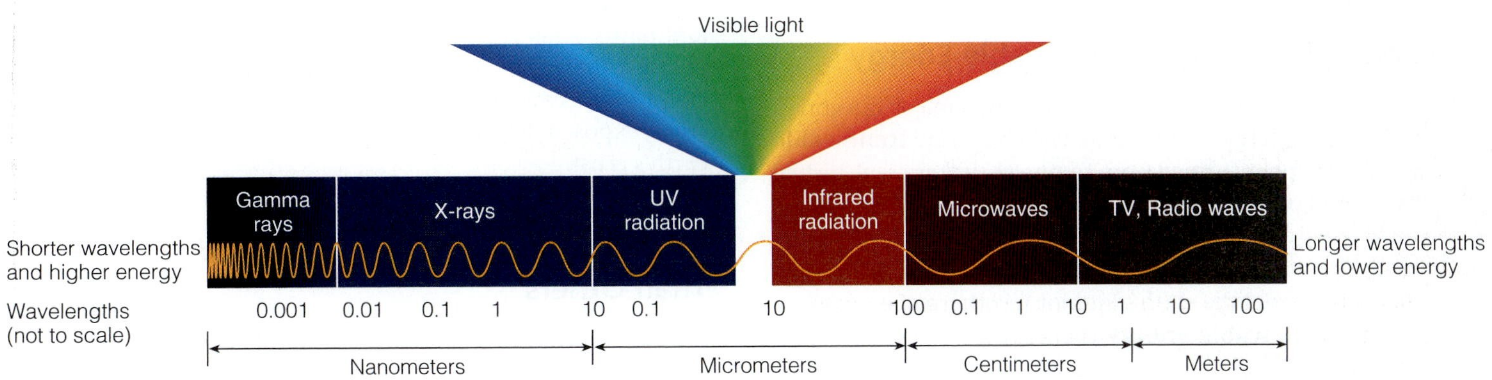

Animated Figure 2-11 The electromagnetic spectrum consists of a range of electromagnetic waves, which differ in wavelength (the distance between successive peaks or troughs) and energy content.

Figure 2-12 The water stored in this reservoir behind a dam in the U.S. state of Tennessee has potential energy, which becomes kinetic energy when the water flows through channels built into the dam where it spins a turbine and produces electricity—another form of kinetic energy.

Bryan Busovicki/Shutterstock.com

infrared (IR) radiation. Visible light makes up most of the spectrum of electromagnetic radiation emitted by the sun.

The other major type of energy is **potential energy**, which is stored and potentially available for use. Examples of this type of energy include a rock held in your hand, the water in a reservoir behind a dam, and the chemical energy stored in the carbon atoms of coal or in molecules of food that you eat.

We can change potential energy to kinetic energy. If you hold this book in your hand, it has potential energy. However, if you drop it on your foot, the book's potential energy changes to kinetic energy. When a car engine burns gasoline, the potential energy stored in the chemical bonds of the gasoline molecules changes into kinetic energy that propels the car, and into heat that flows into the environment. When water in a reservoir flows through channels in a dam (Figure 2-12), its potential energy becomes kinetic energy that we can use to spin turbines in the dam to produce electricity—another form of kinetic energy.

Renewable and Nonrenewable Energy

Scientists divide energy resources into two major categories: renewable energy and nonrenewable energy. **Renewable energy** is energy gained from resources that are replenished by natural processes in a relatively short time. Examples are solar energy, available somewhere on the earth all of the time, firewood from trees, wind, moving water, and heat that comes from the earth's interior (geothermal energy).

Nonrenewable energy is energy from resources that can be depleted and are not replenished by natural processes within a human time scale. Examples are energy produced by the burning of oil, coal, and natural gas, and nuclear energy released when the nuclei of atoms of uranium fuel are split apart.

About 99% of the energy that we use to survive—the energy that keeps us warm and supports the plants that we and other organisms eat—comes from the sun at no cost to us. This is in keeping with the solar energy **principle of sustainability** (see Figure 1-2, p. 6 or back cover). Without this essentially inexhaustible solar energy, the earth would be frozen and life as we know it would not exist.

SUSTAINABILITY

This direct input of solar energy produces several other indirect forms of renewable solar energy. Examples are *wind* (Figure 2-10), *hydropower* (falling and flowing water, Figure 2-12), and *biomass* (solar energy converted to chemical energy and stored in the tissues of trees and other plants) that can be burned to provide heat.

Commercial energy—energy that is sold in the marketplace—makes up the remaining 1% of the energy we use to supplement the earth's direct input of solar energy. About 87% of the commercial energy used in the world and 87% of that used in the United States come from burning nonrenewable **fossil fuels**, or oil, coal, and natural gas (Figure 2-13). They are called fossil fuels because they were formed over millions of years as layers of the decaying remains of ancient plants and animals were exposed to intense heat and pressure within the earth's crust.

Some Types of Energy Are More Useful Than Others

Energy quality is a measure of the capacity of a type of energy to do useful work. **High-quality energy** is concentrated energy that has a high capacity to do useful work. Examples are very high-temperature heat, concentrated sunlight, high-speed wind, and the energy released when we burn gasoline or coal.

Figure 2-13 **Fossil fuels:** Nonrenewable oil, coal, and natural gas (left, center, and right, respectively) supply most of the commercial energy that we use to supplement renewable energy from the sun.

Left: Andrea Danti/Shutterstock.com. Middle: Tom Mc Nemar/Shutterstock.com. Right: Olga Utlyakova/Shutterstock.com.

By contrast, **low-quality energy** is energy that is so dispersed that it has little capacity to do useful work. For example, the enormous number of moving molecules in the atmosphere or in an ocean together have such low-quality energy and such a low temperature that we cannot use them to move things or to heat things to high temperatures.

Energy Changes Are Governed by Two Scientific Laws

After observing and measuring energy being changed from one form to another in millions of physical and chemical changes, scientists have summarized their results in the **first law of thermodynamics**, also known as the **law of conservation of energy**. According to this scientific law, whenever energy is converted from one form to another in a physical or chemical change, no energy is created or destroyed (Concept 2-3A).

This scientific law tells us that no matter how hard we try or how clever we are, we cannot get more energy out of a physical or chemical change than we put in. This is one of nature's basic rules that we cannot violate.

Because the first law of thermodynamics states that energy cannot be created or destroyed, but only converted from one form to another, you may be tempted to think we will never have to worry about running out of energy. Yet if you fill a car's tank with gasoline and drive around or run your computer battery down, something has been lost. What is it? The answer is *energy quality,* the amount of energy available for performing useful work.

Thousands of experiments have shown that whenever energy is converted from one form to another in a physical or chemical change, we end up with lower-quality or less useable energy than we started with (Concept 2-3B). This is a statement of the **second law of thermodynamics**. The resulting low-quality energy usually takes the form of heat that flows into the environment. In the environment, the random motion of air or water molecules further disperses this heat, decreasing its temperature to the point where its energy quality is too low to do much useful work.

In other words, *when energy is changed from one form to another, it always goes from a more useful form to a less useful form.* No one has ever found a violation of this fundamental scientific law.

CONSIDER THIS. . .

CONNECTIONS Can We Recycle or Reuse Energy?

We can recycle and reuse various forms of matter such as paper and aluminum. However, because of the second law of thermodynamics, we can never recycle or reuse high-quality energy to perform useful work. Once the concentrated, high-quality energy in a serving of food, a full tank of gasoline, or a chunk of uranium nuclear fuel is released, it is degraded to low-quality heat and dispersed into the environment.

Scientists estimate that about 84% of the energy used in the United States is either unavoidably wasted because of the second law of thermodynamics (41%) or unnecessarily wasted (43%). Thus, thermodynamics teaches us an important lesson: the cheapest and quickest way to get more energy is to stop wasting almost half the energy we use. One way to do this is to improve our *energy efficiency,* which means getting more work out of the energy we use.

For example, only 5% of the electrical energy used by most incandescent lightbulbs (Figure 2-14, left) produces light, while the other 95% ends up as low-quality waste heat in the environment. There are much more efficient alternatives (Figure 2-14, center and right). We could also use more energy-efficient motor vehicle engines, power plants, and appliances, as we discuss in Chapter 16.

Figure 2-14 Three generations of increasingly energy-efficient lightbulbs: an incandescent bulb (left), a fluorescent bulb (center), and an LED bulb (right).

2-4 What Are Systems and How Do They Respond to Change?

CONCEPT 2-4

Systems have inputs, flows, and outputs of matter and energy, and feedback can affect their behavior.

Systems Respond to Change through Feedback Loops

A **system** is a set of components that function and interact in some regular way. The human body, a river, an economy, and the earth are all systems.

Most systems have the following key components: **inputs** of matter and energy from the environment, **flows** or **throughputs** of matter and energy within the system, and **outputs** of matter and energy to the environment (Figure 2-15) (**Concept 2-4**).

A system can become unsustainable if the throughputs of matter and energy resources exceed the abilities of the system's environment to provide the required resource inputs and to absorb or dilute the system's outputs of matter and energy. One of the most powerful tools used by environmental scientists to study how the components of systems interact is computer modeling (Science Focus 2.2).

When people ask you for *feedback,* they are usually seeking your response to something they said or did. They might feed your response back into their mental processes to help them decide whether and how to change what they are saying or doing.

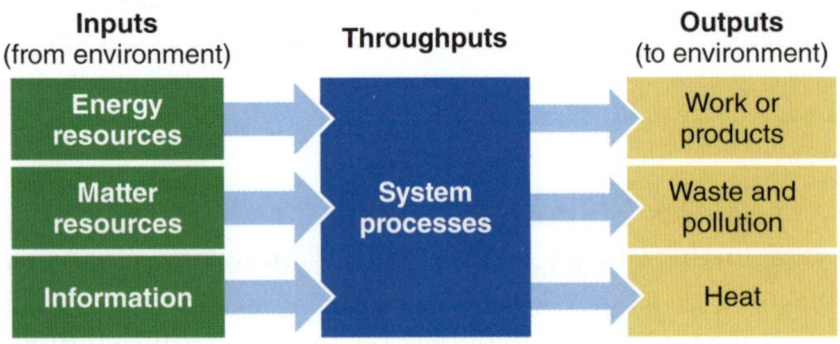

Figure 2-15 A greatly simplified model of a system.

Similarly, most systems are affected in one way or another by **feedback**, any process that increases (positive feedback) or decreases (negative feedback) a change to a system (**Concept 2-4**). Such a process, called a **feedback loop**, occurs when an output of matter, energy, or information is fed back into the system as an input and leads to changes in that system. Note that, unlike the human brain, most systems do not consciously decide how to respond to feedback. Nevertheless, feedback can affect the behavior of systems.

A **positive feedback loop** causes a system to change further in the same direction (Figure 2-16). In the Hubbard Brook experiments, for example (**Core Case Study**), researchers found that when vegetation was removed from a stream valley, flowing water from precipitation caused erosion and losses of nutrients, which caused more vegetation to die. With even less vegetation to hold soil in place, flowing water caused even more erosion and nutrient loss, which caused even more plants to die.

Such accelerating positive feedback loops are of great concern in several areas of environmental science. One of the most alarming is the melting of polar ice, which has occurred as the temperature of the atmosphere has risen during the past few decades. As that ice melts, there is less of it to reflect sunlight, and more water that is exposed to sunlight. Because water is darker than ice, it absorbs more solar energy, making the polar areas warmer and causing the ice to melt faster, thus exposing more water. The melting of polar ice is therefore accelerating, causing a number of serious problems that we explore further in Chapter 19. If a system gets locked into an accelerating positive feedback loop, it can reach a breaking point that can destroy the system or suddenly change its behavior.

A **negative**, or **corrective, feedback loop** causes a system to change in the opposite direction from which it

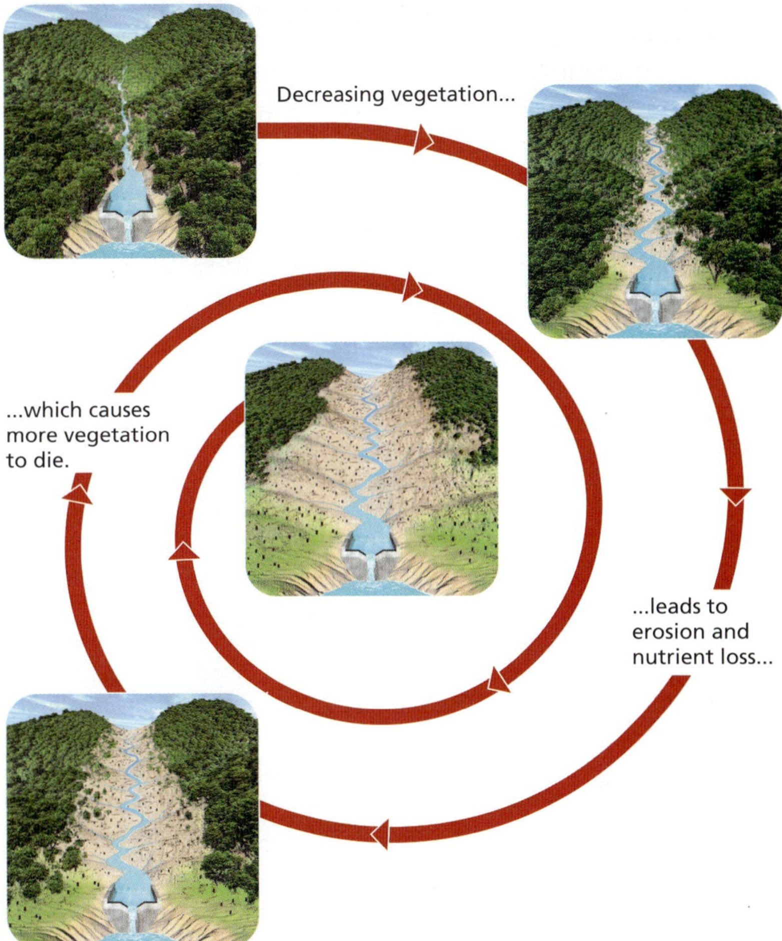

Figure 2-16 A *positive feedback loop.* Decreasing vegetation in a valley causes increasing erosion and nutrient losses that in turn cause more vegetation to die, resulting in more erosion and nutrient losses. ***Question:*** Can you think of another positive feedback loop in nature?

is moving. A simple example is a thermostat, a device that controls how often and how long a heating or cooling system runs (Figure 2-17). When the furnace in a house turns on and begins heating the house, we can set the thermostat to turn the furnace off when the temperature in the house reaches the set number. The house then stops getting warmer and starts to cool.

CONSIDER THIS. . .

THINKING ABOUT The Hubbard Brook Experiments and Feedback Loops

How might experimenters have employed a negative feedback loop to stop, or correct, the positive feedback loop that resulted in increasing erosion and nutrient losses in the Hubbard Brook experimental forest (**Core Case Study**)?

An important example of a negative feedback loop is the recycling and reuse of some resources such as aluminum. For example, an aluminum can is an output of a mining and manufacturing system. When we recycle the can, that output becomes an input. This reduces the amount of aluminum ore that we must mine and process to make aluminum cans. It also reduces the harmful environmental impacts of the mining and processing of aluminum ore. Such a negative feedback loop therefore can help reduce the harmful environmental impacts of human activities by decreasing the use of matter and energy resources and the amount of pollution and solid waste produced by the use of such resources. It is an example of applying the chemical cycling **principle of sustainability** (see Figure 1-2, p. 6 or back cover).

It Can Take a Long Time for a System to Respond to Feedback

A complex system will often show a **time delay**, or a lack of response during a period of time between the input of a feedback stimulus and the system's response to it. For example, scientists could plant trees in a degraded area such as the Hubbard Brook experimental forest to slow erosion and nutrient losses (**Core Case Study**). But it would take years for the trees and other vegetation to grow in order to accomplish this purpose.

Time delays can allow an environmental problem to build slowly until it reaches a *threshold level,* or **tipping point**—the point at which a fundamental shift in the behavior of a system occurs. Reaching a tipping point is somewhat like stretching a rubber band. We can get away with stretching it to several times its original length. But at some point, we reach an irreversible tipping point where the rubber band breaks.

Prolonged delays dampen the negative feedback mechanisms that might slow, prevent, or halt environmental problems. In the Hubbard Brook example (**Core Case Study**), if soil erosion and nutrient losses had reached a certain point where the land could no longer support vegetation, then a tipping point would have been reached and it would have been too late to plant trees in order to try to restore the system. In Chapter 3, we discuss several major environmental tipping points that we may already have exceeded or will likely exceed in the near future.

System Effects Can Be Amplified through Synergy

A **synergistic interaction**, or **synergy**, occurs when two or more processes interact so that the combined effect is greater than the sum of their separate effects. For example, scientific studies reveal such an interaction between smoking and inhaling asbestos particles. Nonsmokers who

SCIENCE FOCUS 2.2

THE USEFULNESS OF MODELS

Scientists use *models,* or simulations, to learn how systems work. Mathematical models are especially useful when there are many interacting variables, when the time frame of events being modeled is long, and when controlled experiments are impossible or too expensive to conduct. One of our most powerful and useful technologies is mathematical modeling with high-speed supercomputers.

Making a mathematical model usually requires that the modelers go through three steps many times. *First,* they identify the major components of the system and how they interact, and develop mathematical equations that summarize this information. In succeeding runs, these equations are steadily refined. *Second,* modelers use a high-speed computer to describe the likely behavior of the system based on the equations. *Third,* they compare the system's projected behavior with known information about its actual behavior. They keep doing this until the model mimics the past and current behavior of the system.

After building and testing a mathematical model, scientists can use it to project what is *likely* to happen under a variety of conditions. In effect, they use mathematical models to answer *if–then* questions: "*If* we do such and such, *then* what is likely to happen now and in the future?" This process can give us a variety of projections or scenarios of possible outcomes based on different assumptions. Mathematical models (like all other models) are no better than the assumptions on which they are built and the data we feed into them.

Scientists applied this process of model-building to the data collected by researchers Bormann and Likens in their Hubbard Brook experiments (**Core Case Study**). These scientists created mathematical models based on the Hubbard Brook data to describe a forest and to project what might happen to soil nutrients and other variables if the forest were disturbed or cut down.

Other areas of environmental science in which computer modeling is becoming increasingly important include studies of the complex systems that govern climate change, deforestation, biodiversity loss, and the oceans.

Critical Thinking

What are two limitations of computer models? Does the existence of limitations mean that we should not rely on such models? Explain. What are the alternatives?

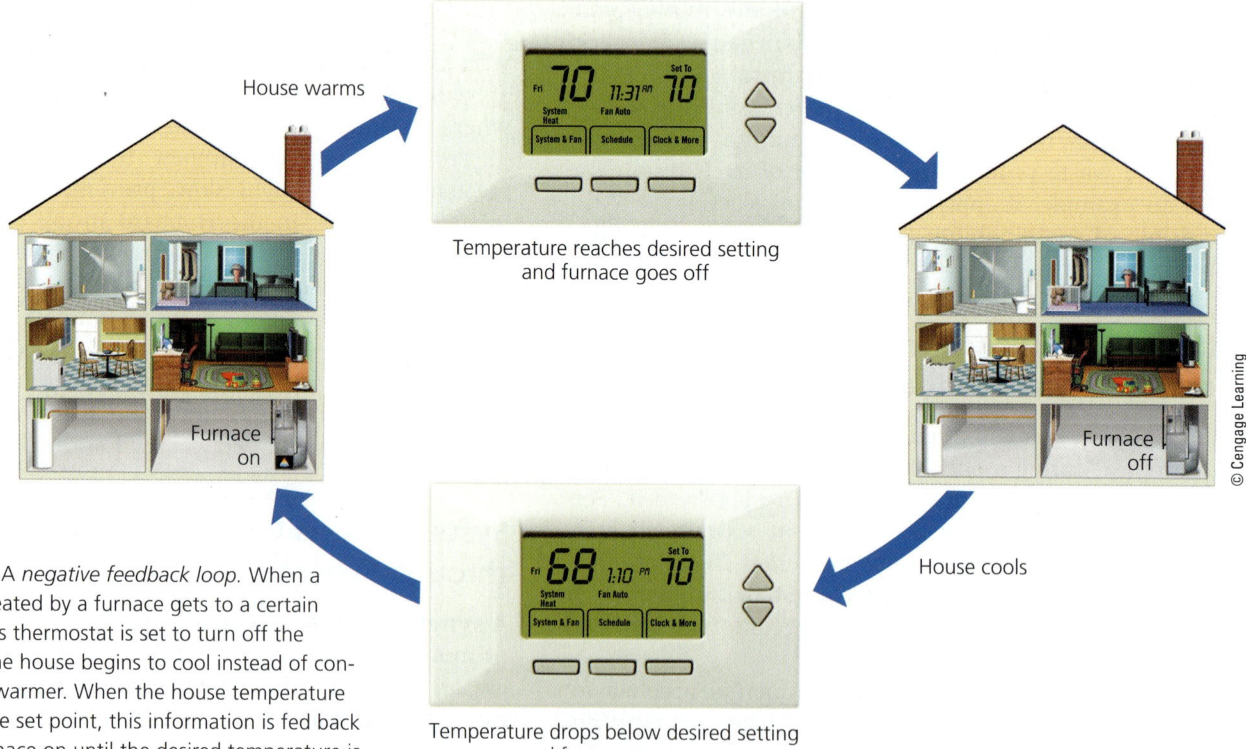

Figure 2-17 A *negative feedback loop.* When a house being heated by a furnace gets to a certain temperature, its thermostat is set to turn off the furnace, and the house begins to cool instead of continuing to get warmer. When the house temperature drops below the set point, this information is fed back to turn the furnace on until the desired temperature is reached again.

are exposed to asbestos particles for long periods of time increase their risk of getting lung cancer fivefold. But people who smoke and are exposed to asbestos have 50 times the risk that nonsmokers have of getting lung cancer.

On the other hand, synergy can be helpful. You may find that you are able to study longer or run farther if you do these activities with a studying or running partner. Your physical and mental systems can do a certain amount of work on their own. But the synergistic effect of you and your partner working together can make your individual systems capable of accomplishing more in the same amount of time. When individuals work together to find and implement win-win solutions to environmental problems, they are applying one of the social science **principles of sustainability** (see Figure 1-5, p. 9 or back cover).

Big Ideas

- **You cannot really throw anything away. According to the *law of conservation of matter,* no atoms are created or destroyed whenever matter undergoes a physical or chemical change. Thus, we cannot do away with matter; we can only change it from one physical state or chemical form to another.**
- **You cannot get something for nothing. According to the *first law of thermodynamics,* or the *law of conservation of energy,* whenever energy is converted from one form to another in a physical or chemical change, no energy is created or destroyed. This means that in causing such changes, we cannot get more energy out than we put in.**
- **You cannot break even. According to the *second law of thermodynamics,* whenever energy is converted from one form to another in a physical or chemical change, we always end up with lower-quality or less usable energy than we started with.**

TYING IT ALL TOGETHER The Hubbard Brook Forest Experiment and Sustainability

steve estvanik/Shutterstock.com

The controlled experiment discussed in the **Core Case Study** that opened this chapter revealed that clearing a mature forest degrades some of its natural capital (see Figure 1-3, p. 7, and photo at left). Specifically, the loss of trees and vegetation altered the ability of the forest to retain and recycle water and other critical plant nutrients—a crucial ecological function based on one of the three scientific **principles of sustainability** (see Figure 1-2, p. 6 or back cover). In other words, the uncleared forest (Figure 2-1, left) was a more sustainable system than a similar area of cleared forest (Figure 2-1, right).

This clearing of vegetation also violated the other two scientific **principles of sustainability**. For example, the cleared forest lost most of its plants that had produced food for the forest's animals by using solar energy and that supplied nutrients to the soil when they died. As a result, many of the forest's key nutrients were lost instead of being recycled. And the loss of plants and the resulting loss of animals reduced the life-sustaining biodiversity of the cleared forest.

Humans clear forests to harvest timber, grow crops, build settlements, and expand cities. The key question is, how far can we go in expanding our ecological footprints (see Figure 1-13, p. 14) without threatening the quality of life for our own species and for the other species that help to keep us alive and support our economies? To live more sustainably, we need to find and maintain a balance between preserving undisturbed natural systems and the natural resources and ecosystem services they provide and modifying other natural systems for our use.

Chapter Review

Core Case Study

1. Describe the controlled scientific experiment carried out in the Hubbard Brook Experimental Forest.

Section 2-1

2. What is the key concept for this section? What is **science**? List the steps involved in a scientific process. What is **data**? What is a **model**? Distinguish among a **scientific hypothesis**, a **scientific theory**, and a **scientific law (law of nature)**. Summarize Jane Goodall's scientific and educational achievements. What is **peer review** and why is it important?

3. Explain why scientific theories are not to be taken lightly and why people often use the term *theory* incorrectly. Explain why scientific theories and laws are the most important and most certain results of science.

4. Distinguish among **tentative science (frontier science)**, **reliable science**, and **unreliable science**. What are four limitations of science in general and environmental science in particular?

Section 2-2

5. What are the two key concepts for this section? What is **matter**? Distinguish between an **element** and a **compound** and give an example of each. Define **atoms**, **molecules**, and **ions** and give an example of each. What is the **atomic theory**? Distinguish among **protons (p)**, **neutrons (n)**, and **electrons (e)**. What is the **nucleus** of an atom? Distinguish between the **atomic number** and the **mass number** of an element. What is an **isotope**? What is **acidity**? What is **pH**?

6. What is a **chemical formula**? Distinguish between **organic compounds** and **inorganic compounds** and give an example of each. Distinguish among complex carbohydrates, proteins, nucleic acids, and lipids. What is a **cell**? Define **gene**, **trait**, and **chromosome**.

7. Define and distinguish between a **physical change** and a **chemical change (chemical reaction)** and give an example of each. What is a **nuclear change**? Define and explain the differences among natural **radioactive decay**, **nuclear fission**, and **nuclear fusion**. What is the **law of conservation of matter** and why is it important?

Section 2-3

8. What are the two key concepts for this section? What is **energy**? Distinguish between **kinetic energy** and **potential energy** and give an example of each. What is **heat (thermal energy)**? Define and give two examples of **electromagnetic radiation**. Define and distinguish between **renewable energy** and **nonrenewable energy**. What are **fossil fuels** and how are they formed? Why are they nonrenewable? What is **energy quality**? Distinguish between **high-quality energy** and **low-quality energy** and give an example of each. What is the **first law of thermodynamics (law of conservation of energy)** and why is it important? What is the **second law of thermodynamics** and why is it important? Explain why the second law means that we can never recycle or reuse high-quality energy.

9. Define and give an example of a **system**. Distinguish among the **inputs, flows (throughputs)**, and **outputs** of a system. Why are scientific models useful? What is **feedback**? What is a **feedback loop**? Distinguish between a **positive feedback loop** and a **negative (corrective) feedback loop** in a system, and give an example of each. Define **time delay** and **synergistic interaction (synergy)**, give an example of each, and explain how they can affect systems. What is a **tipping point**?

10. What are this chapter's *three big ideas*? Explain how the Hubbard Brook Experimental Forest controlled experiments illustrated the three scientific **principles of sustainability**.

Note: Key terms are in bold type.

Critical Thinking

1. What ecological lesson can we learn from the controlled experiment on the clearing of forests described in the **Core Case Study** that opened this chapter?

2. Suppose you observe that all of the fish in a pond have disappeared. Describe how you might use the scientific process described in the **Core Case Study** and in Figure 2-2 to determine the cause of this fish kill.

3. Respond to the following statements:
   a. Scientists have not absolutely proven that anyone has ever died from smoking cigarettes.
   b. The *natural greenhouse effect theory*—that certain gases such as water vapor and carbon dioxide help to warm the lower atmosphere—is not a reliable idea because it is just a scientific theory.

4. A tree grows and increases its mass. Explain why this is not a violation of the law of conservation of matter.
5. If there is no "away" where organisms can get rid of their wastes, due to the law of conservation of matter, why is the world not filled with waste matter?
6. Suppose someone wants you to invest money in an automobile engine, claiming that it will produce more energy than is found in the fuel used to run it. What would be your response? Explain.
7. Use the second law of thermodynamics to explain why we can use oil only once as a fuel, or in other words, why we cannot recycle its high-quality energy.
8. Imagine that for one day **(a)** you have the power to revoke the law of conservation of matter, and **(b)** you have the power to violate the first law of thermodynamics. For each of these scenarios, list three ways in which you would use your new power. Explain your choices.

Doing Environmental Science

Find **(a)** a newspaper or magazine article or a report on the Web that attempts to discredit a scientific hypothesis because it has not been proven, or **(b)** a report of a new scientific hypothesis that has the potential be controversial. Analyze the piece by doing the following: **(1)** determine its source (author or organization); **(2)** detect an alternative hypothesis, if any, that is offered by the author; **(3)** determine the primary objective of the author (for example, to debunk the original hypothesis, to state an alternative hypothesis, or to raise new questions, and so on); **(4)** summarize the evidence given by the author(s) for his or her position; and **(5)** compare the authors' evidence with the evidence for the original hypothesis. Write a report summarizing your analysis and compare it with those of your classmates.

Global Environment Watch Exercise

Search *Easter Island* and under the "News" section, click on "View All." Read current articles on what happened on Easter Island and explain in your own words whether or not the Easter Islanders were living sustainably. How could they have changed their ways in order to live more sustainably in their environment?

Data Analysis

Consider the graph on the right that shows loss of calcium from the experimental cutover site of the Hubbard Brook Experimental Forest (**Core Case Study**) compared with that of the control site. Note that this figure is very similar to Figure 2-5, which compares loss of nitrates from the two sites. After studying this graph, answer the questions below.

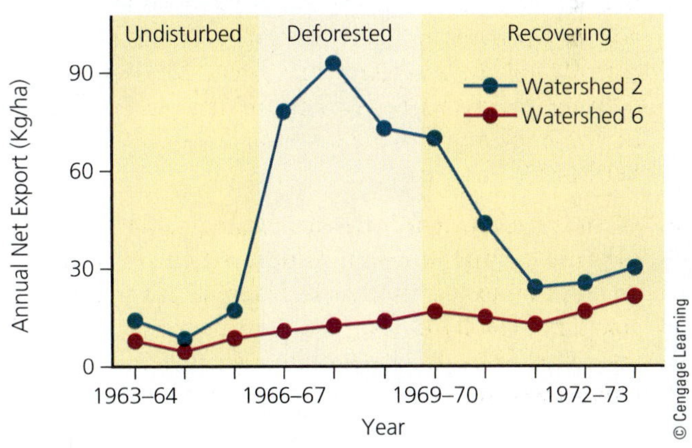

1. In what year did the calcium loss from the experimental site begin a sharp increase? In what year did it peak? In what year did it again level off?
2. In what year were the calcium losses from the two sites closest together? In the span of time between 1963 and 1972, did they ever get that close again?
3. Does this graph support the hypothesis that cutting the trees from a forested area causes the area to lose nutrients more quickly than leaving the trees in place? Explain.

3 Ecosystems: What Are They and How Do They Work?

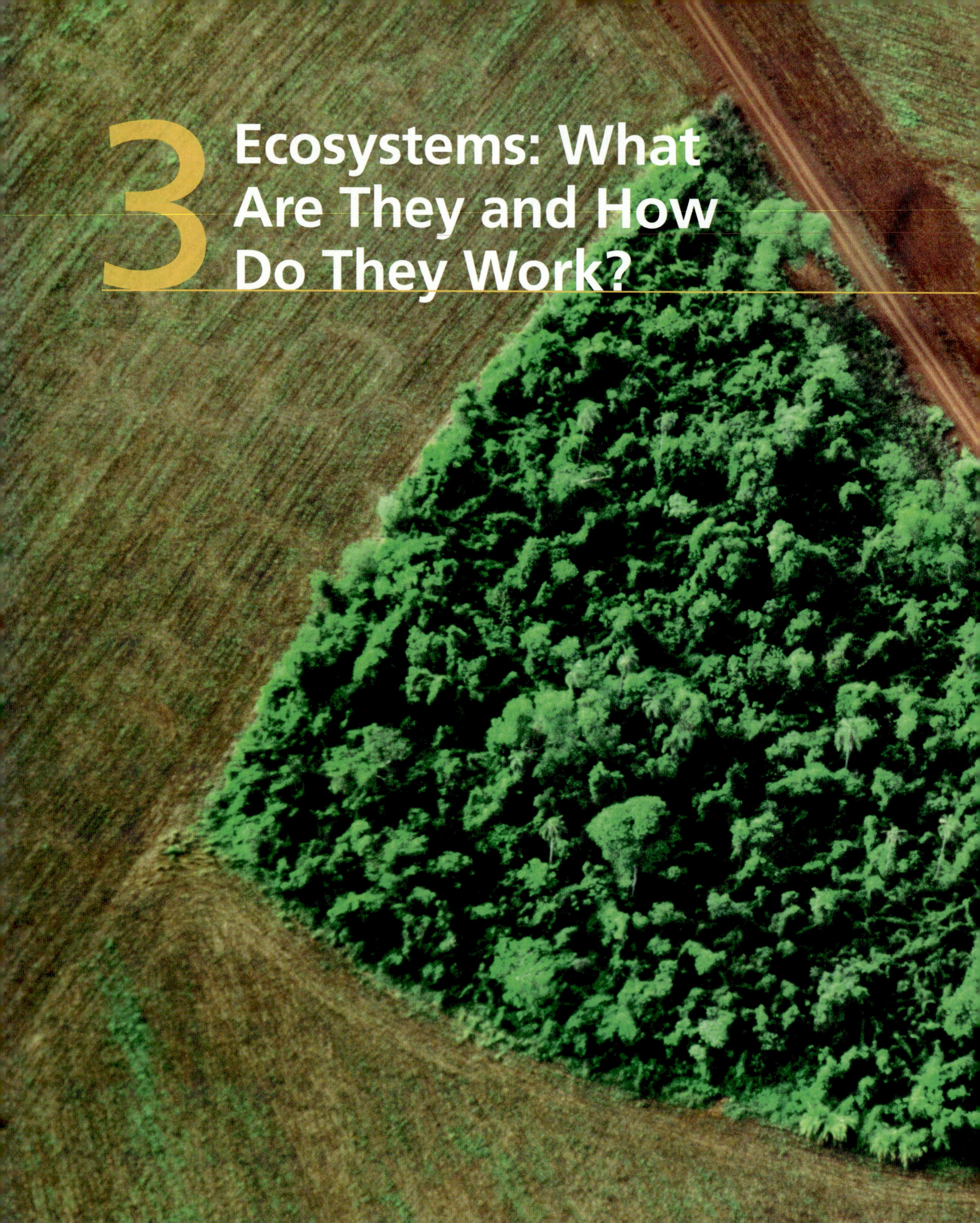

To halt the decline of an ecosystem, it is necessary to think like an ecosystem.

DOUGLAS WHEELER

Key Questions

3-1 How does the earth's life-support system work?

3-2 What are the major components of an ecosystem?

3-3 What happens to energy in an ecosystem?

3-4 What happens to matter in an ecosystem?

3-5 How do scientists study ecosystems?

Tropical rain forest remnant surrounded by farmland near Iguacu National Park, Brazil.

Frans Lanting/National Geographic Creative

CORE CASE STUDY Tropical Rain Forests Are Disappearing

Tropical rain forests are found near the earth's equator and contain an incredible variety of life. These lush forests are warm year-round and have high humidity and heavy rainfall almost daily. Although they cover only about 2% of the earth's land surface, studies indicate that they contain up to half of the world's known terrestrial plant and animal species. For these reasons, they make an excellent natural laboratory for the study of *ecosystems*—communities of organisms interacting with one another and with the physical environment of matter and energy in which they live.

So far, at least half of these forests have been destroyed or disturbed by humans cutting down trees, growing crops, grazing cattle, and building settlements (see chapter-opening photo and Figure 3-1), and the degradation of these centers of biodiversity is increasing. Ecologists warn that without strong protective measures, most of these forests will probably be gone or severely degraded by the end of this century.

So why should we care that tropical rain forests are disappearing? Scientists give three reasons. *First,* clearing these forests will reduce the earth's vital biodiversity by destroying or degrading the habitats of many of the unique plant and animal species that live in them, which could lead to their early extinction.

Second, the destruction of these forests is helping to accelerate atmospheric warming, and thus projected climate change (as discussed in more detail in Chapter 19). The reason is that eliminating large areas of trees faster than they can grow back decreases the forests' ability to remove human-generated emissions of the gas carbon dioxide (CO_2) from the atmosphere, which contributes to atmospheric warming and projected climate change.

Third, large-scale rain forest loss can change regional weather patterns in ways that can prevent the return of diverse tropical rain forests in cleared or severely degraded areas. When this irreversible *ecological tipping point* is reached, tropical rain forests in such areas become much less diverse tropical grasslands.

Ecologists study an ecosystem to learn how its variety of organisms interact with their living *(biotic)* environment of other organisms and with their nonliving *(abiotic)* environment of soil, water, other forms of matter, and energy, mostly from the sun. In effect, ecologists study *connections in nature*. For example, they have found that tropical rain forests and other ecosystems recycle nutrients and provide humans and other organisms with essential natural services and natural resources, in keeping with the chemical cycling **principle of sustainability** (see Figure 1-2, p. 6 or back cover). In this chapter, we look more closely at how tropical rain forests, and ecosystems in general, work. We also examine how human activities such as the clear-cutting of forests can disrupt the cycling of nutrients within ecosystems and the flow of energy through them.

Photos: United Nations Environment Programme

Figure 3-1 **Natural capital degradation:** Satellite image of the loss of tropical rain forest, cleared for farming, cattle grazing, and settlements, near the Bolivian city of Santa Cruz between June 1975 (left) and May 2003 (right).

3-1 How Does the Earth's Life-Support System Work?

CONCEPT 3-1A
The four major components of the earth's life-support system are the atmosphere (air), the hydrosphere (water), the geosphere (rock, soil, and sediment), and the biosphere (living things).

CONCEPT 3-1B
Life is sustained by the flow of energy from the sun through the biosphere, the cycling of nutrients within the biosphere, and gravity.

Earth's Life-Support System Has Four Major Components

The earth's life-support system consists of four main spherical systems (Figure 3-2) that interact with one another—the atmosphere (air), the hydrosphere (water), the geosphere (rock, soil, and sediment), and the biosphere (living things) (Concept 3-1A).

The **atmosphere** is a thin spherical envelope of gases surrounding the earth's surface. Its inner layer, the **troposphere**, extends only about 17 kilometers (11 miles) above sea level at the tropics and about 7 kilometers (4 miles) above the earth's north and south poles. It contains the air we breathe, consisting mostly of nitrogen (78% of the total volume) and oxygen (21%). Most of the remaining 1% of the air consists of water vapor, carbon dioxide, and methane. The next layer, reaching from 17 to 50 kilometers (11–31 miles) above the earth's surface, is called the **stratosphere**. Its lower portion holds enough ozone (O_3) gas to filter out about 95% of the sun's harmful *ultraviolet (UV) radiation*. This global sunscreen allows life to exist on the surface of the planet.

The **hydrosphere** is made up of all of the water on or near the earth's surface. It is found as *water vapor* in the atmosphere, as *liquid water* on the surface and underground, and as *ice*—polar ice, icebergs, glaciers, and ice in frozen soil-layers called *permafrost*. The oceans, which cover about 71% of the globe, contain about 97% of the earth's water.

The **geosphere** consists of the earth's intensely hot *core*, a thick *mantle* composed mostly of rock, and a thin outer *crust*. Most of the geosphere is located in the earth's interior. Its upper portion contains nonrenewable *fossil fuels*—coal, oil, and natural gas—and minerals that we use, as well as renewable soil chemicals (nutrients) that organisms need in order to live, grow, and reproduce.

The **biosphere** consists of the parts of the atmosphere, hydrosphere, and geosphere where life is found. If the earth were an apple, the biosphere would be no thicker than the apple's skin. One important goal of environmental science is to understand the key interactions that occur within this thin layer of air, water, soil, and organisms.

Three Factors Sustain the Earth's Life

Life on the earth depends on three interconnected factors (Concept 3-1B):

1. The *one-way flow of high-quality energy* from the sun, through living things in their feeding interactions, into the environment as low-quality energy (mostly heat dispersed into air or water at a low temperature), and eventually to outer space as heat (Figure 3-3). This is in keeping with the solar energy **principle of sustainability** (see Figure 1-2, p. 6 or back cover). No round-trips are allowed because high-quality energy cannot be recycled, according to the second law of thermodynamics (see Chapter 2, p. 43). As this solar energy interacts with carbon dioxide (CO_2) and other gases in the troposphere, it warms the troposphere—a process known as the **greenhouse effect**. Without this natural process, the earth would be too cold to support the forms of life we find here today.
2. The *cycling of nutrients* (the atoms, ions, and molecules needed for survival by living organisms) through parts of the biosphere. Because the earth does not get

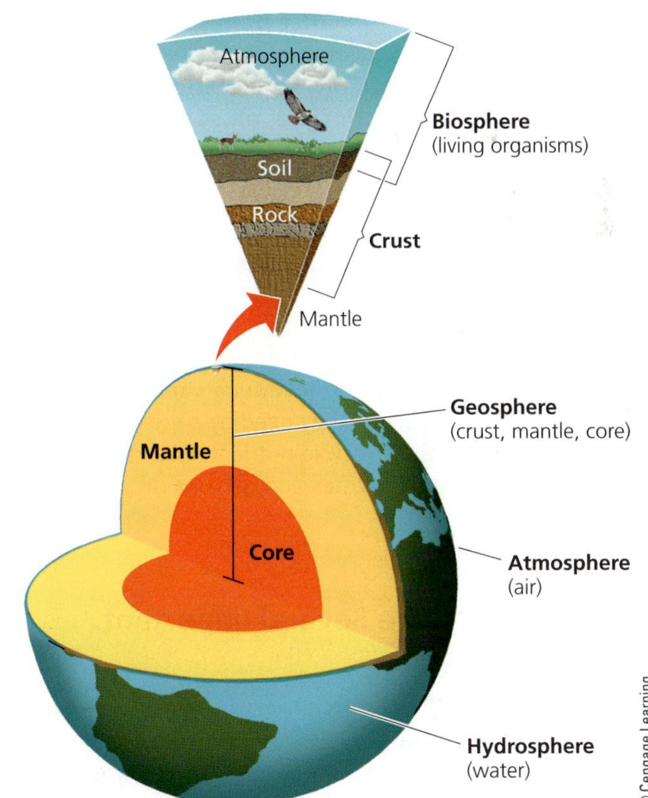

Figure 3-2 **Natural capital:** The earth consists of a land sphere *(geosphere)*, an air sphere *(atmosphere)*, a water sphere *(hydrosphere)*, and a life sphere *(biosphere)* (Concept 3-1A).

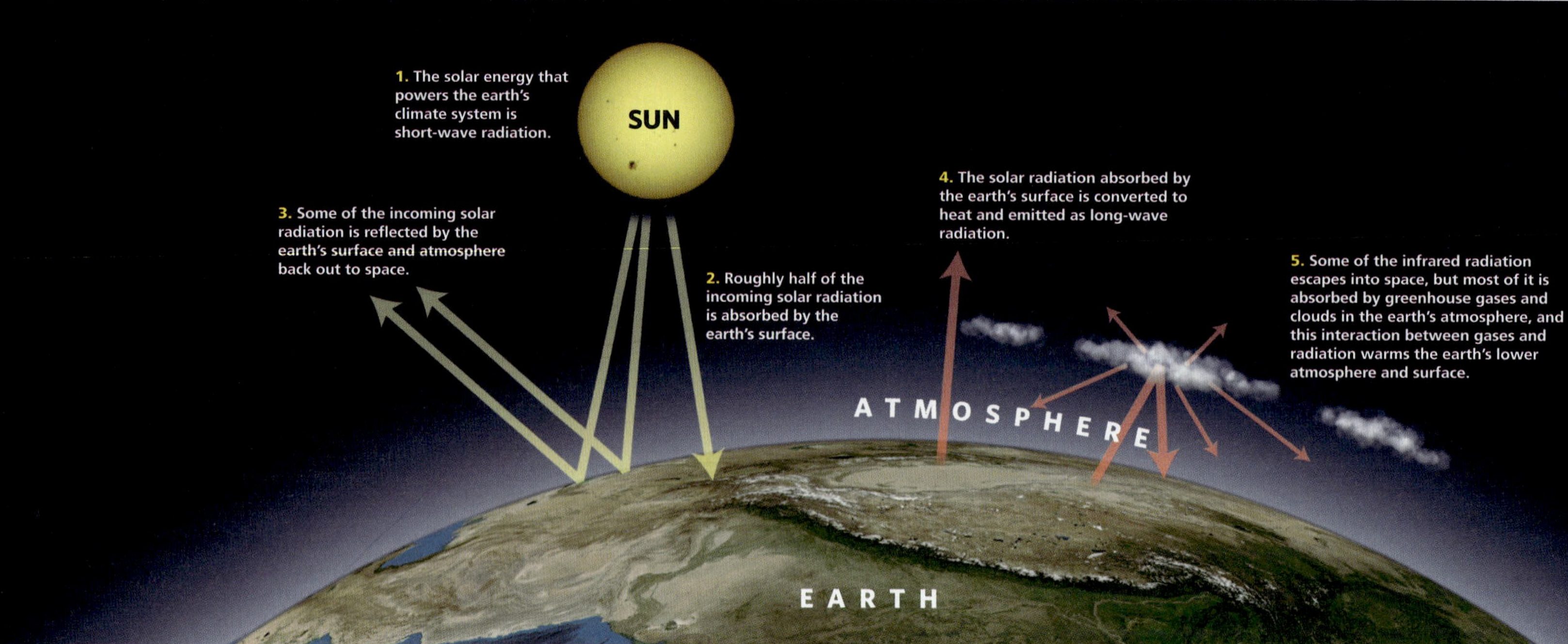

Animated Figure 3-3 Greenhouse Earth. High-quality solar energy flows from the sun to the earth. It is degraded to lower-quality energy (mostly heat) as it interacts with the earth's air, water, soil, and life-forms, and eventually returns to space. Certain gases in the earth's atmosphere retain enough of the sun's incoming energy as heat to warm the planet in what is known as the *greenhouse effect.*

significant inputs of matter from space, its essentially fixed supply of nutrients must be continually recycled to support life in keeping with the chemical cycling **principle of sustainability** (see Figure 1-2, p. 6 or back cover). The law of conservation of matter (see Chapter 2, p. 40) governs this nutrient cycling process.

3. *Gravity,* which allows the planet to hold onto its atmosphere and helps to enable the movement and cycling of chemicals through air, water, soil, and organisms.

3-2 What Are the Major Components of an Ecosystem?

CONCEPT 3-2

Some organisms produce the nutrients they need, others get the nutrients they need by consuming other organisms, and some recycle nutrients back to producers by decomposing the wastes and remains of other organisms.

Ecosystems Have Several Important Components

Ecology is the science that focuses on how organisms interact with one another and with their nonliving environment of matter and energy. Scientists classify matter into levels of organization ranging from atoms to galaxies. Ecologists study interactions within and among five of these levels—**organisms**, **populations**, **communities**, **ecosystems**, and the **biosphere**, which are illustrated and defined in Figure 3-4.

The biosphere and its ecosystems are made up of living *(biotic)* and nonliving *(abiotic)* components. Examples of nonliving components are water, air, nutrients, rocks, heat, and solar energy. Living components include plants, animals, microbes, and all other organisms. Figure 3-5 is a greatly simplified diagram of some of the living and nonliving components of a terrestrial ecosystem.

Ecologists assign every type of organism in an ecosystem to a *feeding level,* or **trophic level**, depending on its source of food or nutrients. We can broadly classify the living organisms that transfer energy and nutrients from one trophic level to another within an ecosystem as *producers* and *consumers.*

Producers, sometimes called **autotrophs** (self-feeders), make the nutrients they need from compounds

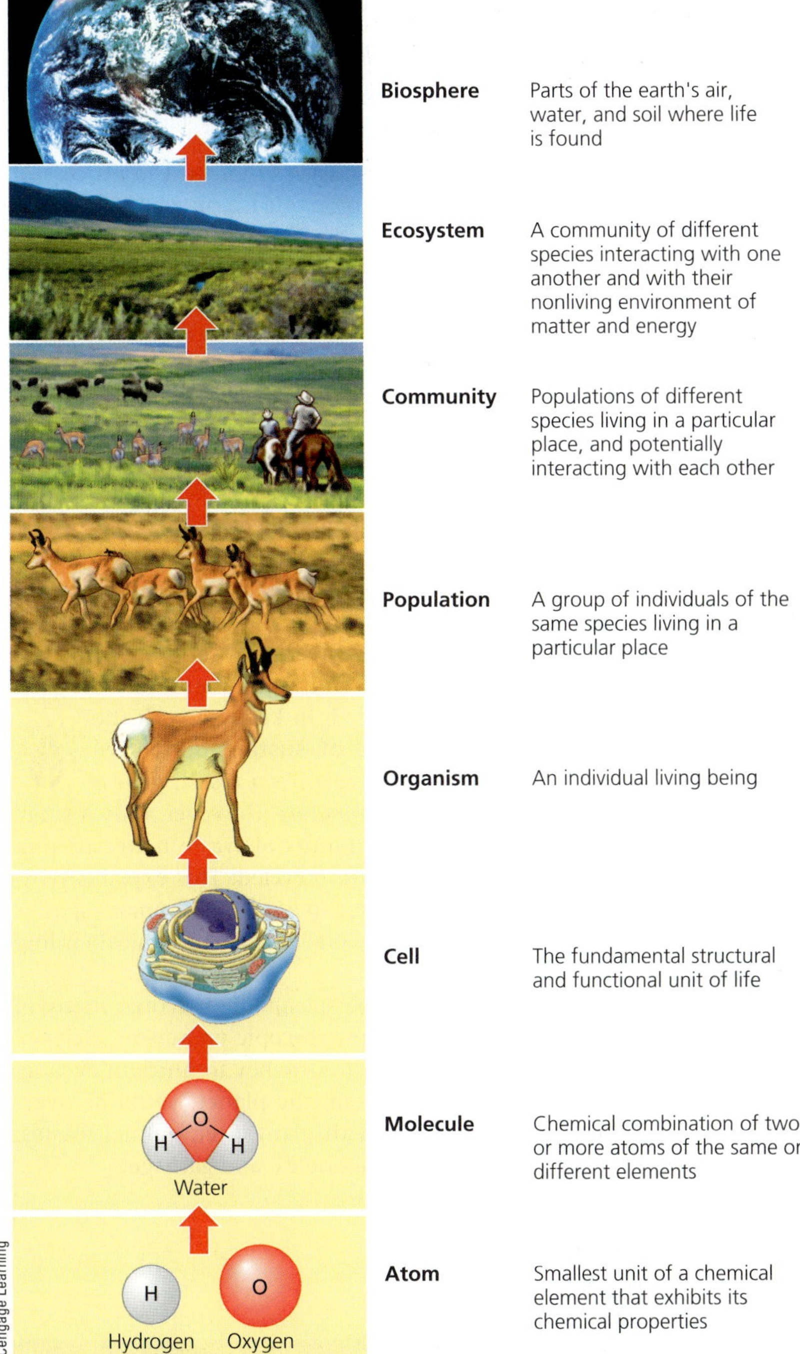

Animated Figure 3-4 This diagram illustrates some levels of the organization of matter in nature. Ecology focuses on the top five of these levels.

and energy obtained from their environment (**Concept 3-2**). In a process called **photosynthesis**, plants typically capture about 1% of the solar energy that falls on their leaves and use it in combination with carbon dioxide and water to form organic molecules, including energy-rich carbohydrates (such as glucose, $C_6H_{12}O_6$), which store the chemical energy plants need. Although hundreds of chemical changes take place during photosynthesis, we can summarize the overall reaction as follows:

carbon dioxide + water + **solar energy** → glucose + oxygen

$$6\ CO_2 + 6\ H_2O + \textbf{solar energy} \rightarrow C_6H_{12}O_6 + 6\ O_2$$

(See Supplement 4, p. S17, for information on how to balance chemical equations such as this one.)

On land, most producers are trees and other green plants (Figure 3-6, left). In freshwater and ocean ecosystems, algae and aquatic plants growing near shorelines are the major producers (Figure 3-6, right). In open water, the dominant producers are *phytoplankton*—mostly microscopic organisms that float or drift in the water. The map in Figure 5 on p. S30 of Supplement 6 shows the distribution of green plants, or *plant biomass,* over the earth's land and water.

Most of the earth's producers are photosynthetic organisms that get their energy indirectly from the sun and convert it to chemical energy stored in their cells. However, some producer bacteria live in the dark in extremely hot water around fissures on the ocean floor. Their source of energy is heat from the earth's interior, or *geothermal energy,* and they represent an exception to the solar energy **principle of sustainability**.

Besides producers, all other organisms in an ecosystem are **consumers**, or **heterotrophs** ("other-feeders"), that cannot produce the nutrients they need through photosynthesis or other processes (**Concept 3-2**). They get their nutrients by feeding on other organisms (producers or other consumers) or their remains. In other words, all consumers (including humans) depend on producers for their nutrients.

There are several types of consumers. **Primary consumers**, or **herbivores** (plant eaters), are animals that eat mostly green plants. Examples are caterpillars, giraffes, and *zooplankton,* or tiny sea animals that feed on phytoplankton. **Carnivores** (meat eaters) are animals that feed on the flesh of other animals. Some carnivores such as spiders, lions (Figure 3-7), and most small fishes are **secondary consumers** that feed on the flesh of herbivores. Other carnivores such as tigers, hawks, and killer whales (orcas) are **tertiary** (or higher-level) **consumers** that feed on the flesh of other carnivores. Some of these relationships are shown in Figure 3-5. **Omnivores** such as pigs, rats, and humans eat plants and other animals.

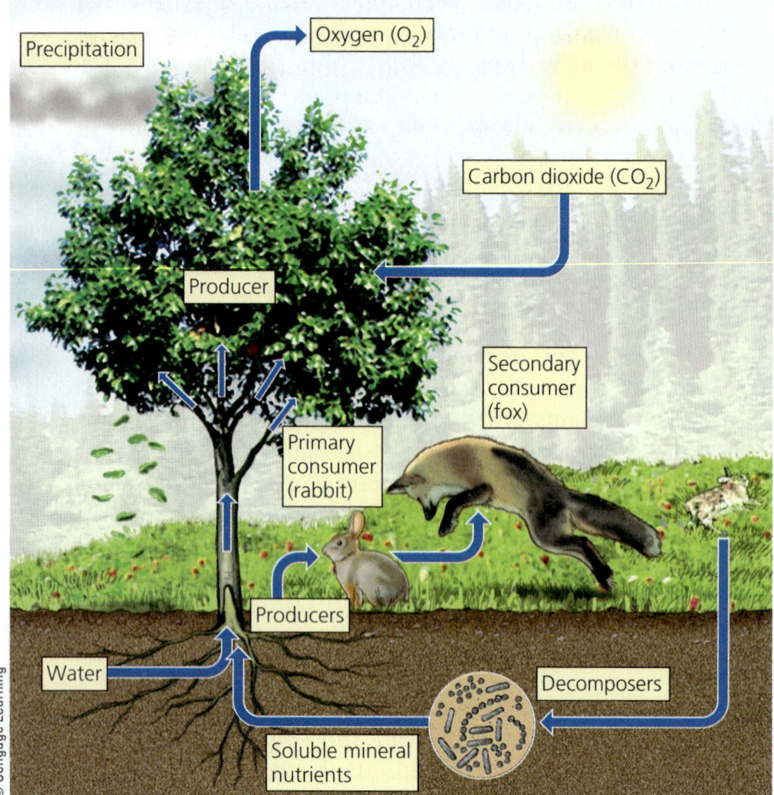

Animated Figure 3-5 Key living and nonliving components of an ecosystem in a field.

CONSIDER THIS. . .

THINKING ABOUT What You Eat

When you ate your most recent meal, were you an herbivore, a carnivore, or an omnivore?

Decomposers are consumers that, in the process of obtaining their own nutrients, release nutrients from the wastes or remains of plants and animals and then return those nutrients to the soil, water, and air for reuse by producers (**Concept 3-2**). Most decomposers are bacteria and fungi (Figure 3-8, left). Other consumers, called **detritus feeders**, or **detritivores**, feed on the wastes or dead bodies of other organisms; these wastes are called *detritus* (dee-TRI-tus), which means debris. Examples are earthworms, some insects, and vultures (Figure 3-8, right).

Hordes of detritivores and decomposers can transform a fallen tree trunk into wood particles and, finally, into simple inorganic molecules that plants can absorb as nutrients (Figure 3-9). Thus, in natural ecosystems the wastes and dead bodies of organisms serve as resources for other organisms, as the nutrients that make life possible are continually recycled, in keeping with the chemical cycling **principle of sustainability** (see Figure 1-2, p. 6 or back cover). As a result, *there is very little waste of nutrients in nature.* However, only a small percentage of the huge amounts of waste materials that human societies produce are recycled. This explains why, unlike other species, the human species is a major violator of the chemical cycling principle.

Decomposers and detritivores, many of which are microscopic organisms (Science Focus 3.1), are the key to nutrient cycling. Without them, the planet would be overwhelmed with plant litter, animal wastes, dead animal bodies, and garbage.

Aleksander Bolbot/Shutterstock.com

Lance Rider/Shutterstock.com

Figure 3-6 Some producers such as trees and other plants (left) live on land. Others such as green algae (right) live in water.

Natasha Chamberlain/Photos.com

Figure 3-7 This lioness (a carnivore) is feeding on a freshly killed zebra (an herbivore) in Kenya, Africa.

Organisms Get Their Energy in Different Ways

Producers, consumers, and decomposers use the chemical energy stored in glucose and other organic compounds to fuel their life processes. In most cells, this energy is released by **aerobic respiration**, which uses oxygen to convert glucose (or other organic nutrient molecules) back into carbon dioxide and water. The net effect of the hundreds of steps in this complex process is represented by the following chemical reaction:

glucose + oxygen → carbon dioxide + water + **energy**

$C_6H_{12}O_6 + 6\,O_2 \rightarrow 6\,CO_2 + 6\,H_2O$ + **energy**

Notice that, although the detailed steps differ, the net chemical change for aerobic respiration is the opposite of that for photosynthesis (p. 55).

Some decomposers get the energy they need by breaking down glucose (or other organic compounds) in the *absence* of oxygen. This form of cellular respiration is called **anaerobic respiration**, or **fermentation**. Instead of carbon dioxide and water, the end products of this process are compounds such as methane gas (CH_4, the main component of natural gas), ethyl alcohol (C_2H_6O), acetic acid ($C_2H_4O_2$, the key component of vinegar), and hydrogen sulfide (H_2S, a highly poisonous gas that smells like rotten eggs). Note that all organisms get their energy from aerobic or anaerobic respiration but only plants carry out photosynthesis.

To summarize, ecosystems and the biosphere are sustained through a combination of *one-way energy flow* from the sun through these systems and the *nutrient cycling* of key materials within them (**Concept 3-1B**)—in keeping with two of the scientific **principles of sustainability** (Figure 3-10).

javaman/Shutterstock.com

Mitrofanov Alexander/Shutterstock.com

Figure 3-8 These sulfur tuft fungi feeding on a tree stump (left) are decomposers. The vultures and Marabou storks (above), eating the carcass of an animal that was killed by another animal, are detritivores.

SCIENCE FOCUS 3.1

MANY OF THE WORLD'S MOST IMPORTANT ORGANISMS ARE INVISIBLE TO US

They are everywhere. Trillions can be found inside your body, on your skin, in a handful of soil, and in a cup of ocean water. They are *microbes*, or *microorganisms*, catchall terms for many thousands of species of bacteria, protozoa, fungi, and floating phytoplankton. Though most of them are too small to be seen with the naked eye, they are the biological rulers of the earth.

Microbes do not get the respect they deserve. Most of us view them primarily as threats to our health in the form of infectious bacteria or fungi that cause athlete's foot and other skin diseases, and protozoa that cause diseases such as malaria. But these harmful microbes are in the minority.

In 2012 scientist Philip Tarr and other researchers identified more than 10,000 species of bacteria, fungi, and other microbes that live in or on our bodies. Many of them provide us with vital services. Bacteria in our intestinal tracts help break down the food we eat, and microbes in our noses help prevent harmful bacteria from reaching our lungs. According to recent research, the greater the diversity of the bacterial zoo in our stomachs, the better our health is likely to be. We are alive largely because of these multitudes of microbes toiling away completely out of sight.

Bacteria and other microbes help to purify the water we drink by breaking down any plant and animal wastes in the water. Bacteria and fungi (such as yeast) also help to produce foods such as bread, cheese, yogurt, soy sauce, beer, and wine. Bacteria and fungi in the soil decompose organic wastes into nutrients that can be taken up by plants that are then eaten by humans and many other plant eaters. Without these tiny creatures, we would go hungry and be up to our necks in waste matter.

Microbes, particularly phytoplankton in the ocean, also provide much of the planet's oxygen and help to regulate the earth's temperature by removing some of the carbon dioxide produced when we burn coal, natural gas, and gasoline. Scientists are working on using microbes to develop new medicines and fuels. Genetic engineers are inserting genetic material into existing microorganisms to convert them to microbes that can be used to clean up polluted water and soils.

Some microorganisms assist us in controlling plant diseases and populations of insect species that attack our food crops. By relying more on these microbes for pest control, we could reduce the use of potentially harmful chemical pesticides. In other words, microbes are a vital part of the earth's natural capital.

Critical Thinking

What are two advantages that microbes have over humans for thriving in the world?

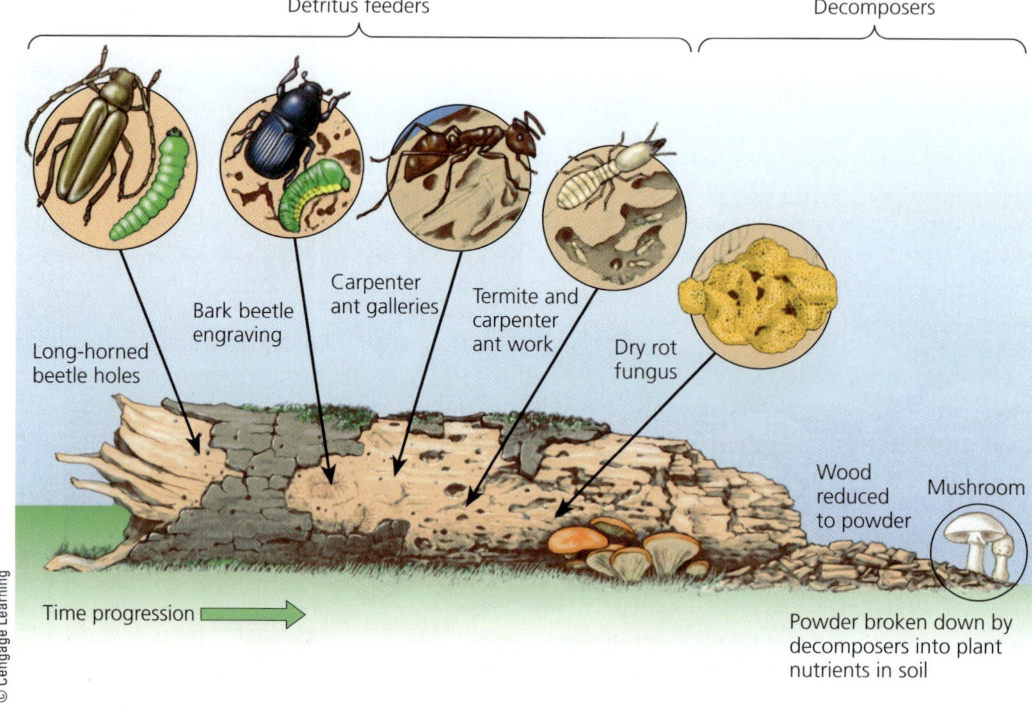

Figure 3-9 Various detritivores and decomposers (mostly fungi and bacteria) can "feed on" or digest parts of a log and eventually convert its complex organic chemicals into simpler inorganic nutrients that can be taken up by producers.

Animated Figure 3-10 **Natural capital:** The main components of an ecosystem are energy, chemicals, and organisms. Nutrient cycling and the flow of energy—first from the sun, then through organisms, and finally into the environment as low-quality heat—link these components.

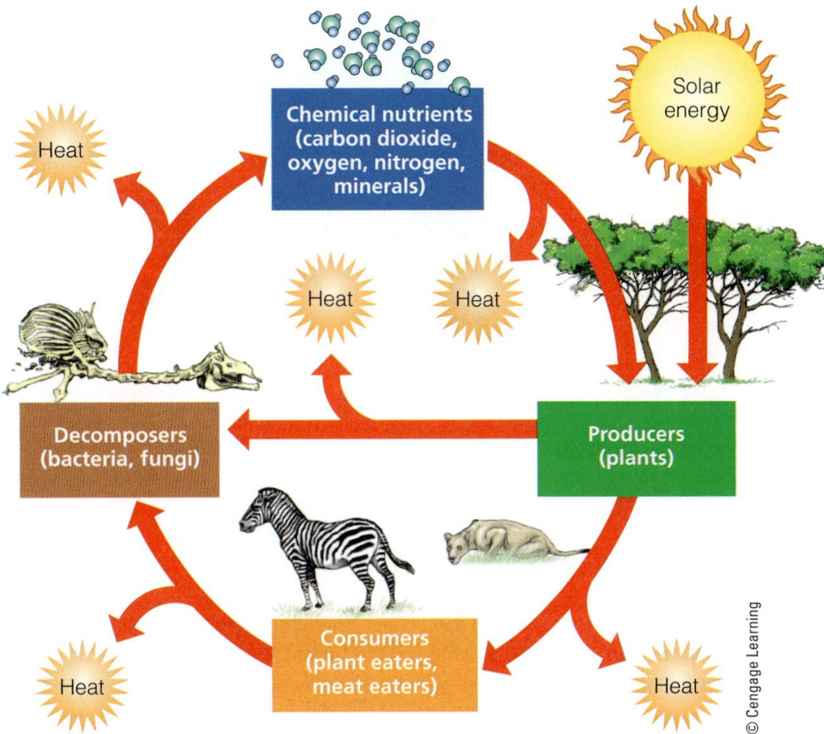

CONSIDER THIS. . .

THINKING ABOUT Chemical Cycling and the Law of Conservation of Matter

Explain the relationship between chemical cycling in ecosystems and the law of conservation of matter (see Chapter 2, p. 40).

3-3 What Happens to Energy in an Ecosystem?

CONCEPT 3-3

As energy flows through ecosystems in food chains and webs, the amount of chemical energy available to organisms at each successive feeding level decreases.

Energy Flows through Ecosystems in Food Chains and Food Webs

The chemical energy stored as nutrients in the bodies and wastes of organisms flows through ecosystems from one trophic (feeding) level to another. For example, a plant uses solar energy to store chemical energy in a leaf. A caterpillar eats the leaf, a robin eats the caterpillar, and a hawk eats the robin. Decomposers and detritus feeders consume the wastes and remains of all members of this and other food chains and return their nutrients to the soil for reuse by producers.

A sequence of organisms, each of which serves as a source of nutrients or energy for the next, is called a **food chain**. It determines how chemical energy and nutrients move along the same pathways from one organism to another through the trophic levels in an ecosystem—primarily through photosynthesis, feeding, and decomposition—as shown in Figure 3-11. Every use and transfer of energy by organisms involves a loss of some degraded high-quality energy to the environment as heat, in accordance with the second law of thermodynamics.

In natural ecosystems, most consumers feed on more than one type of organism, and most organisms are eaten or decomposed by more than one type of consumer. Because of this, organisms in most ecosystems form a complex network of interconnected food chains called a **food web** (Figure 3-12). We can assign trophic levels in food webs just as we can in food chains. Food chains and webs show how producers, consumers, and decomposers are connected to one another as energy flows through trophic levels in an ecosystem.

Usable Energy Decreases with Each Link in a Food Chain or Web

Each trophic level in a food chain or web contains a certain amount of **biomass**, the dry weight of all organic matter contained in its organisms. In a food chain or web, chemical energy stored in biomass is transferred from one trophic level to another.

As a result of the second law of thermodynamics, energy transfer through food chains and food webs is not very efficient because, with each energy transfer from one trophic level to another, some usable chemical energy is degraded and lost to the environment as low-quality heat. In other words, as energy flows through ecosystems in food chains and webs, there is a decrease in the amount of high-quality chemical energy available to organisms at each succeeding feeding level (Concept 3-3).

The percentage of usable chemical energy transferred as biomass from one trophic level to the next varies, depend-

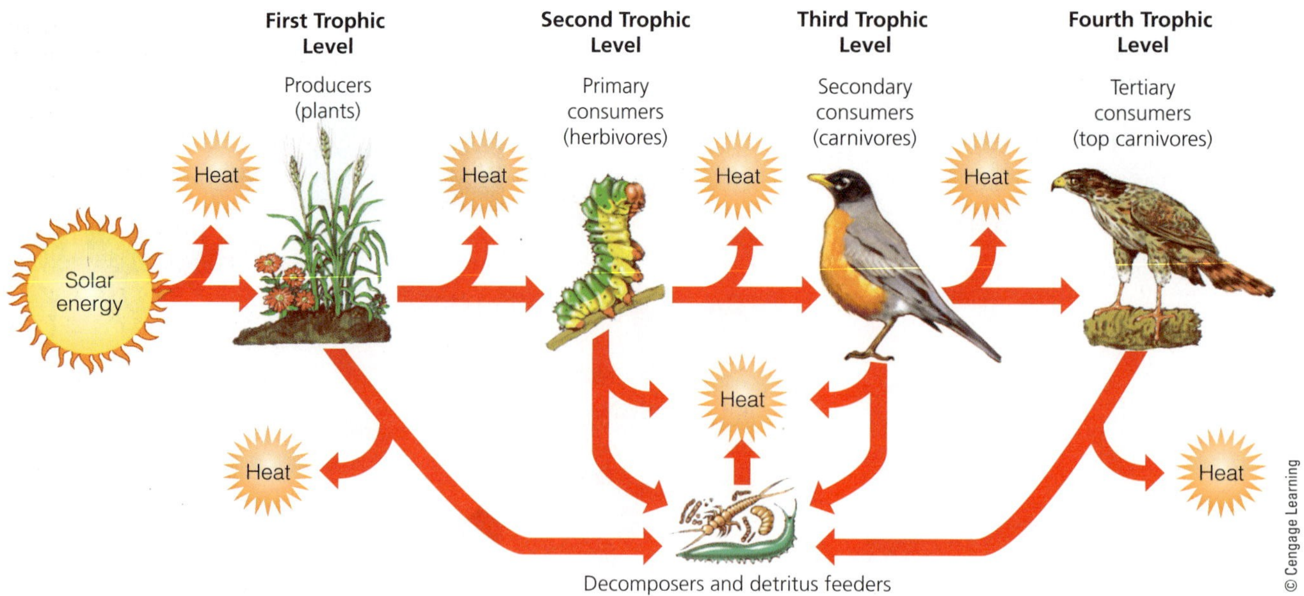

Animated Figure 3-11 In a food chain, chemical energy in nutrients flows through various trophic levels. ***Question:*** Think about what you ate for breakfast. At what level or levels on a food chain were you eating?

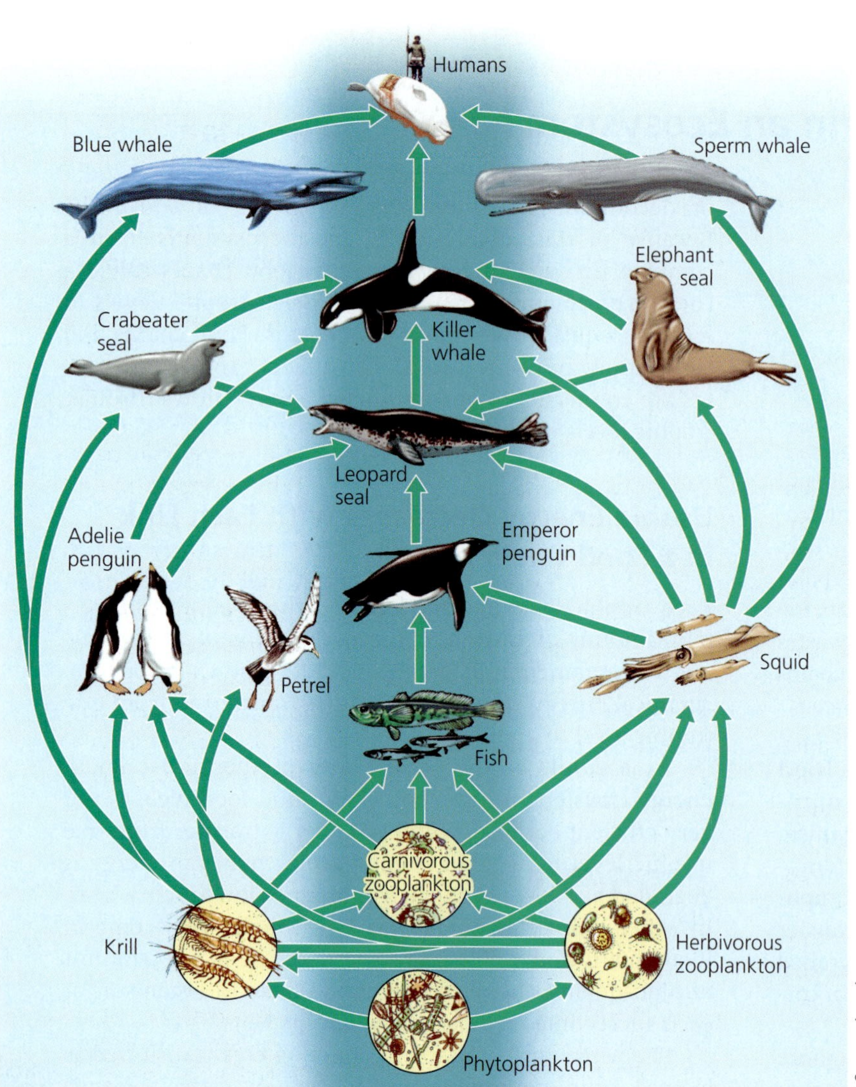

ing on what types of species and ecosystems are involved. The more trophic levels there are in a food chain or web, the greater is the cumulative loss of usable chemical energy as it flows through the trophic levels. The **pyramid of energy flow** in Figure 3-13 illustrates this energy loss for a simple food chain, assuming a 90% energy loss with each transfer.

CONSIDER THIS. . .

THINKING ABOUT Energy Flow and the Second Law of Thermodynamics

Explain the relationship between the second law of thermodynamics (see Chapter 2, p. 43) and the flow of energy through a food chain or web.

The large loss in chemical energy between successive trophic levels explains why food chains and webs rarely have more than four or five trophic levels. In most cases, too little chemical energy is left after four or five transfers to support organisms feeding at these high trophic levels. Thus, there are far fewer tigers in tropical rain forests (**Core Case Study**) and other areas than there are insects.

Animated Figure 3-12 This is a greatly simplified *food web* found in the southern hemisphere. The shaded middle area shows a simple food chain that is part of these complex interacting feeding relationships. Many more participants in the web, including an array of decomposer and detritus feeder organisms, are not shown here. ***Question:*** Can you imagine a food web of which you are a part? Try drawing a simple diagram of it.

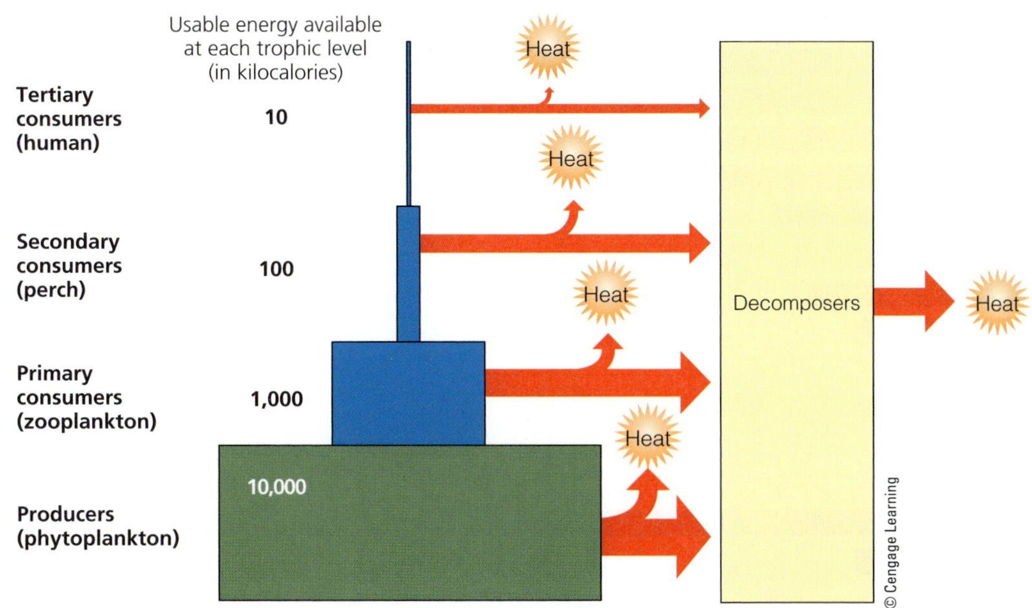

Animated Figure 3-13 This model is a generalized *pyramid of energy flow* that shows the decrease in usable chemical energy available at each succeeding trophic level in a food chain or web. The model assumes that with each transfer from one trophic level to another, there is a 90% loss of usable energy to the environment in the form of low-quality heat. Calories and joules are used to measure energy. 1 kilocalorie = 1,000 calories = 4,184 joules. ***Question:*** Why is a vegetarian diet more energy efficient than a meat-based diet?

CONSIDER THIS. . .

CONNECTIONS Energy Flow and Feeding People

Energy flow pyramids explain why the earth could support more people if they all ate at lower trophic levels by consuming grains, vegetables, and fruits directly rather than passing such crops through another trophic level and eating herbivores such as cattle. About two-thirds of the world's people survive primarily by eating wheat, rice, and corn at the first trophic level because most of them cannot afford to eat much meat.

Some Ecosystems Produce Plant Matter Faster Than Others Do

The amount, or mass, of living organic material (biomass) that a particular ecosystem can support is determined by how much solar energy its producers can capture and store as chemical energy and by how rapidly they can do so. **Gross primary productivity (GPP)** is the *rate* at which an ecosystem's producers (usually plants) convert solar energy into chemical energy in the form of biomass found in their tissues. It is usually measured in terms of energy production per unit area over a given time span, such as kilocalories per square meter per year (kcal/m^2/yr). The map in Figure 7 on p. S33 of Supplement 6 shows the gross primary productivity across North America.

To stay alive, grow, and reproduce, producers must use some of the chemical energy stored in the biomass they make for their own respiration. **Net primary productivity (NPP)** is the *rate* at which producers use photosynthesis to produce and store chemical energy *minus* the *rate* at which they use some of this stored chemical energy through aerobic respiration. NPP measures how fast producers can make the chemical energy that is stored in their tissues and that is potentially available to other organisms (consumers) in an ecosystem.

Gross primary productivity is similar to the *rate* at which you make money, or the number of dollars you earn per year. Net primary productivity is similar to the amount of money earned per year that you can spend after subtracting your work expenses such as the costs of transportation, clothes, food, and supplies.

Ecosystems and aquatic life zones differ in their NPP, as illustrated in Figure 3-14. Despite its low NPP, the open ocean produces more of the earth's biomass per year than any other ecosystem or life zone. This occurs because the enormous volume of the global ocean, which covers 71% of the earth's surface, contains huge numbers of producers, including phytoplankton.

Tropical rain forests have a very high NPP because they have a large number and variety of producer trees and other plants. When such forests are cleared (see chapter-opening photo and **Core Case Study**) or burned to make way for crops or for grazing cattle, there is a sharp drop in the NPP and a loss of many of the diverse array of plant and animal species.

As we have seen, producers are the source of all nutrients in an ecosystem that are available for the producers themselves and for the consumers and decomposers that feed on them. Only the biomass represented by NPP is available as nutrients for consumers, and they use only a portion of it. Thus, *the planet's NPP ultimately limits the number of consumers (including humans) that can survive on the earth*. This is an important lesson from nature.

CONSIDER THIS. . .

CONNECTIONS Humans and Earth's Net Primary Productivity

Peter Vitousek, Stuart Rojstaczer, and other ecologists estimate that humans now use, waste, or destroy about 38% of the earth's total potential NPP. This is a remarkably high value, considering that the human population makes up less than 1% of the total biomass of all of the earth's consumers that depend on producers for their nutrients.

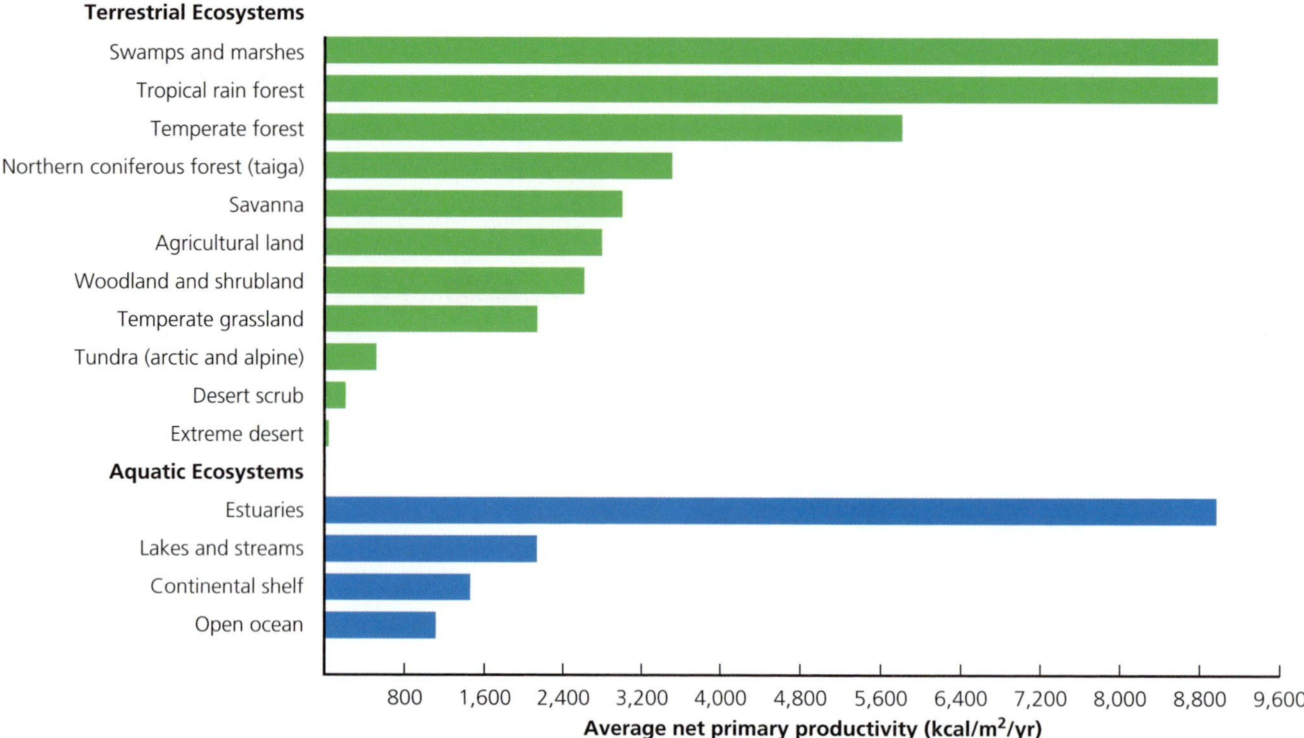

Figure 3-14 The estimated annual average *net primary productivity* in major life zones and ecosystems is expressed in this graph as kilocalories of energy produced per square meter per year (kcal/m^2/yr). ***Question:*** What are the three most productive and the three least productive systems?

(Compiled by the authors using data from R. H. Whittaker, *Communities and Ecosystems,* 2nd ed., New York: Macmillan, 1975.)

3-4 What Happens to Matter in an Ecosystem?

> **CONCEPT 3-4**
> Matter, in the form of nutrients, cycles within and among ecosystems and the biosphere, and human activities are altering these chemical cycles.

Nutrients Cycle within and among Ecosystems

The elements and compounds that make up nutrients move continually through air, water, soil, rock, and living organisms within ecosystems, as well as in the biosphere in cycles called **nutrient cycles**, or *biogeochemical cycles* (literally, life-earth-chemical cycles). This is in keeping with the chemical cycling **principle of sustainability** (see Figure 1-2, p. 6 or back cover). These cycles, which are driven directly or indirectly by incoming solar energy and by the earth's gravity, include the hydrologic (water), carbon, nitrogen, phosphorus, and sulfur cycles. They are important components of the earth's natural capital (see Figure 1-3, p. 7), and human activities are altering them (Concept 3-4).

As nutrients move through their biogeochemical cycles, they may accumulate in certain portions of the cycles and remain there for varying periods of time. These temporary storage sites such as the atmosphere, the oceans and other bodies of water, and underground deposits are called *reservoirs.*

CONSIDER THIS. . .

CONNECTIONS Nutrient Cycles and Life

Nutrient cycles connect past, present, and future forms of life. Some of the carbon atoms in your skin may once have been part of an oak leaf, a dinosaur's skin, or a layer of limestone rock. Your grandmother, rock star Bono, or a hunter–gatherer who lived 25,000 years ago may have inhaled some of the nitrogen molecules you just inhaled.

The Water Cycle

Water (H_2O) is an amazing substance (Science Focus 3.2) that is necessary for life on the earth. The **hydrologic cycle**, or **water cycle**, collects, purifies, and distributes the earth's fixed supply of water, as shown in Figure 3-15.

The water cycle is powered by energy from the sun and involves three major processes—evaporation, precipitation, and transpiration. Incoming solar energy causes *evaporation,* or the conversion of water from liquid to vapor from the earth's oceans, lakes, rivers, and soil. This

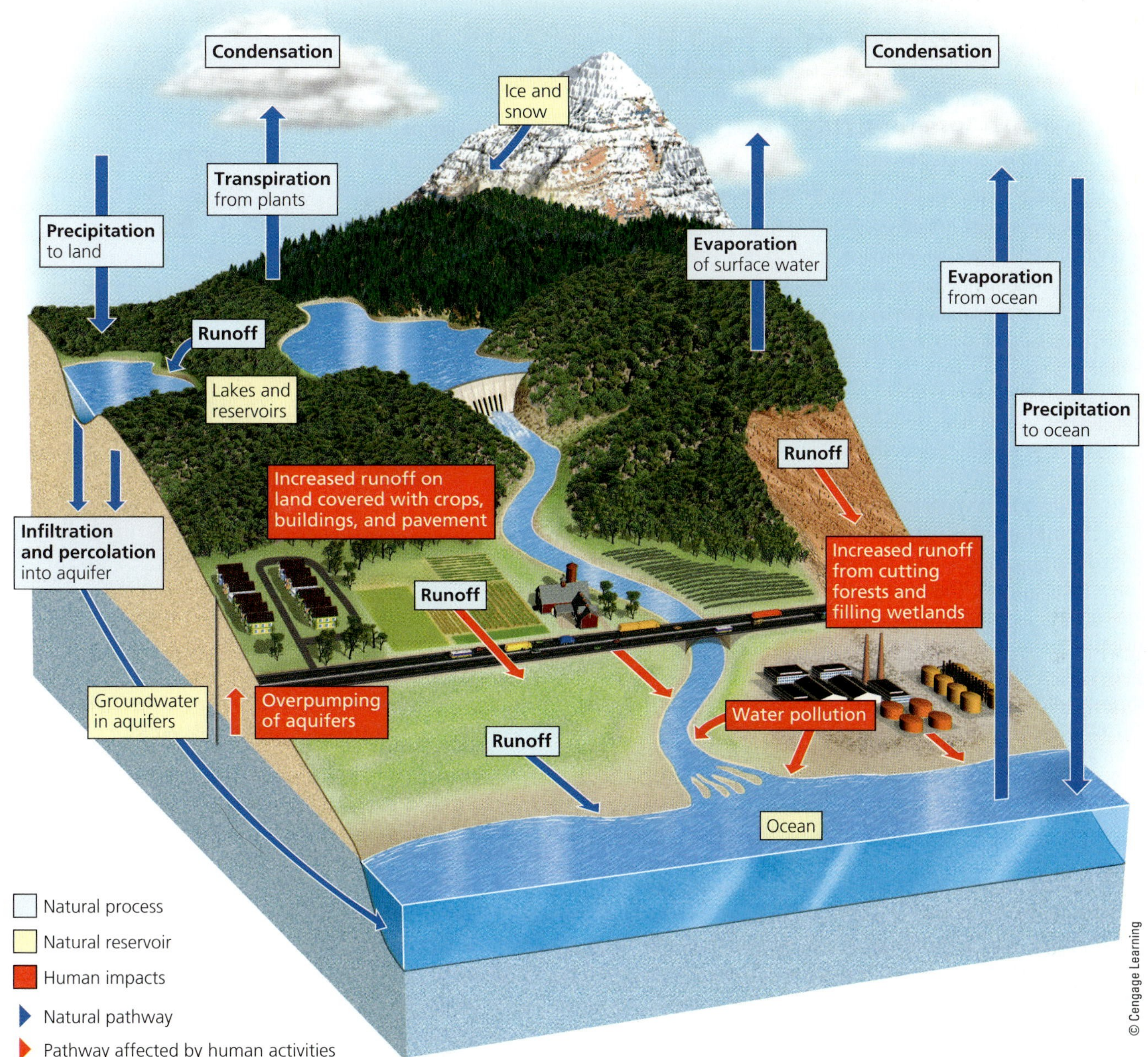

Animated Figure 3-15 **Natural capital:** This is a simplified model of the *water cycle*, or *hydrologic cycle*, in which water circulates in various physical forms within the biosphere. The red arrows and boxes identify major effects of human activities on this cycle. ***Question:*** What are three ways in which your lifestyle directly or indirectly affects the hydrologic cycle?

water vapor rises into the atmosphere, where it condenses into droplets, and gravity then draws the water back to the earth's surface as *precipitation* (rain, snow, sleet, and dew). Over land, about 90% of the water that reaches the atmosphere evaporates from the surfaces of plants, through a process called *transpiration,* and from the soil.

Water that returns to the earth's surface as precipitation takes various paths. Most precipitation falling on terrestrial ecosystems becomes **surface runoff**. This water flows into streams, which eventually carry water to lakes and oceans, from which it can evaporate to repeat the cycle. Some surface water also seeps into the upper layers of soils where it is used by plants, and some evaporates from the soils back into the atmosphere.

Some precipitation is converted to ice that is stored in *glaciers* (Figure 3-16), usually for long periods of time. Some precipitation sinks through soil and permeable rock formations to underground layers of rock, sand, and gravel called **aquifers**, where it is stored as **groundwater**.

A small amount of the earth's water ends up in the living components of ecosystems. As producers, plants absorb some of this water through their roots, most of which evaporates from plant leaves back into the atmosphere during transpiration. Some of the water that plants absorb combines

SCIENCE FOCUS 3.2

WATER'S UNIQUE PROPERTIES

Water (H_2O) is a remarkable substance with a unique combination of properties:

- *Forces of attraction,* called *hydrogen bonds* (Figure 3-A), *hold water molecules together*—the major factor determining water's distinctive properties.
- *Water exists as a liquid over a wide temperature range because of the hydrogen bonds between its molecules.* If water did not have a high boiling point, the oceans would have evaporated long ago.
- *Liquid water changes temperature slowly because it can store a large amount of heat without a large change in its own temperature.* This high heat storage capacity helps protect living organisms from temperature changes, moderates the earth's climate, and makes water an excellent coolant for car engines and power plants.
- *It takes a large amount of energy to evaporate water because of its hydrogen bonds.* Water absorbs large amounts of heat as it changes into water vapor and releases this heat as the vapor condenses back to liquid water. This helps to distribute heat throughout the world and to determine regional and local climates. It also makes evaporation a cooling process—explaining why you feel cooler when perspiration evaporates from your skin.
- *Liquid water can dissolve a variety of compounds* (see Figure 2, p. S13, in Supplement 4). It carries dissolved nutrients into the tissues of living organisms, flushes waste products out of those tissues, serves as an all-purpose cleanser, and helps to remove and dilute the water-soluble wastes of civilization. This property also means that water-soluble wastes can easily pollute water.
- *Water filters out wavelengths of the sun's ultraviolet radiation that would harm some aquatic organisms.* However, down to a certain depth, it is transparent to sunlight that is necessary for photosynthesis.
- *The hydrogen bonds between water molecules also allow liquid water to adhere to a solid surface.* This enables narrow columns of water to rise through a plant from its roots to its leaves (a process called *capillary action*).
- *Unlike most liquids, water expands when it freezes.* This means that ice floats on water because it has a lower density (mass per unit of volume) than liquid water has. Otherwise, lakes and streams in cold climates would freeze solid, losing most of their aquatic life. Because water expands on freezing, it can break pipes, crack a car's engine block (if it doesn't contain antifreeze), break up pavement, and fracture rocks.

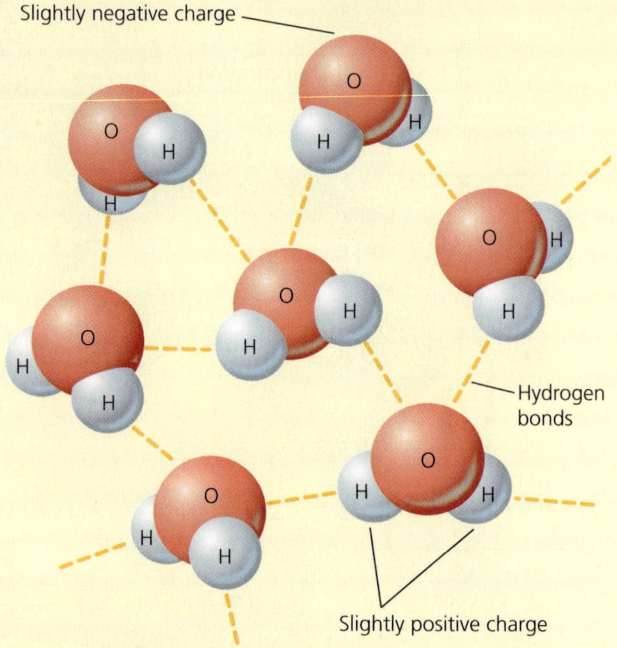

Figure 3-A *Hydrogen bond:* Because of the slightly unequal sharing of electrons in an H_2O molecule, its oxygen atoms hold a slightly negative charge and its hydrogen atoms hold a slightly positive charge. Since opposite electrical charges attract one another, the hydrogen atoms of one water molecule are attracted to the oxygen atoms in other water molecules. These fairly weak forces of attraction *between* molecules (represented by the dashed lines) are called *hydrogen bonds.*

Critical Thinking

Pick two of the special properties listed above and, for each property, explain how life on the earth would be different if it did not exist.

with carbon dioxide during photosynthesis to produce high-energy organic compounds such as carbohydrates. Eventually these compounds are broken down in plant cells, which release the water back into the environment. Consumers get their water from their food and by drinking it.

Because water dissolves many nutrient compounds, it is a major medium for transporting nutrients within and between ecosystems. Water is also the primary sculptor of the earth's landscape as it flows over and wears down rock over millions of years with results such as the Grand Canyon in the southwestern United States.

Throughout the hydrologic cycle, many natural processes purify water. Evaporation and subsequent precipitation act as a natural distillation process that removes impurities dissolved in water. Water flowing aboveground through streams and lakes, and underground in aquifers, is naturally filtered and partially purified by chemical and biological processes—mostly by the actions of decomposer

Pablo H Caridad/Shutterstock.com

Figure 3-16 This glacier in Patagonia, Argentina, stores water for long periods of time as part of the hydrologic cycle.

bacteria—as long as these natural processes are not overloaded. Thus, *the hydrologic cycle can be viewed as a cycle of natural renewal of water quality.*

Only about 0.024% of the earth's vast water supply is available to humans and other species as liquid freshwater in accessible groundwater deposits and in lakes, rivers, and streams (see Figure 25, p. S50, in Supplement 6). The rest is too salty for us to use, is stored as ice, or is too deep underground to extract at affordable prices using current technology.

Humans alter the water cycle in three major ways (see the red arrows and boxes in Figure 3-15). *First,* we withdraw large quantities of freshwater from rivers, lakes, and aquifers sometimes faster than nature can replace it. As a result, some aquifers are being depleted and some rivers no longer flow to the ocean. *Second,* we clear vegetation from land for agriculture, mining, road building, and other activities, and cover much of the land with buildings, concrete, and asphalt. This increases runoff and reduces infiltration that would normally recharge groundwater supplies. *Third,* we drain and fill wetlands for farming and urban development. Left undisturbed, wetlands provide the natural service of flood control, acting like sponges to absorb and hold overflows of water from drenching rains or rapidly melting snow.

CONSIDER THIS. . .

CONNECTIONS Clearing a Rain Forest Can Affect Local Weather and Climate

Clearing vegetation can alter weather patterns by reducing transpiration, especially in dense tropical rain forests (**Core Case Study**). Because so many plants in such a forest transpire water into the atmosphere, vegetation is the primary source of local rainfall. Cutting down large areas of forest raises ground temperatures (because it reduces shade) and can reduce local rainfall so much that the forest cannot grow back. If this occurs over a large area for three or more decades, the climate of the affected area can change, and much less diverse tropical grasslands can replace biologically diverse forests.

The Carbon Cycle

Carbon is the basic building block of the carbohydrates, fats, proteins, DNA, and other organic compounds necessary for life. Various compounds of carbon circulate through the biosphere, the atmosphere, and parts of the hydrosphere, in the **carbon cycle** shown in Figure 3-17.

The carbon cycle is based on carbon dioxide (CO_2) gas, which makes up about 0.039% of the volume of the earth's atmosphere and is also dissolved in water. Carbon dioxide (along with water vapor in the water cycle) is a key component of the atmosphere's thermostat. If the carbon cycle removes too much CO_2 from the atmosphere, the atmosphere will cool, and if it generates too much CO_2, the atmosphere will get warmer. Thus, even slight changes in this cycle caused by natural or human factors can affect the earth's climate and ultimately help to determine the types of life that can exist in various places.

Terrestrial producers remove CO_2 from the atmosphere and aquatic producers remove it from the water. These producers then use the CO_2 and photosynthesis to produce complex carbohydrates such as glucose ($C_6H_{12}O_6$).

The cells in oxygen-consuming producers, consumers, and decomposers then carry out aerobic respiration. This process breaks down glucose and other complex organic compounds to produce CO_2 in the atmosphere and water for reuse by producers. Because of this linkage between *photosynthesis* in producers and *aerobic respiration* in producers, consumers, and decomposers, these processes circulate carbon in the biosphere. Oxygen and hydrogen—the other elements in carbohydrates—cycle almost in step with carbon.

Some carbon atoms take a long time to recycle. Decomposers release the carbon stored in the bodies of dead organisms on land back into the air as CO_2, which can remain in the atmosphere for 100 years or more. And in water, decomposers release carbon that can be stored as insoluble carbonates in bottom sediment for very long periods of time. Indeed, marine sediments are the earth's largest store of carbon. Over millions of years, buried deposits of dead plant matter and bacteria were compressed between layers of sediment, where high pressure and heat converted them to carbon-containing *fossil fuels* such as coal, oil, and natural gas (see Figure 2-13, p. 43, and Figure 3-17).

This long-stored carbon was not released to the atmosphere as CO_2 until fossil fuels were extracted and burned. Small portions of these deposits have also been exposed to air by long-term geological processes that can take place over millions of years. However, in only a few hundred years, we have extracted and burned huge quantities of fossil fuels that took millions of years to form. This is why, on a human time scale, fossil fuels are nonrenewable resources.

We are altering the carbon cycle (see the red arrows in Figure 3-17), mostly by adding large amounts of carbon dioxide to the atmosphere (see Figure 14, p. S70, Supplement 7) when we burn carbon-containing fossil fuels (especially coal to produce electricity). We also alter the cycle by clearing carbon-absorbing vegetation from forests,

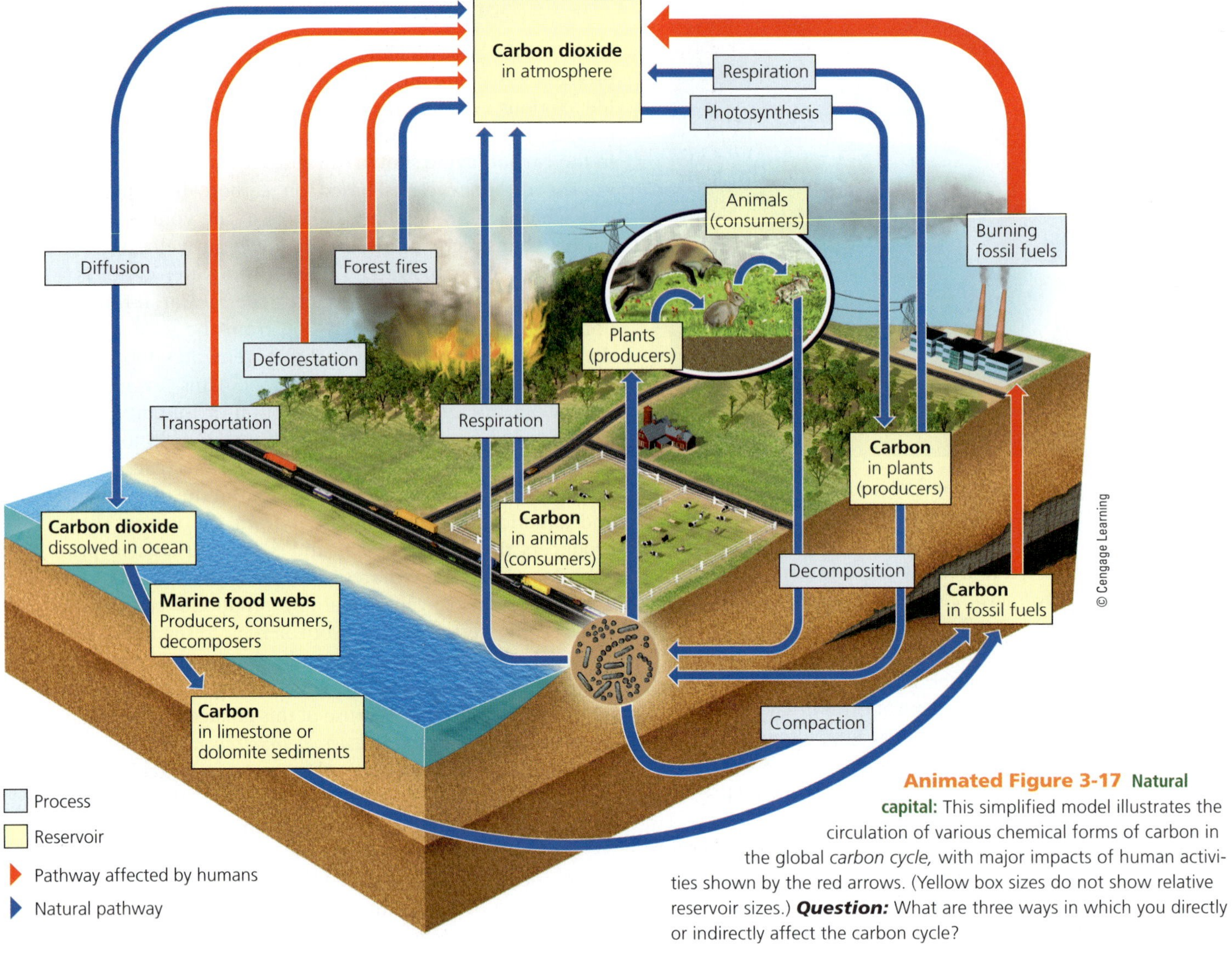

Animated Figure 3-17 Natural capital: This simplified model illustrates the circulation of various chemical forms of carbon in the global *carbon cycle,* with major impacts of human activities shown by the red arrows. (Yellow box sizes do not show relative reservoir sizes.) ***Question:*** What are three ways in which you directly or indirectly affect the carbon cycle?

especially tropical forests, faster than it can grow back (**Core Case Study**). Human activities are altering both the rate of energy flow and the cycling of nutrients within the carbon cycle. In other words, humanity has a large and growing *carbon footprint* that makes up a significant part of our overall ecological footprints (see Figure 1-13, p. 14).

Computer models of the earth's climate systems indicate that increased concentrations of atmospheric CO_2 and other greenhouse gases including methane (CH_4) are very likely to warm the atmosphere by enhancing the planet's natural greenhouse effect (which is why they are called greenhouse gases), and thus to change the earth's climate during this century, as we discuss in Chapter 19.

The Nitrogen Cycle: Bacteria in Action

The major reservoir for nitrogen is the atmosphere. Chemically unreactive nitrogen gas (N_2) makes up 78% of the volume of the atmosphere. Nitrogen is a crucial component of proteins, many vitamins, and nucleic acids such as DNA (see Figure 8, p. S17, in Supplement 4). However, N_2 cannot be absorbed and used directly as a nutrient by multicellular plants or animals.

Two natural processes convert, or *fix,* N_2 into compounds that plants and animals can use as nutrients. One such process is electrical discharges, or lightning, taking place in the atmosphere. The other takes place in aquatic systems, in soil, and in the roots of some plants, where specialized bacteria, called *nitrogen-fixing bacteria,* complete this conversion as part of the **nitrogen cycle**, which is depicted in Figure 3-18.

The nitrogen cycle consists of several major steps. In *nitrogen fixation,* specialized bacteria in soil, as well as blue-green algae (cyanobacteria) in aquatic environments, combine gaseous N_2 with hydrogen to make ammonia (NH_3). The bacteria use some of the ammonia they produce as a nutrient and excrete the rest into the soil or water. Some of the ammonia is converted to ammonium ions (NH_4^+) that plants can use as a nutrient.

Ammonia that is not taken up by plants may undergo *nitrification.* In this process, specialized soil bacteria such as

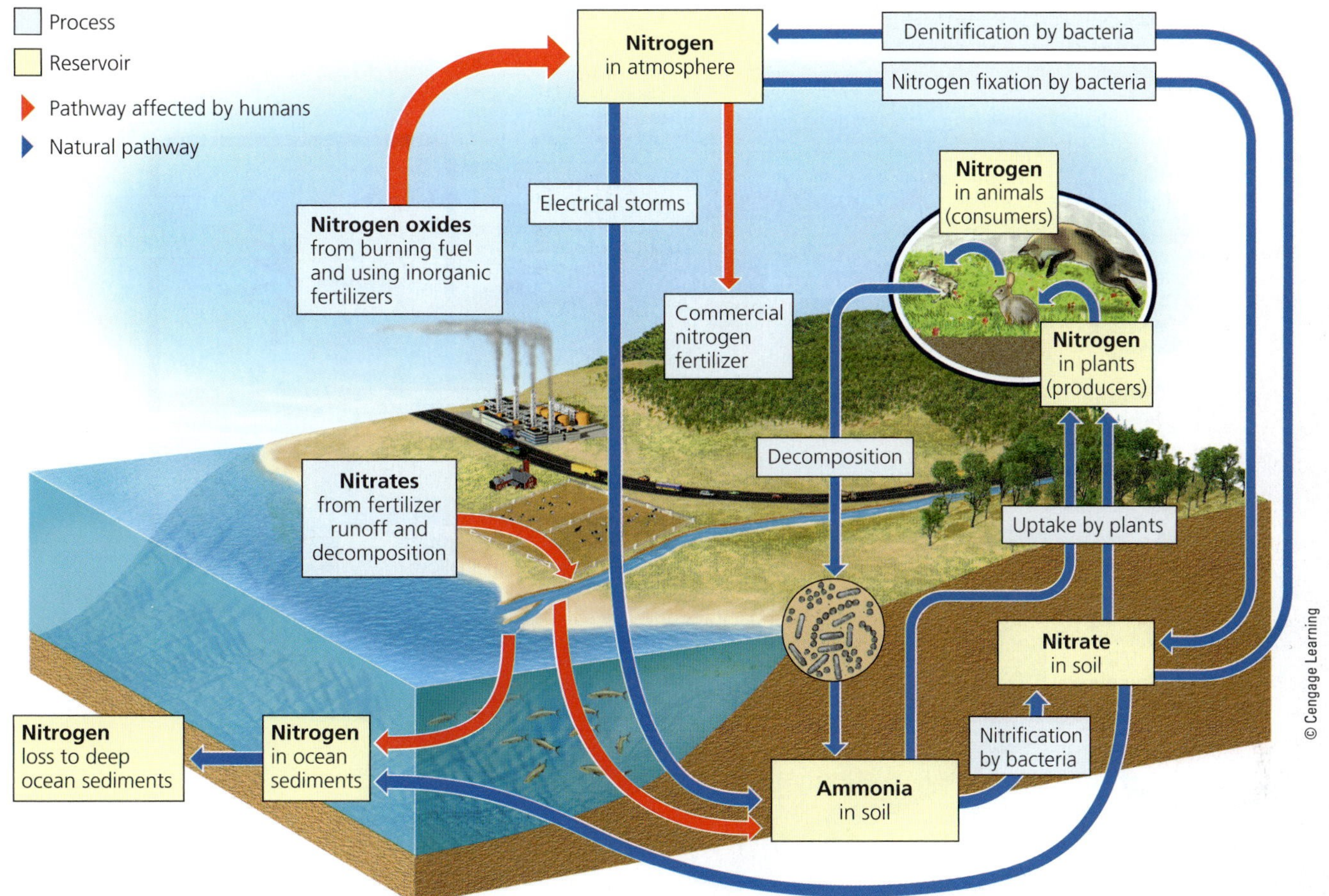

Animated Figure 3-18 **Natural capital:** This is a simplified model of the circulation of various chemical forms of nitrogen in the global *nitrogen cycle,* with major human impacts shown by the red arrows. (Yellow box sizes do not show relative reservoir sizes.) ***Question:*** What are two ways in which the carbon cycle and the nitrogen cycle are linked?

the Rhizobium bacteria convert most of the NH_3 and NH_4^+ in the soil to *nitrate ions* (NO_3^-), which are easily taken up by the roots of plants. The plants then use these forms of nitrogen to produce various amino acids, proteins, nucleic acids, and vitamins. Animals that eat plants eventually consume these nitrogen-containing compounds, as do detritus feeders and decomposers.

Plants and animals return nitrogen-rich organic compounds to the environment as both wastes and cast-off particles of tissues such as leaves, skin, or hair, and through their bodies when they die and are decomposed or eaten by detritus feeders. Vast armies of specialized decomposer bacteria convert this detritus into simpler nitrogen-containing inorganic compounds such as ammonia (NH_3) and water-soluble salts containing ammonium ions (NH_4^+).

In *denitrification,* specialized bacteria in waterlogged soil and in the bottom sediments of lakes, oceans, swamps, and bogs convert NH_3 and NH_4^+ back into nitrate ions, and then into nitrogen gas (N_2), which is released to the atmosphere to begin the nitrogen cycle again.

We intervene in the nitrogen cycle in several ways (see the red arrows in Figure 3-18) that affect what goes on in the atmosphere and in aquatic systems. We add large amounts of nitrogen oxides to the atmosphere when we burn gasoline and other fuels and when we use commercial nitrate fertilizers. For example, we add nitric oxide (NO) into the atmosphere when N_2 and O_2 combine as we burn any fuel at high temperatures, such as in car, truck, and jet engines. In the atmosphere, this gas can be converted to nitrogen dioxide gas (NO_2) and nitric acid vapor (HNO_3), which can return to the earth's surface as damaging *acid deposition,* commonly called *acid rain.*

We also add nitrous oxide (N_2O) to the atmosphere through the action of anaerobic bacteria on commercial nitrogen-containing fertilizer or organic animal manure applied to the soil. This greenhouse gas can warm the atmosphere and take part in reactions that deplete stratospheric ozone, which keeps most of the sun's harmful ultraviolet radiation from reaching the earth's surface (as we discuss in Chapter 18).

We are also removing large amounts of nitrogen (N_2) from the atmosphere faster than the cycle can replace it. This N_2 is removed primarily for use in industrial processes that convert it to ammonia (NH_3) and ammonium ions (NH_4^+) used in fertilizers.

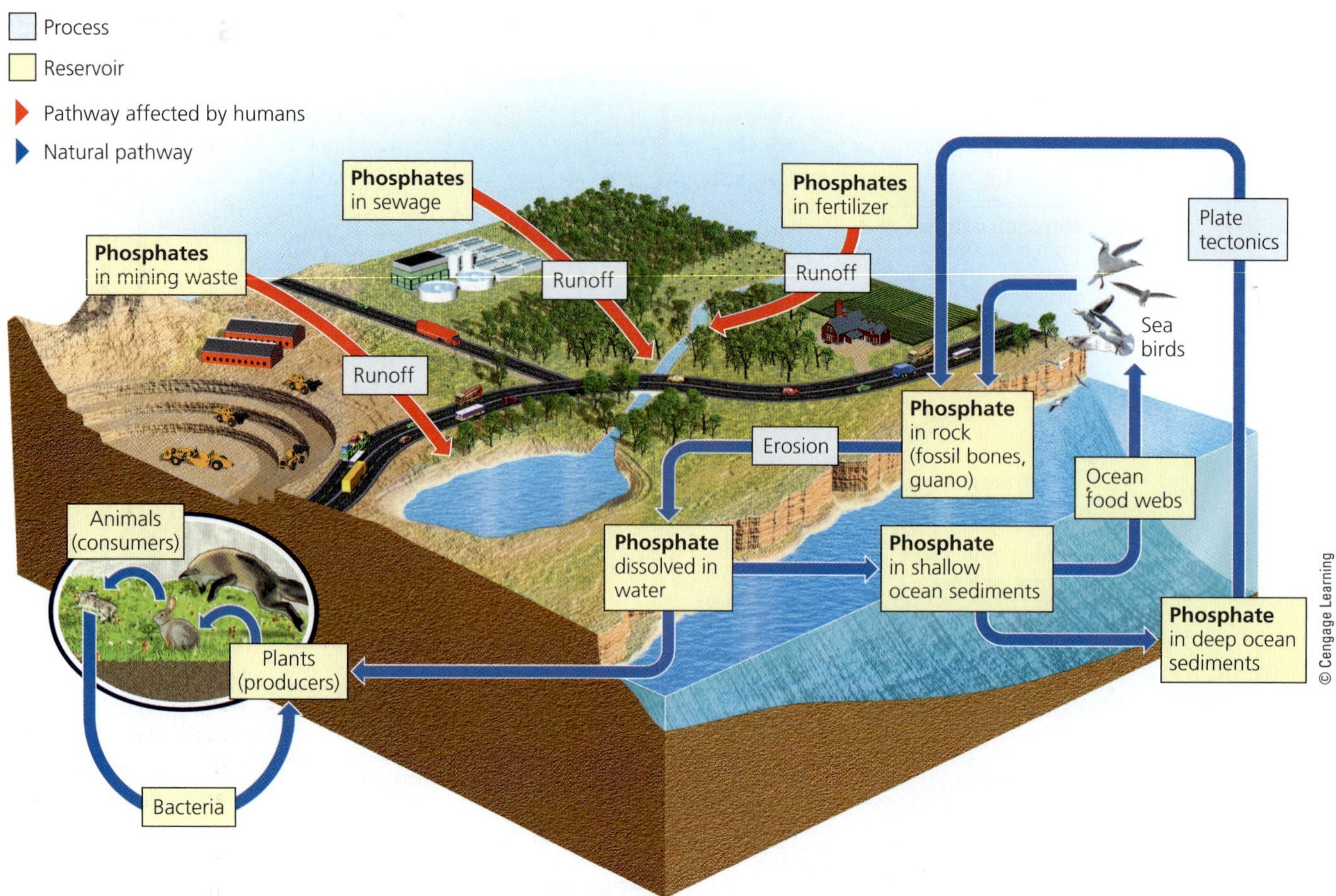

Figure 3-19 **Natural capital:** This is a simplified model of the circulation of various chemical forms of phosphorus (mostly phosphates) in the *phosphorus cycle*, with major human impacts shown by the red arrows. (Yellow box sizes do not show relative reservoir sizes.) ***Questions:*** What are two ways in which the phosphorus cycle and the nitrogen cycle are linked? What are two ways in which the phosphorus cycle and the carbon cycle are linked?

We upset the nitrogen cycle in aquatic ecosystems by adding excess nitrates (NO_3^-) to bodies of water through agricultural runoff of fertilizers and animal manure and through discharges from municipal sewage treatment systems. This can cause excessive growth of algae that can disrupt aquatic systems.

According to the 2005 Millennium Ecosystem Assessment, since 1950, human activities have more than doubled the annual release of nitrogen from the land into the rest of the environment. This is mostly from the greatly increased use of inorganic fertilizers to grow crops. The amount released is projected to double again by 2050 (see Figure 16, p. S70, in Supplement 7).

This excessive input of nitrogen into the air and water contributes to pollution and other problems to be discussed in later chapters. Nitrogen overload is a serious and growing local, regional, and global environmental problem that has attracted little attention. Princeton University physicist Robert Socolow calls for countries around the world to work out some type of nitrogen management agreement to help prevent this problem from reaching crisis levels.

CONSIDER THIS. . .

THINKING ABOUT The Nitrogen Cycle and Tropical Deforestation

What effects might the clearing and degrading of tropical rain forests (**Core Case Study**) have on the nitrogen cycle in these forest ecosystems and on any nearby aquatic systems? (See Figure 2-1, p. 30, and Figure 2-5, p. 37.)

The Phosphorus Cycle

Compounds of phosphorus (P) circulate through water, the earth's crust, and living organisms in the **phosphorus cycle**, depicted in Figure 3-19. Most of these compounds contain *phosphate* ions (PO_4^{3-}), which serve as an important nutrient. In contrast to the cycles of water, carbon, and nitrogen, the phosphorus cycle does not include the atmosphere. The major reservoir for phosphorus is phosphate salts containing PO_4^{3-} in terrestrial rock formations and ocean-bottom sediments. The phosphorus cycle is slow compared to the water, carbon, and nitrogen cycles.

As water runs over exposed rocks, it slowly erodes away inorganic compounds that contain phosphate ions.

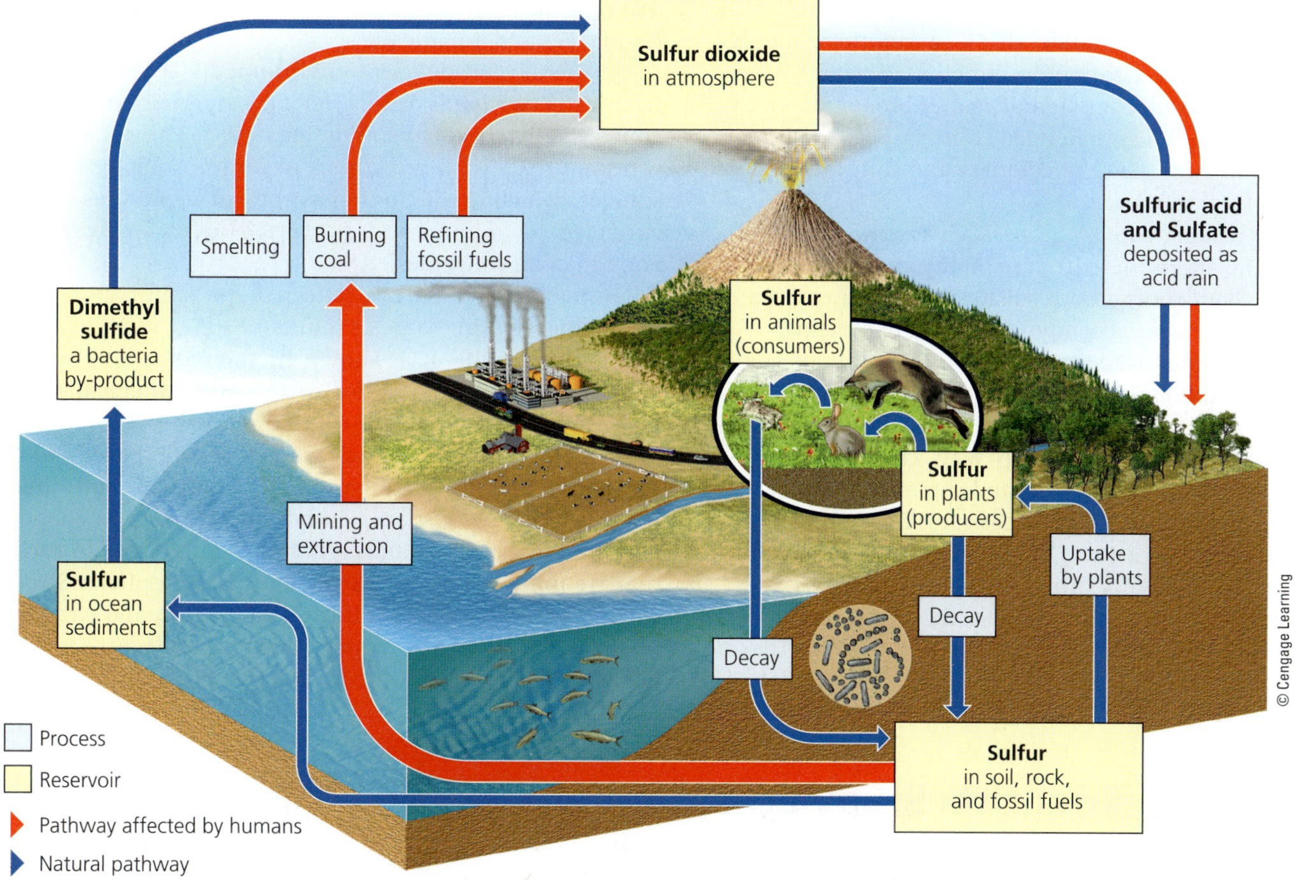

Animated Figure 3-20 **Natural capital:** This is a simplified model of the circulation of various chemical forms of sulfur in the *sulfur cycle*, with major impacts of human activities shown by the red arrows. (Yellow box sizes do not show relative reservoir sizes.) ***Question:*** What are two ways in which the sulfur cycle is linked to each of the phosphorus, nitrogen, and carbon cycles?

The running water carries these phosphate ions into the soil where they can be absorbed by the roots of plants and by other producers. Phosphate compounds are also transferred by food webs from producers to consumers, eventually including detritus feeders and decomposers. In both producers and consumers, phosphates are a component of biologically important molecules such as nucleic acids (see Figure 7, p. S16, in Supplement 4) and energy transfer molecules such as ADP and ATP (see Figure 11, p. S18, in Supplement 4). Phosphate is also a major component of vertebrate bones and teeth.

Phosphate can be lost from the cycle for long periods of time when it is washed from the land into streams and rivers and is carried to the ocean. There it can be deposited as marine sediment and remain trapped for millions of years. Over time, geological processes can uplift and expose these seafloor deposits, from which phosphate can be eroded to start the cycle again.

Because most soils contain little phosphate, the lack of it often limits plant growth on land unless phosphorus (as phosphate salts mined from the earth) is applied to the soil as a fertilizer. Lack of phosphorus also limits the growth of producer populations in many freshwater streams and lakes because phosphate salts are only slightly soluble in water and thus do not release many phosphate ions that producers need as nutrients.

Human activities are affecting the phosphorus cycle (as shown by the red arrows in Figure 3-19). One such activity is the removal of large amounts of phosphate from the earth to make fertilizer. Also, by clearing tropical forests (**Core Case Study**), we reduce phosphate levels in tropical soils. Topsoil that is eroded from fertilized crop fields, lawns, and golf courses carries large quantities of phosphate ions into streams, lakes, and oceans. There they stimulate the growth of producers such as algae and various aquatic plants. Phosphate-rich runoff from the land often produces huge populations of algae, which then upset chemical cycling and other processes in bodies of water.

The Sulfur Cycle

Sulfur circulates through the biosphere in the **sulfur cycle**, shown in Figure 3-20. Much of the earth's sulfur is stored underground in rocks and minerals and in the form of sulfate (SO_4^{2-}) salts buried deep under ocean sediments.

Sulfur also enters the atmosphere from several natural sources. Hydrogen sulfide (H_2S)—a colorless, highly poisonous gas with a rotten-egg smell—is released from active volcanoes and from organic matter broken down by anaerobic decomposers in flooded swamps, bogs, and tidal flats. Sulfur dioxide (SO_2), a colorless and suffocating gas, also comes from volcanoes.

Particles of sulfate (SO_4^{2-}) salts, such as ammonium sulfate, enter the atmosphere from sea spray, dust storms, and forest fires. Plant roots absorb sulfate ions and incorporate the sulfur as an essential component of many proteins.

Certain marine algae produce large amounts of volatile dimethyl sulfide, or DMS (CH_3SCH_3). Tiny droplets of DMS serve as nuclei for the condensation of water into droplets found in clouds. In this way, changes in DMS emissions can affect cloud cover and climate.

In the atmosphere, DMS is converted to sulfur dioxide, some of which in turn is converted to sulfur trioxide gas (SO_3) and to tiny droplets of sulfuric acid (H_2SO_4). DMS also reacts with other atmospheric chemicals such as ammonia to produce tiny particles of sulfate salts. These droplets and particles fall to the earth as components of *acid deposition,* which along with other air pollutants can harm trees and aquatic life.

In the oxygen-deficient environments of flooded soils, freshwater wetlands, and tidal flats, specialized bacteria convert sulfate ions to sulfide ions (S^{2-}). The sulfide ions can then react with metal ions to form insoluble metallic sulfides, which are deposited as rock or metal ores (often extracted by mining and converted to various metals), and the cycle continues.

Human activities have affected the sulfur cycle primarily by releasing large amounts of sulfur dioxide (SO_2) into the atmosphere (as shown by the red arrows in Figure 3-20). We release sulfur to the atmosphere in three ways. *First,* we burn sulfur-containing coal and oil to produce electric power. *Second,* we refine sulfur-containing oil (petroleum) to make gasoline, heating oil, and other useful products. *Third,* we extract metals such as copper, lead, and zinc from sulfur-containing compounds in rocks that are mined for these metals. In the atmosphere, SO_2 is converted to droplets of sulfuric acid (H_2SO_4) and particles of sulfate (SO_4^{2-}) salts, which return to the earth as acid deposition, which in turn can damage ecosystems.

3-5 How Do Scientists Study Ecosystems?

CONCEPT 3-5

Scientists use both field research and laboratory research, as well as mathematical and other models, to learn about ecosystems.

Some Scientists Study Nature Directly

Scientists use both field and laboratory research and mathematical and other models to learn about ecosystems (**Concept 3-5**). *Field research,* sometimes called "muddy-boots biology," involves going into forests and other natural settings to observe and measure the structure of ecosystems and what happens in them (see Chapter 2 opening photo). Most of what we know about ecosystems has come from such research.

Scientists have been particularly creative in studying tropical forests (**Core Case Study**). In a few cases, ecologists have erected tall construction cranes that provide them access to the canopies of tropical forests. This, along with rope walkways between treetops (Figure 3-21), has helped them to identify and observe the rich diversity of species living or feeding in these treetop habitats.

Sometimes ecologists carry out controlled experiments by isolating and changing a variable in part of an area and comparing the results with nearby unchanged areas. A good example of this is reported in the Core Case Study of Chapter 2 (p. 30).

Scientists also use aircraft and satellites equipped with sophisticated cameras and other *remote sensing* devices to scan and collect data on the earth's surface. Then they use *geographic information system (GIS)* software to capture, store, analyze, and display such information. Such software can store geographic and ecological spatial data electronically as numbers or as images. For example, a GIS can convert digital satellite images into global, regional, and local maps showing variations in vegetation, gross primary productivity, air pollution emissions, and many other variables.

Scientists can also attach tiny radio transmitters to animals and use global positioning systems (GPS) to learn about the animals by tracking where they go and how far they go. This technology is very important for studying endangered species. **Green Careers:** GIS analyst; remote sensing analyst

Some Scientists Study Ecosystems in the Laboratory

Since the 1960s, ecologists have increasingly supplemented field research by using *laboratory research*—setting up, observing, and making measurements of model ecosystems and populations under laboratory conditions (Figure 3-22). They have created such simplified systems in containers such as culture tubes, bottles, aquariums, and greenhouses, and in indoor and outdoor chambers where they can control temperature, light, CO_2, humidity, and other variables.

Such systems make it easier for scientists to carry out controlled experiments. In addition, laboratory experiments often are quicker and less costly than similar exper-

Bill Hatcher/National Geographic Creative

Figure 3-21 Scientists learn about tropical forests by using a system of ropes and pulleys to move among the treetops.

iments in the field. But there is a catch: scientists must consider how well their scientific observations and measurements in a simplified, controlled system under laboratory conditions reflect what actually takes place under more complex and dynamic conditions found in nature. Thus, the results of laboratory research must be coupled with and supported by field research.

CONSIDER THIS. . .

THINKING ABOUT Greenhouse Experiments and Tropical Rain Forests

How would you design an experiment that includes an experimental group and a control group and uses a greenhouse to determine how clearing a patch of tropical rain forest vegetation might affect the temperature above the cleared patch?

Some Scientists Use Models to Simulate Ecosystems

Since the late 1960s, ecologists have developed mathematical and other types of models that simulate ecosystems. Computer simulations can help scientists understand large and very complex systems, such as lakes, oceans, forests, and the earth's climate system, that cannot be adequately studied and modeled in field and laboratory research. Scientists are learning a lot about how the earth works by feeding data into increasingly sophisticated models of the earth's systems and running them on high-speed supercomputers. **Green Careers:** ecosystem modeler

Of course, simulations and projections made with ecosystem models are no better than the data and assumptions used to develop the models. Ecologists must determine the relationships among key variables that they will use to develop and test ecosystem models. They must also do careful field and laboratory research to get *baseline data,* or beginning measurements, of the variables they are studying in the world's major terrestrial and aquatic ecosystems.

JOEL SARTORE/National Geographic Creative

Figure 3-22 These biologists are studying frogs in a laboratory in Ecuador.

SCIENCE FOCUS 3.3

TESTING PLANETARY BOUNDARIES: FROM HOLOCENE TO ANTHROPOCENE

For most of the past 10,000 years, we have been living in an era called the **Holocene**—a period of relatively stable climate and other environmental conditions following a glacial period. This general stability has allowed the human population to grow and to develop agriculture and to flourish within societies that have taken over a large and growing share of the earth's resources.

According to Will Steffen, Paul J. Crutzen, Johan Rockström, and other scientists, since the industrial revolution began around 1750, we have entered an era called the **Anthropocene**. In this new era, humans have become major agents of change in the functioning of the earth's life-support system as their ecological footprints have spread over the earth (Figure 1-12, p. 13,and Figure 1-13, p. 14).

In 2009, a group of 28 internationally renowned scientists, led by Johan Rockström of the Stockholm Resilience Centre, designated nine major *planetary boundaries,* or ecological tipping points (see p. 45). In 2012, ecologist Steven Running suggested adding a tenth planetary boundary based on the growing use of the global net primary productivity by humans. These scientists have hypothesized that if we exceed these boundaries, we could trigger abrupt and long-lasting or irreversible environmental changes within ten major systems that play key roles in earth's ability to sustain life (Figure 3-B). Exceeding some of these boundaries will have local and regional effects. However, exceeding any of the climate change, ocean acidification, or stratospheric ozone boundaries would likely have long-lasting global effects.

So far, the limit values of these planetary boundaries are uncertain, and this presents a major problem. We need to do much more research to fill in the missing data on these planetary boundaries and on how our exceeding them could affect the health of humans, other species, and the earth's life-support systems, as well as human economies and political systems. This research could help us to develop strategies for staying within planetary boundaries and for dealing with the effects on systems where we have exceeded such boundaries.

In 2009, the U.S. National Aeronautics and Space Administration and Cisco, a large computer networking company, teamed up to develop a *planetary skin*—a sort of global nervous system that will integrate land, air, and space-based sensors throughout the world. This pilot project will provide scientists, the public, government agencies, and private businesses with valuable baseline data on the state of the earth, which we can use to help us make decisions on how to reduce our environmental impacts. GOOD NEWS

We Need to Learn More about the Health of the World's Ecosystems

We need baseline data on the condition of the world's ecosystems to see how they are changing, to develop effective strategies for preventing or slowing their degradation, and to avoid ecological tipping points beyond which these systems will be disrupted or destroyed. By analogy, your doctor needs baseline data on your blood pressure, weight, and the functioning of your organs and other systems, which can be obtained through basic tests. If your health declines in some way, the doctor can run new tests and compare the results with the baseline data to identify any changes and to determine any necessary treatments.

According to the 2005 Millennium Ecosystem Assessment, scientists have less than half of the basic ecological data they need in order to evaluate the status of ecosystems in the United States. Even fewer data are available for most other parts of the world. Ecologists have called for a massive program to develop baseline data for the world's ecosystems in order to try to identify and avoid certain *planetary boundaries,* or ecological tipping points related to the earth's major natural systems (Science Focus 3.3).

Big Ideas

- **Life is sustained by the flow of energy from the sun through the biosphere, the cycling of nutrients within the biosphere, and gravity.**
- **Some organisms produce the nutrients they need, others survive by consuming other organisms, and still others live on the wastes and remains of organisms while recycling nutrients that are used again by producer organisms.**
- **Human activities are altering the flow of energy through food chains and webs and the cycling of nutrients within ecosystems and the biosphere.**

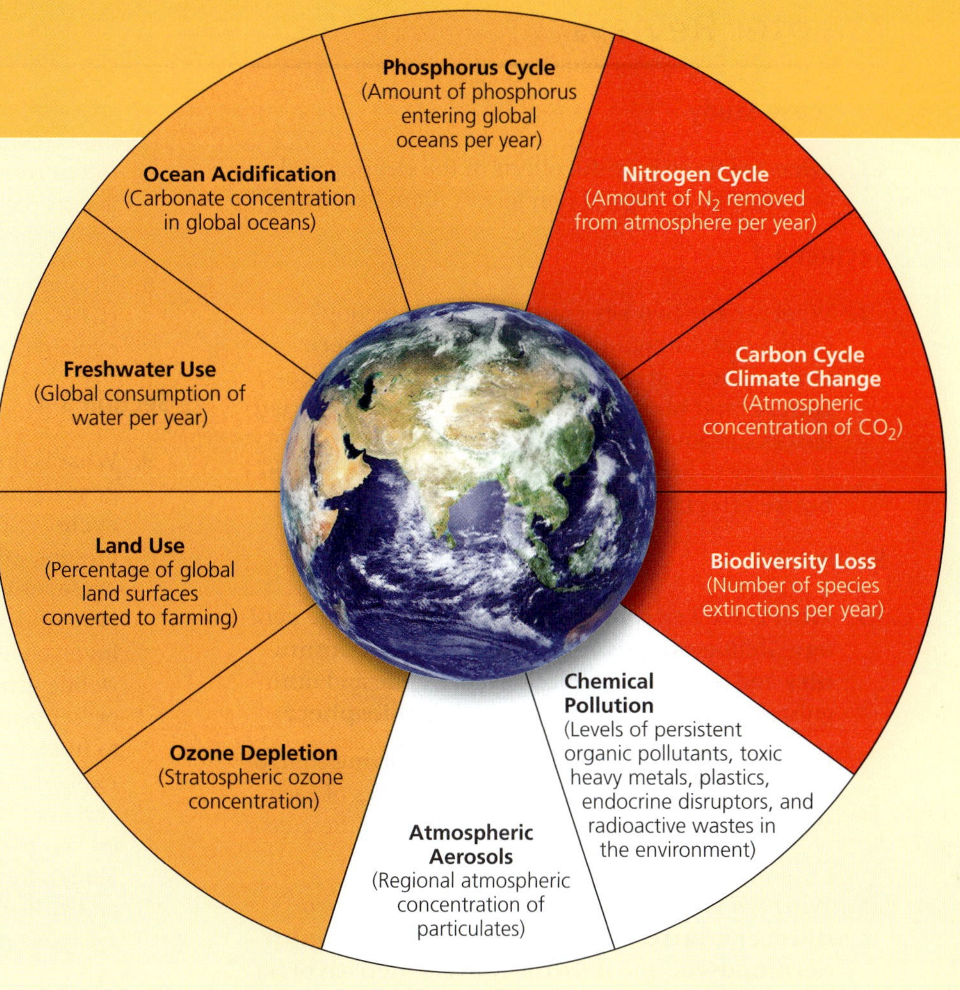

Figure 3-B Planetary boundaries for ten major components of the earth's life-support system. A team of scientists estimated that human activities have exceeded the boundary limits for three systems (shown in red) and are close to the limits for five other systems (shown in orange). There is not enough information to evaluate the other two systems (shown in white).

(Compiled by the authors using data from Johan Rockström, Paul Crutzen, and James Hansen, et al., 2009, "Planetary Boundaries: Exploring the Safe Operating Space for Humanity," *Ecology and Society*, vol. 14, no. 2, p. 32.)

Photo: Sailorr/Shutterstock.com

TYING IT ALL TOGETHER Tropical Rain Forests and Sustainability

Anneka/Shutterstock.com

This chapter began with a discussion of the importance of the world's incredibly diverse tropical rain forests (**Core Case Study**). These ecosystems showcase the functioning of the three scientific **principles of sustainability**, which apply as well to the world's other ecosystems.

First, producers within rain forests rely on *solar energy* to produce a vast amount of biomass through photosynthesis. *Second,* species living in the forests take part in, and depend on, the *cycling of nutrients* and the flow of energy within the forests and throughout the biosphere. *Third,* tropical rain forests contain a huge and vital part of the earth's *biodiversity,* and interactions among species living in these forests help to sustain these complex ecosystems.

We also discussed recent research on the dangers of our exceeding key planetary boundaries through activities that could leave many natural systems, including tropical rain forests, unstable and unsustainable. In particular, exceeding the climate change, ocean acidity, and stratospheric ozone boundaries could have long-lasting global effects. In many of the chapters to follow, we examine these threats further, and we consider ways in which we can apply the six **principles of sustainability** in seeking to stay within the key planetary boundaries and to live more sustainably.

Chapter Review

Core Case Study

1. What are three harmful effects of the clearing and degradation of tropical rain forests (**Core Case Study**)?

Section 3-1

2. What are the two key concepts for this section? Define and distinguish among the **atmosphere**, **troposphere**, **stratosphere**, **hydrosphere**, **geosphere**, and **biosphere**. What three interconnected factors sustain life on the earth? Describe the flow of energy to and from the earth. What is the **greenhouse effect** and why is it important?

Section 3-2

3. What is the key concept for this section? Define **ecology**. Define **organism**, **population**, and **community** and give an example of each. Define and distinguish between an **ecosystem** and the **biosphere**.

4. Distinguish between the living and nonliving components in ecosystems and give two examples of each. What are two other terms used to describe these ecosystem components?

5. What is a **trophic level**? Distinguish among **producers (autotrophs)**, **consumers (heterotrophs)**, **decomposers**, and **detritus feeders (detritivores)**, and give an example of each in an ecosystem. Describe the processes of **photosynthesis**. Distinguish among **primary consumers (herbivores)**, **carnivores**, **secondary consumers**, **tertiary consumers**, and **omnivores**, and give an example of each. Explain why there is little waste in nature and why humans are a major exception to this natural process.

6. Explain the importance of microbes. Distinguish between **aerobic respiration** and **anaerobic respiration (fermentation)**. What two processes sustain ecosystems and the biosphere and how are they linked?

Section 3-3

7. What is the key concept for this section? Define and distinguish between a **food chain** and a **food web**. Explain what happens to energy as it flows through food chains and webs. What is **biomass**? What is the **pyramid of energy flow**? Why are there more insects than tigers in the world? Why don't most of the world's people eat meat?

8. Distinguish between **gross primary productivity (GPP)** and **net primary productivity (NPP)**, and explain their importance.

Section 3-4

9. What is the key concept for this section? What happens to matter in an ecosystem? What is a **nutrient cycle**? Explain how nutrient cycles connect past, present, and future life. Describe the **hydrologic cycle**, or **water cycle**. Summarize the unique properties of water. What three major processes are involved in the water cycle? What is **surface runoff**? Define **groundwater**. What is an **aquifer**? What percentage of the earth's water supply is available to humans and other species as liquid freshwater? Explain how human activities are affecting the water cycle in three ways. Explain how clearing a rain forest can affect local weather and climate (**Core Case Study**). Describe the **carbon**, **nitrogen**, **phosphorus**, and **sulfur cycles**, and explain how human activities are affecting each cycle.

Section 3-5

10. What is the key concept for this section? Describe three ways in which scientists study ecosystems. Explain why we need much more basic data about the structure and condition of the world's ecosystems. Distinguish between the **Holocene** and **Anthropocene** eras. List ten planetary boundaries designated by scientists. Which three of these boundaries have already been exceeded, according to the scientists? What are this chapter's *three big ideas*? How are the three scientific **principles of sustainability** showcased in tropical rain forests? SUSTAINABILITY

Note: Key terms are in bold type.

Critical Thinking

1. How would you explain the importance of tropical rain forests (**Core Case Study**) to people who think that such forests have no connection to their lives?

2. Explain **(a)** why the flow of energy through the biosphere depends on the cycling of nutrients, and **(b)** why the cycling of nutrients depends on gravity (**Concept 3-1B**).

3. Explain why microbes are so important. List two beneficial effects and two harmful effects of microbes on your health. Write a brief description of what you

think would happen to you if microbes were eliminated from the earth.

4. Make a list of the foods you ate for lunch or dinner today. Trace each type of food back to a particular producer species. Describe the sequence of feeding levels that led to your feeding.

5. Use the second law of thermodynamics (see Chapter 2, p. 43) to explain why many poor people in less-developed countries live on a mostly vegetarian diet.

6. What changes might take place in the hydrologic cycle if the earth's climate becomes **(a)** hotter or **(b)** cooler? In each case, what are two ways in which these changes might affect your life?

7. What would happen to an ecosystem if **(a)** all of its decomposers and detritus feeders were eliminated, **(b)** all of its producers were eliminated, and **(c)** all of its insects were eliminated? Could a balanced ecosystem exist with only producers and decomposers and no consumers such as humans and other animals? Explain.

8. For each of the earth's ten major planetary boundaries (Figure 3-B), describe how our exceeding that boundary might affect **(a)** you, **(b)** any child you might have, and **(c)** any grandchild you might have.

Doing Environmental Science

Visit a nearby terrestrial ecosystem or aquatic life zone and try to identify major producers, primary and secondary consumers, detritus feeders, and decomposers. Take notes and describe at least one example of each of these types of organisms. Make a simple sketch showing how these organisms might be related to each other or to other organisms in a food chain or food web.

Global Environment Watch Exercise

Search for *Nitrogen Cycle* and look for information on how humans are affecting the nitrogen cycle. Specifically, look for impacts on the atmosphere and on human health from emissions of nitrogen oxides, and look for the harmful ecological effects of the runoff of nitrate fertilizers into rivers and lakes. Make a list of these impacts and use this information to review your daily activities. Find three things that you do regularly that contribute to these impacts.

Data Analysis

Recall that net primary productivity (NPP) is the *rate* at which producers can make the chemical energy that is stored in their tissues and that is potentially available to other organisms (consumers) in an ecosystem. In Figure 3-14, it is expressed as units of energy (kilocalories, or *kcal*) produced in a given area (square meters, or m^2) over a period of time (a year). Look again at Figure 3-14 and consider the differences in NPP among various ecosystems. Then answer the following questions:

1. What is the approximate NPP of a tropical rain forest in kcal/m^2/yr? Which terrestrial ecosystem produces about one-third of that rate? Which aquatic ecosystem has about the same NPP as a tropical rain forest?

2. Early in the 20th century, large areas of temperate forestland in the United States were cleared to make way for agricultural land. For each unit of this forest area that was cleared and replaced by farmland, about how much NPP was lost?

3. Why do you think deserts and grasslands have dramatically lower NPP than swamps and marshes?

4. About how many times more NPP do estuaries produce, compared to lakes and streams? Why do you think this is so?

4 Biodiversity and Evolution

Few problems are less recognized, but more important than, the accelerating disappearance of the earth's biological resources. In pushing other species to extinction, humanity is busy sawing off the limb on which it is perched.

PAUL EHRLICH

Key Questions

4-1 What is biodiversity and why is it important?

4-2 How does the earth's life change over time?

4-3 How do geological processes and climate change affect evolution?

4-4 How do speciation, extinction, and human activities affect biodiversity?

4-5 What is species diversity and why is it important?

4-6 What roles do species play in ecosystems?

Monarch butterfly on a flower.

Ariel Bravy/Shutterstock.com

CORE CASE STUDY Why Are Amphibians Vanishing?

Amphibians—frogs, toads, and salamanders—were among the earliest vertebrates (animals with backbones) to emerge from the earth's waters and live on the land. Historically, they have been able to adjust to and survive environmental changes more effectively than many other species. However, the amphibian world is changing rapidly.

An amphibian lives part of its life in water and part on land. Now, many of the 6,700 or more amphibian species are having difficulty adapting to rapid changes that have taken place in their water and land habitats during the past few decades. Such changes have resulted primarily from human activities such as use of pesticides and other chemicals that become water pollutants.

Since 1980, populations of hundreds of amphibian species throughout the world have declined or vanished (Figure 4-1). According to the International Union for Conservation of Nature (IUCN), about 33% of all known amphibian species are threatened with extinction, and more than 40% of all known amphibian species are declining.

No single cause has been identified to explain the declines of many amphibian species. However, scientists have identified a number of factors that affect amphibians at various points in their life cycles. For example, frog eggs have no shells to protect frog embryos from water pollutants, and adult frogs are often exposed to insecticides contained in the many insects they eat. We explore these and other factors later in this chapter.

Why should we care if some amphibian species become extinct? Scientists give three reasons. *First,* amphibians are sensitive biological *indicators* of changes in environmental conditions such as habitat loss, air and water pollution, ultraviolet (UV) radiation, and climate change. The growing threats to their survival indicate that environmental conditions are deteriorating in many parts of the world.

Second, adult amphibians play important ecological roles in biological communities. For example, amphibians eat more insects (including mosquitoes) than do birds. In some habitats, the extinction of certain amphibian species could lead to extinction of certain species of other amphibians, aquatic insects, reptiles, birds, fish, and mammals that feed on amphibians or their larvae.

Third, amphibians are a genetic storehouse of pharmaceutical products waiting to be discovered. For example, compounds in secretions from the skin of certain amphibians have been isolated and used as painkillers and antibiotics, and in treatments for burns and heart disease.

Many scientists believe that the threats to amphibians present a warning about a number of environmental threats to biodiversity. In this chapter, we will learn about biodiversity, about how it has arisen and why it is important, and about how it is threatened. We will also look at some possible solutions to these problems.

Joel Sartore/National Geographic Creative

Figure 4-1 These are specimens of some of the nearly 200 amphibian species that have gone extinct since 1970.

4-1 What Is Biodiversity and Why Is It Important?

CONCEPT 4-1
The biodiversity found in genes, species, ecosystems, and ecosystem processes is vital to sustaining life on earth.

Biodiversity Is a Crucial Part of the Earth's Natural Capital

Biological diversity, or **biodiversity**, is the variety of the earth's *species,* or varying life-forms, the genes they contain, the ecosystems in which they live, and the ecosystem processes such as energy flow and nutrient cycling that sustain all life (Figure 4-2). Acting together, these four components of the earth's biodiversity provide us with the ecosystem services (Figure 1-3, orange items, p. 7) that sustain us and our economies.

Recall that a **species** is a group of organisms with a set of characteristics that distinguish it from other groups of organisms, and in sexually reproducing organisms, individuals must be able to mate and produce fertile offspring in order to be grouped within a species.

We do not know how many species there are on the earth. Estimates range from 8 million to 100 million. In 2011, a team of biologists led by Camilo Mora and Boris Worm estimated that there are about 7–10 million species. Other scientific estimates put the number much higher.

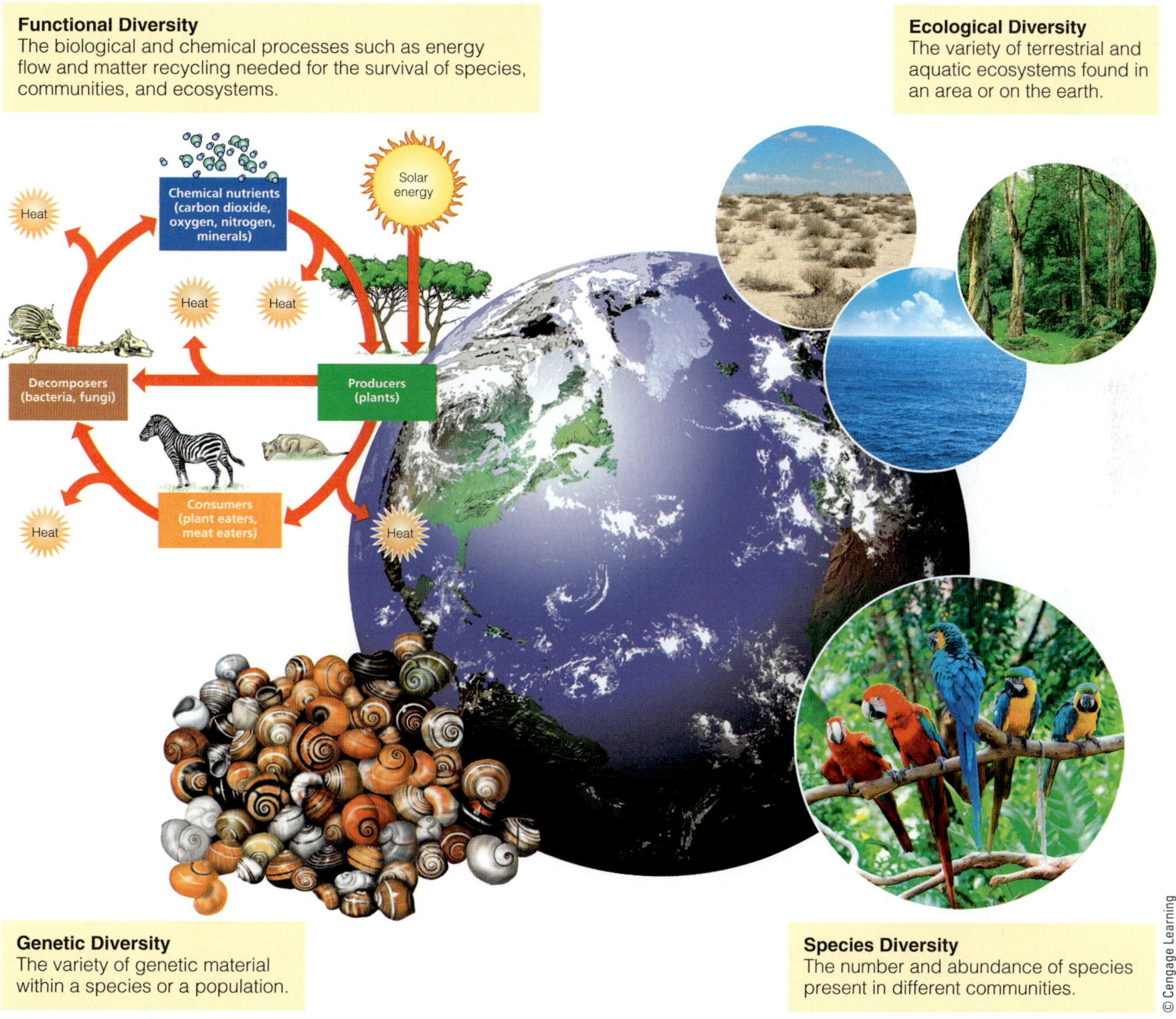

Figure 4-2 **Natural capital:** The major components of the earth's *biodiversity*—one of the planet's most important renewable resources and a key component of its natural capital (see Figure 1-3, p. 7).
Question: Why do you think we should protect the earth's biodiversity from our actions?

Right side, Top left: Laborant/Shutterstock.com; Right side, Top right: leungchopan/Shutterstock.com; Right side, Top center: ©Elenamiv/Shutterstock .com; Bottom right: ©Juriah Mosin/Shutterstock.com

SCIENCE FOCUS 4.1

INSECTS PLAY A VITAL ROLE IN OUR WORLD

We classify many insect species as *pests* because they compete with us for food, spread human diseases such as malaria, bite or sting us, and invade our lawns, gardens, and houses. Some people fear insects and many think the only good bug is a dead bug. They fail to recognize the vital roles insects play in helping to sustain life on earth.

For example, *pollination* is a vital ecosystem service that allows flowering plants to reproduce sexually when pollen grains are transferred from the flower of one plant to a receptive part of the flower of another plant of the same species. Many of the earth's plant species depend on insects to pollinate their flowers (chapter-opening photo and Figure 4-A, left).

Insects that eat other insects—such as the praying mantis (Figure 4-A, right)—help to control the populations of at least half the species of insects we call pests. This free pest control service is an important part of the earth's natural capital. Some insects also play a key role in loosening and renewing the topsoil that supports plant life on land. Others such as the dung beetle recycle animal wastes (dung) often by rolling it into balls (Figure 4-B) and burying it for use as food.

Insects have been around for at least 400 million years—about 2,000 times longer than the latest version of the human species. Some species reproduce at an astounding rate and can rapidly develop new genetic traits such as resistance to pesticides. They also have an exceptional ability to evolve into new species when faced with changing environmental conditions, and many species are now being challenged to do so because of environmental changes brought about by the rapidly expanding human population and its growing resource use per person.

Nicole Duplaix/National Geographic Creative

Figure 4-B A dung beetle rolls a ball of dung. Dung beetles can roll up to 10 times their body weight—equivalent to a 68-kilogram (150-pound) person rolling a 680-kilogram (1,500-pound) boulder.

The scientific study of insects is called *entomology*—a field that includes a vast number of subspecialties and applications. Entomologists are expanding their research in areas related to environmental threats to insect populations. Environmental changes, many of them caused by human activities, are making such threats clearer every year. For example, entomologist Diana Cox-Foster of Pennsylvania State University (USA) is studying the decline of honeybees, which are extremely important pollinators. Her lab work focuses on *colony collapse disorder,* the name given to the disappearances of many bee colonies in recent years. This disorder is threatening to disrupt whole ecosystems that depend on bees for pollination, as well as much of the human food supply. We discuss this serious environmental problem more fully in Chapter 9.

Darlyne A. Murawski/National Geographic Creative

Dr. Morley Read/Shutterstock.com

Figure 4-A *Importance of insects:* Bees (left) and numerous other insects pollinate flowering plants that serve as food for many plant eaters, including humans. This praying mantis, which is eating a moth (right), and many other insect species help to control the populations of most of the insect species we classify as pests.

Critical Thinking

Identify three insect species not discussed above that benefit your life.

So far, biologists have identified about 2 million species—most of them being insects (Science Focus 4.1). Up to half of the world's land-based plant and animal species live in tropical rain forests. Scientists believe that most of the unidentified species live in the planet's rain forests and in the largely unexplored oceans.

Species diversity, the number and variety of the species present in any biological community, is the most obvious component of biodiversity. We discuss it in greater detail in Section 4-5. Another important component is *genetic diversity,* the variety of genes found in a population or in a species (Figure 4-3). The earth's many spe-

© Cengage Learning

Figure 4-3 *Genetic diversity* among individuals in this population of a species of Caribbean snail is reflected in the variations in shell color and banding patterns. Genetic diversity can also include other variations such as slight differences in chemical makeup, sensitivity to various chemicals, and behavior.

cies contain a vast variety of genes, which enable life on the earth to survive and adapt to dramatic environmental changes.

Ecosystem diversity—the earth's variety of deserts, grasslands, forests, mountains, oceans, lakes, rivers, and wetlands—is another major component of biodiversity. Each of these ecosystems is a storehouse of genetic and species diversity. Biologists have classified the terrestrial (land) ecosystems into **biomes**—large regions such as forests, deserts, and grasslands with distinct climates and certain species (especially vegetation) adapted to them. Figure 4-4 shows different major biomes along the 39th parallel spanning the United States. We discuss biomes in more detail in Chapter 7.

Yet another important component of biodiversity is *functional diversity*—the variety of processes such as energy flow and matter cycling that occur within ecosystems (see Figure 3-10, p. 59) as species interact with one another in food chains and webs.

The earth's biodiversity is a vital part of the natural capital (see Figure 1-3, p. 7) that helps to keep us alive and supports our economies. With the help of technology, we use biodiversity to provide us with food, wood, fibers, energy from wood and biofuels, and medicines. Biodiversity also plays critical roles in providing us with the ecosystem services that preserve the quality of the air and water, maintain the fertility of topsoil, decompose and recycle wastes, and control populations of species that we call pests. The four components of biodiversity also increase the stability of ecosystems and increase the resistance of ecosystems to harmful invasive species. We owe much of what we know about biodiversity to a fairly small number of researchers, such as Edward O. Wilson (Individuals Matter 4.1).

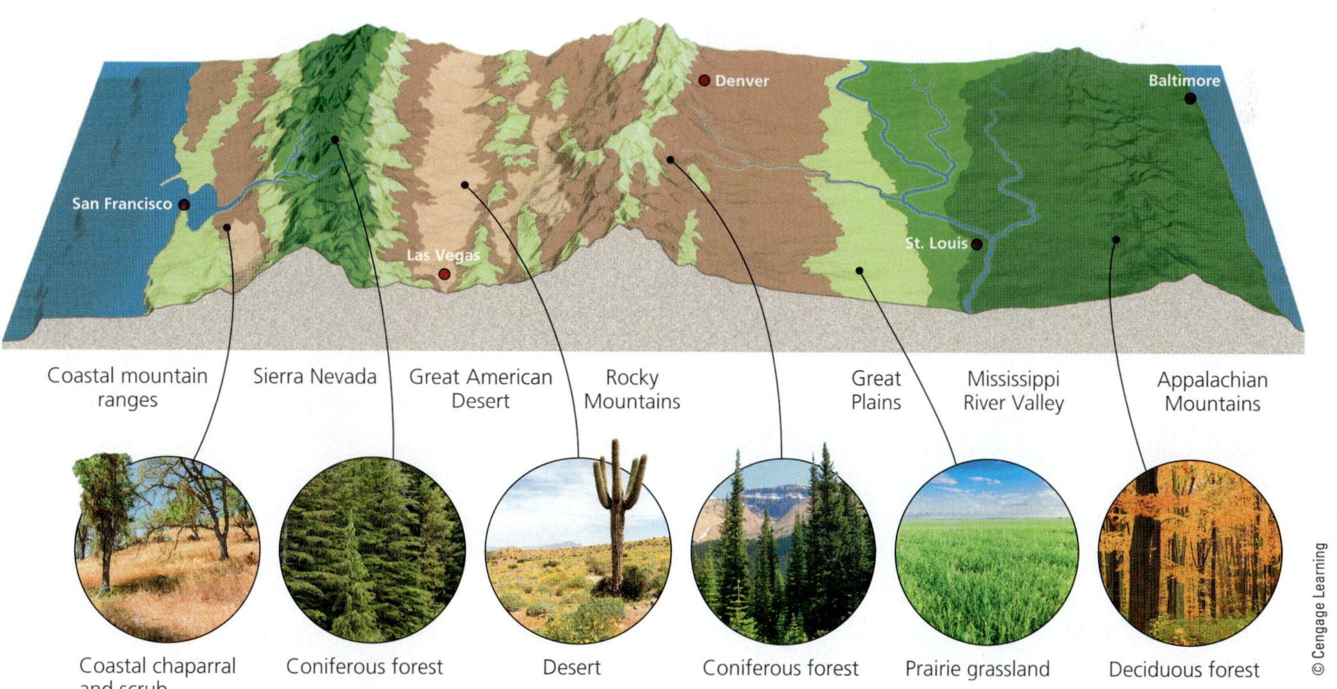

© Cengage Learning

Figure 4-4 The major biomes found along the 39th parallel across the United States show a variety of ecosystems. The differences in tree and other plant species reflect changes in climate, mainly differences in average annual precipitation and temperature.

First: ©Zack Frank/Shutterstock.com; Second: Robert Crum/Shutterstock.com; Third: Joe Belanger/Shutterstock.com; Fourth: ©Protasov AN/Shutterstock.com; Fifth: ©Maya Kruchankova/Shutterstock.com; Sixth: © Marc von Hacht/Shutterstock.com

individuals matter 4.1

Edward O. Wilson: A Champion of Biodiversity

Jim Harrison

As a boy growing up in the southeastern United States, Edward O. Wilson became interested in insects at the age of nine. He has said, "Every kid has a bug period. I never grew out of mine."

Before entering college, Wilson had decided he would specialize in the study of ants. After he grew fascinated with that tiny organism, and throughout his long career, he steadily widened his focus to include the entire biosphere. He now spends much of his time studying, writing, and speaking about biodiversity and working on Harvard University's *Encyclopedia of Life,* an online database for the planet's known and named species.

During his long career, Wilson has taken on many challenges. He and other researchers working together discovered how ants communicate using chemicals called pheromones. He also studied the complex social behavior of ants and has compiled much of his work into the epic volume, *The Ants,* published in 1990. His ant research has been applied to the study and understanding of other social organisms, including humans. He has also proposed the hypothesis that humans have a natural affinity for wildlife and wild places—a concept he calls *biophilia* (or love of life).

One of Wilson's landmark works is *The Diversity of Life,* published in 1992, in which he put together the principles and practical issues of biodiversity more completely than anyone had to that point. Wilson is now deeply involved in writing and lecturing about the need for global conservation efforts and is promoting the goal of completing a global survey of biodiversity. He has won more than 100 national and international awards and has written 25 books, two of which won the Pulitzer Prize for General Nonfiction. About the importance of biodiversity, he writes:

> "Until we get serious about exploring biological diversity. . . science and humanity at large will be flying blind inside the biosphere. . . . How can we save Earth's life forms from extinction if we don't even know what most of them are?"

Background photo: Christian Musat/Shutterstock.com

4-2 How Does the Earth's Life Change over Time?

CONCEPT 4-2A
The scientific theory of evolution explains how life on the earth changes over time due to changes in the genes of populations.

CONCEPT 4-2B
Populations evolve when genes mutate and give some individuals genetic traits that enhance their abilities to survive and to produce offspring with these traits (natural selection).

Biological Evolution by Natural Selection Explains How Life Changes over Time

Most of what we know about the long history of life on the earth comes from **fossils**: mineralized or petrified replicas of skeletons, bones, teeth, shells, leaves, and seeds, or impressions of such items found in rocks (Figure 4-5). Scientists also drill core samples from glacial ice at the earth's poles and on mountaintops, and examine the signs of ancient life found at different layers in these cores.

The entire body of evidence gathered using these methods, which is called the *fossil record*, is uneven and incomplete. Some forms of life left no fossils, and some fossils have decomposed. The fossils found so far represent probably only 1% of all species that have ever lived. Trying to reconstruct the development of life with so little evidence is the work of *paleontology*—a challenging scientific detective game. **Green Careers:** paleontologist

How did we end up with such an amazing array of species? The scientific answer involves **biological evolution** (or simply **evolution**): the process whereby the earth's life changes over time through changes in the genes of populations of organisms in succeeding generations (**Concept 4-2A**). According to the **theory of evolution**, all species evolved from earlier, ancestral species. In other words, life comes from life.

The idea that organisms change over time and are descended from a single common ancestor has been around in one form or another since the early Greek philosophers. But no one had developed a convincing explanation of how this could happen until 1858 when naturalists Charles Darwin (1809–1882) and Alfred Russel Wallace (1823–1913) independently proposed the concept of *natural selection* as a mechanism for biological evolution. Darwin meticulously gathered evidence for this idea and published it in 1859 in his book, *On the Origin of Species by Means of Natural Selection.*

Darwin and Wallace observed that individual organisms must struggle constantly to survive by getting enough food, water, and other resources, to avoid being eaten, and to reproduce. They also observed that individuals in a population with a specific advantage over other individuals in that population were more likely to survive and produce offspring that had the same specific advantage. The advantage was due to a characteristic, or *trait*, possessed by these individuals but not by others of their kind.

Based on these observations, Darwin and Wallace described a process called **natural selection**, in which individuals with certain traits are more likely to survive and reproduce under a particular set of environmental conditions than are those without the traits (**Concept 4-2B**). The scientists concluded that these survival traits would become more prevalent in future populations of the species as individuals with those traits became more numerous and passed their traits on to their offspring.

A huge body of evidence supports this idea. As a result, *biological evolution through natural selection* has become an important scientific theory that generally explains how life has changed over the past 3.5 billion years and why life is so diverse today. However, there are still many unanswered questions that generate scientific debate about the details of evolution by natural selection.

Ira Block/National Geographic Creative

Figure 4-5 This fossil shows the mineralized remains of an early ancestor of the present-day horse. It roamed the earth more than 35 million years ago. Note that you can also see fish skeletons on this fossil.

Mutations and Changes in the Genetic Makeup of Populations Lead to Biological Evolution by Natural Selection

The process of biological evolution by natural selection involves changes in a population's genetic makeup through successive generations. Note that *populations—not individuals—evolve by becoming genetically different.*

The first step in this process is the development of *genetic variability,* or variety in the genetic makeup of individuals in a population. This occurs through **mutations**: changes in the DNA molecules of a gene in any cell that can be inherited by offspring. Most mutations result from random changes that occur in coded genetic instructions when DNA molecules (see Figure 9, p. S17, in Supplement 4) are copied each time a cell divides and whenever an organism reproduces. Some mutations also occur from exposure to external agents such as radioactivity and natural and human-made chemicals (called *mutagens*).

Mutations can occur in any cell, but only those that take place in genes of reproductive cells are passed on to offspring. Sometimes, such a mutation can result in a new genetic trait, called a *heritable trait,* which can be passed from one generation to the next. In this way, populations develop differences among individuals, including genetic variability.

The next step in biological evolution is *natural selection,* in which environmental conditions favor some individuals over others. The favored individuals possess heritable traits that give them some advantage over other individuals in a given population. Such a trait is called an **adaptation**, or **adaptive trait**—any heritable trait that improves the ability of an individual organism to survive and to reproduce at a higher rate than other individuals in a population are able to do under prevailing environmental conditions. For example, in the face of snow and cold, a few gray wolves in a population that have thicker fur might live longer and thus produce more offspring than do those without thicker fur. As those longer-lived wolves mate, genes for thicker fur spread throughout the population and individuals with those genes increase in number and pass this helpful trait on to more offspring. Thus, the scientific concept of natural selection explains how populations adapt to changes in environmental conditions.

Another important example of natural selection at work is the evolution of *genetic resistance*—the ability of one or more organisms in a population to tolerate a chemical designed to kill it. Such resistance develops fairly quickly in populations of organisms that produce large numbers of offspring, such as many species of bacteria and insects.

For example, certain bacteria (Figure 4-6a) have developed genetic resistance to widely used antibacterial drugs, or *antibiotics,* which have become a force of natural selection. Often, when such drugs are used (Figure 4-6b), a few bacteria that are genetically resistant to them survive and rapidly produce more offspring than the bacteria that were killed by the drug could have produced (Figure 4-6c). Thus, the antibiotic eventually loses its effectiveness as genetically resistant bacteria keep reproducing while those that are susceptible to the drug die off (Figure 4-6d). (We discuss this form of genetic resistance further in Chapter 17.)

One way to summarize the process of biological evolution by natural selection is: *Genes mutate, individuals are selected, and populations evolve such that they are better adapted to survive and reproduce under existing environmental conditions* (**Concept 4-2B**).

A remarkable example of species evolution by natural selection is *Homo sapiens sapiens.* We have evolved certain traits that have allowed us to dominate most of the earth (Case Study that follows).

CASE STUDY

How Did Humans Become Such a Powerful Species?

Like many other species, humans have survived and thrived because we have certain traits that allow us to adapt to and modify parts of the environment to increase our chances of surviving and reproducing.

Evolutionary biologists attribute our success to three adaptations: *strong opposable thumbs* that allowed us to grip

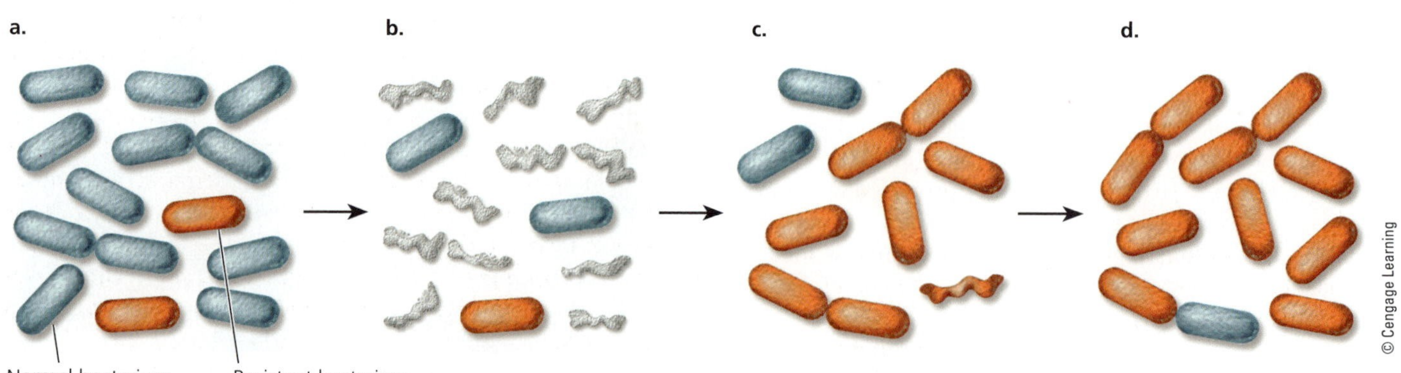

Figure 4-6 *Evolution by natural selection.* A population of bacteria **(a)** is exposed to an antibiotic, which **(b)** kills most individuals, but none of those possessing a trait that makes them resistant to the drug (shown in red). The resistant bacteria multiply **(c)** and eventually, **(d)** replace all or most of the nonresistant bacteria.

and use tools better than the few other animals that have thumbs could do; an *ability to walk upright,* which gave us agility and freed up our hands for many uses; and a *complex brain,* which allowed us to develop many skills, including the ability to use speech and to read and write to transmit complex ideas (Figure 4-7).

These adaptations have helped us to develop tools, weapons, protective devices, and technologies that extend our limited senses of sight, hearing, and smell. Thus, in an eye-blink of the 3.5-billion-year history of life on earth, we have developed powerful technologies and taken over much of the earth's net primary productivity for our own use. At the same time, we have degraded much of the planet's life-support system as our ecological footprints have grown (see Figure 1-13, p. 14).

However, adaptations that make a species successful during one period of time may not be enough to ensure the species' survival when environmental conditions change. This is no less true for humans, and some environmental conditions are now changing rapidly, largely due to our own actions. (We focus on several such changes in later chapters.)

One of our adaptations—our powerful brain—may enable us to live more sustainably by understanding and copying the ways in which nature has sustained itself for billions of years, despite major changes in environmental conditions.

Adaptation through Natural Selection Has Limits

In the not-too-distant future, will adaptations to new environmental conditions through natural selection allow our skin to become more resistant to the harmful effects of UV radiation, our lungs to cope with air pollutants, and our livers to better detoxify pollutants in our bodies?

According to scientists in this field, the answer is *no* because of two limitations on adaptation through natural selection. *First,* a change in environmental conditions can lead to such an adaptation only for genetic traits already present in a population's gene pool or for traits resulting from mutations, which occur randomly.

Second, even if a beneficial heritable trait is present in a population, the population's ability to adapt may be limited by its reproductive capacity. Populations of genetically diverse species that reproduce quickly—such as weeds, mosquitoes, rats, bacteria, and cockroaches—often adapt to a change in environmental conditions in a short time (days to years). By contrast, species that cannot produce large numbers of offspring rapidly—such as elephants, tigers, sharks, and humans—take a much longer time (typically thousands or even millions of years) to adapt through natural selection.

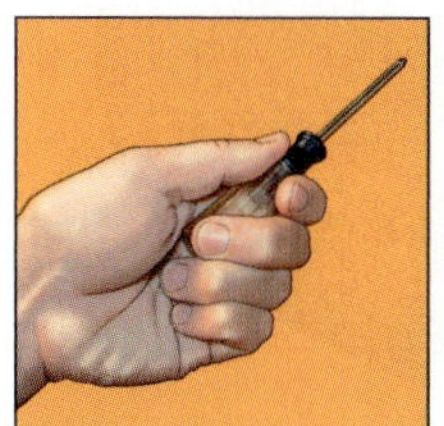

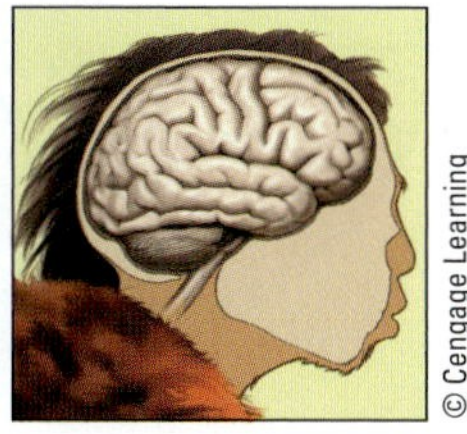

© Cengage Learning

Figure 4-7 *Homo sapiens sapiens* has had three advantages over other mammals that helped the species to survive certain pressures of natural selection and become a dominant species on planet Earth.

Mary Lane/Shutterstock.com

Figure 4-8 One type of carnivorous plant is the Venus flytrap. A fly or other small insect, entering a hinged opening on the plant's leaf, touches trigger hairs that cause the opening to snap shut in less than a second and trap its prey. The plant then uses enzymes to digest its prey over a period of 1–2 weeks.

Three Common Myths about Evolution through Natural Selection

Evolution experts have identified three common misconceptions about biological evolution through natural selection. One is that "survival of the fittest" means "survival of the strongest." To biologists, *fitness* is a measure of reproductive success, not strength. Thus, the fittest individuals are those that leave the most descendants.

Another misconception is that organisms develop certain traits because they need them. For example, certain plants, called *carnivorous plants* (Figure 4-8), feed on insects not because they once needed to in order to survive. Rather, some ancestors of these plants had characteristics that enabled them to trap insects and to draw nutrients from them. This trait gave them an advantage over other plants in an environment where there were lots of insects available, and this enabled them to produce more offspring that had such characteristics. Thus, over time, their populations grew and continued to evolve in this way.

A third misconception is that evolution by natural selection involves some grand plan of nature in which species become more perfectly adapted. From a scientific standpoint, no plan or goal for genetic perfection has been identified in the evolutionary process.

4-3 How Do Geological Processes and Climate Change Affect Evolution?

CONCEPT 4-3
Tectonic plate movements, volcanic eruptions, earthquakes, and climate change have shifted wildlife habitats, wiped out large numbers of species, and created opportunities for the evolution of new species.

Geological Processes Affect Natural Selection

The earth's surface has changed dramatically over its long history. Scientists have discovered that huge flows of molten rock within the earth's interior have broken its surface into a series of gigantic solid plates, called *tectonic plates*. For hundreds of millions of years, these plates have drifted slowly on the planet's mantle (Figure 4-9).

Rock and fossil evidence indicates that 200–250 million years ago, all of the earth's present-day continents were connected in a super continent called Pangaea (Figure 4-9, left). About 135 million years ago, Pangaea began splitting apart as the earth's tectonic plates moved, eventually resulting in the present-day locations of the continents (Figure 4-9, right).

The fact that tectonic plates drift has had two important effects on the evolution and distribution of life on the earth. *First,* the locations of continents and oceanic basins have greatly influenced the earth's climate and thus have helped to determine where plants and animals can live. *Second,* the movement of continents has allowed species to move, adapt to new environments, and form new species through natural selection. When continents join together, populations can disperse to new areas and adapt to new environmental conditions. When continents separate and when islands are formed, populations must evolve under isolated conditions or become extinct.

Adjoining tectonic plates that are grinding along slowly next to one another sometimes shift quickly. Such sudden movement of tectonic plates can cause *earthquakes,* which can also affect biological evolution by causing fissures in the earth's crust that can separate and isolate populations of species. Over long periods of time, this can lead to the formation of new species as each isolated population changes genetically in response to new environmental conditions. *Volcanic eruptions* that occur along the boundaries of tectonic plates can also affect biological evolution by destroying habitats and reducing, isolating, or wiping out populations of species (**Concept 4-3**).

Climate Change and Catastrophes Affect Natural Selection

Throughout its history, the earth's climate has changed drastically. At times, it has cooled and covered much of the earth with glacial ice. At other times it has warmed, melted that ice, and drastically raised sea levels, which in turn increased the total area covered by the oceans and reduced the earth's total land area. Such alternating periods of cooling and heating have led to the advance and retreat of ice sheets at high latitudes over much of the northern hemisphere, most recently about 18,000 years ago (Figure 4-10).

These long-term climate changes have had a major effect on biological evolution by determining where different types of plants and animals can survive and thrive, and by changing the locations of different types of ecosystems such as deserts, grasslands, and forests (**Concept 4-3**). Some species became extinct because the climate changed too rapidly for them to adapt and survive, and new species evolved to take over their ecological roles.

Another force affecting natural selection has been catastrophic events such as collisions between the earth and large asteroids. There may have been many of these collisions during the 3.5 billion years of life on earth.

Figure 4-9 Over millions of years, the earth's continents have moved very slowly on several gigantic tectonic plates. ***Question:*** How might an area of land splitting apart cause the extinction of a species?

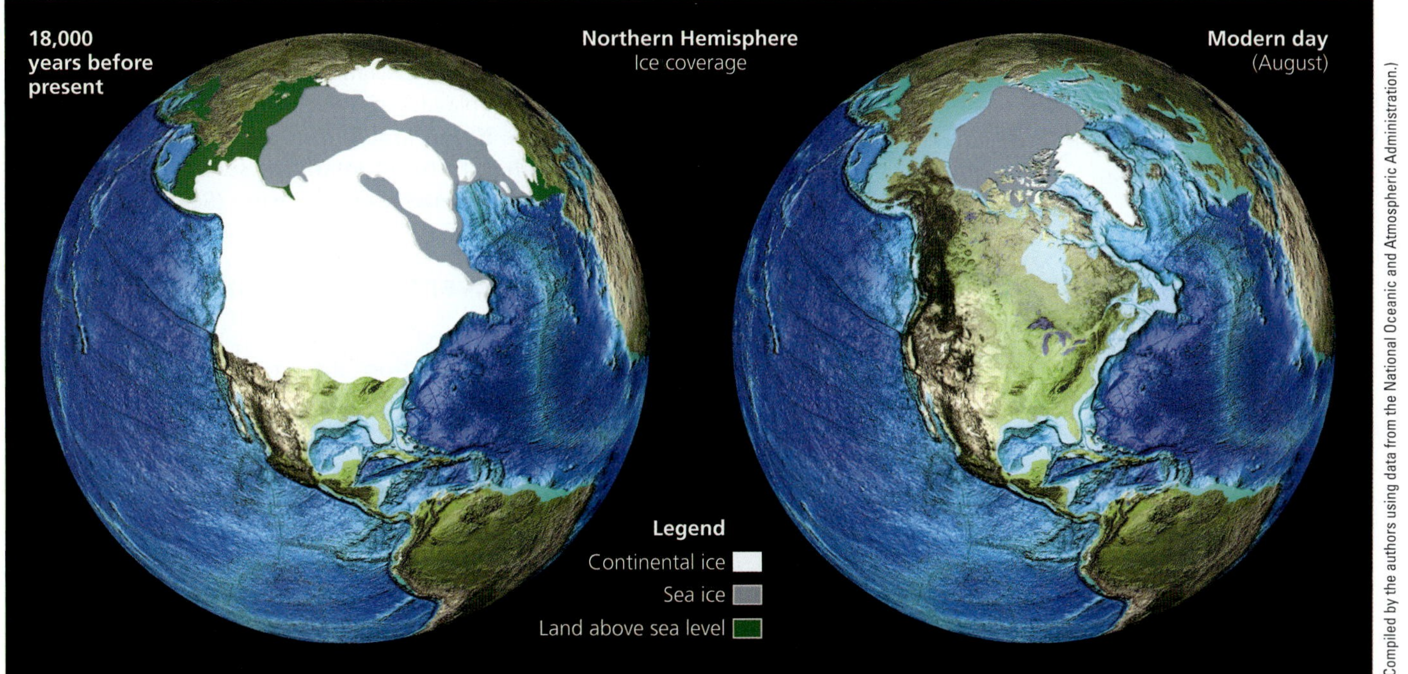

Figure 4-10 These maps of North America show the large-scale changes in glacial ice coverage during the past 18,000 years. ***Question:*** What are two characteristics of an animal and two characteristics of a plant that natural selection would have favored as these ice sheets (left) advanced?

Such impacts have caused widespread destruction of ecosystems and wiped out large numbers of species. On the other hand, they have also caused shifts in the locations of ecosystems and created opportunities for the evolution of new species.

On a long-term basis, the three scientific **principles of sustainability** (see Figure 1-2, p. 6 or back cover), especially the biodiversity principle (Figure 4-2), have enabled life on the earth to adapt to drastic changes in environmental conditions.

4-4 How Do Speciation, Extinction, and Human Activities Affect Biodiversity?

CONCEPT 4-4A
As environmental conditions change, the balance between the formation of new species and the extinction of existing species determines the earth's biodiversity.

CONCEPT 4-4B
Human activities are decreasing biodiversity by causing the extinction of many species and by destroying or degrading habitats needed for the development of new species.

How Do New Species Evolve?

Under certain circumstances, natural selection can lead to an entirely new species. In this process, called **speciation**, one species splits into two or more different species. For sexually reproducing organisms, a new species forms when one population of a species has evolved to the point where its members can no longer breed and produce fertile offspring with members of another population that did not change or that evolved differently.

The most common way in which speciation occurs, especially among sexually reproducing species, is when a barrier or distant migration separates two or more populations of a species and prevents the flow of genes between them. This happens in two phases: first geographic isolation, and then reproductive isolation.

Geographic isolation occurs when different groups of the same population of a species become physically isolated from one another for a long period of time. For example, part of a population may migrate in search of food and then begin living as a separate population in another area with different environmental conditions. Populations can also be separated by a physical barrier (such as a mountain range, stream, or road), a volcanic

Figure 4-11 These poison dart frogs vary in coloration, partly because they were exposed to different environmental conditions.

eruption, tectonic plate movements, or winds or flowing water that carry a few individuals to a distant area. These separated populations can develop quite different characteristics. For example, populations of poison dart frogs (**Core Case Study**) living on different islands or in different parts of a region can have dramatic differences in coloration, as shown in Figure 4-11.

In **reproductive isolation**, mutation and change by natural selection operate independently in the gene pools of geographically isolated populations. If this process continues long enough, members of the geographically and reproductively isolated populations of sexually reproducing species may become so different in genetic makeup that they cannot produce live, fertile offspring if they are rejoined and attempt to interbreed. As a result, one species has become two, and speciation has occurred (Figure 4-12).

Humans are playing an increasing role in the process of speciation. We have learned to shuffle genes from one species to another through **artificial selection** and, more recently, through **genetic engineering** (Science Focus 4.2).

All Species Eventually Become Extinct

Another process affecting the number and types of species on the earth is **extinction**, the process in which an entire species ceases to exist (also referred to as *biological extinction;* when a species becomes extinct over a large region, but not globally, it is called *local extinction*). When environmental conditions change dramatically or rapidly, a population of a species faces three possible futures: *adapt* to the new conditions through natural selection, *migrate* (if possible) to another area with more favorable conditions, or *become extinct.*

Species that are found in only one area, called **endemic species**, are especially vulnerable to extinction. They exist on islands and in other unique areas, especially in tropical rain forests where most species have highly specialized roles. For these reasons, they are unlikely to be able to migrate or adapt in the face of rapidly changing environmental conditions. Many of these endangered species are amphibians (**Core Case Study**). One example is the golden toad (Figure 4-13), which apparently became extinct in 1989 even though it lived in the well-protected Monteverde Cloud Forest Reserve in the mountains of Costa Rica.

Fossils and other scientific evidence indicate that all species eventually become extinct, but drastic changes in environmental conditions can eliminate large groups of species relatively rapidly. Throughout most of the earth's long history, species have disappeared at a low rate, called the **background extinction rate**. Based on the fossil record and analysis of ice cores, biologists estimate that

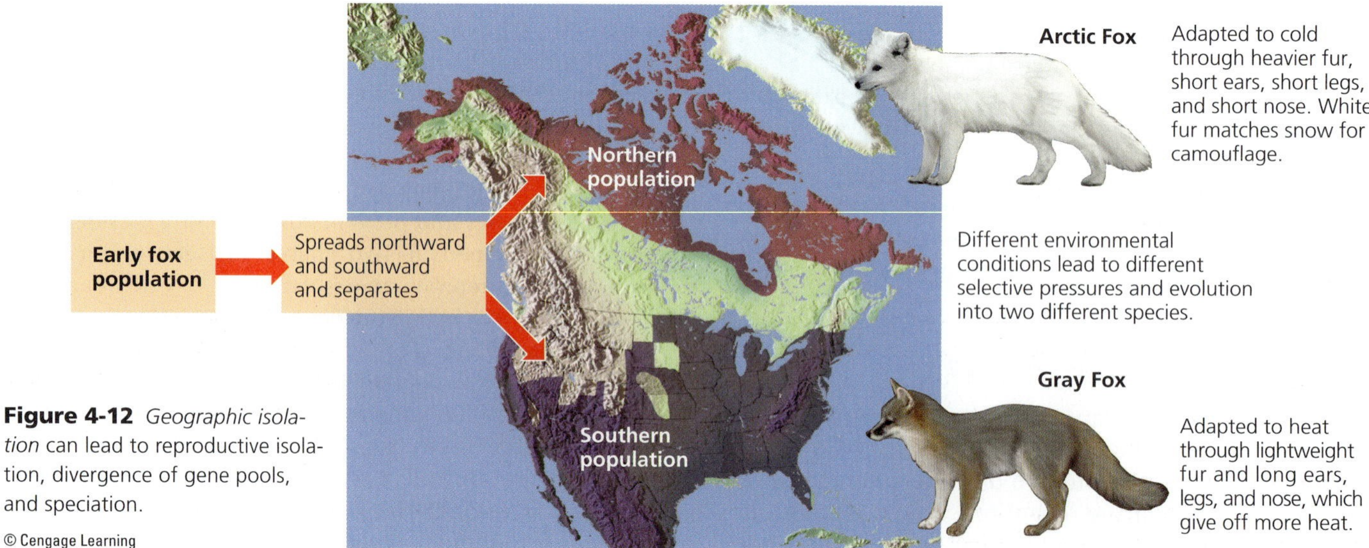

Figure 4-12 *Geographic isolation* can lead to reproductive isolation, divergence of gene pools, and speciation.

SCIENCE FOCUS 4.2

CHANGING THE GENETIC TRAITS OF POPULATIONS

We have used artificial selection to change the genetic characteristics of populations with similar genes. In this process, we select one or more desirable genetic traits in the population of a plant or animal such as a type of wheat, fruit (Figure 4-C), or dog. Then we use *selective breeding,* or *crossbreeding,* to generate populations of the species containing large numbers of individuals with the desired traits.

Note that artificial selection involves crossbreeding between genetic varieties of the same species or between species that are genetically close to one another, and thus it is not a form of speciation. Most of the grains, fruits, and vegetables we eat are produced by artificial selection. It has also given us food crops with higher yields, cows that give more milk, trees that grow faster, and many different types of dogs and cats. But traditional crossbreeding is a slow process and it can be used only on species that are close to one another genetically.

Pear
Apple
Desired trait (color)
Crossbreeding
Offspring
Best result
Crossbreeding
New offspring
Desired result

Figure 4-C Artificial selection involves the crossbreeding of species that are close to one another genetically. In this example, similar fruits are being crossbred.

Now scientists are using genetic engineering to speed up our ability to manipulate genes. In this process, scientists alter an organism's genetic material by adding, deleting, or changing segments of its DNA to produce desirable traits or to eliminate undesirable ones. It enables scientists to transfer genes between different species that would not interbreed in nature. For example, we can put genes from a cold-water fish species into a tomato plant to give it properties that enable it to resist cold weather.

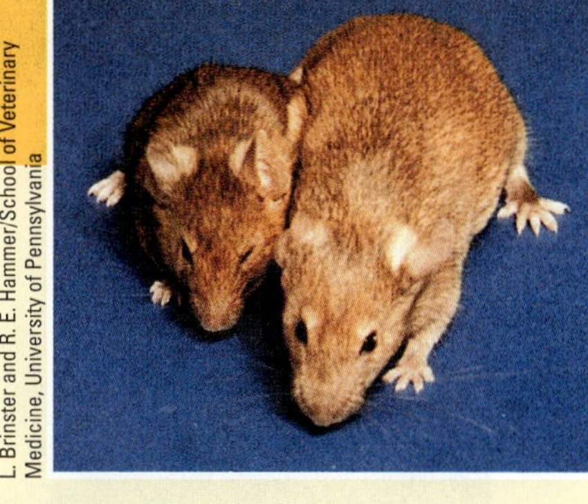

L. Brinster and R. E. Hammer/School of Veterinary Medicine, University of Pennsylvania

Figure 4-D These mice are both 6 months old. The one on the left is normal, while the mouse on the right has had a human growth hormone gene inserted into its cells. Mice with this gene grow 2–3 times faster than, and twice as large as, mice without it. ***Question:*** How do you think the creation of such species might change the process of evolution by natural selection?

Scientists have used genetic engineering to develop modified crop plants, new drugs, pest-resistant plants, and animals that grow rapidly (Figure 4-D). They have also created genetically engineered bacteria to extract minerals such as copper from their underground ores and to clean up spills of oil and other toxic pollutants.

Critical Thinking

If genetic engineering were more widely applied to plants and animals in the future, how might it affect the evolutionary process? What benefits and harms might result?

the average annual background extinction rate has been about 0.0001% of all species per year, which amounts to 1 species lost for every million species on the earth per year. At this rate, if there were 10 million species on the earth, about 10 of them, on average, would go extinct every year.

In contrast to the background extinction rate, a **mass extinction** is a significant rise in extinction rates above the background level. In such a catastrophic, widespread, and often global event, large groups of species (25–95% of all species) are wiped out. Fossil and geological evidence indicate that there have been at least three and probably five mass extinctions (at intervals of 20–60 million years) during the past 500 million years. Scientists point out that there have been numerous other times when extinction rates have gone well above the background extinction rate.

A mass extinction provides an opportunity for the evolution of new species that can fill unoccupied ecological roles or newly created ones. As a result, evidence indicates that each occurrence of mass extinction has been followed by an increase in species diversity over several million years as new species have arisen to occupy new habitats or to exploit newly available resources.

As environmental conditions change, the balance between formation of new species (speciation) and

Figure 4-13 This male golden toad lived in Costa Rica's high-altitude Monteverde Cloud Forest Reserve. The species became extinct in 1989 apparently because its habitat dried up.

extinction of existing species determines the earth's biodiversity (**Concept 4-4A**). The existence of millions of species today means that speciation, on average, has kept ahead of extinction. However, many scientists argue that higher extinction rates and other evidence indicate that we are experiencing the beginning of a new mass extinction.

There is also considerable evidence that much of the current rise in the extinction rate and the resulting loss of biodiversity are primarily due to human activities (**Concept 4-4B**), as our ecological footprints spread over the planet (see Figure 1-12, p. 13, and Figure 1-13, p. 14). Research indicates that the largest cause of the rising rate of species extinctions is the loss, fragmentation, and degradation of habitats. These losses occur as we cultivate more land to grow crops and clear more forestland for farming, ranching, and settlement (see Figure 3-1, p. 52). We examine this issue further in Chapters 9 and 10.

4-5 What Is Species Diversity and Why Is It Important?

> **CONCEPT 4-5**
> Species diversity is a major component of biodiversity and tends to increase the sustainability of some ecosystems.

Species Diversity Includes the Variety and Abundance of Species in a Particular Place

An important characteristic of a community and the ecosystem to which it belongs is its **species diversity**, or the number and variety of species it contains. One important component of species diversity is *species richness*, the number of different species in a given area. For example, an area of natural temperate rain forest (Figure 4-14, left) has a much higher species richness than a tree farm (Figure 4-14, right), which is usually planted with one species and hosts just a few other species that migrate in.

For any given community, another component of species diversity is *species evenness*, a measure of the relative abundance, or the comparative numbers of individuals of each species present. The more even the numbers of individuals in each species in a community, the higher the species evenness in that community. For example, a tree farm (Figure 4-14, right) with a large number of individuals of one species of trees and low numbers of individuals from the other plant species has low species evenness. However, most temperate rain forests (Figure 4-14, left) have high species evenness, because they contain similar numbers of individuals from each of many different species.

The species diversity of communities varies with their geographical location. For most terrestrial plants and animals, species diversity (primarily species richness) is highest in the tropics and declines as we move from the equator toward the poles. The most species-rich environments are tropical rain forests, large tropical lakes, coral reefs, and the ocean-bottom zone.

Species-Rich Ecosystems Tend to Be Productive and Sustainable

Ecologists have been conducting research to answer two important questions: *First,* is plant productivity higher in species-rich ecosystems? *Second,* does species richness enhance the *stability,* or *sustainability,* of an ecosystem?

Research suggests that the answers to both questions may be *yes,* and from that research, two hypotheses have emerged. According to the first, the more diverse an ecosystem is, the more productive it will be. That is, with a greater variety of producer species, an ecosystem will produce more plant biomass, which in turn will support a greater variety of consumer species.

The second hypothesis is that greater species richness and productivity will make an ecosystem more stable or sustainable. According to this hypothesis, a complex ecosystem, with many different species (high species richness) and the resulting variety of feeding paths, has more ways to respond to most environmental disturbances and stresses because it does not have all its eggs in one basket. According to biologist Edward O. Wilson, "There's a common sense element to this: the more species you have, the more likely you're going to have an insurance policy for the whole ecosystem."

Ecologist David Tilman and his colleagues at the University of Minnesota did research that supported both hypotheses. They found that communities with high plant species richness produced a certain amount of biomass more consistently than did communities with fewer spe-

Figure 4-14 This area of natural temperate rain forest in Washington's Olympic National Park (left) has a much higher number of species (higher species richness) and higher species evenness than this tree farm planted in Oregon (right) has.

Left: Natalia Bratslavsky/Shutterstock.com; Right: Knowlesgal. . . | Dreamstime.com

cies. The species-rich communities were also less affected by drought and more resistant to invasions by insect species. Later laboratory studies involved setting up artificial ecosystems in growth chambers where researchers could control and manipulate key variables such as temperature, light, and atmospheric gas concentrations. These studies have supported Tilman's findings.

Ecologists hypothesize that in a species-rich ecosystem, each species can exploit a different portion of the resources available. For example, some plants will bloom early and others will bloom late. Some have shallow roots to absorb water and nutrients in topsoil, and others use longer roots to tap into deeper soils. A number of studies support this hypothesis, although some do not.

There is debate among scientists about how much species richness is needed to help sustain various ecosystems. Some research suggests that the average annual net primary productivity of an ecosystem reaches a peak with 10–40 producer species. Many ecosystems contain more than 40 producer species, but do not necessarily produce more biomass or reach a higher level of stability. While there is more research to do in this area, most ecologists now accept as a useful hypothesis the idea that species richness appears to increase the productivity and stability, or sustainability, of an ecosystem (Concept 4-5).

4-6 What Roles Do Species Play in Ecosystems?

CONCEPT 4-6A
Each species plays a specific ecological role called its *niche*.

CONCEPT 4-6B
Any given species may play one or more of four important roles—native, nonnative, indicator, or keystone—in a particular ecosystem.

Each Species Plays a Role in Its Ecosystem

An important principle of ecology is that *each species has a specific role to play in the ecosystems where it is found* (Concept 4-6A). Scientists describe the role that a species plays in its ecosystem as its **ecological niche**, or simply **niche**. It is a species' way of life in a community and includes everything that affects its survival and reproduction, such as how much water and sunlight it needs, how much space it requires, what it feeds on, what feeds on it, and the temperatures and other conditions it can tolerate. A species' niche should not be confused with its **habitat**, which is the place where it lives. Its niche is its pattern of living.

Scientists use the niches of species to classify them mostly as *generalists* or *specialists*. **Generalist species** have broad niches (Figure 4-15, right curve). They can live in many different places, eat a variety of foods, and often tolerate a wide range of environmental conditions. Flies, cockroaches, mice, rats, white-tailed deer, and humans are generalist species.

In contrast, **specialist species** occupy narrow niches (Figure 4-15, left curve). They may be able to live in only

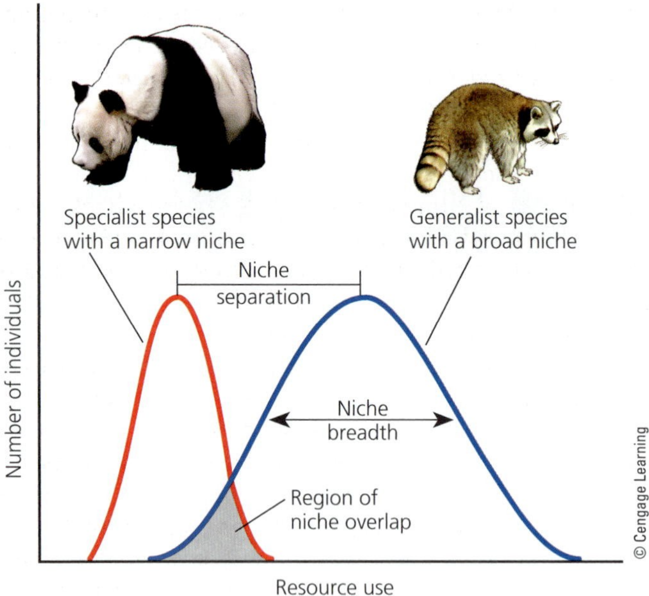

Figure 4-15 Specialist species such as the giant panda have a narrow niche (left curve) and generalist species such as the raccoon have a broad niche (right curve).

one type of habitat, use just one or only a few types of food, or tolerate a narrow range of climatic and other environmental conditions. For example, some shorebirds occupy specialized niches, feeding on crustaceans, insects, and other organisms found on sandy beaches and their adjoining coastal wetlands (Figure 4-16).

Because of their narrow niches, specialists are more prone to extinction when environmental conditions change. For example, China's *giant panda* (Case Study that follows) is highly endangered because of a combination of habitat loss, low birth rate, and its specialized diet consisting mostly of bamboo.

Is it better to be a generalist or a specialist? It depends. When environmental conditions are fairly constant, as in a tropical rain forest, specialists have an advantage because they have fewer competitors. But under rapidly changing environmental conditions, the more adaptable generalist usually is better off than the specialist.

CONSIDER THIS. . .

THINKING ABOUT Amphibians' Niches

Do you think that most amphibian species (**Core Case Study**) occupy specialist or generalist niches? Explain.

CASE STUDY

The Giant Panda—A Highly Endangered Specialist

The *giant panda* (Figure 4-17) is among the most threatened of all species, rated as *endangered* by the IUCN. According to the World Wildlife Fund (WWF) there are 1,600 to 3,000 giant pandas left in the wild, most of them in China.

Pandas evolved to live in the forests of China where bamboo, their main food source, grows thickly among evergreen trees. A typical panda spends 12 hours a day eating bamboo stalks and leaves.

The need for bamboo makes the panda a specialist species. For this reason, the main threat to pandas is destruction and fragmentation of their forest habitat where the bamboo grows. The pandas are now limited to six remaining mountain forests of central and southwestern China. Some scientists warn that projected climate change due to atmospheric warming will limit the range of bamboo forests, making the problem even worse.

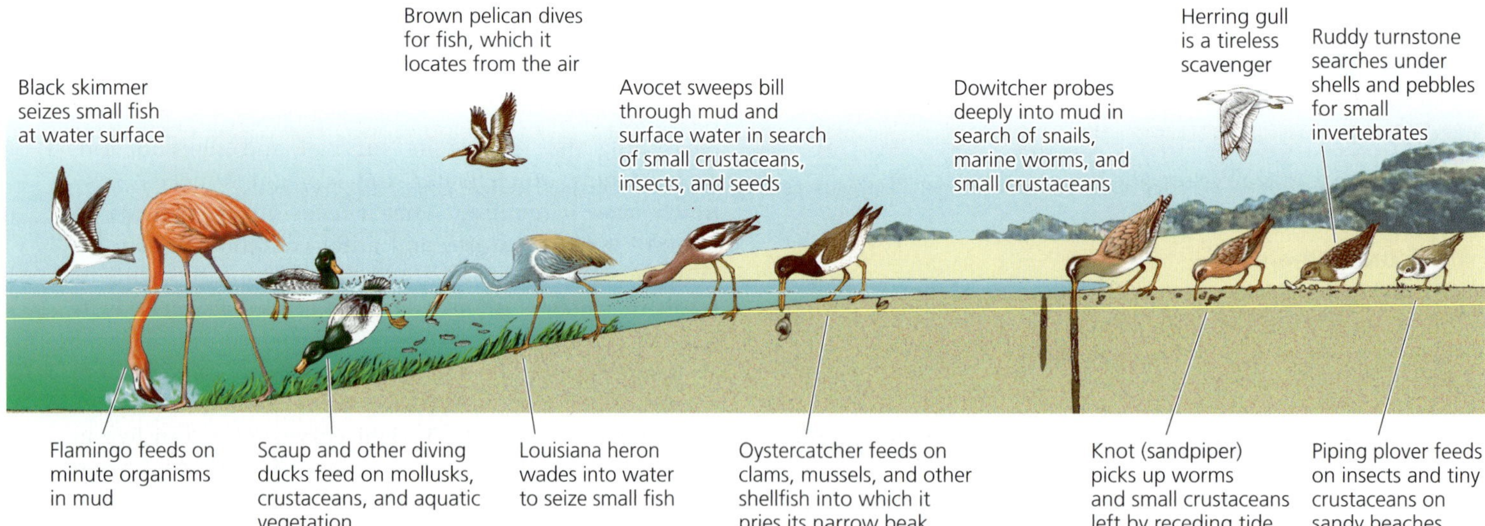

Figure 4-16 Various bird species in a coastal wetland occupy specialized feeding niches. This specialization reduces competition and allows for sharing of limited resources.

© Cengage Learning

Figure 4-17 This giant panda is eating a stalk from a bamboo tree. Bamboo stalks and leaves make up about 95% of the diet of this specialist species.

Mike Flippo/Shutterstock.com

The giant panda is also endangered because of its low reproductive rate. Pandas have a very short breeding season, and females give birth to only one or two cubs every 2–3 years. If twins are born, usually only one survives, because the mother usually cannot produce enough milk for two cubs. Another threat is poaching—the illegal killing of giant pandas for their fur.

GOOD NEWS

The government of China, along with international conservation groups, has been trying to address these threats to pandas. By 2011, the government had set aside more than 50 panda reserves in about half of the remaining panda habitat area. The government estimated that about 60% of the remaining wild panda population was thus protected. In addition, according to the WWF, there were more than 300 pandas in captivity on mainland China and in 30 other countries.

Groups like the IUCN and WWF call for protecting larger areas of forest and for connecting fragmented areas with forested corridors through which the pandas could move between seasons to gain more access to bamboo. They also call for more patrolling to control poaching and illegal logging in protected areas.

Species Can Play Four Major Roles within Ecosystems

Niches can be classified further in terms of specific roles that certain species play within ecosystems. Ecologists describe *native, nonnative, indicator,* and *keystone* roles. Any given species may play one or more of these four roles in a particular ecosystem (**Concept 4-6B**).

Native species are those species that normally live and thrive in a particular ecosystem. Other species that migrate into, or are deliberately or accidentally introduced into, an ecosystem are called **nonnative species**, also referred to as *invasive, alien,* and *exotic species.*

People often think of nonnative species as threatening. In fact, most introduced and domesticated plant species such as food crops and flowers and animals such as chickens, cattle, and fish from around the world are beneficial to us. However, some nonnative species can compete with and reduce a community's native species, causing unintended and unexpected consequences. In 1957, for example, Brazil imported wild African honeybees (Figure 4-18) to help increase honey production. Instead of helping, the bees displaced some native honeybee populations, which had the effect of reducing the honey supply.

Since then, these nonnative honeybee species—popularly known as "killer bees"—have moved northward into Central America and parts of the southwestern and southeastern United States. The wild African bees are not the fearsome killers portrayed in some horror movies. However, they are aggressive and unpredictable and have killed thousands of domesticated animals and an estimated 1,000 people in the western hemisphere, many of whom were allergic to bee stings.

Nonnative species can spread rapidly if they find a new location with favorable conditions. In their new niches, these species often do not face the predators and diseases they face in their native niches, or they may be able to out-compete some native species in their new locations. We will examine this environmental threat in greater detail in Chapter 9.

Indicator Species Serve as Biological Smoke Alarms

Species that provide early warnings of damage to a community or an ecosystem are called **indicator species**. For example, in the **Core Case Study** that opened this chapter, we learned that some amphibians are classified as indicator species. The decline of an amphibian population can indicate the presence of parasites, disease-causing

John Lindsay-Smith/Shutterstock.com

Figure 4-18 Wild African honeybees, popularly known as "killer bees," were imported into Brazil and their populations have since spread widely.

SCIENCE FOCUS 4.3

SCIENTISTS ARE SEARCHING FOR THE CAUSES OF AMPHIBIAN DECLINES

Herpetologists, the scientists who study frogs and other amphibians (Figure 4-E), have identified a number of factors—both natural and human-caused—that affect these species at various points in their life cycles. One of the natural causes is *parasites* such as flatworms, which feed on the amphibian eggs laid in water and apparently have caused an increase in the number of births of amphibians with limbs missing or with extra limbs.

Some herpetologists hypothesize that *viral and fungal diseases,* especially the chytrid fungus that attacks the skin of frogs, are reducing the frogs' ability to take in water through their skin. This leads to death from dehydration. Such diseases can spread fairly easily, because adults of many amphibian species congregate in large numbers to breed.

As with other threatened species, one of the major threats to amphibians is *habitat loss and fragmentation.* This is almost completely a human-caused problem resulting from the clearing of forests and the draining and filling of freshwater wetlands for farming and urban development.

Another threat is *prolonged drought,* which can dry up breeding pools that frogs and other amphibians depend on for reproduction and in early life. There is evidence that human activities are accelerating climate change by adding to atmospheric warming (as we discuss in Chapter 19) and this is helping to prolong natural droughts in many areas of the world. A 2005 study found an apparent correlation between climate change caused by atmospheric warming and the extinction of about two-thirds of the 110 known species of harlequin frogs in tropical forests of Central and South America.

Another human-influenced problem is *higher levels of UV radiation,* which can harm embryos of amphibians in shallow ponds, as well as adults basking in the sun for warmth. Historically, such radiation has been screened by ozone in the stratosphere, but during the past few decades, ozone-depleting chemicals released into the atmosphere from human sources have destroyed some of the protective ozone. International action has been taken to reduce the threat of stratospheric ozone depletion, but it will take about 50 years for ozone levels to recover to those of 1960.

Pollution and *overhunting* are two other human-caused threats to amphibians. Frogs and other species are increasingly exposed to pesticides in ponds and in the bodies of insects that they consume. This can make them more vulnerable to bacterial, viral, and fungal diseases and to some parasites. Also, frogs are hunted for their leg meat, especially in Asia and France, where they are overhunted in many areas.

Yet another threat to amphibians being studied by scientists is the invasion of their habitats by *nonnative predators and competitors,* such as certain fish species. Some of this immigration is natural, but humans accidentally or deliberately transport many species to amphibian habitats, another problem that we discuss further in Chapter 9.

George Grall/National Geographic Creative

Figure 4-E This researcher is watching frogs mating in a tropical forest in Brazil.

Most herpetologists believe that a combination of these factors, which vary from place to place, probably is responsible for most of the decline and disappearances among amphibian species.

Critical Thinking

Of the factors listed above, which three do you think could be most effectively controlled by human efforts?

microbes, or pollution in the local environment (Science Focus 4.3), as well as habitat destruction and fragmentation. It can also be a sign of the effects of climate change (see Figure 4-13).

Birds are excellent biological indicators because they are found almost everywhere and are affected quickly by environmental changes such as the loss or fragmentation of their habitats and the introduction of chemical pesticides. Some butterflies are also indicator species because their association with various plant species makes them vulnerable to habitat loss and fragmentation. The populations of many bird and butterfly species are declining—a problem we explore more fully in Chapter 9.

Keystone Species Play Critical Roles in Their Ecosystems

A keystone is the wedge-shaped stone placed at the top of a stone archway. Remove this stone and the arch collapses. In some communities and ecosystems, ecologists hypothesize that certain species play a similar role. **Keystone spe-**

Figure 4-19 *Keystone species:* The American alligator plays an important ecological role in its marsh and swamp habitats in the southeastern United States.

Martha Marks/Shutterstock.com

cies are species whose roles have a large effect on the types and abundance of other species in an ecosystem.

Keystone species often exist in relatively limited numbers in their ecosystems, but the effects that they have there are often much larger than their numbers would suggest. And because of their often-smaller numbers, some keystone species are more vulnerable to extinction than other species are.

Keystone species can play several critical roles in helping to sustain ecosystems. One such role is the *pollination* of flowering plant species by butterflies (see chapter-opening photo), honeybees (Figure 4-A), hummingbirds, bats, and other species. In addition, *top predator* keystone species feed on and help to regulate the populations of other species. Examples are the wolf, leopard, lion, some shark species, and the American alligator (see Case Study that follows).

The loss of a keystone species can lead to population crashes and extinctions of other species in a community that depends on them for certain ecosystem services. This is why it so important for scientists to identify and protect keystone species.

CASE STUDY

The American Alligator—A Keystone Species That Almost Went Extinct

The American alligator (Figure 4-19) is a keystone species because it plays a number of important roles in the ecosystems where it is found in the southeastern United States. This species has outlived the dinosaurs and survived many challenges to its existence.

In the 1930s, hunters began killing large numbers of these animals for their exotic meat and their soft belly skin, used to make expensive shoes, belts, and pocketbooks. Other people hunted alligators for sport or out of dislike for the large reptile. By the 1960s, hunters and poachers had wiped out 90% of the alligators in the state of Louisiana, and the alligator population in the Florida Everglades was also near extinction.

Those who did not care much for the alligator were probably not aware of its important ecological role—its *niche*—in subtropical wetland ecosystems. Alligators dig deep depressions, or gator holes. These depressions hold freshwater during dry spells, serve as refuges for aquatic life, and supply freshwater and food for fishes, insects, snakes, turtles, birds, and other animals.

The large nesting mounds that alligators build provide nesting and feeding sites for some herons and egrets, and red-bellied turtles lay their eggs in old gator nests. In addition, alligators eat large numbers of gar, a predatory fish, which helps to maintain populations of game fish such as bass and bream that the gar eat.

As alligators create gator holes and nesting mounds, they help to keep shore and open water areas free of invading vegetation. Without this free ecosystem service, freshwater ponds and coastal wetlands where alligators live would be filled in with shrubs and trees, and dozens of species would disappear from these ecosystems.

In 1967, the U.S. government placed the American alligator on the endangered species list. By 1977, because it was protected, its populations had made a strong enough comeback to be removed from the endangered species list.

Today, there are well over a million alligators in Florida. In fact, the population has recovered to the point where the state allows property owners to kill alligators that stray onto their land. To conservation biologists, the comeback of the American alligator is an important success story in wildlife conservation.

CONSIDER THIS. . .

THINKING ABOUT The American Alligator and Biodiversity

What are two ways in which the American alligator supports one or more of the four components of biodiversity (Figure 4-2) within its environment?

Big Ideas

- **Populations evolve when genes mutate and give some individuals genetic traits that enhance their abilities to survive and to produce offspring with these traits (natural selection).**
- **Human activities are degrading the earth's vital biodiversity by hastening the extinction of species and by disrupting habitats needed for the development of new species.**
- **Each species plays a specific ecological role (its ecological niche) in the ecosystem where it is found.**

TYING IT ALL TOGETHER Amphibians and Sustainability

Robert King/Shutterstock.com

The **Core Case Study** on amphibians at the beginning of this chapter shows that the importance of a species does not always match the public's perception of it. While many people worry about the fates of the giant panda and polar bears, most are not aware that many species of amphibians are declining dramatically or that they play key roles in their ecosystems, such as controlling insect populations. Their extinction would soon be followed by the disappearances of many species of birds, fish, and mammals that feed on them.

In this chapter, we studied the importance of biodiversity—the numbers and varieties of species found in different parts of the world, along with genetic, ecosystem, and functional diversity. We also studied the process whereby all species came to be, according to the scientific theory of biological evolution through natural selection. Taken together, biodiversity and evolution represent two vital and irreplaceable forms of natural capital.

Finally, we examined the variety of roles played by species in ecosystems. For example, we saw that some species, including many amphibians, are indicator species that clue us in to the presence of threats to biodiversity, to ecosystems, and to the biosphere. Others such as the American alligator are keystone species that play vital roles in sustaining the ecosystems where they live.

Ecosystems and the variety of species they contain are functioning examples of the three scientific **principles of sustainability** (see Figure 1-2, p. 6 or back cover) in action. They depend on solar energy and on the cycling of nutrients. Ecosystems also help to sustain biodiversity.

Chapter Review

Core Case Study

1. Describe the threats to many of the world's amphibian species (**Core Case Study**) and explain why we should avoid hastening the extinction of any amphibian species through our activities.

Section 4-1

2. What is the key concept for this section? Define **biodiversity (biological diversity)** and list and describe its four major components. What is the importance of biodiversity? Define **species**. Summarize the importance of insects. Define and give three examples of **biomes**. Summarize the scientific contributions of Edward O. Wilson.

Section 4-2

3. What are the two key concepts for this section? What is a **fossil** and why are fossils important for understanding the history of life? What is **biological evolution** (or **evolution**)? State the **theory of evolution**. What is **natural selection**? What is a **mutation** and what role do mutations play in evolution by natural selection? What is an **adaptation (adaptive trait)**? How did humans become such a powerful species? What are two limitations on evolution by natural selection? What are three myths about evolution through natural selection?

Section 4-3

4. What is the key concept for this section? Describe how geologic processes can affect natural selection. How can climate change and catastrophes such as asteroid impacts affect natural selection?

Section 4-4

5. What are the two key concepts for this section? Define **speciation**. Distinguish between **geographic isolation** and **reproductive isolation**, and explain how they can lead to the formation of a new species. Define and distinguish between **artificial selection** and **genetic engineering** and give an example of each.

6. What is **extinction**? What is an **endemic species** and why can such a species be vulnerable to extinction? Define and distinguish between **background extinction rate** and **mass extinction**.

Section 4-5

7. What is the key concept for this section? Define **species diversity** and distinguish between species richness and species evenness. Explain why species-rich ecosystems tend to be productive and sustainable.

Section 4-6

8. What are the two key concepts for this section? Define and distinguish between an **ecological niche** (or **niche**) and a **habitat**. Distinguish between **generalist species** and **specialist species** and give an example of each. Why has the fact that the giant panda is a specialist species led to its classification as an endangered species?

9. Define and distinguish among **native**, **nonnative**, **indicator**, and **keystone species** and give an example of each. What major ecological role do many amphibian species play (**Core Case Study**)? List six factors that contribute to the threats of extinction for frogs and other amphibians. Describe the role of the American alligator as a keystone species.

10. What are this chapter's *three big ideas*? How are ecosystems and the variety of species they contain related to the three scientific **principles of sustainability**?

Note: Key terms are in bold type.

Critical Thinking

1. How might we and other species be affected if all amphibians (**Core Case Study**) were to go extinct?

2. What role does each of the following processes play in helping to implement the three scientific **principles of sustainability**: **(a)** natural selection, **(b)** speciation, and **(c)** extinction?

3. How would you respond to someone who tells you that:
   - **a.** he or she does not believe in biological evolution because it is "just a theory"?
   - **b.** we should not worry about air pollution because natural selection will enable humans to develop lungs that can detoxify pollutants?

4. Is the human species a keystone species? Explain. If humans were to become extinct, what are three species that might also become extinct and three species whose populations would probably grow?

5. How would you respond to someone who says that because extinction is a natural process, we should not worry about the loss of biodiversity when species become extinct largely as a result of our activities?

6. List three ways in which you could apply **Concept 4-4B** to making your lifestyle more environmentally sustainable.

7. Congratulations! You are in charge of the future evolution of life on the earth. What are the three things that you would consider to be the most important to do?

8. If you were forced to choose between saving the giant panda from extinction and saving amphibians, which would you choose? Explain.

Doing Environmental Science

Study an ecosystem of your choice, such as a meadow, a patch of forest, a garden, or an area of wetland. (If you cannot do this physically, do so virtually by reading about an ecosystem online or in a library.) Determine and list five major plant species and five major animal species in your ecosystem. Write hypotheses about **(a)** which of these species, if any, are indicator species and **(b)** which of them, if any, are keystone species. Explain how you arrived at your hypotheses. Then design an experiment to test each of your hypotheses, assuming you would have unlimited means to carry them out.

Global Environment Watch Exercise

Search for *Amphibians* to find out more about the current state of these species with regard to threats to their existence (**Core Case Study**). What actions are being taken by various nations and organizations to protect amphibians? Write a short summary report on your research.

Data Analysis

The following table is a sample of a very large body of data reported by R. A. Alford and S. J. Richards in their book *Extinction in our Times–Global Amphibian Decline.* It compares various areas of the world in terms of the number of amphibian species found and the number of amphibian species that were endemic, or unique to each area. Scientists like to know these percentages because endemic species tend to be more vulnerable to extinction than do nonendemic species. Study the table below and then answer the questions that follow it.

Area	Number of Species	Number of Endemic Species	Percentage Endemic
Pacific/Cascades/Sierra Nevada Mountains—North America	52	43	
Southern Appalachian Mountains—USA	101	37	
Southern Coastal Plain—USA	68	27	
Southern Sierra Madre—Mexico	118	74	
Highlands of Western Central America	126	70	
Highlands of Costa Rica and Western Panama	133	68	
Tropical Southern Andes Mountains—Bolivia and Peru	132	101	
Upper Amazon Basin—Southern Peru	102	22	

1. Fill in the fourth column to calculate the percentage of amphibian species that are endemic to each area.
2. What two areas have the highest numbers of endemic species? Name the two areas with the highest percentages of endemic species.
3. What two areas have the lowest numbers of endemic species? What two areas have the lowest percentages of endemic species?
4. What two areas have the highest percentages of non-endemic species?

5 Biodiversity, Species Interactions, and Population Control

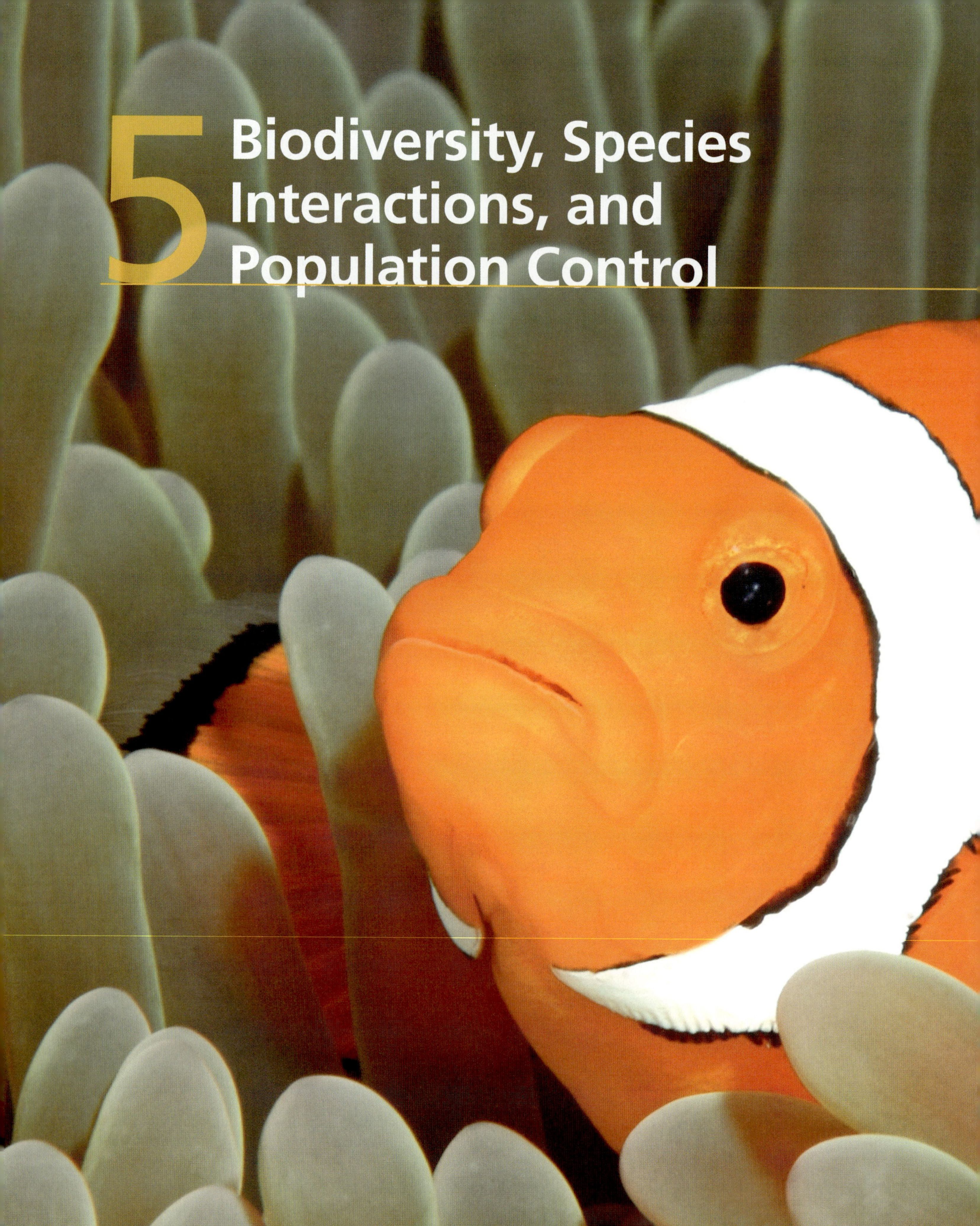

In looking at nature, never forget that every single organic being around us may be said to be striving to increase its numbers.

CHARLES DARWIN, 1859

Key Questions

5-1 How do species interact?

5-2 How do communities and ecosystems respond to changing environmental conditions?

5-3 What limits the growth of populations?

A clownfish gains protection by living among sea anemones and helps protect the anemones from some of their predators.

cbpix/Shutterstock.com

CORE CASE STUDY The Southern Sea Otter: A Species in Recovery

Southern sea otters (Figure 5-1, left) live in giant kelp forests (Figure 5-1, right) in shallow waters along parts of the Pacific coast of North America. Most of the remaining members of this endangered species are found off the western coast of the United States between the cities of Santa Cruz and Santa Barbara, California.

Southern sea otters are fast and agile swimmers that dive to the ocean bottom looking for shellfish and other prey, including sea urchins. On the surface they swim on their backs and feed on their prey, using their bellies as a table (Figure 5-1, left). Each day, a sea otter consumes 20–35% of its weight in clams, mussels, crabs, sea urchins, abalone, and about 40 other species of bottom-dwelling organisms.

It is estimated that between 13,000 and 20,000 southern sea otters once lived in California's coastal waters. By the early 1900s, they were hunted almost to extinction in this region by fur traders who killed them for their thick, luxurious fur. Commercial fisherman also killed otters because they viewed them as competitors in the hunt for valuable abalone and other shellfish.

Since that time, this population has generally grown from a low of about 50 in 1938 to an estimated 2,800 in 2012. Their partial recovery got a boost in 1977 when the U.S. Fish and Wildlife Service declared the species endangered in most of its range, with a total population of only 1,850 individuals. Despite such progress, the population has a long way to go to justify removing it from the endangered species list.

Why should we care about the southern sea otters of California? One reason is *ethical:* many people believe it is wrong to allow human activities to cause the extinction of a species. Another reason is that people love to look at these appealing and highly intelligent animals as they play in the water. As a result, they help to generate millions of dollars a year in tourism revenues. A third reason—and a key reason in our study of environmental science—is that biologists classify them as a *keystone species* (p. 94). Scientists hypothesize that in the absence of southern sea otters, sea urchins and other kelp-eating species would probably destroy the Pacific coast kelp forests and much of the rich biodiversity they support.

Biodiversity is an important part of the earth's natural capital and is the focus of one of the three scientific **principles of sustainability** (see Figure 1-2, p. 6 or back cover). In this chapter, we will look at two factors that affect biodiversity: how species interact and help control one another's population sizes and how communities, ecosystems, and populations of species respond to changes in environmental conditions.

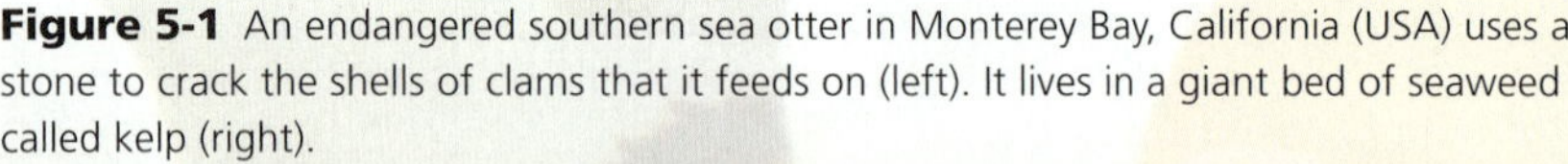

Figure 5-1 An endangered southern sea otter in Monterey Bay, California (USA) uses a stone to crack the shells of clams that it feeds on (left). It lives in a giant bed of seaweed called kelp (right).

Left: © Xfkirsten | Dreamstime.com; Right: Paul Whitted/Shutterstock.com

5-1 How Do Species Interact?

CONCEPT 5-1

Five types of interactions among species—interspecific competition, predation, parasitism, mutualism, and commensalism—affect the resource use and population sizes of species.

Most Species Compete with One Another for Certain Resources

Ecologists have identified five basic types of interactions among species as they share limited resources such as food, shelter, and space. These types of interactions are called *interspecific competition, predation, parasitism, mutualism,* and *commensalism,* and they all have significant effects on the population sizes of the species in an ecosystem and on how they use resources (Concept 5-1).

The most common interaction among species is **interspecific competition**, which occurs when members of two or more species interact to use the same limited resources such as food, water, light, and space. While fighting for resources does occur, most interspecific competition involves the ability of one species to become more efficient than another species in obtaining the resources it needs.

Recall that each species plays a role in its ecosystem called its *ecological niche* (p. 91 and Figure 4-15, p. 92). When two species compete with one another for the same resources, their niches *overlap*. The greater this overlap, the more intense is their competition for key resources. If one species can take over the largest share of one or more key resources, each of the other competing species must move to another area (if possible), adapt by shifting its feeding habits or behavior through natural selection to reduce or alter its niche, suffer a sharp population decline, or become extinct in that area.

Humans compete with many other species for space, food, and other resources. As our ecological footprints grow and spread (see Figure 1-13, p. 14), we are taking over or degrading the habitats of many other species and depriving them of resources they need in order to survive.

Some Species Evolve Ways to Share Resources

Over a time scale long enough for natural selection to occur, populations of some species develop adaptations that allow them to reduce or avoid competition with other species for resources. One way this happens is through **resource partitioning**. It occurs when species competing for similar scarce resources evolve specialized traits that allow them to share resources by using parts of them, using them at different times, or using them in different ways.

Figure 5-2 shows resource partitioning by some insect-eating bird species. In this case, their adaptations allow them to reduce competition by feeding in different portions of certain spruce trees and by feeding on different insect species.

Figure 5-2 *Sharing the wealth:* Resource partitioning among five species of insect-eating warblers in the spruce forests of the U.S. state of Maine. Each species spends at least half its feeding time in its associated yellow-highlighted areas of these spruce trees.

(Based on R. H. MacArthur, "Population Ecology of Some Warblers in Northeastern Coniferous Forests," *Ecology* 36 (1958): 533–536, 1958.)

Figure 5-3 *Specialist species of honeycreepers:* Through natural selection, different species of honeycreepers have shared resources by evolving specialized beaks to take advantage of certain types of food such as insects, seeds, fruits, and nectar from certain flowers. ***Question:*** Look at each bird's beak and take a guess at what sort of food that bird might eat.

Another example of resource partitioning through natural selection involves birds called honeycreepers that live in the U.S. state of Hawaii (Figure 5-3). And Figure 4-16 (p. 92) shows how the evolution of specialized feeding niches has reduced competition for resources among bird species in a coastal wetland.

Consumer Species Feed on Other Species

In **predation**, a member of one species (the **predator**) feeds directly on all or part of a living organism (the **prey**) as part of a food web. Together, the two different species, such as a brown bear (the predator, or hunter) and a salmon (the prey, or hunted), form a **predator–prey relationship** (Figure 5-4). Such relationships are also shown in Figure 3-7 (p. 57).

Figure 5-4 *Predator–prey relationship:* This brown bear (the predator) in the U.S. state of Alaska has captured and will feed on this salmon (the prey).

CONSIDER THIS. . .

CONNECTIONS Grizzly Bears and Moths

During the summer months, the grizzly bears of the Greater Yellowstone ecosystem in the western United States eat huge amounts of army cutworm moths, which huddle in masses high on remote mountain slopes. In this predator–prey interaction, one grizzly bear can dig out and lap up as many as 40,000 cutworm moths in a day. Consisting of 50–70% fat, the moths offer a nutrient that the bear can store in its fatty tissues and draw on during its winter hibernation.

In a giant kelp forest ecosystem, sea urchins prey on kelp, a form of seaweed (Science Focus 5.1). However, as a keystone species (see Chapter 4, p. 94), southern sea otters (**Core Case Study**) prey on the sea urchins and thus help keep them from destroying the kelp forests.

Predators have a variety of methods that help them to capture prey. *Herbivores* can simply walk, swim, or fly to the plants they feed on. *Carnivores* feeding on mobile prey have two main options: *pursuit* and *ambush*. Some, such as the cheetah, catch prey by running fast. Others, such as the American bald eagle, can fly and have keen eyesight. Still others cooperate in capturing their prey. For example, female African lions often hunt together to prey on zebras (Figure 3-7, p. 57), wildebeest, antelopes, and other fast-running large animals of the open savanna grasslands.

Other predators use *camouflage* to hide in plain sight and ambush their prey. For example, praying mantises (see Figure 4-A, right, p. 80) sit on flowers or plants of a color similar to their own and ambush visiting insects. White ermines (a type of weasel), snowy owls, and arctic foxes (Figure 5-5) hunt their prey in snow-covered areas.

SCIENCE FOCUS 5.1

THREATS TO KELP FORESTS

A kelp forest is composed of large concentrations of a seaweed called *giant kelp.* Anchored to the ocean floor, its long blades grow toward the sunlit surface waters (Figure 5-1, right). Under good conditions, the blades can grow 0.6 meter (2 feet) in a day and the plant can grow as high as a ten-story building. The blades are very flexible and can survive all but the most violent storms and waves.

Kelp forests are one of the most biologically diverse ecosystems found in marine waters, supporting large numbers of marine plants and animals. These forests help reduce shore erosion by blunting the force of incoming waves and trapping some of the outgoing sand.

Kelp plants are prey to sea urchins (Figure 5-A). Large populations of these predators can rapidly devastate a kelp forest because they eat the bases of young kelp plants. Scientific studies by biologists including James Estes of the University of California at Santa Cruz indicate that the southern sea otter is a keystone species (**Core Case Study**) that helps to sustain kelp forests by controlling populations of sea urchins. An adult southern sea otter (Figure 5-1, left) can eat as many as 1,500 sea urchins a day. That is equivalent to about 200 quarter-pound hamburgers a day for a 70-kilogram (150-pound) person. Also, a 2012 study by Estes and ecologist Chris Wilmers indicated that kelp forests absorb 12 times more CO_2 when otters are present than when they are not.

A second threat to kelp forests is polluted water running off the land and into the coastal waters where kelp forests grow. The pollutants in this runoff include pesticides and herbicides, which can kill kelp plants and other kelp forest species and upset the food webs in these aquatic forests. Another runoff pollutant is fertilizer. Its plant nutrients (mostly nitrates) can cause excessive growth of algae and other plants, which block some of the sunlight needed to support the growth of giant kelp.

Deborah Meeks/SuperStock

Figure 5-A This purple sea urchin inhabits the coastal waters of the U.S. state of California and feeds on kelp.

Critical Thinking

List three ways to protect giant kelp forests and southern sea otters.

Paul Nicklen/National Geographic Creative

Figure 5-5 A white arctic fox hunts its prey by blending into its snowy background to try to avoid being detected.

People camouflage themselves to hunt wild game and use camouflaged traps to capture wild animals. Some predators use *chemical warfare* to attack their prey. For example, some spiders and poisonous snakes use venom to paralyze their prey and to deter their predators.

Prey species have evolved many ways to avoid predators, including abilities to run, swim, or fly fast, and some have highly developed senses of sight, sound, or smell that alert them to the presence of predators. Other avoidance adaptations include protective shells (as on armadillos and turtles), thick bark (on giant sequoias), spines (on porcupines), and thorns (on cacti and rose bushes). Many lizards have brightly colored tails that break off when they are attacked, often giving them enough time to escape.

Other prey species use the camouflage of certain shapes or colors. Some insect species have shapes that look like twigs (Figure 5-6a), or bird droppings on leaves. A leaf insect can be almost invisible against its background (Figure 5-6b), as can an arctic hare in its white winter fur.

Chemical warfare is another common strategy. Some prey species discourage predators with chemicals that are *poisonous* (oleander plants), *irritating* (stinging nettles and bombardier beetles, Figure 5-6c), *foul smelling* (skunks and stinkbugs), or *bad tasting* (buttercups and monarch butterflies, Figure 5-6d). When attacked, some species of squid and octopus emit clouds of black ink, allowing them to escape by confusing their predators.

Many bad-tasting, bad-smelling, toxic, or stinging prey species have evolved *warning coloration*, brightly colored advertising that helps experienced predators to recognize and avoid them. They flash a warning: "Eating me is risky." Examples are the brilliantly colored, foul-tasting monarch butterflies (Figure 5-6d) and poisonous frogs (Figure 5-6e and Figure 4-11, p. 88). When a bird such as a blue jay eats a monarch butterfly, it usually vomits and learns to avoid them.

CONSIDER THIS. . .

CONNECTIONS Coloration and Dangerous Species

Biologist Edward O. Wilson gives us two rules, based on coloration, for evaluating possible danger from any unknown animal species we encounter in nature. *First,* if it is small and strikingly beautiful, it is probably poisonous. *Second,* if it is strikingly beautiful and easy to catch, it is probably deadly.

Some butterfly species gain protection by looking and acting like other, more dangerous species, a protective device known as *mimicry*. For example, the nonpoisonous viceroy butterfly (Figure 5-6f) mimics the monarch butterfly. Other prey species use *behavioral strategies* to avoid predation. Some attempt to scare off predators by puffing up (blowfish), spreading their wings (peacocks), or mimicking a predator (Figure 5-6h). Some moths have wings that look like the eyes of much larger animals (Figure 5-6g). Other prey species gain some protection by living in large groups such as schools of fish and herds of antelope.

At the individual level, members of the predator species benefit from their predation and members of the prey species are harmed. At the population level, predation plays a role in evolution by natural selection. Animal predators, for example, tend to kill the sick, weak, aged, and least fit members of a prey population because they are the easiest to catch. Individuals with better defenses against predation thus tend to survive longer and leave more offspring with adaptations that can help them avoid predation.

Figure 5-6 These prey species have developed specialized ways to avoid their predators: (a, b) *camouflage,* (c, d, e) *chemical warfare,* (d, e, f) *warning coloration,* (f) *mimicry,* (g) *deceptive looks,* and (h) *deceptive behavior.*

©Minden Pictures/SuperStock

Figure 5-7 *Coevolution:* This bat is using ultrasound to hunt a moth. As the bats evolve traits to increase their chance of getting a meal, the moths evolve traits to help them avoid being eaten.

©Great Lakes Fishery Commission

Figure 5-8 *Parasitism:* This blood-sucking, parasitic sea lamprey has attached itself to an adult lake trout from the Great Lakes (USA, Canada).

Some people view certain animal predators with contempt. When a hawk tries to capture and feed on a rabbit, some root for the rabbit. Yet the hawk, like all predators, is merely trying to get enough food for itself and its young. In doing so, it plays an important ecological role in controlling rabbit populations.

Interactions between Predator and Prey Species Can Drive Each Other's Evolution

Predator and prey populations can exert intense natural selection pressures on one another. Over time, as a prey species develops traits that make it more difficult to catch, its predators face selection pressures that favor traits increasing their ability to catch their prey. Then the prey species must get better at eluding the more effective predators.

When populations of two different species interact in such a way over a long period of time, changes in the gene pool of one species can lead to changes in the gene pool of the other. Such changes can help both competing species to become more competitive or to avoid or reduce competition. Biologists call this natural selection process **coevolution**.

For example, bats prey on certain species of moths (Figure 5-7), and they hunt at night using *echolocation* to navigate and to locate their prey. They emit pulses of extremely high-frequency, high-intensity sound that bounce off objects, and they capture the returning echoes that tell them where their prey is located. As a countermeasure, certain moth species have evolved ears that are especially sensitive to the sound frequencies that bats use to find them. When they hear the bat frequencies, they try to escape by dropping to the ground or flying evasively. Some bat species have evolved ways to counter this defense by changing the frequency of their sound pulses. In turn, some moths have evolved their own high-frequency clicks to jam the bats' echolocation systems. Some bat species have then adapted by turning off their echolocation systems and using the moths' clicks to locate their prey.

Some Species Feed off Other Species by Living on or inside Them

Parasitism occurs when one species (the *parasite*) feeds on another organism (the *host*), usually by living on or inside the host. In this relationship, the parasite benefits and the host is often harmed but not immediately killed.

A parasite usually is much smaller than its host and rarely kills it. However, most parasites remain closely associated with their hosts, draw nourishment from them, and may gradually weaken them over time.

Some parasites such as tapeworms live inside their hosts. Other parasites such as mistletoe plants and blood-sucking sea lampreys (Figure 5-8) attach themselves to the outsides of their hosts. Some parasites, including fleas and ticks, move from one host to another while others, such as tapeworms, spend their adult lives within a single host.

From the host's point of view, parasites are harmful. But from the population perspective, parasites can promote biodiversity by helping to keep the populations of their hosts in check.

In Some Interactions, Both Species Benefit

In **mutualism**, two species behave in ways that benefit both by providing each with food, shelter, or some other resource. One example is pollination of flowering plants by species such as honeybees, hummingbirds, and butterflies (see Chapter 4 opening photo, p. 76) that feed on the nectar of flowers.

Figure 5-9 shows an example of a mutualistic relationship that combines *nutrition* and *protection*. It involves

Villiers/Dreamstime.com

Figure 5-9 *Mutualism:* Oxpeckers feed on parasitic ticks that infest animals such as this impala and warn of approaching predators.

Frans Lanting/National Geographic Creative

Figure 5-10 In an example of *commensalism,* this pitcher plant is attached to a branch of a tree without penetrating or harming the tree. This carnivorous plant feeds on insects that become trapped inside it.

birds that ride on the backs of large animals such as African buffalo, elephants, rhinoceroses, and impalas (Figure 5-9). The birds remove and eat parasites and pests (such as ticks and flies) from the animals' bodies and often make noises warning the larger animals when predators are approaching.

Another example of mutualism involves clownfish, which usually live within sea anemones (see chapter-opening photo), whose tentacles sting and paralyze most fish that touch them. The clownfish, which are not harmed by the tentacles, gain protection from predators and feed on the waste matter left from the anemones' meals. The sea anemones benefit because the clownfish protect them from some of their predators and parasites.

In *gut inhabitant mutualism,* armies of bacteria in the digestive systems of animals help to break down (digest) the animals' food. In turn, the bacteria receive a sheltered habitat and food from their hosts. Hundreds of millions of bacteria in your gut secrete enzymes that help you digest the food you eat.

It is tempting to think of mutualism as an example of cooperation between species. In reality, the species in a mutualistic interaction benefit one another unintentionally and are in it for themselves.

In Some Interactions, One Species Benefits and the Other Is Not Harmed

Commensalism is an interaction that benefits one species but has little, if any, beneficial or harmful effect on the other. One example involves plants called *epiphytes* (air plants), which attach themselves to the trunks or branches of trees (Figure 5-10) in tropical and subtropical forests. Epiphytes benefit by having a solid base on which to grow. They also live in an elevated spot that gives them better access to sunlight, water from the humid air and rain, and nutrients falling from the tree's upper leaves and limbs. Their presence apparently does not harm the tree. Similarly, birds benefit by nesting in trees, generally without harming them.

5-2 How Do Communities and Ecosystems Respond to Changing Environmental Conditions?

CONCEPT 5-2
The structure and species composition of communities and ecosystems change in response to changing environmental conditions through a process called *ecological succession.*

Communities and Ecosystems Change over Time: Ecological Succession

The types and numbers of species in biological communities and ecosystems change in response to changing environmental conditions such as a fires, volcanic eruptions, climate change, and the clearing of forests to plant crops. The normally gradual change in species composition in a given area is called **ecological succession** (Concept 5-3).

Ecologists recognize two main types of ecological succession, depending on the conditions present at the beginning of the process. **Primary ecological succession** involves the gradual establishment of communities of different species in lifeless areas where there is no soil in a terrestrial ecosystem or no bottom sediment in an aquatic ecosystem. Examples include bare rock exposed by a retreating glacier (Figure 5-11), newly cooled lava, an abandoned highway or parking lot, and a newly created shallow pond or reservoir. Primary succession usually takes hundreds to thousands of years because of the need to build up fertile soil or aquatic sediments to provide the nutrients needed to establish a plant community.

The other, more common type of ecological succession is called **secondary ecological succession**, in which a series of communities or ecosystems with different species develop in places containing soil or bottom sediment. This type of succession begins in an area where an ecosystem has been disturbed, removed, or destroyed, but some soil or bottom sediment remains. Candidates for secondary succession include abandoned farmland (Figure 5-12), burned or cut forests, heavily polluted streams, and flooded land. Because some soil or sediment is present, new vegetation can begin to grow, usually within a few weeks. It begins with the germination of seeds already in the soil and seeds imported by wind or in the droppings of birds and other animals.

Ecological succession is an important ecosystem service that tends to increase the biodiversity of communities and ecosystems by increasing species richness and interactions among species. Such interactions in turn enhance sustainability by promoting population control and by increasing the complexity of food webs, which enhances energy flow and nutrient cycling. As part of the earth's natural capital, both primary and secondary ecological succession are examples of *natural ecological restoration.*

Ecologists have identified three factors that affect how and at what rate succession occurs. One is *facilitation,* in which one set of species makes an area suitable for spe-

Figure 5-11 *Primary ecological succession:* Over almost a 1,000 years, these plant communities developed, starting on bare rock exposed by a retreating glacier on Isle Royal, Michigan (USA) in western Lake Superior. The details of this process vary from one site to another.

Animated Figure 5-12 *Natural ecological restoration of disturbed land:* This diagram shows the undisturbed secondary ecological succession of plant communities on an abandoned farm field in the U.S. state of North Carolina. It took 150–200 years after the farmland was abandoned for the area to become covered with a mature oak and hickory forest.

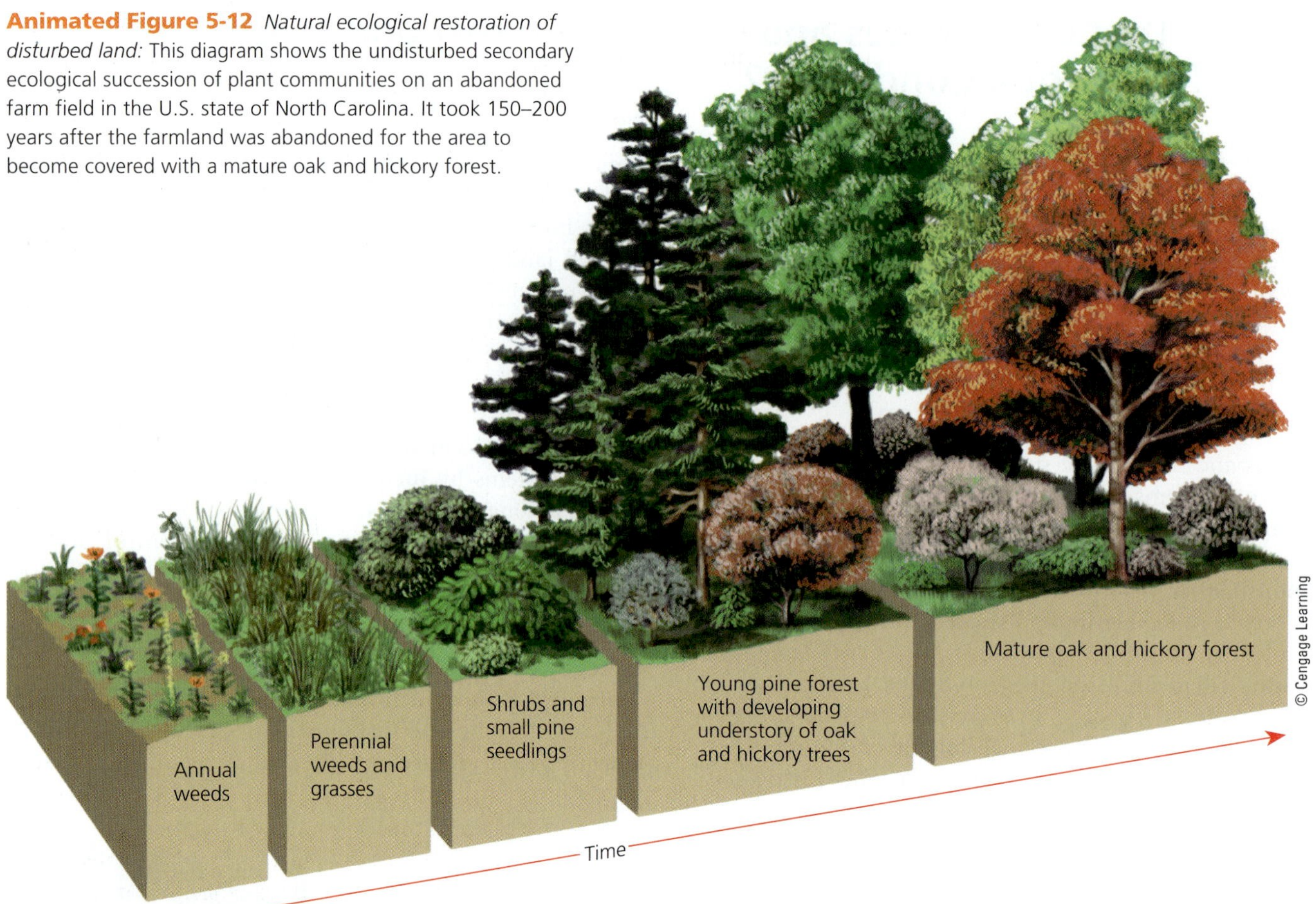

cies with different niche requirements, and often less suitable for itself. For example, as lichens and mosses gradually build up soil on a rock in primary succession, herbs and grasses can move in and crowd out the lichens and mosses.

A second factor is *inhibition,* in which some species hinder the establishment and growth of other species. For example, the needles dropping off some pine trees make the soil beneath the trees too acidic for most other plants to grow there. A third factor is *tolerance,* in which plants in the late stages of succession succeed because they are not in direct competition with other plants for key resources. Shade-tolerant plants, for example, can live in shady forests because they do not need as much sunlight as the trees above them do.

Ecological Succession Does Not Follow a Predictable Path

According to the traditional view, ecological succession proceeds in an orderly sequence along an expected path until a certain stable type of *climax community* occupies an area. On land, such a community is dominated by a few long-lived plant species, often within a mature forest (Figures 5-11 and 5-12), and is in balance with its environment. This equilibrium model of succession is what ecologists once meant when they talked about the *balance of nature.*

Over the last several decades, many ecologists have changed their views about balance and equilibrium in nature. There is a general tendency for succession to lead to more complex, diverse, and presumably sustainable ecosystems. However, a close look at almost any terrestrial community or ecosystem reveals that it consists of an ever-changing mosaic of patches of vegetation in different stages of succession. The current scientific view is that we cannot predict a given course of succession or view it as inevitable progress toward an ideally adapted climax plant community or ecosystem. Rather, ecological succession reflects the ongoing struggle by different species for enough light, water, nutrients, food, space, and other key resources. Most ecologists now recognize that mature, late-successional ecosystems are in a state of continual disturbance and change.

Living Systems Are Sustained through Constant Change

All living systems, from a cell to the biosphere, are constantly changing in response to changing environmental conditions. However, living systems contain complex processes that interact to provide some degree of stability, or sustainability, over each system's expected life span. This *stability,* or capacity to withstand external stress and disturbance, is maintained only by constant change in response to changing environmental conditions. For example, in a mature tropical rain forest, some trees die and others take their places. However, unless the forest is cut, burned, or otherwise destroyed, you would still recognize it as a tropical rain forest 50 or 100 years from now.

It is useful to distinguish between two aspects of stability or sustainability in living systems. One is **inertia**, or **persistence**: the ability of a living system such as a grassland or a forest to survive moderate disturbances. A second factor is **resilience**: the ability of a living terrestrial system to be restored through secondary ecological succession after a more severe disturbance.

Evidence suggests that some ecosystems have one of these properties but not the other. For example, tropical rain forests have high species richness and high inertia and thus are resistant to significant change or damage. But once a large tract of tropical rain forest is cleared or severely damaged, the resilience of the resulting degraded forest ecosystem may be so low that it reaches an ecological tipping point after which it may not be restored by secondary ecological succession. One reason for this is that most of the nutrients in a typical rain forest are stored in its vegetation, not in the topsoil, as in most other terrestrial ecosystems. Once the nutrient-rich vegetation is gone, daily rains can remove most of the remaining soil nutrients and thus prevent the return of a tropical rain forest on a large cleared area.

By contrast, grasslands are much less diverse than most forests, and consequently they have low inertia and can burn easily. However, because most of their plant matter is stored in underground roots, these ecosystems have high resilience and can recover quickly after a fire, as their root systems produce new grasses. Grassland can be destroyed only if its roots are plowed up and something else is planted in its place, or if it is severely overgrazed by livestock or other herbivores.

5-3 What Limits the Growth of Populations?

> **CONCEPT 5-3**
> No population can grow indefinitely because of limitations on resources and because of competition among species for those resources.

Most Populations Live in Clumps

A **population** is a group of interbreeding individuals of the same species (Figure 5-13). Figure 5-14 shows three ways in which the members of a population are typically distributed or dispersed in their habitat. Most populations live together in *clumps* (Figure 5-14a) such as packs of wolves, schools of fish (Figure 5-13), and flocks of birds. Southern sea otters (**Core Case Study**) (Figure 5-1, left), for example, are usually found in groups known as rafts or pods ranging in size from a few to several hundred animals.

Why clumps? Several reasons: *First,* the resources a species needs vary greatly in availability from place to place, so the species tends to cluster where the resources are available. *Second,* individuals moving in groups have a better chance of encountering patches or clumps of resources, such as water and vegetation, than they would searching for the resources on their own. *Third,* living in groups provides some protection from predators. *Fourth,* living in packs gives some predator species a better chance of getting a meal.

Rich Carey/iStockphoto.com

Figure 5-13 A population, or school, of Anthias fish on coral in Australia's Great Barrier Reef.

a. Clumped (elephants)

b. Uniform (creosote bush)

c. Random (dandelions)

Figure 5-14 Three general habitat *dispersion patterns* for a population's individuals.

Left: EcoPrint/Shutterstock.com. Center: kenkistler/Shutterstock.com. Right: Nataly Lukhanina/Shutterstock.com.

Populations Can Grow, Shrink, or Remain Stable

Four variables—*births, deaths, immigration,* and *emigration*—govern changes in population size. A population increases through birth and immigration (arrival of individuals from outside the population) and decreases through death and emigration (departure of individuals from the population):

Population change = (Births + Immigration) − (Deaths + Emigration)

A population's **age structure**—its distribution of individuals among various age groups—can have a strong effect on how rapidly it grows or declines. Age groups are usually described in terms of organisms not mature enough to reproduce (the *prereproductive stage*), those capable of reproduction (the *reproductive stage*), and those too old to reproduce (the *postreproductive stage*).

The size of a population will likely increase if it is made up mostly of individuals in their reproductive stage, or soon to enter this stage. In contrast, a population dominated by individuals in their postreproductive stage will tend to decrease over time.

Some Factors Can Limit Population Size

Different species and their populations thrive under different physical and chemical conditions. Some need bright sunlight; others flourish in shade. Some need a hot environment; others prefer a cool or cold one. Some do best under wet conditions; others thrive in dry conditions.

Each population in an ecosystem has a **range of tolerance** to variations in its physical and chemical environment, as shown in Figure 5-15. Individuals within a population may also have slightly different tolerance ranges for temperature or other physical or chemical factors because of small differences in their genetic makeup, health, and age. For example, a trout population may do best within a narrow band of temperatures (*optimum level* or *range*), but a few individuals can survive above and below that band. Of course, if the water becomes much too hot or too cold, none of the trout can survive.

A number of physical or chemical factors can help to determine the number of organisms in a population. Sometimes one or more factors, known as **limiting factors**, are more important than other factors in regulating population growth. This ecological principle is called the **limiting factor principle**: *Too much or too little of any physical or chemical factor can limit or prevent the growth of a population, even if all other factors are at or near the optimal range of tolerance.*

On land, precipitation often is the limiting factor. For example, low precipitation levels in desert ecosystems limit desert plant growth. Soil nutrients also can act as a limiting factor on land. Suppose a farmer plants corn in phosphorus-poor soil. Even if water, nitro-

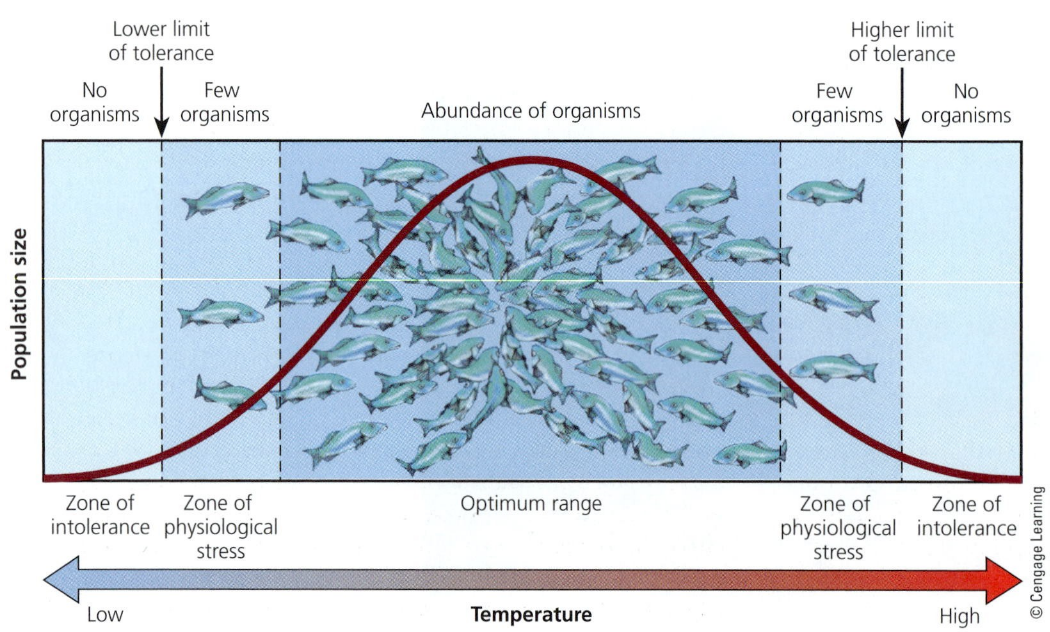

Figure 5-15 Range of tolerance for a population of trout to changes in water temperature.

gen, potassium, and other nutrients are at optimal levels, the corn will stop growing when it uses up the available phosphorus.

Too much of a physical or chemical factor can also be limiting. For example, too much water or fertilizer can kill plants. Temperature can be a limiting factor as well. Both high and low temperatures can limit the survival and population sizes of various terrestrial species, especially plants and many insect species.

Important limiting physical factors for populations in *aquatic life zones*, or water-filled areas that support life, include temperature (Figure 5-15), sunlight, nutrient availability, acidity, and the low levels of oxygen gas in the water *(dissolved oxygen content)*. Another limiting factor in aquatic life zones is *salinity*—the amounts of various inorganic minerals or salts dissolved in a given volume of water.

Another factor that can limit the sizes of some populations is **population density**, the number of individuals in a population found within a defined area or volume. Some limiting factors become more important as a population's density increases. For example, in a dense population, parasites and diseases can spread more easily and have the effect of controlling population growth. On the other hand, a higher population density can help sexually reproducing individuals to find mates more easily in order to produce offspring and increase the size of a population.

Different Species Have Different Reproductive Patterns

Different reproductive patterns help to ensure the long-term survival of species. Some species have many, usually small, offspring and give them little or no parental care or protection. Examples include algae, bacteria, and most insects. These species reproduce at an early age and overcome typically massive losses of offspring by producing so many offspring that a few will likely survive to reproduce many more offspring and keep this reproductive pattern going.

At the other extreme are species that tend to reproduce later in life and have a small number of offspring with longer life spans. Typically, the offspring of mammals with this reproductive strategy develop inside their mothers (where they are safe), and are born fairly large. After birth, they mature slowly and are cared for and protected by one or both parents, and in some cases by living in herds or groups, until they reach reproductive age and begin the cycle again.

Most large mammals (such as elephants, whales, and humans) follow this reproductive pattern. Many of these species—especially those with long times between generations and with low reproductive rates, such as elephants, rhinoceroses, and sharks—are vulnerable to extinction. Most organisms have reproductive patterns between these two extremes.

No Population Can Grow Indefinitely: J-Curves and S-Curves

Some species have an incredible ability to increase their numbers. Members of such populations typically reproduce at an early age, have many offspring each time they reproduce, and reproduce many times, with short intervals between successive generations. For example, with no controls on its population growth, a species of bacteria that can reproduce every 20 minutes would generate enough offspring to form a 0.3-meter-deep (1-foot-deep) layer over the surface of the entire earth in only 36 hours—a dramatic illustration of the potential of exponential growth.

Fortunately, this will not happen. Research reveals that regardless of their reproductive strategy, no population of a species can grow indefinitely because of limitations on resources and competition with populations of other species for those resources (**Concept 5-2**). In the real world, a rapidly growing population of any species eventually reaches some size limit imposed by one or more limiting factors such as the availability of light, water, temperature, space, or nutrients, or by exposure to predators or infectious diseases.

There are always limits to population growth in nature. For example, one reason California's southern sea otters (**Core Case Study**) face extinction is that they cannot reproduce rapidly (Science Focus 5.2).

Environmental resistance is the combination of all factors that act to limit the growth of a population. It largely determines an area's **carrying capacity**: the maximum population of a given species that a particular habitat can sustain indefinitely. The growth rate of a population decreases as its size nears the carrying capacity of its environment because resources such as food, water, and space begin to dwindle.

A population with few, if any, limitations on its resource supplies can grow exponentially at a fixed rate or percentage per year. Exponential growth starts slowly but then accelerates as the population increases, because as its base size increases, so does the number added to the population each year. Plotting the number of individuals against time yields a J-shaped growth curve. Figure 5-16 shows such a curve for sheep on an island south of Australia in the early 19th century.

Exponential growth (left third of the curve in Figure 5-16) occurs when a population has essentially unlimited resources to support its growth. Such exponential growth is eventually converted to *logistic growth* (center of the curve in Figure 5-16), in which the growth rate decreases as the population becomes larger and faces environmental resistance. Over time, the population size typically stabilizes at or near the *carrying capacity* of its environment, which results in a sigmoid (S-shaped) population growth curve. Depending on resource availability, the size of a population often fluctuates around its carrying capacity (right third of the curve in Figure 5-16). Also, unlike in

SCIENCE FOCUS 5.2

WHY DO CALIFORNIA'S SOUTHERN SEA OTTERS FACE AN UNCERTAIN FUTURE?

The southern sea otter (**Core Case Study**) cannot rapidly increase its numbers for several reasons. Female southern sea otters reach sexual maturity between 2 and 5 years of age, can reproduce until age 15, and typically each produce only one pup a year.

The population size of southern sea otters has fluctuated in response to changes in environmental conditions. One such change has been a rise in populations of the orcas (killer whales) that feed on them. Scientists hypothesize that orcas began feeding more on southern sea otters when populations of their normal prey, sea lions and seals, began declining. In 2010 and 2011, the number of southern sea otters killed or injured by sharks increased for reasons that scientists are trying to understand.

Another factor may be parasites known to breed in the intestines of cats. Scientists hypothesize that some southern sea otters may be dying because coastal area cat owners flush feces-laden cat litter down their toilets or dump it in storm drains that empty into coastal waters. The feces contain parasites that then infect the otters.

Otters are also being threatened by blooms of toxic algae that are fed by urea, a key ingredient in fertilizer that washes into coastal waters. Other pollutants released by human activities are PCBs and other fat-soluble toxic chemicals that can accumulate in the tissues of the shellfish on which otters feed, and this proves fatal to otters. The facts that southern sea otters feed at high trophic levels and live close to the shore make them vulnerable to these and other pollutants in coastal waters. Scientists note that as an indicator species, the southern sea otter is revealing the degraded condition of its coastal water habitat.

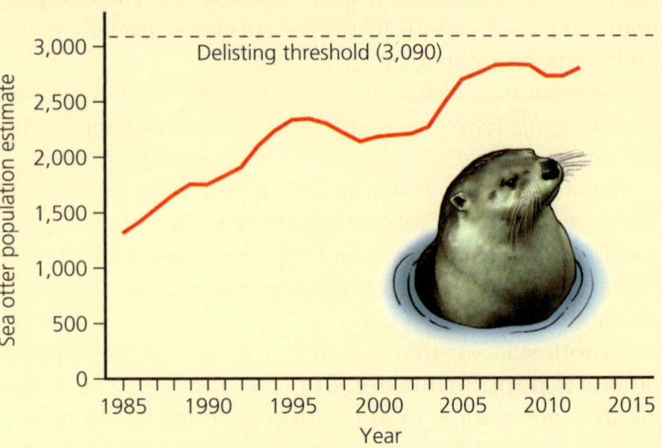

Figure 5-B Changes in the population size of southern sea otters off the coast of the U.S. state of California, 1983–2012.
(Compiled by the authors using data from U.S. Geological Survey.)

Some southern sea otters die when they encounter oil spilled from ships. The entire California southern sea otter population could be wiped out by a large oil spill from a single tanker off the state's central coast or by the rupture of an offshore oil well, should drilling for oil be allowed off this coast.

The factors listed here, mostly resulting from human activities, plus a fairly low reproductive rate, have hindered the ability of the endangered southern sea otter to rebuild its population (Figure 5-B). According to the U.S. Geological Survey, the California southern sea otter population would have to reach at least 3,090 for 3 years in a row before it could be considered for removal from the endangered species list.

Critical Thinking

How would you design a controlled experiment to test the hypothesis that cat litter flushed down toilets might be killing southern sea otters?

---

the simple model in Figure 5-16, the carrying capacity can vary over time due to changing conditions.

Changes in habitat or other environmental conditions can reduce the populations of some species while increasing the populations of others (see the following Case Study).

CASE STUDY

Exploding White-Tailed Deer Populations in the United States

By 1900, habitat destruction and uncontrolled hunting had reduced the white-tailed deer (Figure 5-17) population in the United States to about 500,000 animals. In the 1920s and 1930s, laws were passed to protect the remaining deer. Hunting was restricted and predators, including wolves and mountain lions that preyed on the deer, were nearly eliminated.

These protections worked, and for some suburbanites and farmers, perhaps too well. Today there are over 25 million white-tailed deer in the United States. During the last 50 years, large numbers of Americans have moved into the wooded habitat of deer where suburban areas have expanded. By gardening and landscaping around their homes, people have provided deer with flowers, shrubs, garden crops, and other plants that they like to eat.

Deer prefer to live in the edge areas of forests and woodlots for security, and they go to nearby fields,

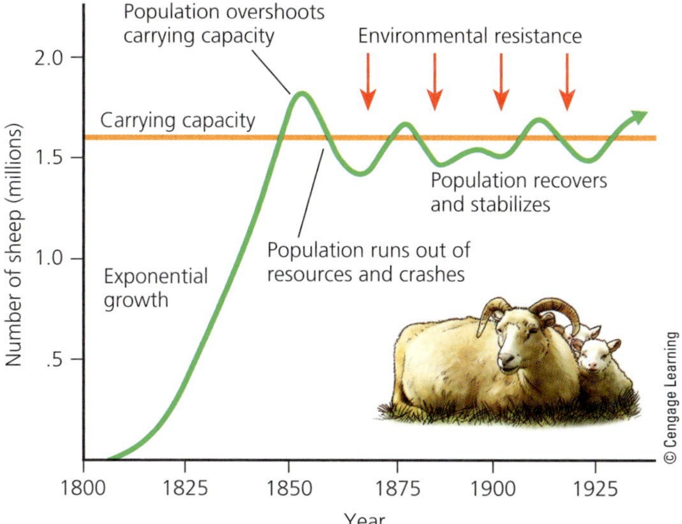

Animated Figure 5-16 Growth of a sheep population on the island of Tasmania between 1800 and 1925. After sheep were introduced in 1800, their population grew exponentially thanks to an ample food supply and few predators. By 1855, they had overshot the land's carrying capacity. Their numbers then stabilized and fluctuated around a carrying capacity of about 1.6 million sheep.

Figure 5-17 White-tailed deer populations in the United States have been growing.

orchards, lawns, and gardens for food. Thus, a suburban neighborhood can be an all-you-can-eat paradise for white-tailed deer. Their populations in such areas have soared. In some woodlands, they are consuming native ground-cover vegetation and this has allowed nonnative weed species to take over. The deer also help to spread Lyme disease (carried by deer ticks) to humans.

In addition, each year about 1.5 million deer–vehicle collisions injure at least 10,000 Americans and kill at least 200—the highest of all death tolls resulting from encounters with wild animals. By comparison, between 2000 and 2010, deer–vehicle collisions killed about 2,000 Americans, compared to 28 killed by bears, and 10 killed by sharks.

There are no easy answers to the deer population problem in the suburbs. Changes in hunting regulations that allow for the killing of more female deer have cut down the overall deer population. However, this has had a limited effect on deer populations in suburban areas, because it is too dangerous to allow widespread hunting with guns in such populated communities. Some areas have hired experienced and licensed archers who use bows and arrows to help reduce deer numbers. To protect nearby residents the archers hunt from elevated tree stands and only shoot their arrows downward.

Some communities spray the scent of deer predators or of rotting deer meat in edge areas to scare off deer. Others use electronic equipment that emits high-frequency sounds, which humans cannot hear, for the same purpose. Some home owners surround their gardens and yards with high, black plastic mesh fencing that is invisible from a distance.

Deer can be trapped and moved from one area to another, but this is expensive and must be repeated whenever they move back into an area. Also, there are questions concerning where to move the deer and how to pay for such programs.

We can also put deer on birth control. Darts loaded with contraceptives can be shot into female deer to hold down their birth rates. However, this is expensive and must be repeated every year. One possibility is an experimental, single-shot contraceptive vaccine that lasts for several years. Another approach is to trap dominant males and use chemical injections to sterilize them. Both of these approaches will require years of testing.

Meanwhile, suburbanites can expect deer to chow down on their shrubs, flowers, and garden plants unless they can protect their properties with high, deer-proof fences, repellents, or other methods. Deer have to eat every day just as we do. Suburban dwellers could consider not planting trees, shrubs, and flowers that attract deer around their homes.

CONSIDER THIS. . .

THINKING ABOUT White-Tailed Deer

Some people blame the white-tailed deer for invading farms and suburban yards and gardens to eat food that humans have made easily available to them. Others say humans are mostly to blame because they have invaded deer territory, eliminated most of the predators that kept deer populations under control, and provided the deer with plenty to eat in their lawns, gardens, and crop fields. Which view do you hold? Why? Do you see a solution to this problem?

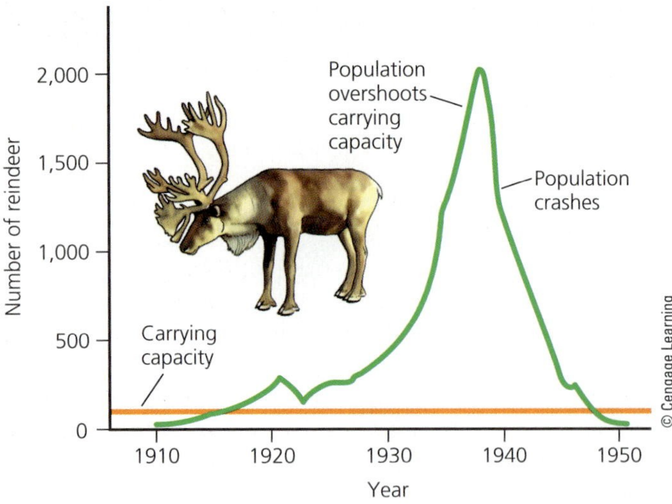

Figure 5-18 Exponential growth, overshoot, and population crash of reindeer introduced onto the small Bering Sea island of St. Paul in 1910.

When a Population Exceeds Its Carrying Capacity It Can Crash

Some populations do not make a smooth transition from exponential growth to logistic growth. Instead, they use up their resource supplies and temporarily *overshoot*, or exceed, the carrying capacity of their environment. In such cases, the population suffers a sharp decline, called *dieback*, or **population crash**, unless part of the population can switch to new resources or move to an area that has more resources. Such a crash occurred when reindeer were introduced onto a small island in the Bering Sea (Figure 5-18).

The carrying capacity of any given area is not fixed. In some areas, it can rise or decline seasonally and from year to year because of variations in weather, such as a drought that causes decreases in available vegetation. Other factors include the presence or absence of predators and an abundance or scarcity of competitors.

Humans Are Not Exempt from Nature's Population Controls

Humans are not exempt from population crashes. Ireland experienced such a crash after a fungus destroyed its potato crop in 1845. About 1 million people died from hunger or diseases related to malnutrition, and 3 million people migrated to other countries, especially the United States.

During the 14th century, the *bubonic plague* spread through densely populated European cities and killed at least 25 million people. The bacterium causing this disease normally lives in rodents. It was transferred to humans by fleas that fed on infected rodents and then bit humans. The disease spread like wildfire through crowded cities, where sanitary conditions were poor and rats were abundant. Today, several antibiotics, not available until recently, can be used to treat bubonic plague.

Currently, the world is experiencing a global epidemic of AIDS, caused by infection with the human immunodeficiency virus (HIV). Between 1981 and 2011, AIDS killed more than 30 million people and continues to claim another 1.7 million lives each year—an average of 3 deaths per minute.

So far, technological, social, and other cultural changes have expanded the earth's carrying capacity for the human species. We have increased food production and used large amounts of energy and matter resources to occupy formerly uninhabitable areas, to expand agriculture, and to control the populations of other species that compete with us for resources. Some say we can keep expanding our ecological footprint in this way indefinitely, mostly because of our technological ingenuity. Others say that sooner or later, we will reach the limits that nature eventually imposes on any population that exceeds or degrades its resource base. We discuss these issues in Chapter 6.

Big Ideas

- Certain interactions among species affect their use of resources and their population sizes.
- Changes in environmental conditions cause communities and ecosystems to gradually alter their species composition and population sizes (ecological succession).
- There are always limits to population growth in nature.

TYING IT ALL TOGETHER Southern Sea Otters and Sustainability

Before the arrival of European settlers on the western coast of North America, the southern sea otter population was part of a complex ecosystem made up of kelp, bottom-dwelling creatures, whales, and other species depending on one another for survival. Giant kelp forests served as food and shelter for sea urchins. Sea otters ate the sea urchins and other kelp eaters. Some species of whales and sharks ate the otters. And detritus from all these species helped to maintain the giant kelp forests. Each of these interacting populations was kept in check by—and helped to sustain—all the others.

When European settlers arrived and began hunting the otters for their pelts, they probably didn't know much about the intricate web of life beneath the ocean surface. But with the effects of overhunting, people realized they had done more than simply take sea otters. They had torn the web, disrupted an entire ecosystem, and triggered a loss of valuable natural resources and ecosystem services, including biodiversity.

Populations of most plants and animals depend, directly or indirectly, on solar energy, and all populations play roles in the cycling of nutrients in the ecosystems where they live. In addition, the biodiversity found in the variety of species in different terrestrial and aquatic ecosystems provides alternative paths for energy flow and nutrient cycling, better opportunities for natural selection as environmental conditions change, and natural population control mechanisms. When we disrupt these paths, we violate all three scientific **principles of sustainability**. In this chapter, we focused on how the biodiversity principle promotes sustainability, provides a variety of species to restore damaged ecosystems through ecological succession, and limits the sizes of populations.

SUSTAINABILITY

Chapter Review

Core Case Study

1. Explain how southern sea otters act as a keystone species in their environment (**Core Case Study**). Explain why we should care about protecting this species from extinction that could result primarily from human activities.

Section 5-1

2. What is the key concept for this section? Define **interspecific competition**. Define and give two examples of **resource partitioning** and explain how it can increase species diversity. Define **predation** and distinguish between a **predator** species and a **prey** species and give an example of each. What is a **predator–prey relationship**?

3. Explain why we should help to preserve kelp forests. Describe three ways in which predators can increase their chances of feeding on their prey and three ways in which prey species can avoid their predators. Define and give an example of **coevolution**.

4. Define **parasitism**, **mutualism**, and **commensalism** and give an example of each. Explain how each of these species interactions, along with predation, can affect the population sizes of species in ecosystems.

Section 5-2

5. What is the key concept for this section? What is **ecological succession**? Distinguish between **primary ecological succession** and **secondary ecological succession** and give an example of each. List three factors that can affect how ecological succession occurs. Explain why succession does not follow a predictable path.

6. Explain how living systems achieve some degree of sustainability by undergoing constant change in response to changing environmental conditions. In terms of the stability of ecosystems, distinguish between **inertia (persistence)** and **resilience** and give an example of each.

Section 5-3

7. What is the key concept for this section? Define **population**. Why do most populations live in clumps? List four variables that govern changes in population size. Write an equation showing how these variables interact. What is a population's **age structure** and what are the three major age groups called? Define **range of tolerance**. Define **limiting factor** and give an example. State the **limiting factor principle**. Define **population density** and explain how some limiting factors can become more important as a population's density increases. Describe two different reproductive strategies that can enhance the long-term survival of a species.
8. Explain why no population can grow indefinitely. Distinguish between the **environmental resistance** and the **carrying capacity** of an environment, and use these concepts to explain why there are always limits to population growth in nature. Why is the recovery of southern sea otters a slow one, and what factors are threatening this recovery? Describe the exploding white-tailed deer population problem in the United States and discuss options for dealing with it.
9. Define and give an example of a **population crash**. Explain why humans are not exempt from nature's population controls.
10. What are this chapter's *three big ideas*? Explain how the interactions of plant and animal species in any ecosystem are related to the scientific **principles of sustainability**.

Note: Key terms are in bold type.

Critical Thinking

1. What difference would it make if the southern sea otter (**Core Case Study**) became extinct primarily because of human activities? What are three things we could do to help prevent the extinction of this species?
2. Use the second law of thermodynamics (see Chapter 2, p. 43) to help explain why predators are generally less abundant than their prey. In your explanation, make use of the pyramid of energy flow (see Figure 3-13, p. 61).
3. How would you reply to someone who argues that we should not worry about the effects that human activities have on natural systems because ecological succession will heal the wounds of such activities and restore the balance of nature?
4. How would you reply to someone who contends that efforts to preserve natural systems are not worthwhile because nature is largely unpredictable?
5. Explain why most species with a high capacity for population growth (such as bacteria, flies, and cockroaches) tend to have small individuals, while those with a low capacity for population growth (such as humans, elephants, and whales) tend to have large individuals.
6. Which reproductive strategy do most species of insect pests and harmful bacteria use? Why does this make it difficult for us to control their populations?
7. List three factors that have limited human population growth in the past and that we have overcome. Explain how we overcame each of these factors. List two factors that may limit human population growth in the future. Do you think that we are close to reaching those limits? Explain.
8. If the human species were to suffer a population crash, what are three species that might move in to occupy part of our ecological niche? Explain why this might happen.

Doing Environmental Science

Visit a nearby land area, such as a partially cleared or burned forest, a grassland, or an abandoned crop field, and record signs of secondary ecological succession. Take notes on your observations and formulate a hypothesis about what sort of disturbance led to this succession. Include your thoughts about whether this disturbance was natural or caused by humans. Study the area carefully to see whether you can find patches that are at different stages of succession and record your thoughts about what sorts of disturbances have caused these differences. You might want to research the topic of ecological succession in such an area.

Global Environment Watch Exercise

Search for *kelp forests,* and use the results to find sources of information about how a warmer ocean, as a result of climate change, might affect California's coastal kelp beds on which the southern sea otters depend (**Core Case Study**). Write a report on what you found. Try to include information on current effects of warmer water on the kelp beds as well as projections about future effects. Also, summarize any information you might find on possible ways to prevent harm to these kelp beds.

Data Analysis

The graph below shows changes in the size of an emperor penguin population in terms of numbers of breeding pairs on the island of Terre Adelie in the Antarctic. Use the graph to answer the questions below.

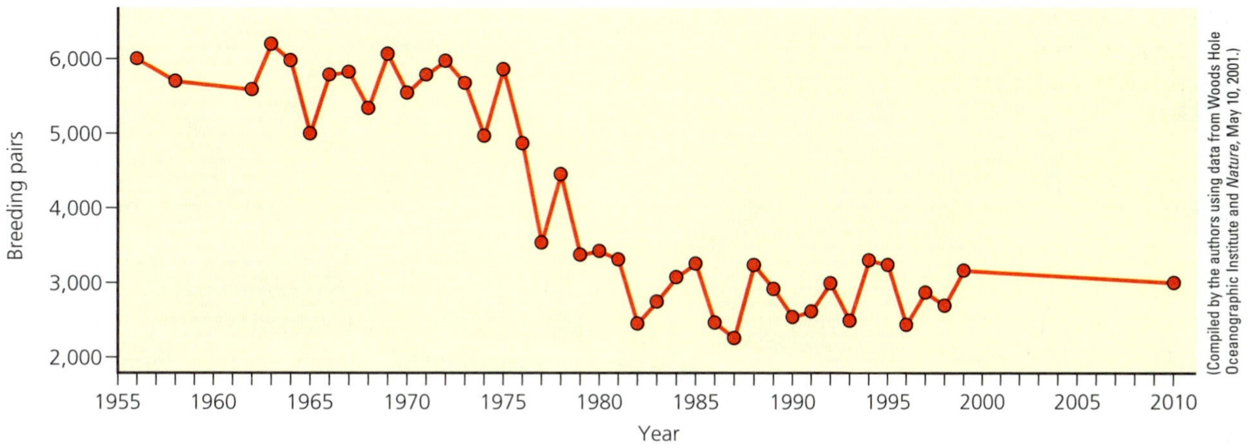

1. Assuming that the penguin population fluctuates around the carrying capacity, what was the approximate carrying capacity of the island for the penguin population from 1960 to 1975? What was the approximate carrying capacity of the island for the penguin population from 1980 to 2010?
2. What was the percentage decline in the penguin population from 1975 to 2010?

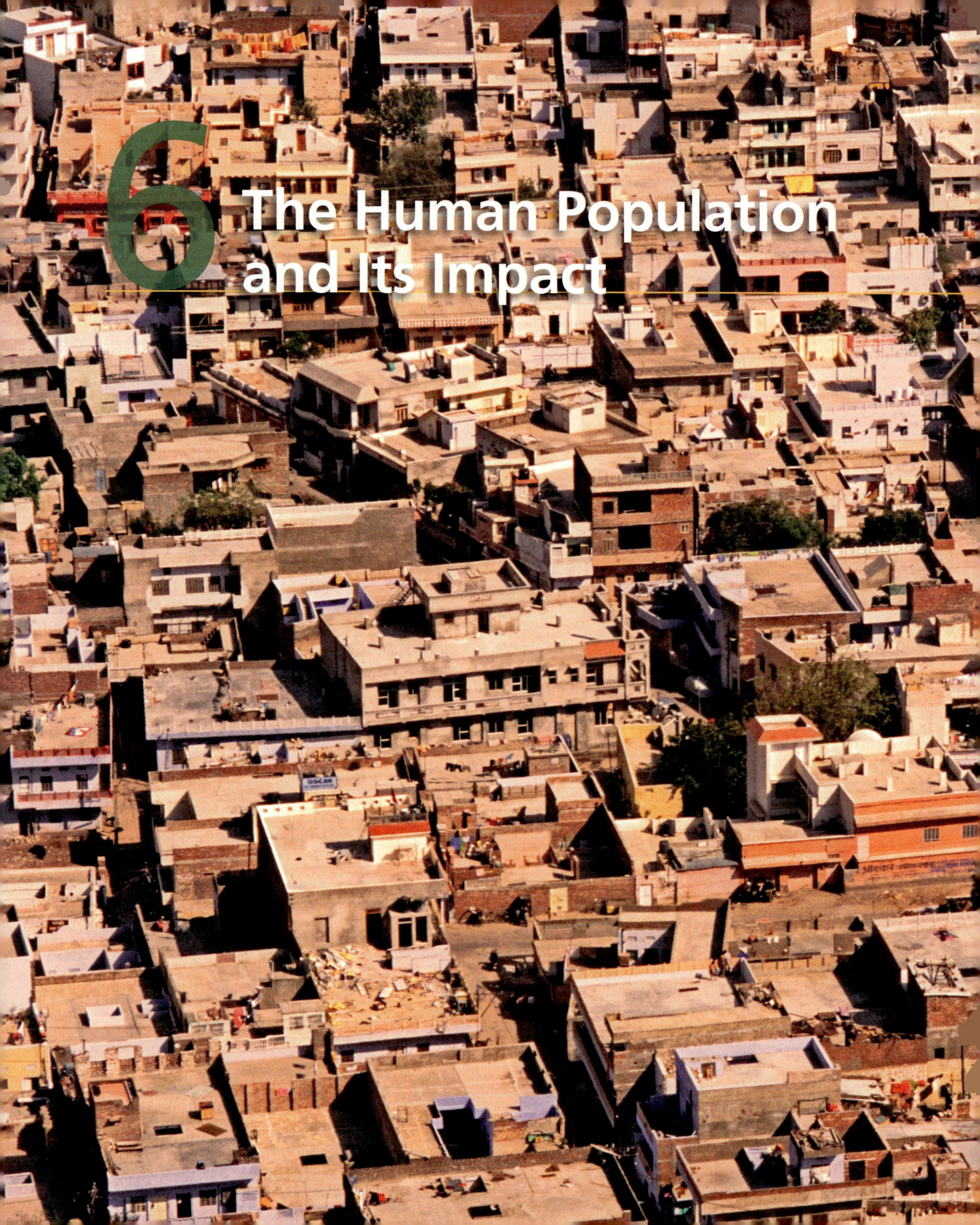

6 The Human Population and Its Impact

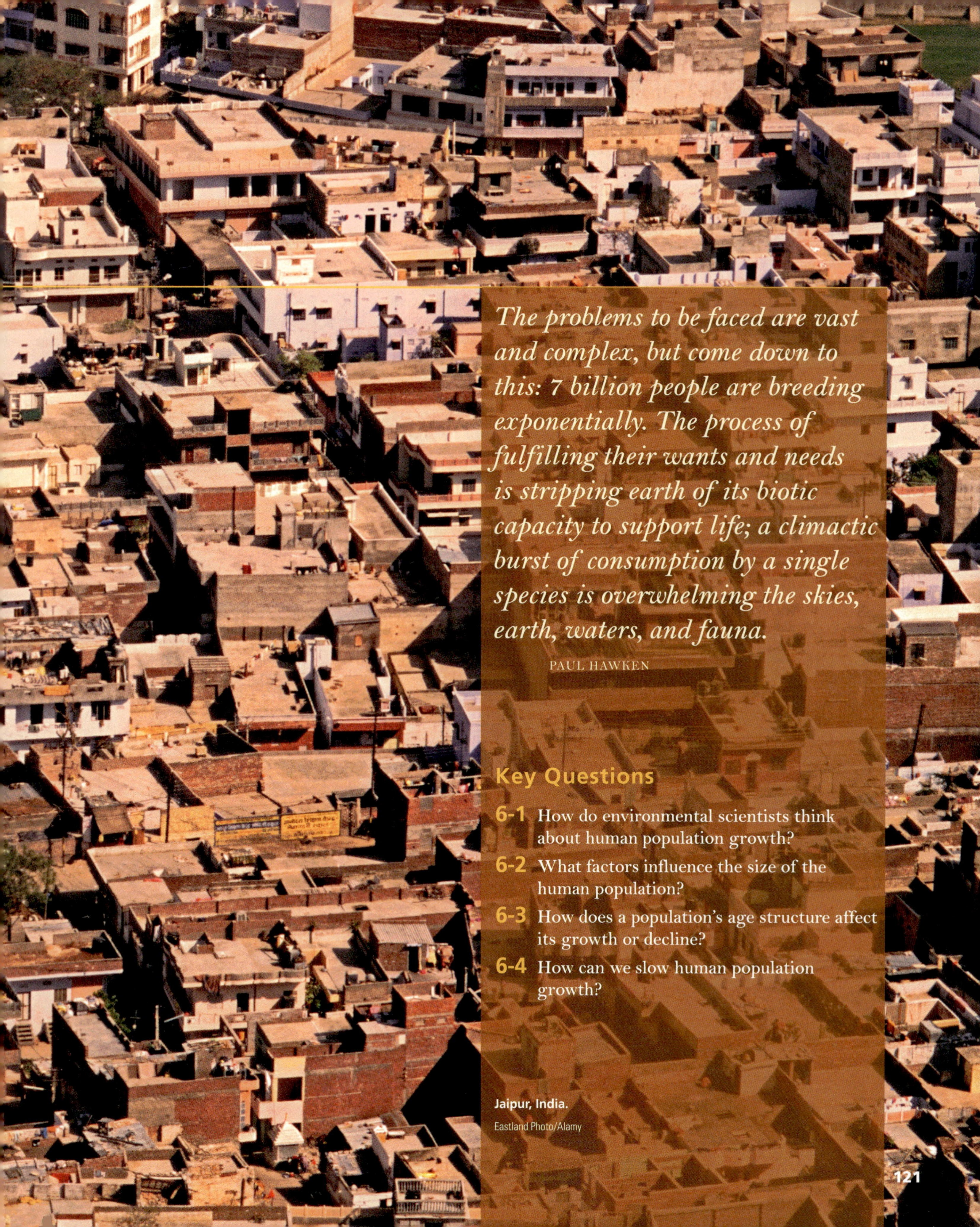

The problems to be faced are vast and complex, but come down to this: 7 billion people are breeding exponentially. The process of fulfilling their wants and needs is stripping earth of its biotic capacity to support life; a climactic burst of consumption by a single species is overwhelming the skies, earth, waters, and fauna.

PAUL HAWKEN

Key Questions

6-1 How do environmental scientists think about human population growth?

6-2 What factors influence the size of the human population?

6-3 How does a population's age structure affect its growth or decline?

6-4 How can we slow human population growth?

Jaipur, India.

Eastland Photo/Alamy

CORE CASE STUDY Planet Earth: Population 7 Billion

It took about 200,000 years—from the time that the latest version of our species *Homo sapiens sapiens* evolved, to the 1920s—for our population to reach an estimated 2 billion. It took less than 50 years to add the second 2 billion people (by about 1974), and 25 years to add the third 2 billion (by 1999). Twelve years later, in late 2011, we topped 7 billion (Figure 6-1).

So why does it matter that there are now 7.1 billion people on the earth—almost 3 times as many as there were in 1950? Some say it doesn't matter and they contend that we can develop new technologies that could easily support 16 billion people. Many scientists disagree and argue that the current exponential growth of the human population is unsustainable. They point out that as our population grows and as we use more of the earth's natural resources, our ecological footprints also expand (see Figure 1-13, p. 14) and degrade the natural capital that keeps us alive and supports our lifestyles and economies.

In 2009, a team of internationally renowned scientists estimated that we have exceeded the global boundary limits for three of the major components of the earth's life-support system and are close to the boundary limits for five other components (Figure 3-B, p. 73). Some scientists say that because of our strong effect on natural systems, we now live in a new *Anthropocene era* in which our collective actions are moving key parts of our life-support system into danger zones.

Three major factors account for the rapid rise of the human population. *First,* the emergence of early and modern agriculture about 10,000 years ago allowed us to feed more people with each passing century. *Second,* we have developed technologies that have helped us to expand into almost all of the planet's climate zones and habitats (see Figure 1-12, p. 13). *Third,* our death rates have dropped sharply because of improved sanitation and health care and the development of antibiotics and vaccines to help control infectious diseases.

So what is a sustainable level for the human population? Population experts have made low, medium, and high projections for how big the human population will be by the end of this century (Figure 6-1). No one knows whether any of these population sizes are sustainable or for how long.

In this chapter, we examine population growth trends, the environmental impacts of the growing population, proposals for dealing with human population growth, and some stories about countries where population growth has been stabilized.

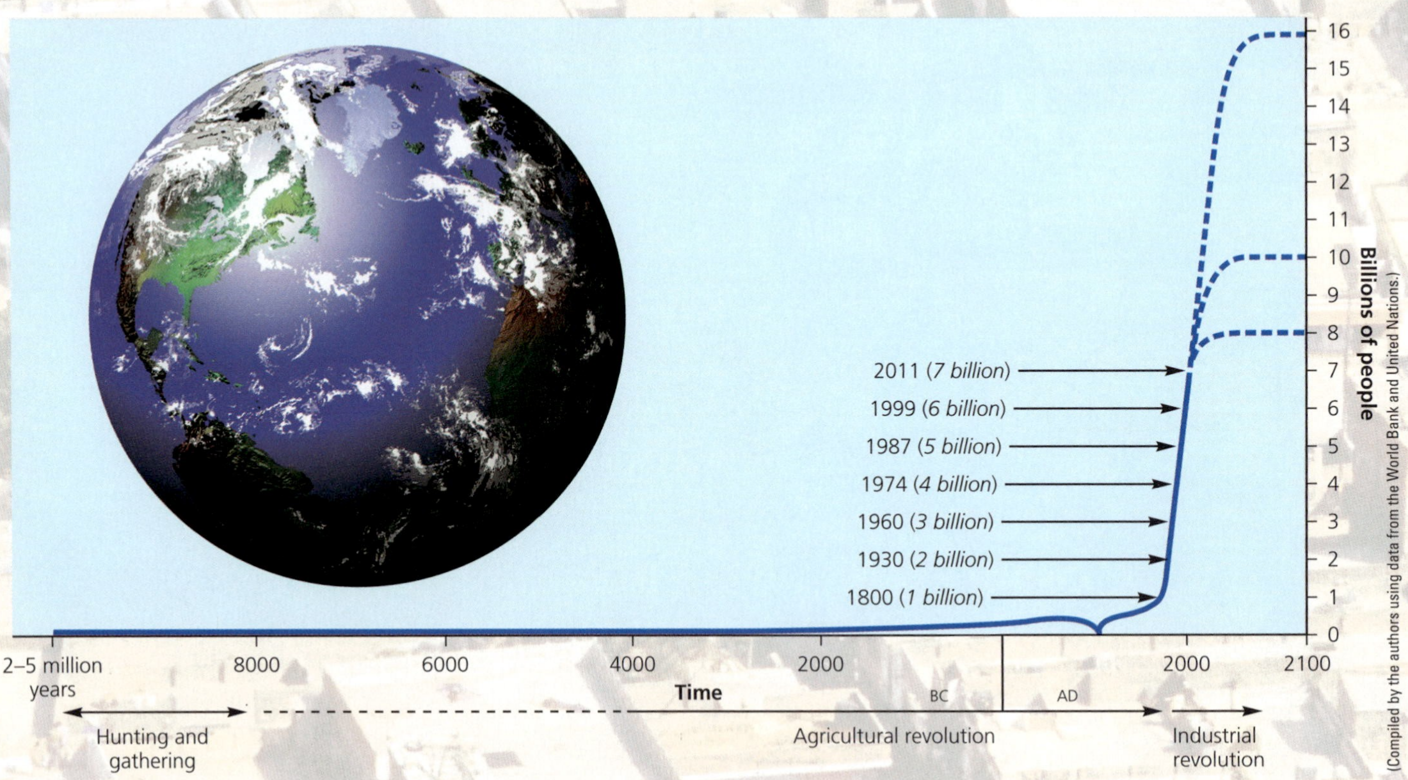

Figure 6-1 The human population has grown exponentially—showing slow growth throughout most of history and shooting up at a rapid rate within the last 200 years. This graph also shows projections to 2100 that range from 8 billion to 16 billion. (This figure is not to scale.)

Photo: ©NASA

6-1 How Do Environmental Scientists Think about Human Population Growth?

CONCEPT 6-1
The continuing rapid growth of the human population and its impacts on natural capital raise questions about how long the human population can keep growing.

Human Population Growth Shows Certain Trends

Estimates of what the global human population is likely to be in 2050 range from 7.8 billion to 10.8 billion people, with a medium projection of 9.6 billion people. For the year 2100, the projected population size ranges from 8 billion to 16 billion (Figure 6-1). These varying estimates depend on a variety of factors that we consider in more detail later in this chapter. However, *demographers*, or population experts, recognize three important growth trends.

First of all, in recent decades, the rate of population growth has slowed (Figure 6-2), but the world's population is still growing (Figure 6-1) at a rate of about 1.2%. This may not seem like much but in 2012 this growth added about 84 million people to the population—an average of more than 230,000 people each day, or almost 3 more people every second.

Second, demographers recognize that, geographically, human population growth is unevenly distributed and this pattern is expected to continue (Figure 6-3). About 2% of the 84 million new arrivals on the planet in 2012 were added to the world's more-developed coun-

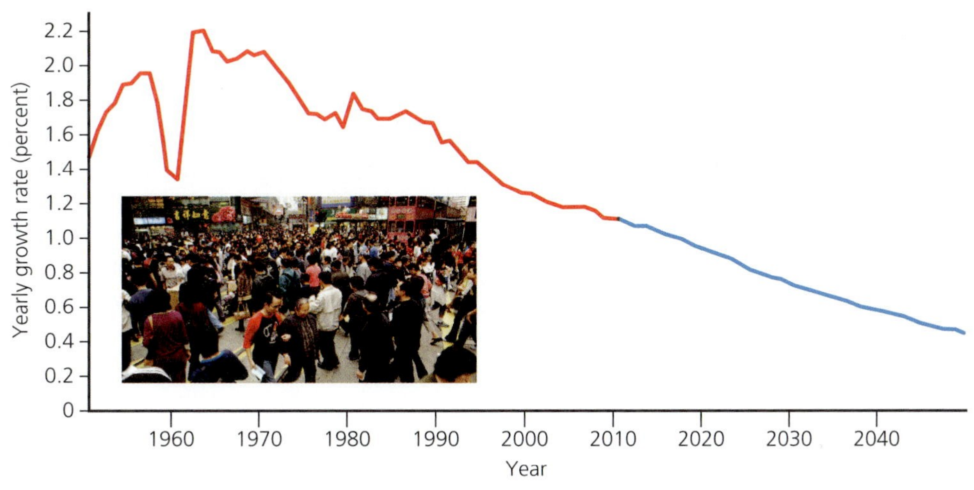

Figure 6-2 The annual growth *rate* of world population has generally dropped since the 1960s, but the population has continued to grow (Figure 6-1).

(Compiled by the authors using data from United Nations Population Division, U.S. Census Bureau, and Population Reference Bureau.)

Photo: JUSTIN GUARIGLIA/National Geographic Creative

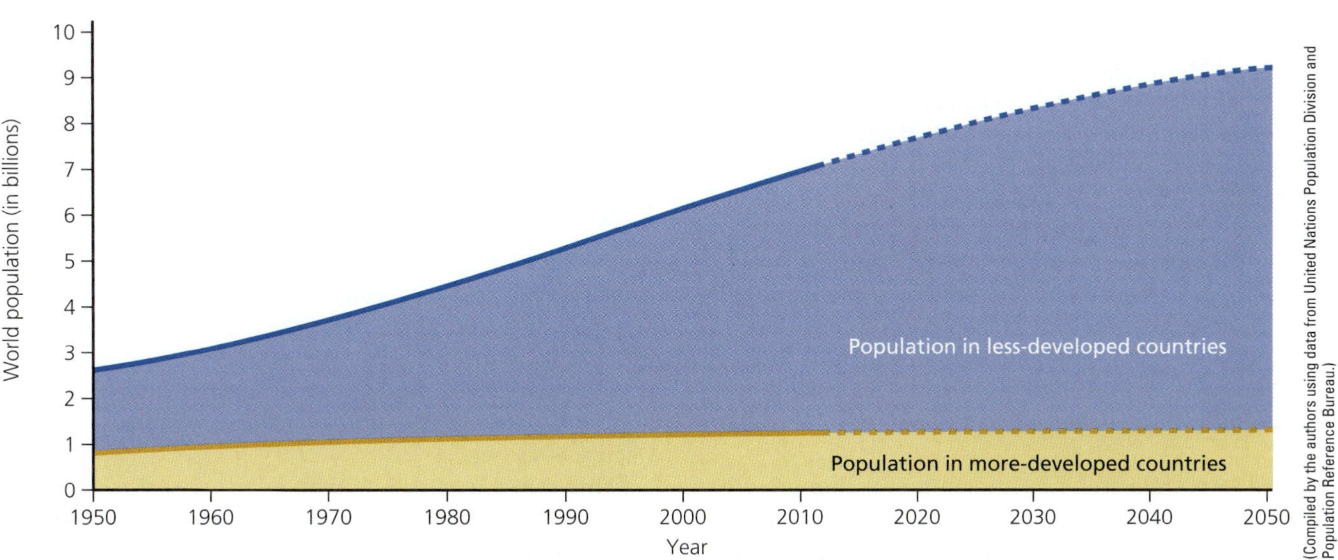

(Compiled by the authors using data from United Nations Population Division and Population Reference Bureau.)

Figure 6-3 Most of the world's population growth between 1950 and 2012 took place in the world's less-developed countries. This gap has been projected to increase between 2012 and 2050.

SCIENCE FOCUS 6.1

HOW LONG CAN THE HUMAN POPULATION KEEP GROWING?

To survive and provide resources for growing numbers of people, humans have modified, cultivated, built on, and degraded a large and increasing portion of the earth's natural systems (Figure 6-A). As a result, our ecological footprints have spread across much of the globe (see Figure 1-12, p. 13) and are projected to expand significantly further by 2030. Some scientists estimate that by then, we would need the equivalent of two planet Earths to sustain the projected population of 9.6 billion people (see Figure 1-13, p. 14).

Scientific studies of the populations of other species tell us that *no population can continue growing indefinitely* (p. 113). How long can we continue to avoid the reality of the earth's carrying capacity for our species by sidestepping many of the factors that sooner or later limit the growth of any population?

The debate over this important question has been going on since 1798 when Thomas Malthus, a British economist, hypothesized that the human population tends to grow exponentially, while food supplies tend to increase more slowly at a linear rate. So far, Malthus has been proven wrong. Food production has grown at an exponential rate instead of at a linear rate because of genetic and other technological advances in industrialized food production.

Environmental scientists are reexamining arguments such as those of Malthus about possible limits to the growth of human populations and economies. One view is that we have exceeded some of those limits, with too many people collectively degrading the earth's life-support system (see Figure 3-B, p. 73). To some scientists and other analysts, the key problem is *overpopulation* because of the sheer number of people in less-developed countries (see Figure 1-14, top, p. 15), which have 82% of the world's population. To others, the key factor is *overconsumption* in affluent, more-developed countries because of their high rates of resource use per person (see Figure 1-14, bottom).

Another view of population growth is that, so far, technological advances have allowed us to overcome the environmental limits that all populations of other species face and that this has had the effect of increasing the earth's carrying capacity for our species. They point out that average life expectancy in most of the world has been steadily rising despite dire warnings from some environmental scientists that we are seriously degrading our life-support system.

Some of these analysts argue that because of our technological ingenuity, there are few, if any, limits to human population growth and resource use per person. They believe that we can continue ever-increasing economic growth and avoid serious damage to our life-support systems by making technological advances in areas such as food production and medicine, and by finding substitutes for resources that we are depleting. As a result, they see no need to slow the world's population growth.

Proponents of slowing and eventually stopping population growth point out that

tries, where population is growing exponentially at about 0.1% a year. The other 98% were added to the world's middle- and low-income, less-developed countries, where the population is growing exponentially, 14 times faster at about 1.4% a year, on average.

At least 95% of the 2.6 billion people projected to be added to the world's population between 2012 and 2050 will be born into the least-developed countries, which are not equipped to deal with the pressures of such rapid growth. The world's three most populous countries are, in order, China, India, and the United States. In 2012, more than one of every three persons on the earth lived in China (with 19% of the world's population) or India (with 18%).

A third important trend in human population growth is the movement of people from rural areas to cities. More than half of the world's people now live in *urban areas,* or cities (see chapter-opening photo) and their surrounding suburbs. The great majority of these urban dwellers live in less-developed countries where resources for dealing with rapidly growing populations are limited. (We cover this and other *urbanization trends* in Chapter 22.)

Human Population Growth Impacts Natural Capital

As the human population grows, so does the global total human ecological footprint (see Figure 1-13, p. 14), and the bigger this footprint, the higher the overall impact on the earth's natural capital. The 2005 Millennium Ecosystem Assessment concluded that human activities have degraded about 60% of the earth's ecosystem services (Figure 6-4).

This raises a question that many scientists are now studying: How many people can the earth support indefinitely (**Concept 6-1**)? Some analysts contend that we need to define the planet's **cultural carrying capacity**: the maximum number of people who could live in reason-we are not providing the basic necessities for about 1.4 billion people—one of every five on the planet—who struggle to survive on the equivalent of about $1.25 per day. This raises a serious question: How will we meet the basic needs of the additional 2.6 billion people projected to be alive in 2050?

Proponents of slowing growth also warn of two potentially serious consequences that we could face if we do not sharply lower birth rates. First, death rates might increase because of declining health and environmental conditions and increasing social disruption in some areas, as is already happening in parts of Africa. A worst-case scenario for such a trend is a crash of the human population to a more sustainable level, perhaps as low as 2 billion. Second, resource use and degradation of normally renewable resources may intensify as more consumers increase their already large ecological footprints in more-developed countries and in rapidly developing countries such as China, India, and Brazil.

So far, advances in food production and health care have staved off widespread population declines. But there is extensive and growing evidence that we are steadily depleting and degrading much of the earth's irreplaceable natural capital (see Figure 1-3, p. 7) that supports us. We can get away with this for awhile, because the earth's life-support system is very complex and resilient, and because of built-in time delays between disturbances to the system and responses to them from within the system. But like unseen termites eating away the foundation of a house, at some point, such disturbances could reach a *tipping* or *breaking point* (see Figure 3-B, p. 73) beyond which there could be damaging and long-lasting change.

No one knows how close we are to environmental limits that, many scientists say, eventually will control the size of the human population. However, these scientists argue that human population growth is a vital scientific, political, economic, and ethical issue that we must confront.

Figure 6-A **Natural capital degradation:** Fertile topsoil is an irreplaceable form of natural capital necessary for supplying the world's people with food. According to a joint survey by the UN Environment Programme (UNEP) and the World Resources Institute, topsoil is eroding faster than it forms on about 38% of the world's cropland, including this farmland in Iowa.

Lynn Betts/USDA Natural Resources Conservation Service

Critical Thinking

How close do you think we are to the environmental limits of human population growth? Very close, moderately close, or far away? Explain.

able freedom and comfort indefinitely, without decreasing the ability of the earth to sustain future generations. (See the online Guest Essay by Garrett Hardin on this topic.) This issue has long been a topic of scientific and political debate (Science Focus 6.1).

Natural Capital Degradation

Altering Nature to Meet Our Needs

Reducing biodiversity

Increasing use of net primary productivity

Increasing genetic resistance in pest species and disease-causing bacteria

Eliminating many natural predators

Introducing harmful species into natural communities

Using some renewable resources faster than they can be replenished

Disrupting natural chemical cycling and energy flow

Relying mostly on polluting and climate-changing fossil fuels

Animated Figure 6-4 We humans have altered the natural systems that sustain our lives and economies in at least eight major ways to meet the increasing needs and wants of our growing population (Concept 6-1). ***Questions:*** In your daily living, do you think you contribute directly or indirectly to any of these harmful environmental impacts? Which ones? Explain.

Top: ©Dirk Ercken/Shutterstock.com. Center: ©Fulcanelli/Shutterstock.com. Bottom: © Werner Stoffberg/Shutterstock.com.

6-2 What Factors Influence the Size of the Human Population?

CONCEPT 6-2A
Population size increases through births and immigration, and decreases through deaths and emigration.

CONCEPT 6-2B
The average number of children born to the women in a population *(total fertility rate)* is the key factor that determines population size.

The Human Population Can Grow, Decline, or Remain Fairly Stable

The basics of global population change are quite simple. If there are more births than deaths during a given period of time, the earth's population increases, and when the opposite is true, it decreases. When the number of births equals the number of deaths during a particular time period, population size does not change.

Instead of using the total numbers of births and deaths per year, demographers use the **crude birth rate** (the number of live births per 1,000 people in a population in a given year) and the **crude death rate** (the number of deaths per 1,000 people in a population in a given year).

Human populations grow or decline in particular countries, cities, or other areas through the interplay of three factors: *births (fertility), deaths (mortality),* and *migration.* We can calculate the **population change** of an area by subtracting the number of people leaving a population (through death and emigration) from the number entering it (through birth and immigration) during a specified period of time (usually 1 year) (Concept 6-2A):

$$\text{Population change} = (\text{Births} + \text{Immigration}) - (\text{Deaths} + \text{Emigration})$$

When births plus immigration exceed deaths plus emigration, a population grows; when the reverse is true, a population declines. (see Figure 18, p. S44, in Supplement 6 for a map showing various percentage rates of population growth in the world's countries in 2012.)

Women Are Having Fewer Babies but the World's Population Is Still Growing

Another measurement used in population studies is the **fertility rate**, a measure of how many children are born in a population over a set period of time. Here, we consider two types of fertility rates, the first being the **replacement-level fertility rate**: the average number of children that couples in a population must bear to replace themselves. It is slightly higher than two children per couple (2.1 in more-developed countries and as high as 2.5 in some less-developed countries), mostly because some children die before reaching their reproductive years. Any fertility rate above the replacement level will cause a population to grow.

If we were to reach the replacement-level fertility rate for the world, would it bring an immediate halt to population growth? No, because the number of *future* parents alive has grown dramatically. If all of today's couples were to have an average of 2.1 children, they would not be contributing to population growth. But if all of today's girl children were to grow up to have an average of 2.1 children as well, the world's population would continue to grow for 50 years or more (assuming death rates do not rise) because there are so many girls under age 15 who will be moving into their reproductive years.

The second type of fertility rate, and the key factor affecting human population growth and size, is the **total fertility rate (TFR)**: the average number of children born to the women in a population during their reproductive years (Concept 6-2B). Between 1955 and 2012, the average TFR for more-developed countries dropped from 2.8 to 1.6, and the average TFR for less-developed countries dropped from 6.2 to 2.6. The latter number is projected to continue dropping. If the global TFR, which was 2.4 in 2012, were to drop to 2.1, the world's population would continue to grow for about 50 years and then would level off and gradually decline.

Between 1955 and 2012, the global TFR dropped from 5 to 2.4. Those who support slowing the world's population growth view this as good news. However, to eventually halt population growth, the global TFR will have to drop to 2.1. (See Figure 19, p. S45, in Supplement 6 for a map showing how TFRs vary globally.)

CASE STUDY

The U.S. Population—Third-Largest and Growing

The population of the United States grew from 76 million in 1900 to 314 million by 2012, despite oscillations in the country's TFR and population growth rate (Figure 6-5). It took the country 139 years to reach a population of 100 million people, 52 years to add another 100 million (by 1967), and only 39 years to add the third 100 million (by 2006). During the period of high birth rates between 1946 and 1964, known as the *baby boom,* 79 million people were added to the U.S. population. At the peak of the baby boom in 1957, the average TFR was 3.7 children per woman. In 2012, and in most years since 1972, it has been at or below 2.1 children per woman (1.9 in 2012), compared to a global TFR of 2.4.

The drop in the TFR has slowed the rate of population growth in the United States. But the country's population is still growing faster than those of most other more-developed countries and it is not close to leveling off. According to the U.S. Census Bureau, about 2.3 mil-

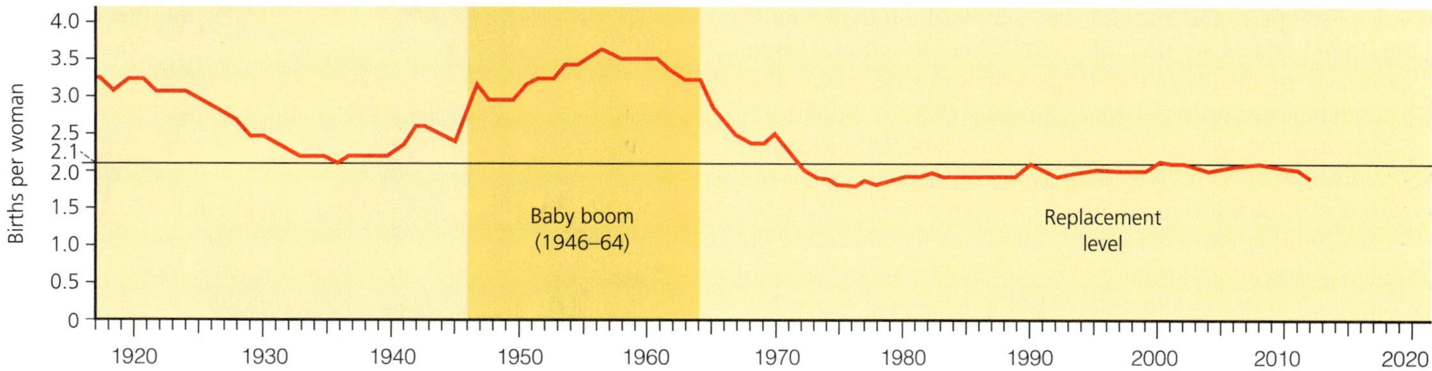

Figure 6-5 The graph shows the total fertility rates for the United States between 1917 and 2012. ***Question:*** The U.S. fertility rate has declined and remained at or below replacement levels since 1972. So why is the population of the United States still increasing?

(Compiled by the authors using data from Population Reference Bureau and U.S. Census Bureau.)

lion people were added to the U.S. population in 2012. About 1.6 million (70% of the total) were added because there were that many more births than deaths, and about 0.7 million (30% of the total) were immigrants.

Immigration has become a political issue in the United States. The country was founded by immigrants and since 1820, it has admitted almost twice as many immigrants and refugees as all other countries combined. The number of legal immigrants (including refugees) has varied during different periods because of changes in immigration laws and rates of economic growth (Figure 6-6).

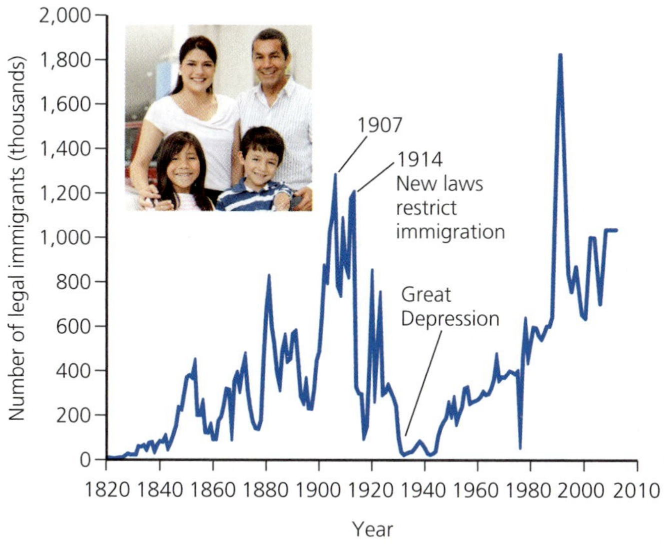

Figure 6-6 Legal immigration to the United States, 1820–2006 (the last year for which data are available). The large increase in immigration since 1989 resulted mostly from the Immigration Reform and Control Act of 1986, which granted legal status to certain illegal immigrants who could show they had been living in the country before January 1, 1982.

Photo: Andresr/Shutterstock.com

(Compiled by the authors using data from U.S. Immigration and Naturalization Service and the Pew Hispanic Center.)

In addition to the fourfold increase in population growth since 1900, some amazing changes in lifestyles took place in the United States during the 20th century (Figure 6-7), which led to Americans living longer. Along with this came dramatic increases in per capita resource use and a much larger total and per capita ecological footprint.

(Compiled by the authors using data from U.S. Census Bureau and Department of Commerce.)

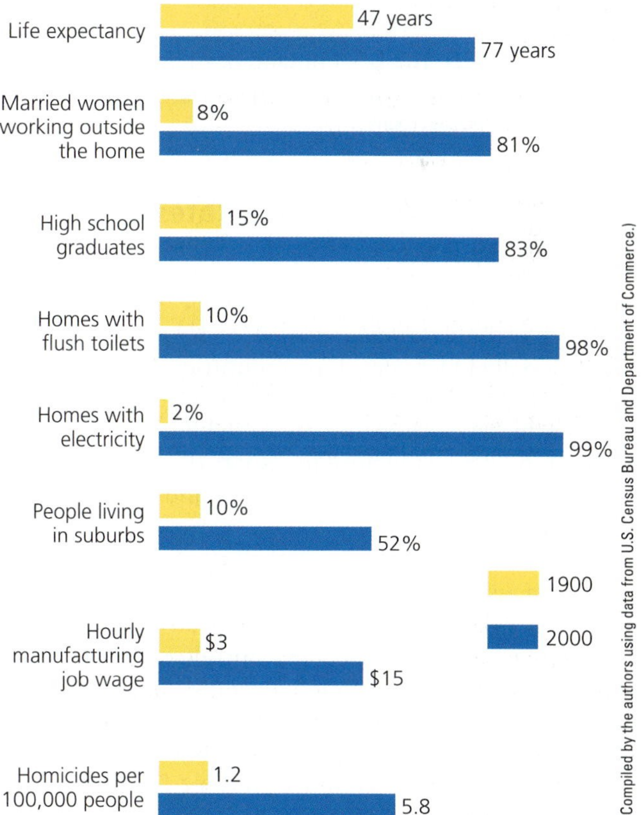

Figure 6-7 Some major changes took place in the United States between 1900 and 2000. ***Question:*** Which two of these changes do you think had the biggest impacts on the U.S. ecological footprint?

Compare the present-day standard of living in the United States to that of 1907 when the three leading causes of death in the United States were pneumonia, tuberculosis, and diarrhea (ailments that are seldom life-threatening now); 90% of U.S. doctors had no college education; one of five adults could not read or write; only 6% of Americans graduated from high school; the average U.S. worker earned a few hundred to a few thousand dollars per year; and there were only 9,000 cars in the country and only 232 kilometers (144 miles) of paved roads.

The U.S. Census Bureau projects that the U.S. population is likely to increase from 314 million in 2012 to 400 million by 2050—an addition of 86 million more Americans over four decades. Because of a high per-person rate of resource use and the resulting waste and pollution, each addition to the U.S. population has an enormous environmental impact (see the map in Figure 8, p. S33, in Supplement 6). Recall that the environmental impact of a population is obtained by multiplying the effects of three factors (see Figure 1-14, p. 15, bottom): population size, affluence (and resulting high rates of resource use per person), and technology. The United States has by far the world's largest total and per capita ecological footprint (see Figure 1-11, p. 13), mostly because of the size of its population multiplied by its very high rate of resource use per person. This explains why some analysts consider the United States to be the world's most overpopulated country.

CONSIDER THIS. . .

THINKING ABOUT The U.S. Population

Considering this information, do you believe that the United States is the world's most overpopulated country? Explain.

Several Factors Affect Birth Rates and Fertility Rate

Many factors affect a country's average birth rate and TFR. One is the *importance of children as a part of the labor force,* especially in less-developed countries. This is a major part of why it makes sense for many poor couples in those countries to have a large number of children. They need help with hauling daily drinking water (Figure 6-8), gathering wood for heating and cooking, and tending crops and livestock.

Another economic factor is the *cost of raising and educating children.* Birth and fertility rates tend to be lower in more-developed countries, where raising children is much more costly because they do not enter the labor force until they are in their late teens or twenties. In the United States, it costs more than $235,000 to raise a middle-class child from birth to age 18. By contrast, many children in poor countries receive little education and instead have to work to help their families survive (Figure 6-9).

The *availability of, or lack of, private and public pension systems* can influence the decisions of some couples on how many children to have, especially the poor in less-developed countries. Pensions reduce a couple's need to have several children to help support them in old age.

Syed Sajjad Ali/Shutterstock.com

Figure 6-8 These young girls in India are carrying water.

Also, there are more *infant deaths* in poorer countries, and over time, this has affected cultural norms related to family size in these countries. The more children a couple has, the more likely it is that at least a few will survive and grow to adulthood.

Urbanization plays a role in birth rate trends. People living in urban areas usually have better access to family planning services and tend to have fewer children than do those living in the rural areas of poorer countries.

Another important factor is the *educational and employment opportunities available for women.* Total fertility rates tend to be low when women have access to education and paid employment outside the home. In less-developed countries, a woman with no education typically has two more children than does a woman with a high school education. In most societies, better-educated women tend to marry later and have fewer children.

Average age at marriage (or, more precisely, the average age at which a woman has her first child) also plays a role. Women normally have fewer children when their average age at marriage is 25 or older.

Birth rates and TFRs are also affected by the *availability of legal abortions.* According to the World Health Organization and the Guttmacher Institute, each year, about 210 million women become pregnant and at least 42 million of them get abortions—about 22 million of them legal and the other 20 million illegal (and often unsafe). The *availability of reliable birth control methods* also allows women to control the number and spacing of the children they have.

Religious beliefs, traditions, and cultural norms also play a role. In some countries, these factors favor large families as many people strongly oppose abortion and some forms of birth control.

PRIT VESILIND/National Geographic Creative

Figure 6-9 This young girl is breaking granite into gravel in the Kerala State of India.

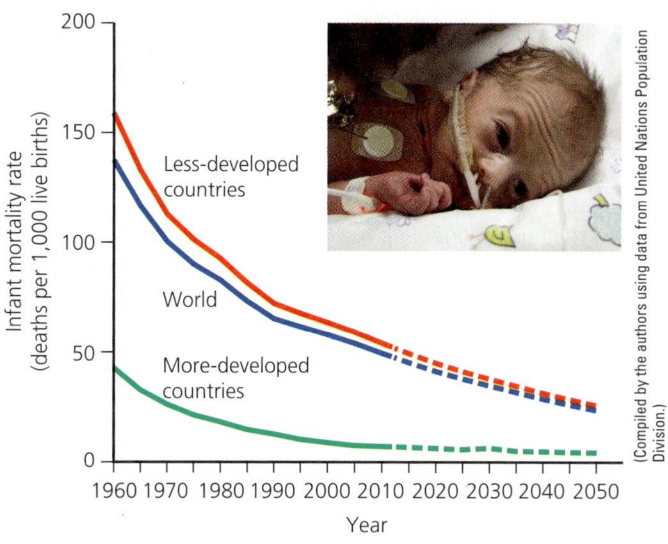

Figure 6-10 Infant mortality rates for the world's more-developed countries and less-developed countries, 1950–2012, with projections to 2050 based on medium population projections.

Photo: Gert Vrey/Shutterstock.com

CONSIDER THIS. . .

CONNECTIONS Preference for Male Children and Social Problems in China

In China, where families are strongly encouraged to have only one child, there is a strong preference for male children, partly because daughters are likely to marry and leave their parents. Some pregnant Chinese women use ultrasound to determine the gender of their fetuses and get an abortion if the child is female. As a result, baby boys are sometimes kidnapped and sold to families that want boys, and some young girls have been kidnapped and sold as brides for single men. The Chinese government estimates that by 2030, about 30 million Chinese men will not be able to find wives and if the economy slows, they might have poor job prospects. Sociologists say this is a recipe for social upheaval.

Several Factors Affect Death Rates

The rapid growth of the world's population over the past 100 years is not primarily the result of a rise in the birth rate. Instead, it has been caused largely by declining death rates, especially in less-developed countries. More people in some of these countries are living longer and fewer infants are dying because of increased food supplies, improvements in food distribution, better nutrition, medical advances such as immunizations and antibiotics, improved sanitation, and safer water supplies.

A useful indicator of the overall health of people in a country or region is **life expectancy**, which for any given year is the average number of years a person born in that year can be expected to live. Between 1955 and 2012, the average global life expectancy increased from 48 years to 70 years. In 2012, Japan had the world's longest life expectancy of 83 years. In the world's poorest countries, life expectancy is 55 years or less.

GOOD NEWS

Between 1900 and 2012, the average U.S. life expectancy increased from 47 years to 79 years. However, the United States ranks 32nd among nations in life expectancy, down from 5th in 1950. Research indicates that a key factor in this ranking is poor health, even though the United States leads the world in health-care costs per person.

Another important indicator of overall health in a population is its **infant mortality rate**, the number of babies out of every 1,000 born who die before their first birthday. It is viewed as one of the best measures of a society's quality of life because it reflects a country's general level of nutrition and health care. A high infant mortality rate usually indicates insufficient food *(undernutrition),* poor nutrition *(malnutrition),* and a high incidence of infectious disease. Infant mortality also affects the TFR. In areas with low infant mortality rates, women tend to have fewer children because fewer of their children die at an early age.

Infant mortality rates in most countries have declined dramatically since 1965 (Figure 6-10) but vary widely among different countries (see Figure 20, p. S45, Supplement 6). Even so, every year, more than 4 million infants (most of them in less-developed countries) die of *preventable* causes during their first year of life. This average of nearly 11,000 mostly unnecessary infant deaths per day is equivalent to 55 jet airliners, each loaded with 200 infants younger than age 1, crashing *every day* with no survivors—a tragedy rarely reported in the news.

In 1900 the U.S. infant mortality rate was 165. In 2012 it was 6.0. This sharp decline was a major factor in the marked increase in U.S. average life expectancy during this period. However, in 2012 the United States ranked 44th among all nations in terms of infant mortality rates for two main reasons: the generally inadequate health care for poor women during pregnancy, as well as for their babies after birth, and drug addiction among many pregnant women.

SCIENCE FOCUS 6.2

PROJECTING POPULATION CHANGE

Estimates of the human population size in 2050 range from 7.8 billion to 10.8 billion people—a difference of 3 billion. Why this wide range of estimates? The answer is that there are countless factors that demographers have to consider when making projections for any country or region of the world.

First, they have to determine the reliability of current population estimates. Many of the more-developed countries such as the United States have fairly good estimates. But many countries have poor knowledge of their population size. Some countries even deliberately inflate or deflate the numbers for economic or political purposes.

Second, demographers make assumptions about trends in fertility—that is, whether women in a country will, on average, have fewer babies or more babies in the future. The demographers might assume that fertility is declining by a certain percentage per year. If the estimate of the rate of declining fertility is off by a few percentage points, the resulting percentage increase in population can be magnified over a number of years, and be quite different from the projected population size increase.

For example, UN demographers assumed that Kenya's fertility rate would decline, and based on that, in 2002, they projected that Kenya's total population would be 44 million by 2050. In reality, the fertility rate rose from 4.7 to 4.8 children per woman. The population momentum was therefore more powerful than expected, and in 2010, the UN revised its projection for Kenya's population in 2050 to 70.8 million, which was 60% higher than its earlier projection.

Third, population projections are made by a variety of organizations employing demographers. UN projections tend to be cited most often. But the U.S. Census Bureau also makes world population projections, as does the International Institute for Applied Systems Analysis (IIASA) and the U.S. Population Reference Bureau. The World Bank has also made projections in the past and maintains a database. All of these organizations use differing sets of data and differing methods, and their projections vary (Figure 6-B).

Critical Thinking

If you were in charge of the world and making decisions about resource use based on population projections, which of the projections in Figure 6-B would you rely on? Explain.

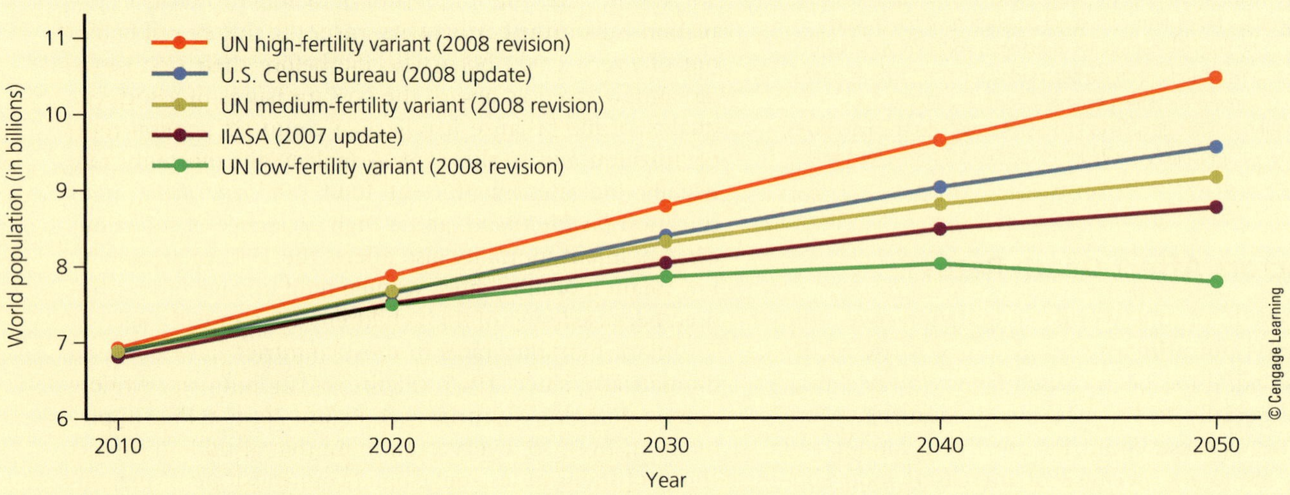

Figure 6-B World population projections to 2050 from three different organizations: the UN, the U.S. Census Bureau, and IIASA. Note that the uppermost, middle, and lowermost curves of these five projections are all from the UN, each assuming a different level of fertility.

Migration Affects an Area's Population Size

The third factor in population change is **migration**: the movement of people into *(immigration)* and out of *(emigration)* specific geographic areas. Most people migrating from one area or country to another seek jobs and economic improvement. But many are driven to migrate by religious persecution, ethnic conflicts, political oppression, or war.

According to a United Nations (UN) study and another study by environmental scientist Norman Myers, in 2008 there were at least 40 million *environmental refugees*—people who had to leave their homes because of water or food shortages, soil erosion, or some other form of environmental degradation or depletion. It is difficult to find more recent, reliable estimates of the number of such refugees, but in 2012, the UN projected that by 2020, there will be 50 million environmental refugees.

Estimates of any population's future numbers can vary considerably. These varying estimates depend mostly on TFR projections. However, demographers also have to make assumptions about death rates, migration, and a number of other variables. If their assumptions are wrong, their population forecasts can be way off the mark (Science Focus 6.2).

6-3 How Does a Population's Age Structure Affect Its Growth or Decline?

CONCEPT 6-3
The numbers of males and females in young, middle, and older age groups determine how fast a population grows or declines.

A Population's Age Structure Helps Us to Make Projections

An important factor determining whether the population of a country increases or decreases is its **age structure**: the numbers or percentages of males and females in young, middle, and older age groups in that population (**Concept 6-3**).

Population experts construct a population *age-structure diagram* by plotting the percentages or numbers of males and females in the total population in each of three age categories: *prereproductive* (ages 0–14), consisting of individuals normally too young to have children; *reproductive* (ages 15–44), consisting of those normally able to have children; and *postreproductive* (ages 45 and older), with individuals normally too old to have children. Figure 6-11 presents generalized age-structure diagrams for countries with rapid, slow, zero, and negative population growth rates.

A country with a large percentage of its people younger than age 15 (represented by a wide base in Figure 6-11, far left) will experience rapid population growth unless death rates rise sharply. Because of this *demographic momentum,* the number of births in such a country will rise for several decades even if women have an average of only one or two children each, due to the large number of girls entering their prime reproductive years.

In 2012, about 26% of the world's population—29% in the less-developed countries and 16% in more-developed countries—was under age 15. By 2025, the world's current 1.8 billion people under age 15—roughly one of every four persons on the planet—will move into their prime reproductive years. The dramatic differences in population age structure between less-developed and more-developed countries (Figure 6-12) show why most future human population growth will take place in less-developed countries (Figure 6-3).

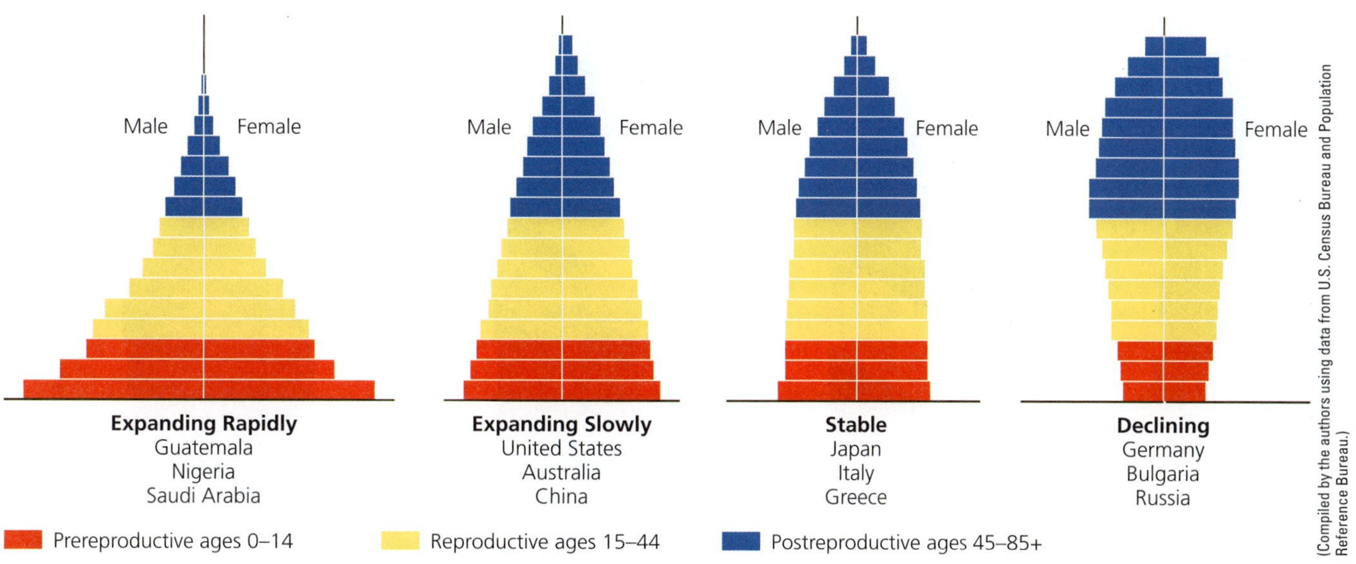

Animated Figure 6-11 Generalized population age-structure diagrams for countries with rapid (1.5–3%), slow (0.3–1.4%), zero (0–0.2%), and negative (declining) population growth rates. ***Question:*** Which of these diagrams best represents the country where you live?

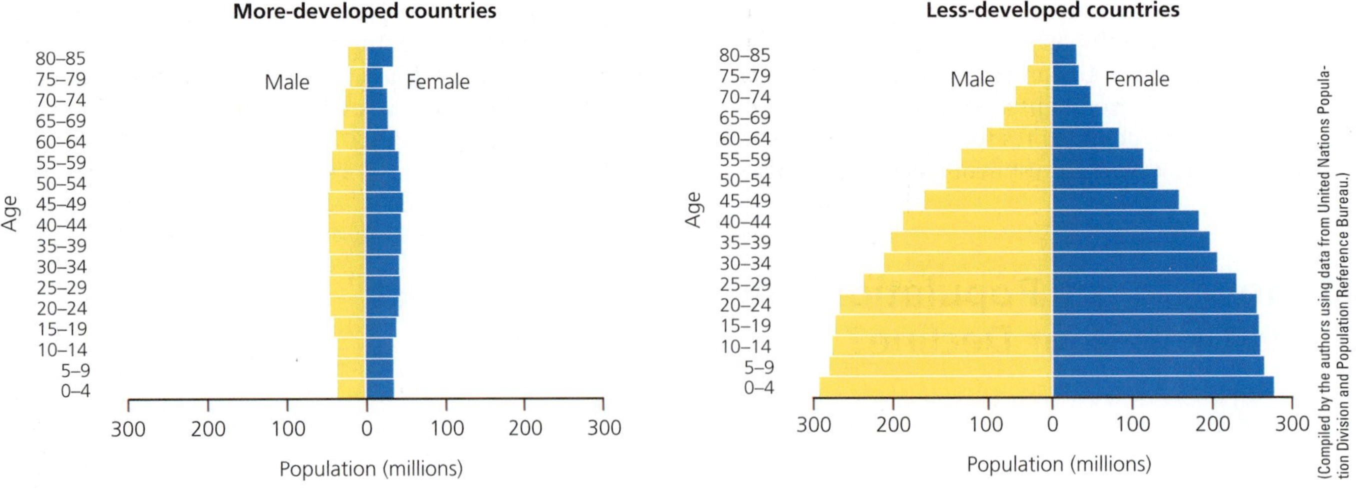

Figure 6-12 Population structure by age and sex in less-developed countries and more-developed countries for 2011. ***Question:*** If all girls under 15 were to have only one child during their lifetimes, how do you think these structures would change over time?

The global population of seniors—people who are 65 and older—is projected to triple by 2050, when one of every six people will be a senior. (See the Case Study that follows.) This graying of the world's population is due largely to declining birth rates and medical advances that have extended life spans. In 2012, the three nations with the largest percentage of their population age 65 or older were, in order, Japan, Germany, and Italy. In such countries, the number of working adults is shrinking in proportion to the number of seniors, which in turn is slowing the growth of tax revenues in these countries. Some analysts worry about how such societies will support their growing populations of seniors.

CASE STUDY

The American Baby Boom

Changes in the distribution of a country's age groups have long-lasting economic and social impacts. For example, consider the American baby boom, which added 79 million people to the U.S. population between 1946 and 1964. Over time, this group looks like a bulge moving up through the country's age structure, as shown in Figure 6-13.

For decades, members of the baby-boom generation have strongly influenced the U.S. economy because they make up about 36% of all adult Americans. Baby boomers created the youth market in their teens and twen-

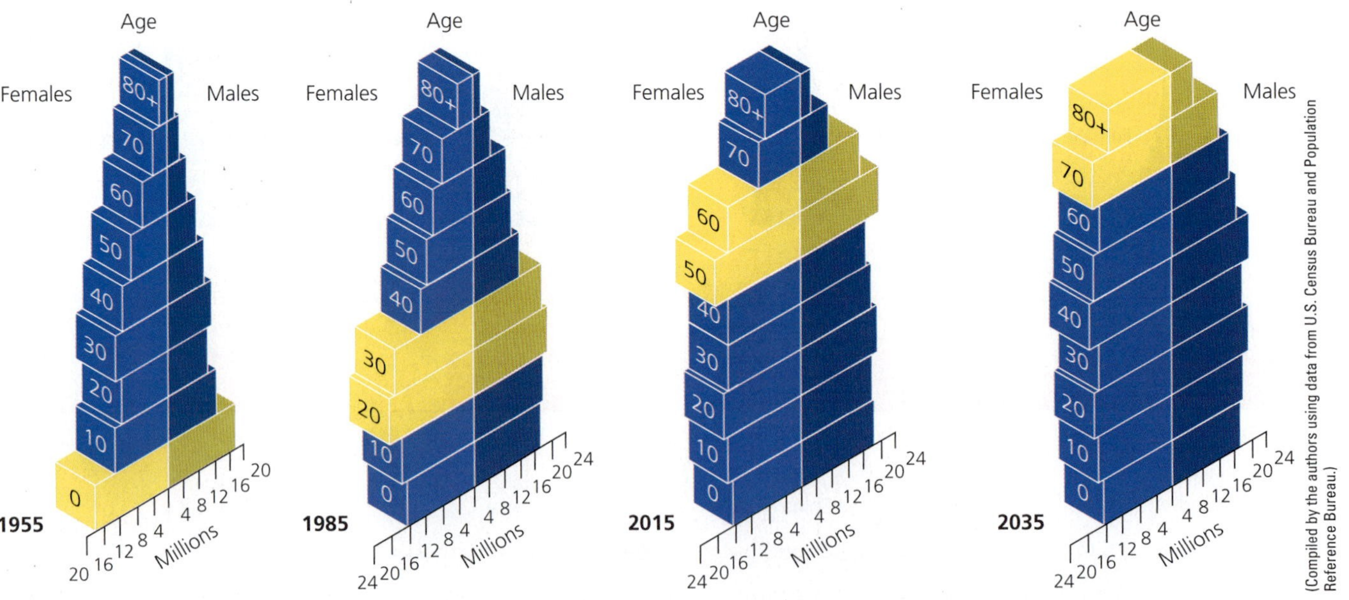

Animated Figure 6-13 Age-structure charts tracking the baby-boom generation in the United States, 1955, 1985, 2015 (projected), and 2035 (projected).

ties and are now creating the late middle age and senior markets. In addition to having this economic impact, the baby-boom generation plays an increasingly important role in deciding who gets elected to public office and what laws are passed or weakened.

Since 2011, when the first baby boomers began turning 65, the number of Americans older than age 65 has grown at the rate of about 10,000 a day and will do so through 2030. This process has been called the *graying of America.* As the number of working adults declines in proportion to the number of seniors, there may be political pressure from baby boomers to increase tax revenues to help support the growing senior population. This could lead to economic and political conflicts between younger and older Americans.

CONSIDER THIS. . .

CONNECTIONS Baby Boomers, the U.S. Work Force, and Immigration

According to the U.S. Census Bureau, after 2020, much higher immigration levels will be needed to supply enough workers as baby boomers retire. According to a recent study by the UN Population Division, if the United States wants to maintain its current ratio of workers to retirees, it will need to absorb an average of 10.8 million immigrants every year—more than 10 times the current immigration level—through 2050.

Populations Made Up Mostly of Older People Can Decline Rapidly

As the percentage of people age 65 or older increases, more countries will begin experiencing population declines. If population decline is gradual, its harmful effects usually can be managed. However, some countries are experiencing fairly rapid declines and feeling such effects more severely.

Japan has the world's highest percentage of elderly people (above age 65) and the world's lowest percentage of young people (below age 15). In 2012, Japan's population was 128 million. By 2050, its population is projected to be 95.5 million, a 25% drop. As its population declines, there will be fewer adults working and paying taxes to support an increasing elderly population. Because Japan discourages immigration, it may face a bleak economic future. As a result, some have called for the country to rely more on robots to do its manufacturing jobs and on selling robots in the global economy to help support its aging population.

In China, the growth in numbers of children has slowed because of its one-child policy. As a result, the average age of China's population has been increasing over the past two decades at one of the fastest rates ever recorded. While China's population is not yet declining, the UN estimates that by 2025, China is likely to have too few young workers to support its rapidly aging population. This graying of the Chinese population could lead to a declining work force, higher wages for workers, limited funds for supporting continued economic development, and fewer children and grandchildren to care for the growing number of elderly people. These concerns and other factors may slow economic growth and have led to some relaxation of China's one-child population control policy.

Figure 6-14 lists some of the problems associated with rapid population decline. Countries currently faced with rapidly declining populations include Japan, Russia, Germany, Bulgaria, Hungary, Ukraine, Serbia, Greece, Portugal, and Italy.

Populations Can Decline Due to a Rising Death Rate: The AIDS Tragedy

A large number of deaths from AIDS can disrupt a country's social and economic structure by removing significant numbers of young adults from its population. According to the World Health Organization, between 1981 and 2012, AIDS killed more than 30 million people (617,000 in the United States).

Unlike hunger and malnutrition, which kill mostly infants and children, AIDS kills primarily young adults and leaves many children orphaned, some of whom are also infected with HIV, the virus that can lead to AIDS. Worldwide, AIDS is the leading cause of death for people of ages 15–49.

This pandemic has had a devastating effect in some countries, and has changed their population age structures

Some Problems with Rapid Population Decline

- Can threaten economic growth
- Labor shortages
- Less government revenues with fewer workers
- Less entrepreneurship and new business formation
- Less likelihood for new technology development
- Increasing public deficits to fund higher pension and health-care costs
- Pensions may be cut and retirement age increased

Figure 6-14 Rapid population decline can cause several problems. ***Question:*** Which two of these problems do you think are the most important?

Top: ©Slavoljub Pantelic/Shutterstock.com. Center: lofoto/Shutterstock.com. Bottom: Yuri Arcurs/Shutterstock.com.

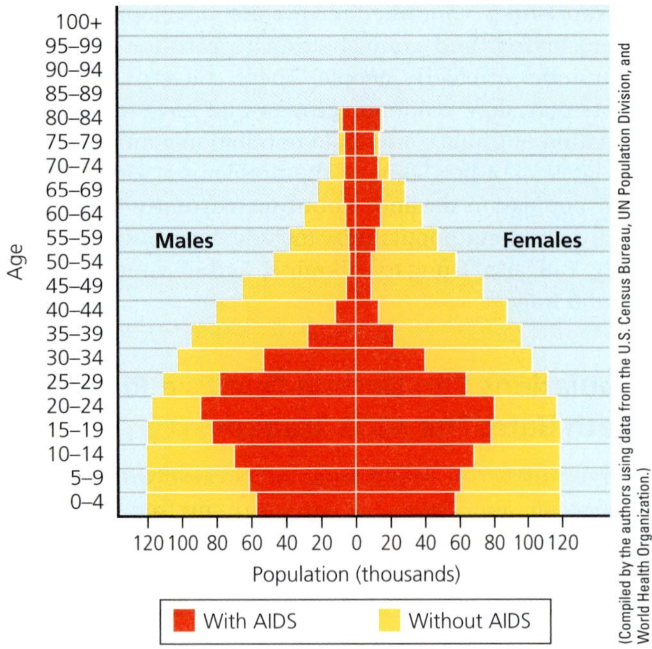

Figure 6-15 In Botswana, more than 25% of people ages 15–49 were infected with HIV in 2011. This figure shows two projected age structures for Botswana's population in 2020—one including the possible effects of the AIDS epidemic (red bars), and the other not including those effects (yellow bars). ***Question:*** How might this affect Botswana's economic development?

(Figure 6-15). This has had a number of harmful effects. One is a sharp drop in average life expectancy, especially in several southern African countries where 15–26% of all people between ages 15 and 49 are infected with HIV. Another is the loss of productive young-adult workers and trained personnel such as scientists, farmers, engineers, and teachers, as well as government, business, and health-care workers. The essential services they could provide are therefore lacking, and there are fewer taxpayers and fewer workers available to support the very young and the elderly. Many experts have called for the creation of a massive international program to help countries ravaged by AIDS.

6-4 How Can We Slow Human Population Growth?

CONCEPT 6-4

We can slow human population growth by reducing poverty, elevating the status of women, and encouraging family planning.

The First Step Is to Promote Economic Development

Scientific studies and experience have shown that the three most effective ways to slow or stop population growth are to reduce poverty, primarily through economic development and universal primary education; to elevate the status of women; and to encourage family planning and reproductive health care (**Concept 6-4**). Let's begin by looking at the role of economic development.

In the world's most desperately poor countries, couples tend to have more children for reasons listed earlier in this chapter (p. 128). Thus, in many less-developed countries, total fertility and population growth rates tend to be high, and large numbers of poor people are increasingly being crowded into unsanitary and difficult living conditions in slums and shantytowns.

Demographers, examining the birth and death rates of western European countries that became industrialized during the 19th century, have developed a hypothesis of population change known as the **demographic transition**: As countries become industrialized and economically developed, their populations tend to grow more slowly. According to the hypothesis, this transition takes place in four stages, as shown in Figure 6-16.

Some analysts believe that most of the world's less-developed countries will make a demographic transition over the next few decades, mostly because newer technologies will help them to develop economically and to raise their per capita incomes. Other analysts fear that rapid population growth, extreme poverty, and increasing environmental degradation and resource depletion in some low-income, less-developed countries could leave these countries stuck in stage 2 of the demographic transition.

Another factor that could hinder the demographic transition in some less-developed countries is that since 1985, economic assistance from more-developed countries has generally dropped. The resulting shortage of funds, coupled with poor economies in some countries, could leave large numbers of people trapped in poverty, which could in turn keep population growth rates high in such countries. Many experts argue that more-developed countries should help less-developed nations to make the demographic transition by aiding them in their economic development. They contend that such an effort could help to stabilize the global population, thereby making for a more sustainable world.

Empowering Women Can Help to Slow Population Growth

A number of studies show that women tend to have fewer children if they are educated, have the ability to control their own fertility, earn an income of their own, and live in societies that do not suppress their rights. Although women make up roughly half of the world's population,

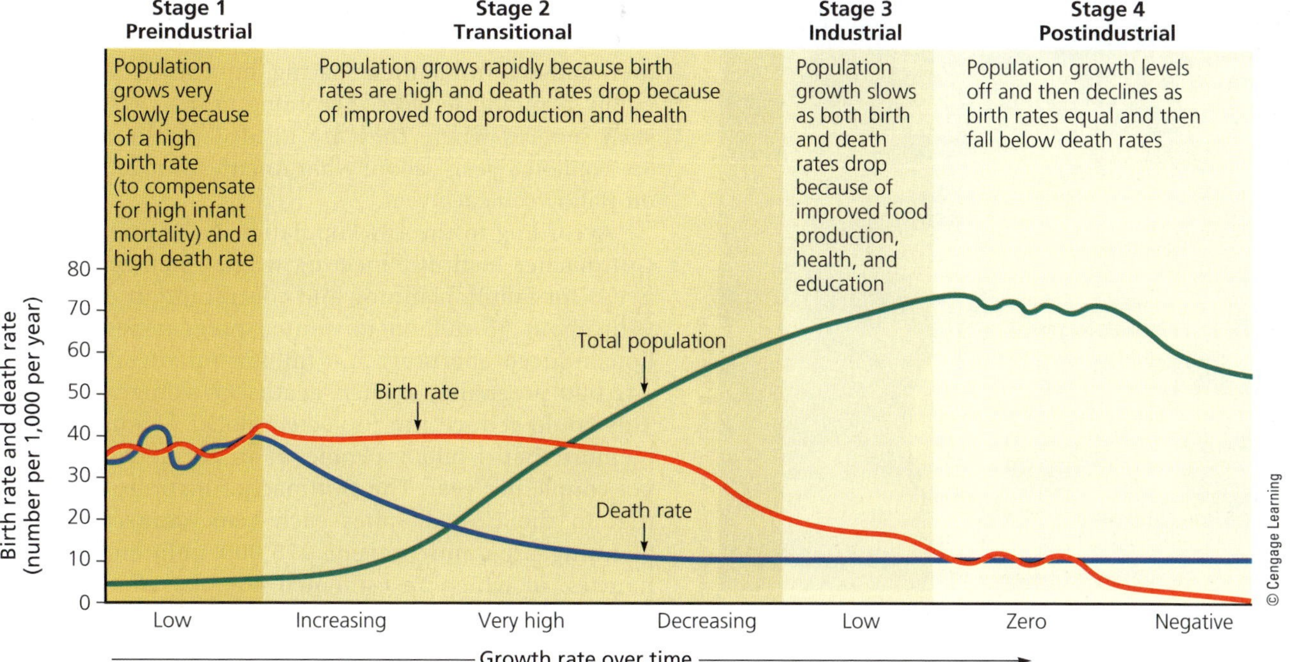

Animated Figure 6-16 The *demographic transition*, which a country can experience as it becomes industrialized and more economically developed, can take place in four stages. ***Question:*** At what stage is the country where you live?

in most societies they have fewer rights and educational and economic opportunities than men have.

Women do almost all of the world's domestic work and child care for little or no pay and provide more unpaid health care (within their families) than do all of the world's organized health-care services combined. In rural areas of Africa, Latin America, and Asia, women do 60–80% of the work associated with growing food, gathering, and hauling wood (Figure 6-17) and animal dung for use as fuel, and hauling water. As one Brazilian woman observed, "For poor women, the only holiday is when you are asleep."

While women account for 66% of all hours worked, they receive only 10% of the world's income and own just 2% of the world's land. They also make up 70% of the world's poor and 66% of its 800 million illiterate adults. Because sons are more valued than daughters in many societies, girls are often kept at home to work instead of being sent to school. Globally, the number of school-age girls who do not attend elementary school is more than 900 million—almost 3 times the entire U.S. population. Poor women who cannot read often have an average of five to seven children, compared to two or fewer children in societies where almost all women can read.

However, an increasing number of women in less-developed countries are taking charge of their lives and reproductive behavior. As it expands, such bottom-up change driven by individual women will play an important role in stabilizing populations, reducing poverty and environmental degradation, and allowing more access to basic human rights.

Family Planning Can Provide Several Benefits

Family planning provides educational and clinical services that help couples choose how many children to have and when to have them. Such programs vary from culture to culture, but most of them provide information on birth spacing, birth control, and health care for pregnant women and infants (Figure 6-18).

Figure 6-17 This woman in Nepal is bringing home firewood. Typically, she spends 2 hours a day, 2 or 3 times a week, on this task.

Helene Rogers/Art Directors & TRIP/Alamy

Figure 6-18 This family planning center is located in Kolkata (Calcutta), India.

Family planning has been a major factor in reducing the number of births throughout most of the world. It has also reduced the number of abortions performed each year and has decreased the numbers of mothers and fetuses dying during pregnancy, according to studies by the UN Population Division and other population agencies.

Such studies indicate that family planning is responsible for a drop of at least 55% in total fertility rates (TFRs) in less-developed countries, from 6.0 in 1960 to 2.6 in 2012. For example, family planning, coupled with economic development, played a major role in the sharp drop in the TFR in Bangladesh from around 6.8 to 2.3 by 2012. Between 1971 and 2012, Thailand also used family planning to cut its annual population growth rate from 3.2% to 0.5%, and to reduce its TFR from 6.4 to 1.6. According to the UN, had there not been the sharp drop in TFRs since the 1970s, with all else being equal, the world's population today would be about 8.5 billion instead of 7 billion.

Family planning also has financial benefits. Studies have shown that each dollar spent on family planning in countries such as Thailand, Egypt, and Bangladesh saves $10–$16 in health, education, and social-service costs by preventing unwanted births.

Despite these successes, certain problems have hindered progress in some countries. There are two major problems. *First,* according to the UN Population Fund, about 42% of all pregnancies in less-developed countries are unplanned and about 26% end with abortion. So ensuring access to voluntary contraception would play a key role in stabilizing the populations and reducing the number of abortions in such countries.

Second, an estimated 215 million couples in less-developed countries want to limit their number of children and determine their spacing, but they lack access to family planning services. According to the UN, providing such services where they are needed would cost about $6.7 billion a year—about what Americans together spend on Halloween each year.

According to the UN Population Fund and the Alan Guttmacher Institute, meeting women's current unmet needs for family planning and contraception could prevent about 53 million unwanted pregnancies, 24 million induced abortions, 1.6 million infant deaths, and 142,000 pregnancy-related deaths of women per year. This could reduce the projected global population size by more than 1 billion people, at an average cost of $20 per couple per year. The Guttmacher Institute estimates that in the United States each year, domestic family planning programs prevent 973,000 unintended pregnancies, of which an estimated 406,000 would end in abortion.

Some analysts call for expanding family planning programs to educate men about the importance of having fewer children and taking more responsibility for raising them. Proponents also call for greatly increased research in order to develop more effective birth control methods for men.

The experiences of countries such as Japan, Thailand, Bangladesh, South Korea, Taiwan, and China show that a country can achieve or come close to replacement-level fertility within a decade or two. Thus, the real story of the past 50 years has been the sharp reduction in the *rate* of population growth (Figure 6-2) resulting from a combination of the reduction of poverty through economic development, empowerment of women, and the promotion of family planning. However, the global population is still growing fast enough to add possibly several billion more people during this century (Figure 6-1).

CASE STUDY

Slowing Population Growth in India

For six decades, India has tried to control its population growth with only modest success. The world's first national family planning program began in India in 1952, when its population was nearly 400 million. In 2012, after 60 years of population control efforts, India had 1.26 billion people—the world's second largest population. Much of this increase occurred because the country's life expectancy rose from 38 years of age in 1952 to 65 in 2012, mostly as a result of declining death rates.

In 1952, India added 5 million people to its population. In 2012, it added 19 million—more than any other country. Also, 31% of India's population is under age 15, which sets the country up for further rapid population growth. The United Nations projects that by 2030, India will be the world's most populous country, and by 2050 it will have a population of 1.69 billion.

Figure 6-19 These people are shopping at Kolkata's (Calcutta's) South City Mall, the largest shopping mall in India.

STEVE RAYMER/National Geographic Creative

India has the world's fourth largest economy and a thriving and rapidly growing middle class of more than 100 million people—a number nearly equal to a third of the U.S. population. This growing class of consumers will enlarge India's ecological footprint, as more Indians use more resources with every passing year (Figure 6-19).

However, the country faces a number of serious poverty, malnutrition, and environmental problems that could worsen as its population continues to grow rapidly. About one-fourth of all people in India's cities live in slums, and prosperity and progress have not touched many of the nearly 650,000 rural villages where more than two-thirds of India's population lives. Nearly half of the country's labor force is unemployed or underemployed and 42% of its population lives in extreme poverty (Figure 6-20).

For decades, the Indian government has provided family planning services throughout the country and has strongly promoted a smaller average family size. Even so, Indian women have an average of 2.5 children.

Two factors help to account for larger families in India. *First,* most poor couples believe they need several children to work and care for them in old age. *Second,* the strong cultural preference in India for male children means that some couples keep having children until they produce one or more boys. The result: even though 90% of Indian couples have access to at least one modern birth control method, only 47% actually use one (compared to 85% in China).

India also faces critical resource and environmental problems. With 18% of the world's people, India has just 2.3% of the world's land resources and 2% of its forests. About half the country's cropland is degraded as a result of soil erosion and overgrazing. In addition, more than two-thirds of its water is seriously polluted, sanitation services often are inadequate, and many of its major cities suffer from serious air pollution.

India is undergoing rapid economic growth, which is expected to accelerate. This not only will help many people in India, but it will also put more pressure on the country's and the earth's natural capital. On the other hand, economic growth may help India to slow its population growth by accelerating its demographic transition.

CASE STUDY

Slowing Population Growth in China: A Success Story

China is the world's most populous country, with 1.35 billion people (Figure 6-21), followed by India with 1.26 billion and the United States with 314 million. In 2011, the U.S. Census Bureau projected that if current trends continue, China's population will increase to about 1.4 billion by 2026 and then will begin a slow decline to as low as 750 million by the end of this century.

In the 1960s, China's large population was growing so rapidly that there was a serious threat of mass starvation. To avoid this, government officials decided to take measures that eventually led to the establishment of the

Samrat35/Dreamstime LLC

Figure 6-20 Homeless people in Kolkata, India in 2011.

JUSTIN GUARIGLIA/National Geographic Creative

Figure 6-21 Thousands of people crowd a street in China, where almost one-fifth of all the people on the planet live.

world's most extensive, intrusive, and strict family planning and birth control program.

China's goal has been to sharply reduce population growth by promoting one-child families. The government provides contraceptives, sterilizations, and abortions for married couples. In addition, married couples pledging to have no more than one child receive a number of benefits, including better housing, more food, free health care, salary bonuses, and preferential job opportunities for their child. Couples who break their pledge lose such benefits.

Since this government-controlled program began, China has made impressive efforts to feed its people and bring its population growth under control. Between 1972 and 2012, the country cut its birth rate in half and reduced the average number of children born to its women from 5.7 to 1.5, compared to 1.9 children per woman in the United States.

Since 1980, China has undergone rapid industrialization and economic growth. According to the Earth Policy Institute, between 1990 and 2010, this reduced the number of people living in extreme poverty by almost 500 million. It also helped at least 300 million Chinese—a number almost equal to the entire U.S. population—to become middle-class consumers. Over time, China's rapidly growing middle class will consume more resources per person, expanding China's ecological footprint (see Figure 1-11, p. 13) within its own borders and in other parts of the world that provide it with resources. This will put a strain on China's and the earth's natural capital unless China steers a course toward more environmentally sustainable economic development.

Some have criticized China for having such a strict population control policy. However, government officials say that the alternative was mass starvation. They estimate that China's one-child policy has reduced its population size by as many as 400 million people.

Big Ideas

- **The human population is growing rapidly and may soon bump up against environmental limits.**
- **Even if population growth is not a problem, the increasing use of resources per person is expanding the overall human ecological footprint and putting a strain on the earth's resources.**
- **We can slow human population growth by reducing poverty, elevating the status of women, and encouraging family planning.**

TYING IT ALL TOGETHER World Population Growth and Sustainability

Jeremy Richards/Shutterstock.com

This chapter began with a discussion of the fact that the world's human population has now passed 7 billion (**Core Case Study**). We noted that this is a result of exponential population growth and that many environmental scientists believe such growth to be unsustainable for the long run. We briefly considered some of the environmental problems brought on by exponential human population growth. We looked at factors that influence the growth of populations, as well as at how some countries have made progress in controlling population growth.

In the first six chapters of this book, you have learned how ecosystems and species have been sustained throughout the earth's history, in keeping with the three scientific **principles of sustainability**, by nature's reliance on solar energy, nutrient cycling, and biodiversity (see Figure 1-2, p. 6 or back cover). These three principles can guide us in dealing with the problems brought on by population growth and decline. That is, by employing solar and other renewable-energy technologies more widely, we can cut pollution and emissions of climate-changing gases that are increasing as the population grows. By reusing and recycling more materials, we could cut our waste and reduce our ecological footprints. And in focusing on preserving biodiversity, we could help to sustain the life-support system on which we and all other species depend.

SUSTAINABILITY

Chapter Review

Core Case Study

1. Summarize the story of how human population growth has surpassed 7 billion and explain why this is significant to many environmental scientists (**Core Case Study**). List three factors that account for the rapid increase in the world's human population over the past 200 years.

Section 6-1

2. What is the key concept for this section? What is the range of estimates for the size of the human population in 2050? Summarize the three major population growth trends recognized by demographers. About how many people are added to the world's population each year? What are the world's three most populous countries? List eight major ways in which we have altered natural systems to meet our needs. Define **cultural carrying capacity**. Summarize the debate over whether and how long the human population can keep growing.

Section 6-2

3. What are the two key concepts for this section? Define and distinguish between **crude birth rate** and **crude death rate**. List three variables that affect the growth and decline of human populations. Explain how a given area's **population change** is calculated. Define **fertility rate**, and distinguish between the **replacement-level fertility rate** and the **total fertility rate (TFR)**. How has the global TFR changed since 1955?

4. Summarize the story of population growth in the United States and explain why it is high compared to population growth in most other more-developed countries. About how much of the annual U.S. population growth is due to immigration? List six changes in lifestyles that have taken place in the United States during the 20th century, leading to a rise in per capita resource use. What is the end effect of such changes in terms of the U.S. ecological footprint?

5. List nine factors that affect birth rates and fertility rates. Explain why there are more boys than girls in some countries. Define **life expectancy** and **infant mortality rate** and explain how they affect the population size of a country. Why does the United States have a lower life expectancy and higher infant mortality rate than a number of other more-developed countries?

6. What is **migration**? What are environmental refugees and how quickly are their numbers growing? Describe three major factors that demographers have to consider in making population projections.

Section 6-3

7. What is the key concept for this section? What is the **age structure** of a population? Explain how age structure affects population growth and economic growth. Describe the American baby boom and some of its economic and social effects. What are some problems related to rapid population decline due to an aging population? How has the AIDS epidemic affected the age structure of some countries, especially in Africa?

Section 6-4

8. What is the key concept for this section? What is the **demographic transition** and what are its four stages? What factors could hinder some less-developed countries from making this transition?
9. Explain how the reduction of poverty and empowerment of women can help countries to slow their population growth. What is **family planning** and how can it help to stabilize populations? Describe India's efforts to control its population growth. Describe China's population control program and compare it with that of India.
10. What are this chapter's *three big ideas*? Summarize the story of human population growth and explain how the three scientific **principles of sustainability** (see Figure 1-2, p. 6 or back cover) can guide us in dealing with the problems that stem from population growth and decline.

Note: Key terms are in bold type.

Critical Thinking

1. Do you think that the global population of 7.1 billion (**Core Case Study**) is too large? Explain. If your answer was *yes,* what do you think should be done to slow human population growth? If your answer was *no,* do you believe that there is a population size that would be too big? Explain.
2. If you could say hello to a new person every second without taking a break and working around the clock, how many people could you greet in 1 day? How many in a year? How long would it take you to greet the 84 million people who were added to the world's population this year? How many years would it take you to greet all 7.1 billion people on the planet?
3. Which of the three major environmental worldviews summarized on p. 21 do you believe underlie the two major positions on whether the world is overpopulated (Science Focus 6.1)?
4. Should everyone have the right to have as many children as they want? Explain. Is your belief on this issue consistent with your environmental worldview?
5. Is it rational for a poor couple in a less-developed country such as India to have four or five children? Explain.
6. Identify a major local, national, or global environmental problem, and describe the role that population growth plays in this problem.
7. Some people believe the most important environmental goal is to sharply reduce the rate of population growth in less-developed countries, where at least 92% of the world's population growth is expected to take place between now and 2050. Others argue that the most serious environmental problems stem from high levels of resource consumption per person in more-developed countries, which have much larger ecological footprints per person than do less-developed countries. What is your view on this issue? Explain.
8. Experts have identified population growth as one of the major causes of the environmental problems we face. The population of the United States is growing faster than that of any other more-developed country. However, this fact is rarely discussed, and the U.S. government has no official policy for slowing U.S. population growth. Why do think this is so? Do you think there should be such a policy? If so, explain your thinking and list three steps you would take as a leader to slow U.S. population growth. If not, explain your thinking.

Doing Environmental Science

Prepare an age-structure diagram for your community. You will need to estimate how many people belong in each age category (see p. 131). To do this, interview a randomly drawn sample of the population to find out their ages and then divide your sample into age groups. (Be sure to interview equal numbers of males and females.) Then find out the total population of your community and apply the percentages for each age group from your sample to the whole population in order to make your estimates. Create your diagram and then use it to project future population trends. Write a report in which you discuss some economic, social, and environmental effects that might result from these trends.

Global Environment Watch Exercise

Find three different projections for the size of the global population in 2050 (**Core Case Study**). Explain how the projections were made. To do this, try to find out the assumptions behind each of the projections with regard to total fertility rates, crude death rates, infant mortality rates, life expectancies, and other factors. Based on your reading, choose the projection that you believe to be the closest to reality, and explain why you chose this projection.

Data Analysis

The chart below shows selected population data for two different countries, A and B. Study the chart and answer the questions that follow.

	Country A	Country B
Population (millions)	144	82
Crude birth rate (number of live births per 1,000 people per year)	43	8
Crude death rate (number of deaths per 1,000 people per year)	18	10
Infant mortality rate (number of babies per 1,000 born who die in first year of life)	100	3.8
Total fertility rate (average number of children born to women during their child-bearing years)	5.9	1.3
% of population under 15 years old	45	14
% of population older than 65 years	3.0	19
Average life expectancy at birth	47	79
% urban	44	75

1. Calculate the rates of natural increase (due to births and deaths, not counting immigration) for the populations of country A and country B. Based on these calculations and the data in the table, for each of the countries, suggest whether it is a more-developed country or a less-developed country and explain the reasons for your answers.

2. Describe where each of the two countries may be in the stages of demographic transition (Figure 6-16). Discuss factors that could hinder either country from progressing to later stages in the demographic transition.

3. Explain how the percentages of people under 15 years of age in each country could affect its per capita and total ecological footprints.

7 Climate and Biodiversity

When we try to pick out anything by itself, we find it hitched to everything else in the universe.

JOHN MUIR

Key Questions

7-1 What factors influence climate?

7-2 How does climate affect the nature and location of biomes?

7-3 How have human activities affected the world's terrestrial ecosystems?

Giraffes on tropical grassland (savanna) in Africa.

Oleg Znamenskiy/Shutterstock.com

CORE CASE STUDY A Temperate Deciduous Forest

The earth hosts a great diversity of species and *habitats,* or places where these species can live. Some species live in *terrestrial,* or land, habitats such as deserts, grasslands (see chapter-opening photo), and forests—the three major types of terrestrial ecosystems, also called *biomes.* Other species live in water-covered habitats in *aquatic life zones,* such as rivers, lakes, and oceans. In this chapter, we look at key terrestrial habitats, and in the next chapter, we look at major aquatic habitats.

Why do forests grow on some areas of the earth's land while deserts form in other areas? The answers lie largely in differences in *climate,* the average atmospheric conditions in a given region over a period of time ranging from at least three decades to thousands of years. Differences in climate result mostly from long-term differences in average annual precipitation and temperature. These differences lead to three major types of climate—*tropical* (areas near the equator, receiving the most intense sunlight), *polar* (areas near the earth's poles, receiving the least intense sunlight), and *temperate* (areas between the tropic and polar regions).

Throughout these regions, we find different types of ecosystems, vegetation, and animals adapted to the various climate conditions. For example, scattered throughout the temperate areas of the globe, we find *temperate deciduous forests* (Figure 7-1). Such forests typically see warm summers, cold winters, and abundant precipitation—rain in summer and snow in winter months. They are dominated by a few species of *broadleaf deciduous trees* such as oak, hickory, maple, aspen, and birch. Animal species living in these forests include predators such as wolves, foxes, and wildcats. They feed on herbivores such as white-tailed deer (see Figure 5-17, p. 115), squirrels, rabbits, opossums, and mice. Warblers, robins, and other bird species live in these forests during the spring and summer, mating and raising their young.

These species are adapted to the conditions of temperate forests. For example, most of the trees' leaves, after developing their vibrant colors in the fall (Figure 7-1, left), drop off the trees. This allows the trees to survive the cold winters by becoming dormant (Figure 7-1, right). Each spring, they sprout new leaves and spend their summers growing and producing until the cold weather returns. Some forest mammals, including bears, spend their summers storing fat and then hibernating during winter, sleeping the coldest months away. Many bird species migrate to warmer climates in the late fall to avoid the cold and snow.

Temperate deciduous forests once covered the eastern half of the United States and western Europe. But as these areas became urbanized, most of the original forests were cleared. Today, on a worldwide basis, this terrestrial ecosystem has been disturbed more than any other terrestrial system by human activities.

In this chapter, we examine the key role that climate plays in the location and formation of forests and all other major terrestrial ecosystems that make up an important part of the earth's terrestrial biodiversity.

Figure 7-1 A temperate deciduous forest in fall (left) and in winter (right).

7-1 What Factors Influence Climate?

CONCEPT 7-1

Key factors that influence an area's climate are incoming solar energy, the earth's rotation, global patterns of air and water movement, gases in the atmosphere, and the earth's surface features.

The Earth Has Many Different Climates

It is important to understand the difference between weather and climate. **Weather** is a set of physical conditions of the lower atmosphere, including temperature, precipitation, humidity, wind speed, cloud cover, and other factors, in a given area over a period of hours or days. (Supplement 5, p. S19, introduces you to the basics of weather.)

Weather differs from **climate**, which is the general pattern of atmospheric conditions in a given area over periods ranging from at least three decades to thousands of years. In other words, climate is the sum of weather conditions in a given area, averaged over a long period of time.

We know a lot about the daily weather where we live, but we might not know as much about the climate where we live or of how it has changed over the past 30 to 100 years. To get such an understanding, you would first have to find and plot data on the average temperature and average precipitation in your area from year to year over at least the last three decades. Then you would have to note whether these numbers have generally gone up, stayed the same, or gone down.

Based on this type of analysis, scientists have described the various regions of the earth according to their climates. Figure 7-2 shows these major climate zones along with the major **ocean currents**—mass movements of surface water driven by winds blowing over the oceans. These currents help to determine regional climates and are a key component of the earth's natural capital (see Figure 1-3, p. 7).

Climate varies among the earth's different regions primarily because, over long periods of time, patterns of global air circulation and ocean currents distribute heat and precipitation unevenly between the tropics and other

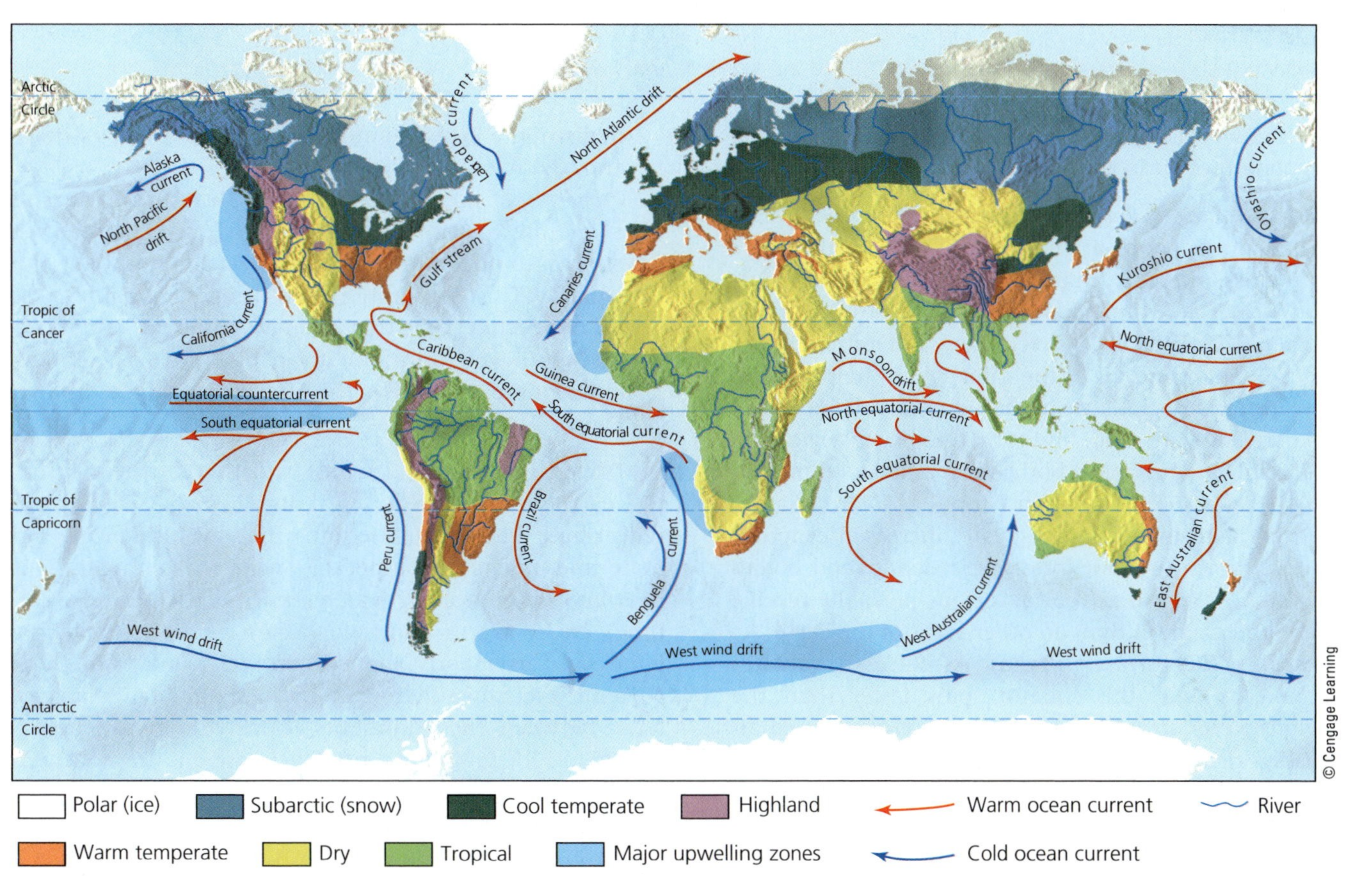

Animated Figure 7-2 **Natural capital:** This generalized map of the earth's current climate zones also shows the major ocean currents and upwelling areas (where currents bring nutrients from the ocean bottom to the surface). ***Question:*** Based on this map, what is the general type of climate where you live?

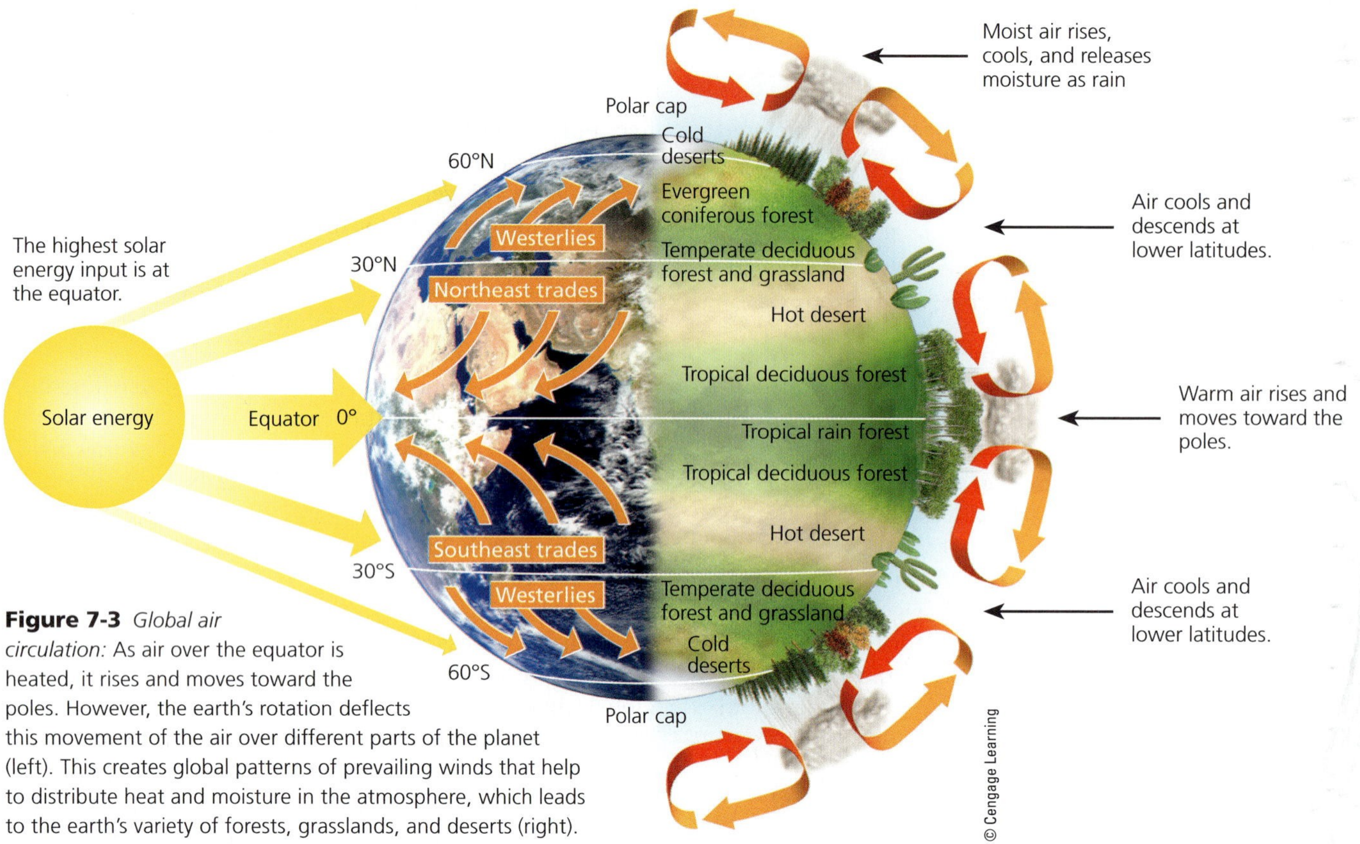

Figure 7-3 *Global air circulation:* As air over the equator is heated, it rises and moves toward the poles. However, the earth's rotation deflects this movement of the air over different parts of the planet (left). This creates global patterns of prevailing winds that help to distribute heat and moisture in the atmosphere, which leads to the earth's variety of forests, grasslands, and deserts (right).

parts of the world (Figure 7-3). Three major factors affect the circulation of air in the lower atmosphere:

1. *Uneven heating of the earth's surface by the sun.* Air is heated much more at the equator, where the sun's rays strike directly, than at the poles, where sunlight strikes at an angle and spreads out over a much greater area (Figure 7-3, left). These differences in the input of solar energy to the atmosphere help explain why tropical regions near the equator are hot, why polar regions are cold, and why temperate regions in between generally have both warm and cool temperatures (Figure 7-2). The intense input of solar radiation in tropical regions leads to greatly increased evaporation of moisture from forests, grasslands, and bodies of water. As a result, tropical regions normally receive more precipitation than do other areas of the earth.
2. *Rotation of the earth on its axis.* As the earth rotates around its axis, the equator spins faster than the regions to its north and south. As a result, heated air masses, rising above the equator and moving north and south to cooler areas, are deflected in different ways over different parts of the planet's surface (Figure 7-3, left). The atmosphere over these different areas is divided into huge regions called *cells,* distinguished by the direction of air movement. The differing directions of air movement are called *prevailing winds*—major surface winds that blow almost continuously and help to distribute heat and moisture over the earth's surface and to drive ocean currents.
3. *Properties of air, water, and land.* Heat from the sun evaporates ocean water and transfers heat from the oceans to the atmosphere, especially near the hot equator. This evaporation of water creates giant cyclical convection cells that circulate air, heat, and moisture both vertically and from place to place in the atmosphere, as shown in Figure 7-3, right, and in Figure 7-4.

Driven by prevailing winds and the earth's rotation, the earth's major ocean currents (Figure 7-2) help to redistribute heat from the sun, thereby influencing climate and vegetation, especially near coastal areas. This heat and differences in water *density* (mass per unit volume) create warm and cold ocean currents. Prevailing winds and irregularly shaped continents interrupt these currents and cause them to flow in roughly circular patterns between the continents, clockwise in the northern hemisphere and counterclockwise in the southern hemisphere.

Water also moves vertically in the oceans as denser water sinks while less dense water rises. This creates a connected loop of deep and shallow ocean currents (which are separate from those shown in Figure 7-2). This loop acts somewhat like a giant conveyor belt that moves heat from the surface to the deep sea and trans-

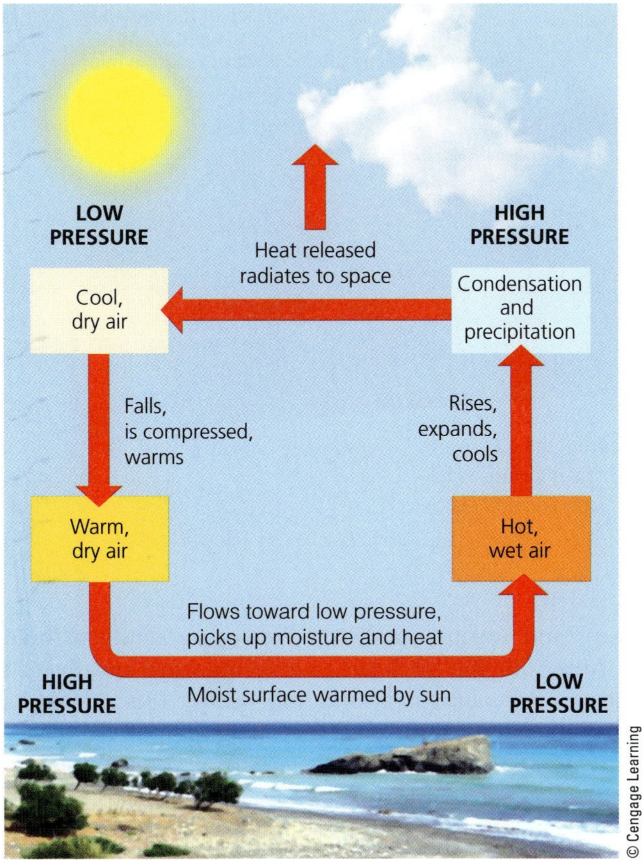

Figure 7-4 Energy is transferred by *convection* in the atmosphere—the process by which warm, wet air rises, then cools and releases heat and moisture as precipitation (right side and top, center). Then the cooler, denser, and drier air sinks, warms up, and absorbs moisture as it flows across the earth's surface (left side and bottom) to begin the cycle again.

fers warm and cold water between the tropics and the poles (Figure 7-5).

The ocean and the atmosphere are strongly linked in two ways: ocean currents are affected by winds in the atmosphere, and heat from the ocean affects atmospheric circulation. One example of the interactions between the ocean and the atmosphere is the *El Niño–Southern Oscillation,* or *ENSO* (see Figure 4, p. S21, in Supplement 5). This large-scale weather phenomenon occurs every few years when prevailing winds in the tropical Pacific Ocean weaken and change direction. The resulting above-average warming of Pacific waters alters the weather over at least two-thirds of the earth for 1 or 2 years (see Figure 5, p. S21, in Supplement 5).

The earth's air circulation patterns, prevailing winds, and configuration of continents and oceans are all factors in the formation of six giant convection cells (like the one shown in Figure 7-4), three of them south of the equator and three north of the equator. These cells lead to an irregular distribution of climates and of the resulting deserts, grasslands, and forests, as shown in Figure 7-3, right (**Concept 7-1**).

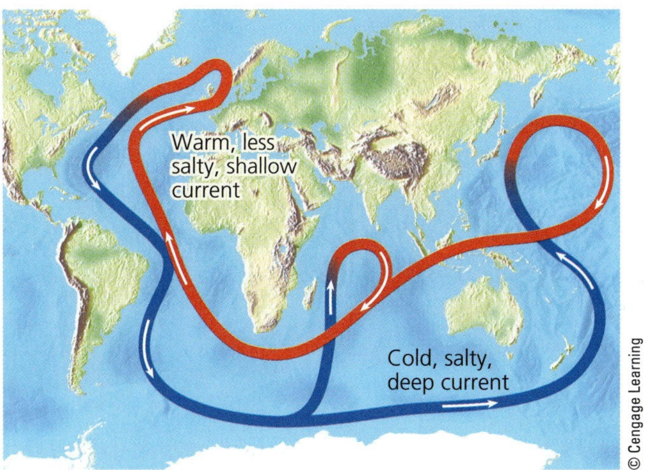

Figure 7-5 A connected loop of deep and shallow ocean currents transports warm and cool water to various parts of the earth.

Greenhouse Gases Warm the Lower Atmosphere

As energy flows from the sun to the earth, some of it is reflected by the earth's surface back into the atmosphere (see Figure 3-3, p. 54). Molecules of certain gases in the atmosphere, including water vapor (H_2O), carbon dioxide (CO_2), methane (CH_4), and nitrous oxide (N_2O), absorb some of this solar energy and release a portion of it as infrared radiation (heat) that warms the lower atmosphere. Thus, these gases, called **greenhouse gases**, play a role in determining the lower atmosphere's average temperatures and thus the earth's climates.

The earth's surface also absorbs much of the solar energy that strikes it and transforms it into longer-wavelength infrared radiation, which then rises into the lower atmosphere. Some of this heat escapes into space, but some is absorbed by molecules of greenhouse gases and emitted into the lower atmosphere as even longer-wavelength infrared radiation (see Figure 2-11, p. 41). Some of this released energy radiates into space, and some adds to the warming of the lower atmosphere and the earth's surface. Together, these processes result in a natural warming of the troposphere, called the **greenhouse effect** (see Figure 3-3, p. 54). Without this natural warming effect, the earth would be a very cold and mostly lifeless planet.

Human activities such as the burning of fossil fuels, clearing of forests, and growing of crops release carbon dioxide, methane, and nitrous oxide into the atmosphere. A considerable body of scientific evidence, combined with climate model projections, indicates that continued inputs of these greenhouse gases into the atmosphere from human activities are likely to enhance the earth's natural greenhouse effect and change the earth's climate during this century. If this occurs, it will likely alter temperature and precipitation patterns, raise average sea levels, and shift areas where we can grow crops and where some

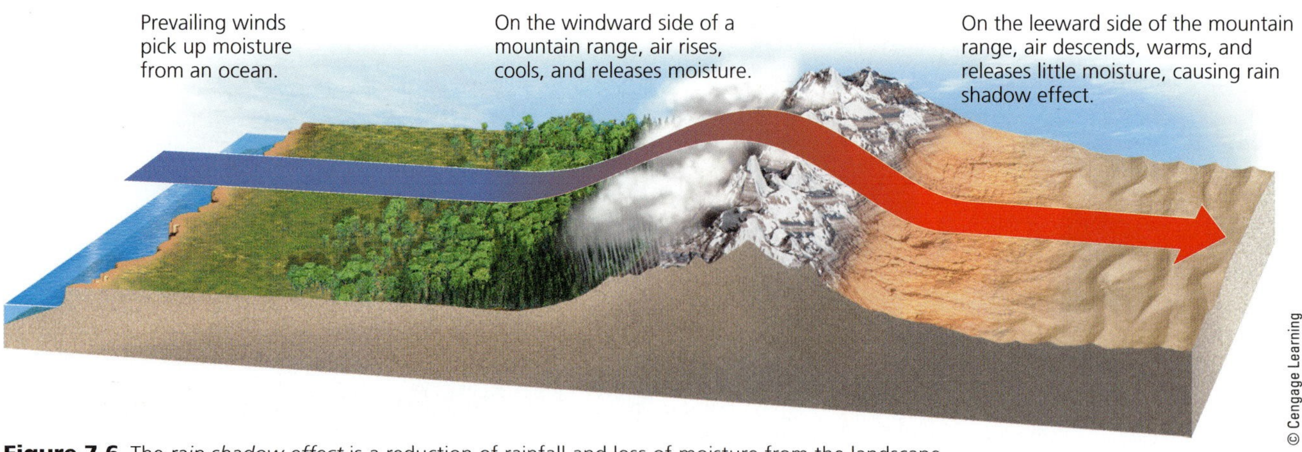

Figure 7-6 The *rain shadow effect* is a reduction of rainfall and loss of moisture from the landscape on the leeward side of a mountain. Warm, moist air in onshore winds loses most of its moisture as rain and snow that fall on the windward slopes of a mountain range. This leads to semiarid and arid conditions on the leeward side of the mountain range and on the land beyond.

types of plants and animals (including humans) can live, as discussed more fully in Chapter 19.

The Earth's Surface Features Affect Local Climates

Various topographic features of the earth's surface can create local and regional climatic conditions that differ from the general climate in some regions. For example, mountains interrupt the flow of prevailing surface winds and the movement of storms. When moist air blowing inland from an ocean reaches a mountain range, it is forced upward. As it rises, it cools and expands, and then loses most of its moisture as rain and snow that fall on the windward slope of the mountain.

As the drier air mass passes over the mountaintops, it flows down the leeward slopes (facing away from the wind), and warms up. This increases its ability to hold moisture, but the air releases little moisture and instead tends to dry out plants and soil below. This process is called the **rain shadow effect** (Figure 7-6), and over many decades, it results in *semiarid* or *arid* conditions on the leeward side of a high mountain range. Sometimes this effect leads to the formation of deserts such as Death Valley, a part of the Mojave Desert found in parts of the U.S. states of California, Nevada, Utah, and Arizona.

Cities also create distinct microclimates. Bricks, concrete, asphalt, and other building materials absorb and hold heat, and buildings block wind flow. Motor vehicles and the heating and cooling systems of buildings release large quantities of heat and pollutants. As a result, cities on average tend to have more haze and smog, higher temperatures that make them *heat islands,* and lower wind speeds than the surrounding countryside.

7-2 How Does Climate Affect the Nature and Location of Biomes?

CONCEPT 7-2

Differences in long-term average annual precipitation and temperature lead to the formation of tropical, temperate, and cold deserts, grasslands, and forests, and largely determine their locations.

Climate Helps to Determine Where Terrestrial Organisms Can Live

Differences in climate (Figure 7-2) help to explain why one area of the earth's land surface is a desert, another a grassland, and another a forest (see Figure 2, p. S26, Supplement 6). Furthermore, different combinations of varying average annual precipitation and temperatures, along with global air circulation and ocean currents, lead to the formation of tropical (hot), temperate (moderate), and polar (cold) deserts, grasslands, and forests, as summarized in Figure 7-7 (Concept 7-2).

Climate and vegetation vary according to *latitude* and also according to *elevation,* or height above sea level. If you climb a tall mountain, from its base to its summit, you can observe changes in plant life similar to those you would encounter in traveling from the equator to the earth's northern polar region (Figure 7-8).

Figure 7-9 shows how scientists have divided the world into several major **biomes**—large terrestrial regions, each characterized by a certain type of climate and a certain combination of dominant plant life. The variety of terrestrial biomes and aquatic systems is one of

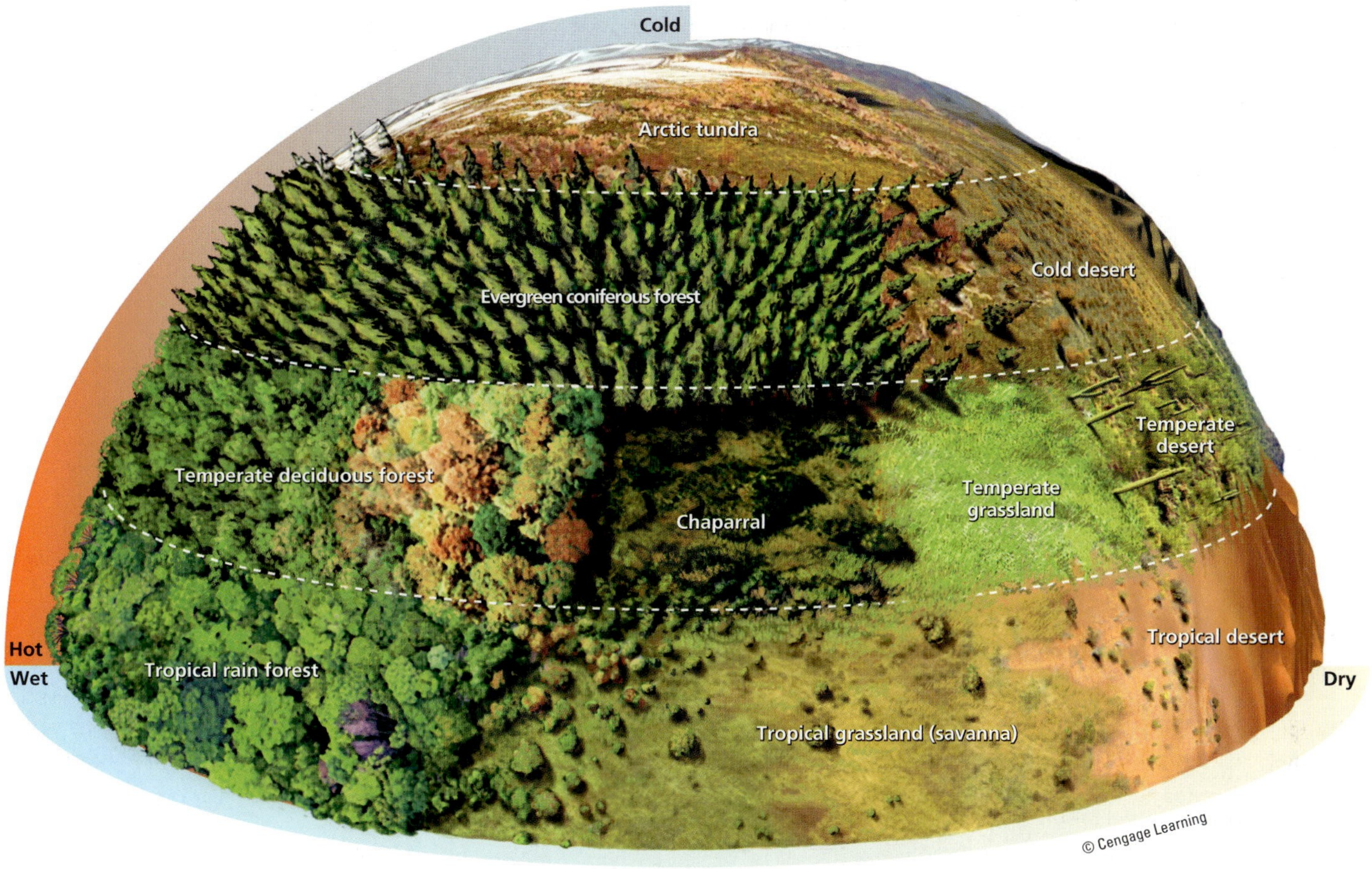

Figure 7-7 **Natural capital:** Average precipitation and average temperature, acting together as limiting factors over a long time, help to determine the type of desert, grassland, or forest in any particular area, and thus the types of plants, animals, and decomposers found in that area (assuming it has not been disturbed by human activities).

Figure 7-8 Biomes and climate both change with elevation (left), as well as with latitude (right).

the four components of the earth's biodiversity (see Figure 4-2, p. 79)—a vital part of the earth's natural capital.

By comparing Figure 7-9 with Figure 7-2, you can see how the world's major biomes vary with climate. Figure 4-4 (p. 81) shows how major biomes along the 39th parallel in the United States are related to different climates. See the maps in Supplement 6 (pp. S32–S39) that show the major biomes in North America (Figure 6), South America (Figure 10), Europe (Figure 11), Asia (Figure 12), Africa (Figure 13), and Oceania and Australia (Figure 14).

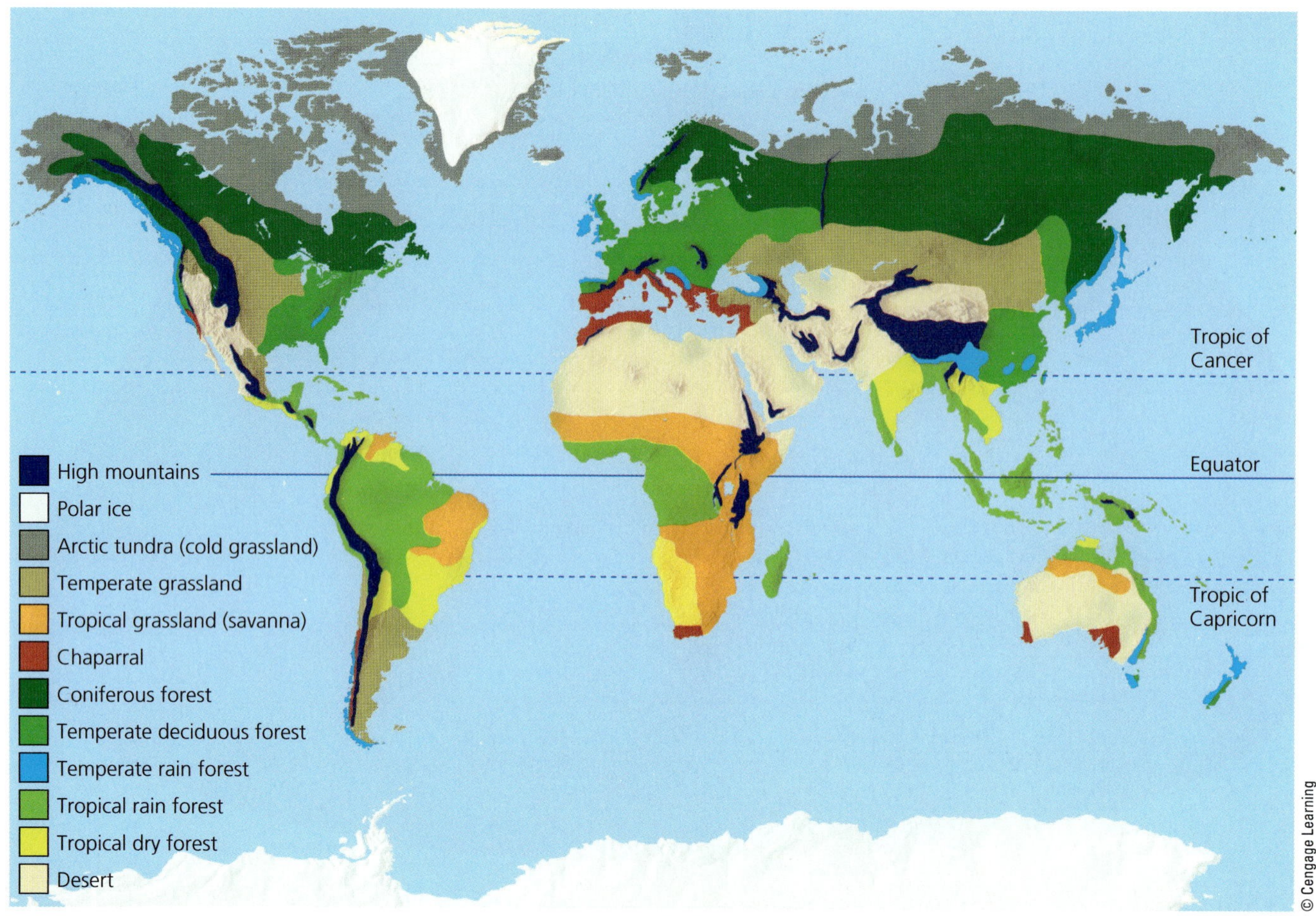

Animated Figure 7-9 **Natural capital:** The earth's major *biomes* result primarily from differences in climate.

On maps such as the one in Figure 7-9, biomes are shown with sharp boundaries, and each biome is covered with one general type of vegetation. In reality, biomes are not uniform. They consist of a *mosaic of patches,* each with somewhat different biological communities but with similarities typical of the biome. These patches occur primarily because of the irregular distribution of the resources needed by plants and animals and because human activities have removed or altered the natural vegetation in many areas.

CONSIDER THIS. . .

THINKING ABOUT Biomes, Climate, and Human Activities

Use Figure 7-2 to determine the general type of climate where you live and Figure 7-9 to determine the general type of biome that should exist where you live. Then use Figure 1-12, p. 13 and Figure 8 in Supplement 6, p. S33 to determine how human ecological footprints have affected the biome where you live.

There Are Three Major Types of Deserts

In a *desert,* annual precipitation is low and often scattered unevenly throughout the year. During the day, the baking sun warms the ground and evaporates water from plant leaves and from the soil. But at night, most of the heat stored in the ground radiates quickly into the atmosphere. Desert soils have little vegetation and moisture to help store the heat and the skies above deserts are usually clear. This explains why in a desert you might roast during the day but shiver at night.

A combination of low rainfall and varying average temperatures creates a variety of desert types—tropical, temperate, and cold (Figures 7-7 and 7-10 and **Concept 7-2**). *Tropical deserts* (Figure 7-10, top photo) such as the Sahara and the Namib of Africa are hot and dry most of the year (Figure 7-10, top graph). They have few plants and a hard, windblown surface strewn with rocks and some sand. They are the deserts we often see in the movies.

In *temperate deserts* (Figure 7-10, center photo) such as the Sonoran Desert in southeastern California, southwestern Arizona, and northwestern Mexico, daytime temperatures are high in summer and low in winter and there is more precipitation than in tropical deserts (Figure 7-10, center graph). The sparse vegetation primarily consists of widely dispersed, drought-resistant shrubs and cacti or other succulents adapted to the lack of water and temperature variations.

In *cold deserts* such as the Gobi Desert in Mongolia, vegetation is sparse (Figure 7-10, bottom photo).

Apdesign/Shutterstock.com

Tropical desert

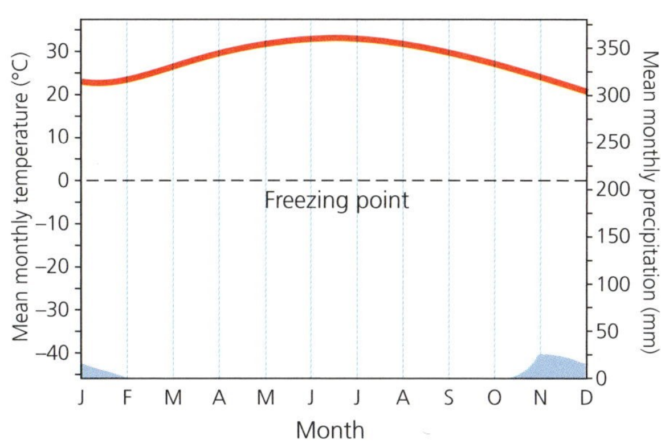

Mike Norton/Shutterstock.com

Temperate desert

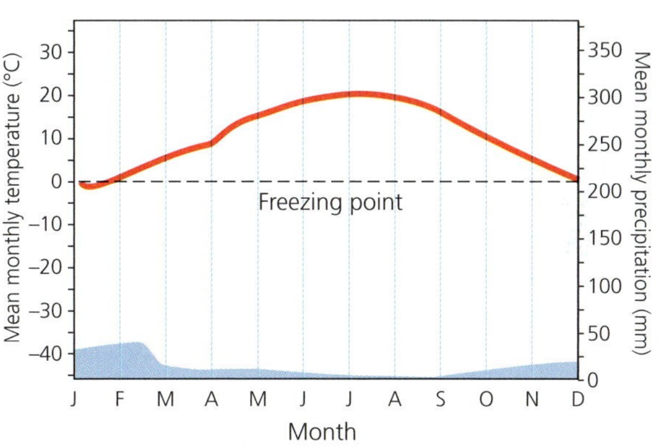

ageFotostock/SuperStock

Cold desert

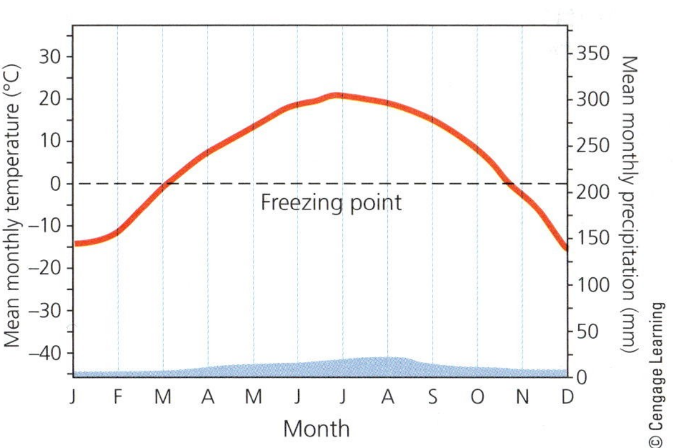

Figure 7-10 These climate graphs track the typical variations in annual temperature (red) and precipitation (blue) in tropical, temperate, and cold deserts. Top photo: a *tropical desert* in Morocco. Center photo: a *temperate desert* in southeastern California, with saguaro cactus, a prominent species in this ecosystem. Bottom photo: a *cold desert,* Mongolia's Gobi Desert, where Bactrian camels live. ***Question:*** Which month of the year has the highest temperature and which month has the lowest rainfall for each of the three types of deserts?

SCIENCE FOCUS 7.1

STAYING ALIVE IN THE DESERT

Adaptations for survival in the desert have two themes: *beat the heat* and *every drop of water counts.*

Desert plants have evolved a number of strategies based on such adaptations. During long hot and dry spells, plants such as mesquite and creosote drop their leaves to survive in a dormant state. *Succulent* (fleshy) *plants* such as the saguaro ("sah-WAH-ro") cactus (Figure 7-A and Figure 7-10, middle photo) have three adaptations: they have no leaves, which can lose water to the atmosphere through *transpiration;* they store water and synthesize food in their expandable, fleshy tissue; and they reduce water loss by opening their pores only at night to take up carbon dioxide (CO_2). The spines of these and many other desert plants guard them from being eaten by herbivores seeking the precious water they hold.

Some desert plants use deep roots to tap into groundwater. Others such as prickly pear and saguaro cacti use widely spread shallow roots to collect water after brief showers and store it in their spongy tissues.

Some desert plants conserve water by having wax-coated leaves that reduce water loss. Others such as annual wildflowers and grasses store much of their biomass in seeds that remain inactive, sometimes for years, until they receive enough water to germinate. Shortly after a rain, these seeds germinate, grow, and carpet some deserts with dazzling arrays of colorful flowers (Figure 7-A) that last for up to a few weeks.

Most desert animals are small. Some beat the heat by hiding in cool burrows or rocky crevices by day and coming out at night or in the early morning. Others become dormant during periods of extreme heat or drought. Some larger animals such as camels (Figure 7-10, bottom photo) can drink massive quantities of water when it is available and store it in their fat for use as needed. Also, the camel's thick fur actually helps it to keep cool because the air spaces in the fur insulate the camel's skin against the outside heat. And camels do not sweat, which reduces their water loss through evaporation. Kangaroo rats never drink water. They get the water they need by breaking down fats in seeds that they consume.

Insects and reptiles such as rattlesnakes have thick outer coverings to minimize water loss through evaporation, and their wastes are dry feces and a dried concentrate of urine. Many spiders and insects get their water from dew or from the food they eat.

Anton Foltin/Shutterstock.com

Figure 7-A After a brief rain, these wildflowers bloomed in this temperate desert in Picacho Peak State Park in the U.S. state of Arizona.

Critical Thinking

What are three steps you would take to survive in the open desert if you had to?

Winters are cold, summers are warm or hot, and precipitation is low (Figure 7-10, bottom graph). In all types of deserts, plants and animals have evolved adaptations that help them to stay cool and to get enough water to survive (Science Focus 7.1).

Desert ecosystems are fragile because they experience slow plant growth, low species diversity, slow nutrient cycling (due to low bacterial activity in the soils), and very little water. Their soils take from decades to centuries to recover from disturbances such as the growing problem of off-road vehicle traffic, which can also destroy the habitats for a variety of animals that live underground. The lack of vegetation, especially in tropical and polar deserts, also makes them vulnerable to heavy erosion from sandstorms.

There Are Three Major Types of Grasslands

Grasslands occur primarily in the interiors of continents in areas that are too moist for deserts to form and too dry for forests to grow (Figure 7-9). Grasslands persist because of a combination of seasonal drought, grazing by large herbivores, and occasional fires—all of which keep shrubs and trees from growing in large numbers.

The three main types of grasslands—tropical, temperate, and cold (arctic tundra)—result from combinations of low average precipitation and varying average temperatures (**Concept 7-2**) (Figures 7-7 and 7-11).

One type of tropical grassland, called a *savanna* (Figure 7-11, top photo), contains widely scattered clumps of trees. This biome usually has warm temperatures year-

erichon/Shutterstock.com

Tropical grassland (savanna)

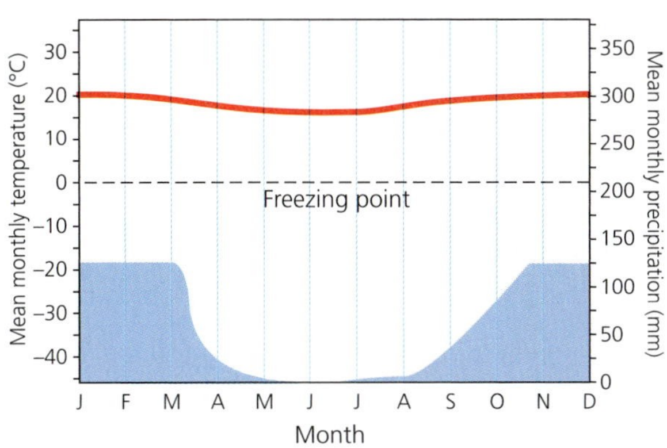

biletskiy/Shutterstock.com

Temperate grassland (prairie)

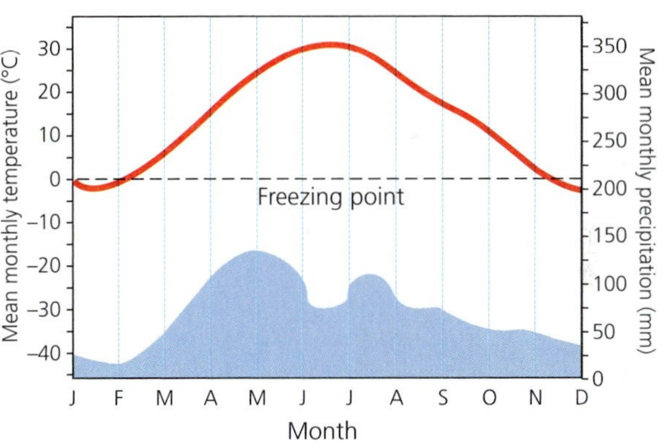

©ARCTIC NATIONAL WILDLIFE REFUGE

Cold grassland (arctic tundra)

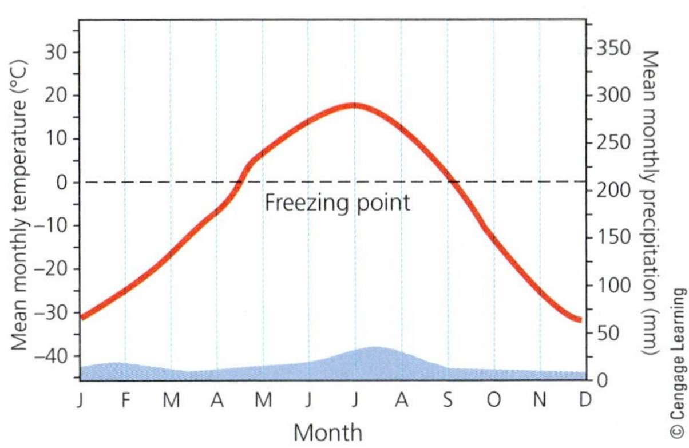

Figure 7-11 These climate graphs track the typical variations in annual temperature (red) and precipitation (blue) in tropical, temperate, and cold (arctic tundra) grasslands. Top photo: *savanna (tropical grassland)* in Kenya, Africa, with zebras grazing. Center photo: *prairie (temperate grassland)* in the U.S. state of Illinois, with wildflowers in bloom. Bottom photo: *arctic tundra (cold grassland)* in Alaska's Arctic National Wildlife Refuge in summer. ***Question:*** Which month of the year has the highest temperature and which month has the lowest rainfall for each of the three types of grasslands?

National Archives/EPA Documerica

Figure 7-12 **Natural capital degradation:** This intensively cultivated cropland is an example of the replacement of biologically diverse temperate grasslands (such as in the center photo of Figure 7-11) with a monoculture crop.

round and alternating dry and wet seasons (Figure 7-11, top graph).

Tropical savannas in East Africa are home to *grazing* (primarily grass-eating) and *browsing* (twig- and leaf-nibbling) hoofed animals, including wildebeests, gazelles, zebras (Figure 7-11, top photo), giraffes (chapter-opening photo), and antelopes, as well as their predators such as lions, hyenas, and humans. Herds of these grazing and browsing animals migrate to find water and food in response to seasonal and year-to-year variations in rainfall (Figure 7-11, blue region in top graph) and food availability. Savanna plants, like those in deserts, are adapted to survive drought and extreme heat; many have deep roots that can tap into groundwater.

CONSIDER THIS. . .

CONNECTIONS Grassland Niches and Feeding Habits

As an example of differing niches, some large herbivores have evolved specialized eating habits that minimize competition among species for the vegetation found on the savanna. For example, giraffes eat leaves and shoots from the tops of trees, elephants eat leaves and branches farther down, wildebeests prefer short grasses, and zebras graze on longer grasses and stems.

In a *temperate grassland,* winters can be bitterly cold, summers are hot and dry, and annual precipitation is fairly sparse and falls unevenly throughout the year (Figure 7-11, center graph). Because the aboveground parts of most of the grasses die and decompose each year, organic matter accumulates to produce deep, fertile topsoil. This topsoil is held in place by a thick network of the grasses' intertwined roots (unless the topsoil is plowed up, which exposes it to high winds found in these biomes). This biome's grasses are adapted to periodic droughts and to fires that burn the plant parts above the ground but do not harm the roots, from which new grass can grow.

In the mid-western and western areas of the United States, we find two types of temperate grasslands depending primarily on average rainfall: *short-grass prairies* (Figure 7-11, center photo) and the *tallgrass prairies* (which get more rain). In all prairies, winds blow almost continuously and evaporation is rapid, often leading to fires in the summer and fall. This combination of winds and fires helps to maintain such grasslands by hindering tree growth. Many of the world's natural temperate grasslands have been converted to farmland, because their fertile soils are useful for growing crops (Figure 7-12) and grazing cattle.

Cold grasslands, or *arctic tundra* (Russian for "marshy plain"), lie south of the arctic polar ice cap (Figure 7-9). During most of the year, these treeless plains are bitterly cold (Figure 7-11, bottom graph), swept by frigid winds, and covered with ice and snow. Winters are long with few hours of daylight, and the scant precipitation falls primarily as snow.

Under the snow, this biome is carpeted with a thick, spongy mat of low-growing plants, primarily grasses, mosses, lichens, and dwarf shrubs. Trees and tall plants cannot survive in the cold and windy tundra because they would lose too much of their heat. Most of the annual growth of the tundra's plants occurs during the 7- to 8-week summer, when there is daylight almost around the clock.

One outcome of the extreme cold is the formation of **permafrost**, underground soil in which captured water stays frozen for more than two consecutive years. During the brief summer, the permafrost layer keeps melted snow and ice from draining into the ground. As a consequence, many shallow lakes, marshes, bogs, ponds, and other seasonal wetlands form when snow and frozen surface soil melt on the waterlogged tundra. Hordes of mosquitoes, black flies, and other insects thrive in these shallow surface pools. They serve as food for large colonies of migratory birds (especially waterfowl) that migrate from the south to nest and breed in the bogs and ponds.

Animals in this biome survive the intense winter cold through adaptations such as thick coats of fur (arctic wolf, arctic fox, and musk oxen) or feathers (snowy owl) or living underground (arctic lemming). In the summer, caribou (often called reindeer) and other types of deer migrate to the tundra to graze on its vegetation.

Tundra is a fragile biome. Most tundra soils formed about 17,000 years ago when glaciers began retreating after the last Ice Age (see Figure 4-10, p. 87). These soils usually are nutrient poor. Because of the short growing season, tundra soil and vegetation recover very slowly from damage or disturbance. Human activities in the arctic tundra—primarily on and around oil drilling sites, pipelines, mines, and military bases—leave scars that persist for centuries.

Another type of tundra, called *alpine tundra,* occurs above the limit of tree growth but below the permanent snow line on high mountains (Figure 7-8, left). The vegetation is similar to that found in arctic tundra, but it receives more sunlight than arctic vegetation gets. During the brief summer, alpine tundra can be covered with an array of beautiful wildflowers.

There Are Three Major Types of Forests

Forests are lands that are dominated by trees. The three main types of forests—*tropical, temperate* (**Core Case Study**), and *cold* (northern coniferous, or boreal)—result from combinations of varying precipitation levels and varying average temperatures (**Concept 7-2**) (Figures 7-7 and 7-13).

Tropical rain forests (Figure 7-13, top photo) are found near the equator (Figure 7-9), where hot, moisture-laden air rises and dumps its moisture (Figure 7-3). These lush forests have year-round, uniformly warm temperatures, high humidity, and almost daily heavy rainfall (Figure 7-13, top graph). This fairly constant warm, wet climate is ideal for a wide variety of plants and animals.

Tropical rain forests are dominated by *broadleaf evergreen plants,* which keep most of their leaves year-round. The tops of the trees form a dense *canopy* (Figure 7-13, top photo) that blocks most light from reaching the forest floor. For this reason, there is little vegetation on the forest floor. Many of the plants that do live at the ground level have enormous leaves to capture what little sunlight filters down to them.

Some trees are draped with vines (called *lianas*) that reach for the treetops to gain access to sunlight. In the canopy, the vines grow from one tree to another, providing walkways for many species living there. When a large tree is cut down, its network of lianas can pull down other trees.

Tropical rain forests have a very high net primary productivity (see Figure 3-14, p. 62). They are teeming with life and possess incredible biological diversity. Although tropical rain forests cover only about 2% of the earth's land surface, ecologists estimate that they contain at least 50% of the known terrestrial plant and animal species. For example, a single tree in these forests may support several thousand different insect species (see Chapter 2 opening photo, p. 28). Plants from tropical rain forests are a source of chemicals, many of which have been used as blueprints for making most of the world's prescription drugs.

Rain forest species occupy a variety of specialized niches in distinct layers, which help to enable these forests' great biodiversity (high species richness). Vegetation layers are structured, for the most part, according to the plants' needs for sunlight, as shown in Figure 7-14. Much of the animal life, particularly insects, bats, and birds, lives in the sunny canopy layer, with its abundant shelter and supplies of leaves, flowers, and fruits. To study life in the treetops, ecologists climb trees and build platforms and boardwalks in the upper canopy.

Dropped leaves, fallen trees, and dead animals decompose quickly in tropical rain forests because of the warm, moist conditions and the hordes of decomposers. About 90% of the nutrients released by this rapid decomposition are quickly taken up and stored by trees, vines, and other plants. Nutrients that are not taken up are soon leached from the thin topsoil by the almost daily rainfall. As a result, very little plant litter builds up on the ground. The resulting lack of fertile soil helps to explain why rain forests are not good places to clear and grow crops or graze cattle on a sustainable basis.

So far, at least half of these forests have been destroyed or disturbed by human activities such as farming and cattle production, and the pace of this destruction and degradation is increasing (see Chapter 3 Core Case Study, p. 52). Ecologists warn that without strong protective

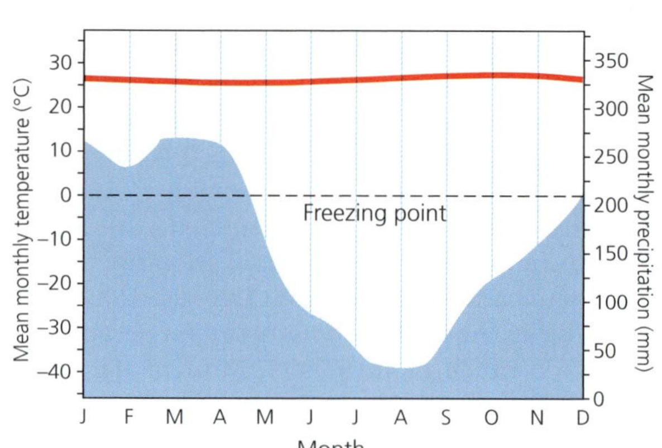

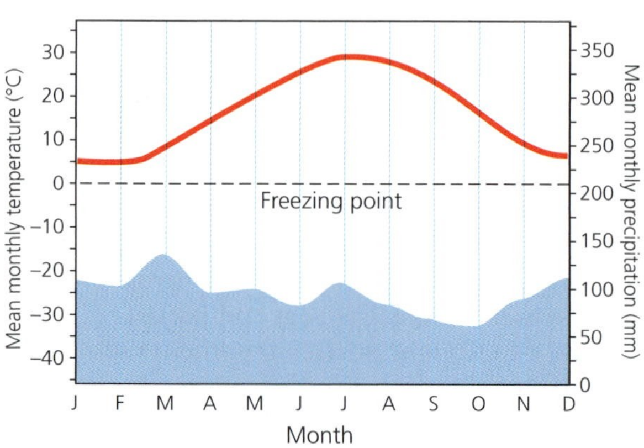

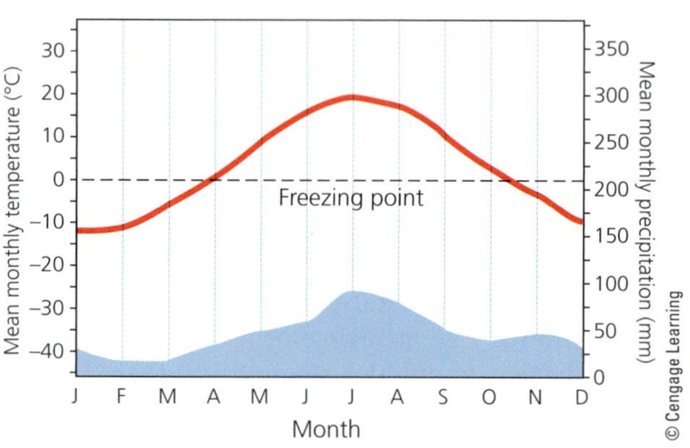

Figure 7-13 These climate graphs track the typical variations in annual temperature (red) and precipitation (blue) in tropical, temperate, and cold (northern coniferous, or boreal) forests. Top photo: the closed canopy of a *tropical rain forest* in Costa Rica. Middle photo: a *temperate deciduous forest* in the autumn near Hamburg, Germany (see also Figure 7-1, left). Bottom photo: a *northern coniferous forest* in Canada's Jasper National Park. ***Question:*** Which month of the year has the highest temperature and which month has the lowest rainfall for each of the three types of forests?

Figure 7-14 Specialized plant and animal niches are *stratified*, or arranged roughly in layers, in a tropical rain forest. Filling such specialized niches enables species to avoid or minimize competition for resources and results in the coexistence of a great variety of species.

measures, most of these forests, along with their rich biodiversity and other very valuable ecosystem services, could be gone by the end of this century.

The second major type of forest, the *temperate deciduous forest,* is the subject of this chapter's **Core Case Study** (see middle photo of Figure 7-13). Because they have cooler temperatures and fewer decomposers than tropical forests have, these forests also have a slower rate of decomposition. As a result, they accumulate a thick layer of slowly decaying leaf litter, which becomes a storehouse of nutrients. Figure 7-15 shows two of the animal species found in this biome.

On a global basis, temperate forests have been degraded by various human activities, especially logging and urban expansion, more than any other terrestrial biome. However, within 100–200 years, forests of this type that have been cleared can return through secondary ecological succession (see Figure 5-12, p. 110).

Cold or *northern coniferous forests* (Figure 7-13, bottom photo) are also called *boreal forests* or *taigas* ("TIE-guhs"). These forests are found just south of the arctic tundra in northern regions across North America, Asia, and Europe (Figure 7-9) and above certain altitudes in the Sierra Nevada and Rocky Mountain ranges of the United States. In this subarctic climate, winters are long and extremely cold; in the northernmost taigas, winter sunlight is available only 6–8 hours per day. Summers are short, with cool to warm temperatures (Figure 7-13, bottom graph), and the sun shines as long as 19 hours a day during mid-summer.

Most boreal forests are dominated by a few species of *coniferous* (cone-bearing) *evergreen trees* such as spruce, fir, cedar, hemlock, and pine that keep most of their leaves (or needles) year-round. Most of these species have small, needle-shaped, wax-coated leaves that can withstand the intense cold and drought of winter, when snow blankets the ground. Plant diversity is low because few species can survive the winters when soil moisture is frozen.

Beneath the stands of trees in these forests is a deep layer of partially decomposed conifer needles. Decomposi-

Figure 7-15 The eastern gray squirrel (left) and the broad-winged hawk (right), which preys on the gray squirrel, are found in North America's temperate deciduous forest ecosystems (**Core Case Study**).

Left: GEORGE GRALL/National Geographic Creative; Right: Pr2is/Dreamstime.com

tion is slow because of low temperatures, the waxy coating on the needles, and high soil acidity. The decomposing conifer needles make the thin, nutrient-poor topsoil acidic, which prevents most other plants (except certain shrubs) from growing on the forest floor.

This biome contains a variety of wildlife. Year-round residents include bears, wolves, moose, lynx, and many burrowing rodent species. Caribou spend the winter in the taiga and the summer in the arctic tundra (Figure 7-11, bottom). During the brief summer, warblers and other insect-eating birds feed on hordes of flies, mosquitoes, and caterpillars.

Coastal coniferous forests or *temperate rain forests* are found in scattered coastal temperate areas with ample rainfall or moisture from dense ocean fogs. Thick stands of these forests with large conifers such as Sitka spruce, Douglas fir, and redwoods once dominated undisturbed areas of these biomes along the coast of North America, from Canada to northern California in the United States.

Mountains Play Important Ecological Roles

Some of the world's most spectacular environments are high on *mountains* (Figure 7-16), steep or high-elevation lands that cover about one-fourth of the earth's land surface (Figure 7-9). Mountains are places where dramatic changes in altitude, slope, climate, soil, and vegetation take place over very short distances (Figure 7-8, left).

About 1.2 billion people (17% of the world's population) live in mountain ranges or in their foothills, and 4 billion people (56% of the world's population) depend on mountain systems for all or some of their water. Because of the steep slopes, mountain soils are easily eroded when the vegetation holding them in place is removed by natural disturbances such as landslides and avalanches, or by human activities such as timber cutting and agriculture. Many mountains are *islands of biodiversity* surrounded by a sea of lower-elevation landscapes transformed by human activities.

Mountains play important ecological roles. They contain the majority of the world's forests, which are habitats for much of the planet's terrestrial biodiversity. They often are habitats for *endemic species,* those that are found nowhere else on earth. They also serve as sanctuaries for animals that are capable of migrating to higher altitudes and surviving in such environments. Every year, more of these animals are driven from lowland areas to mountain habitats by human activities and by a warming climate.

CONSIDER THIS. . .

CONNECTIONS Mountains and Climate

Mountains help to regulate the earth's climate. Many mountaintops are covered with glacial ice and snow that reflect some solar radiation back into space, which helps to cool the earth. However, many of the world's mountain glaciers are melting, primarily because the atmosphere has gotten warmer over the last three decades. While glaciers reflect solar energy, the darker rocks exposed by melting glaciers absorb that energy. This helps to warm the atmosphere above them, which melts more ice and warms the atmosphere more—in an escalating cycle of change.

Finally, mountains play a critical role in the hydrologic cycle (see Figure 3-15, p. 63) by serving as major storehouses of water. During winter, precipitation is stored as ice and snow. In the warmer weather of spring and summer, much of this snow and ice melts, releasing water to streams for use by wildlife and by humans for drinking and for irrigating crops. As the atmosphere has warmed over the last three decades, some mountaintop snow packs and glaciers are melting earlier in the spring each year.

Figure 7-16 Mountains, such as this one in Aspen, Colorado, play important ecological roles.
CHARLES KOGOD/National Geographic Creative

This is lowering food production in certain areas, because much of the water needed throughout the summer to irrigate crops gets released too quickly and too early.

Scientific measurements and climate models indicate that a large number of the world's mountaintop glaciers may disappear during this century if the atmosphere keeps getting warmer as projected. This could force many people to move from their homelands in search of new water supplies and places to grow their crops. Despite the ecological, economic, and cultural importance of mountain ecosystems, protecting them has not been a high priority for governments or for many environmental organizations.

7-3 How Have Human Activities Affected the World's Terrestrial Ecosystems?

CONCEPT 7-3
Human activities are disrupting ecosystem and economic services provided by many of the earth's deserts, grasslands, forests, and mountains.

Humans Have Disturbed Much of the Earth's Land

According to the 2005 Millennium Ecosystem Assessment and later updates of such research, about 60% of the world's major terrestrial ecosystems are being degraded or used unsustainably, as the human ecologi-

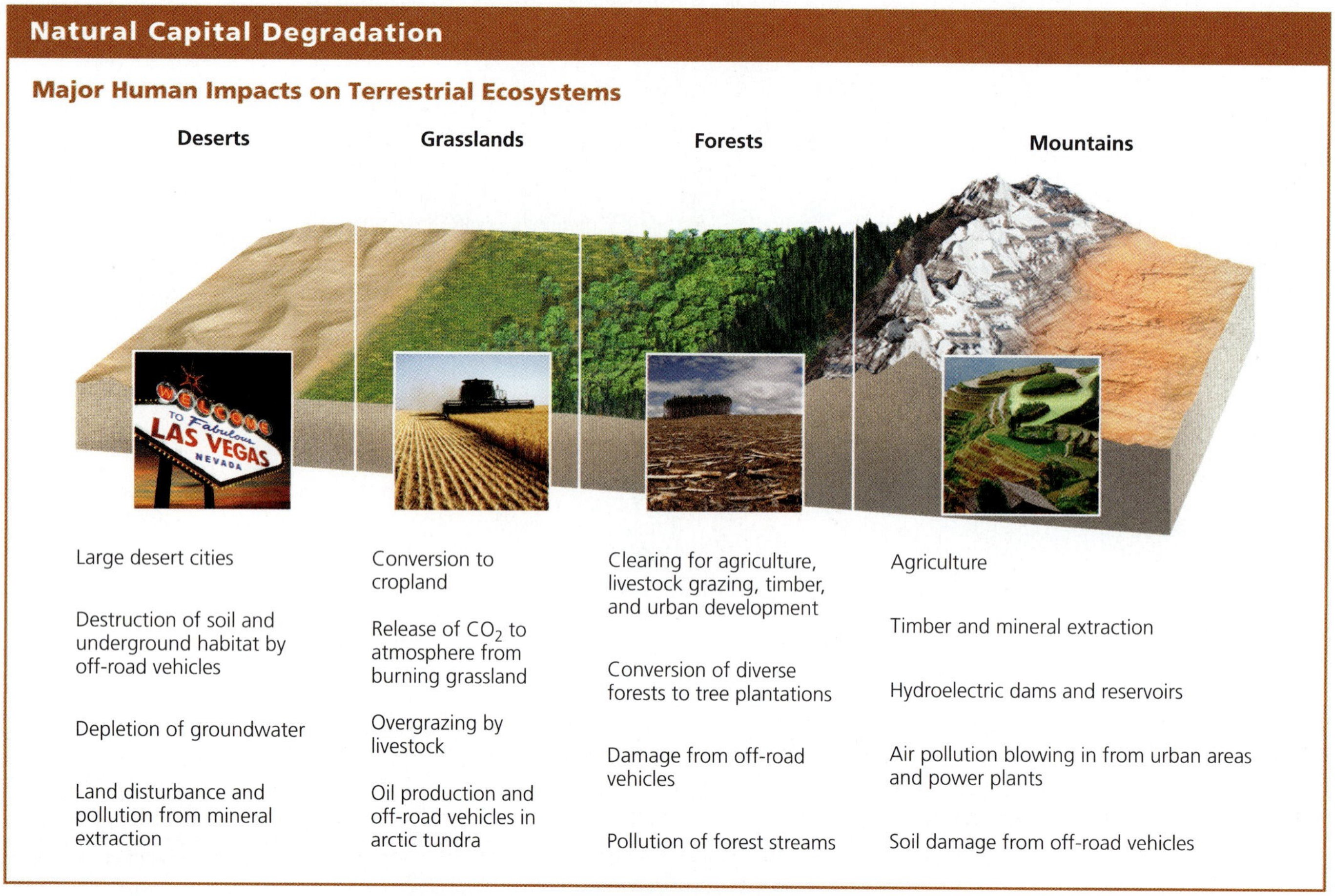

Natural Capital Degradation

Major Human Impacts on Terrestrial Ecosystems

Deserts

- Large desert cities
- Destruction of soil and underground habitat by off-road vehicles
- Depletion of groundwater
- Land disturbance and pollution from mineral extraction

Grasslands

- Conversion to cropland
- Release of CO_2 to atmosphere from burning grassland
- Overgrazing by livestock
- Oil production and off-road vehicles in arctic tundra

Forests

- Clearing for agriculture, livestock grazing, timber, and urban development
- Conversion of diverse forests to tree plantations
- Damage from off-road vehicles
- Pollution of forest streams

Mountains

- Agriculture
- Timber and mineral extraction
- Hydroelectric dams and reservoirs
- Air pollution blowing in from urban areas and power plants
- Soil damage from off-road vehicles

WELCOME TO Fabulous LAS VEGAS NEVADA

Figure 7-17 Human activities have had major impacts on the world's deserts, grasslands, forests, and mountains (**Concept 7-3**), as summarized here. ***Question:*** For each of these biomes, which two of the impacts listed do you think are the most harmful?

Left: ©somchaij/Shutterstock.com. Left center: Orientaly/Shutterstock.com. Right center: © Eppic | Dreamstime.com. Right: ©Vasik Olga/Shutterstock.com.

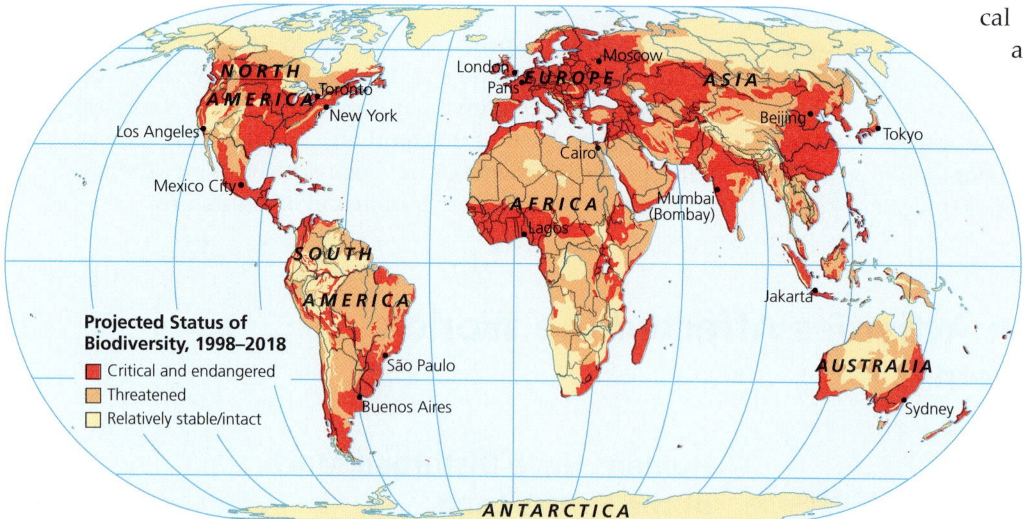

Figure 7-18 **Natural capital degradation:** Projected status of biodiversity in 2018, resulting from ongoing threats of habitat loss and fragmentation, invasions of nonnative species, pollution, climate change, and unsustainable exploitation.

National Geographic; Conservation International

cal footprint gets bigger and spreads across the globe (see Figure 1-12, p. 13, and Figure 1-13, p. 14). Figure 7-17 summarizes some of the human impacts on the world's deserts, grasslands, forests, and mountains (**Concept 7-3**). Figure 7-18 shows the projected status of biodiversity in 2018 due to long-term threats from human activities.

How long can we keep eating away at these terrestrial forms of natural capital without threatening our economies and the long-term survival of our own and many other species? No one knows. But there are increasing signs that we need to come to grips with this vital issue.

individuals matter 7.1

J. Michael Fay: Defender of Threatened Wild Places and National Geographic Explorer-in-Residence

Michael Nichols/National Geographic Creative

Conservationist J. Michael Fay is a biologist with the Wildlife Conservation Society. For much of his life, he has been driven by a passion to explore and preserve some of the earth's vital and threatened wild places. To do this, he has gathered important ecological data about several major ecosystems by flying over and photographing them and by walking through these areas.

In 1996, Fay flew over the forests of the African countries of Gabon and the Congo and studied a vast corridor of forest spanning the two countries. In 1997, he hiked more than 3,200 kilometers (2,000 miles) through this corridor and talked with people to learn more about what he had seen from the air. The data that he collected were used by the government of Gabon to create a system of 13 protected national parks. In 2004, Fay spent 8 months flying over the entire continent of Africa, documenting its ecosystems and the impacts of human activities on them. Many of the 116,000 photographs he took can be seen on Google.

In 2008, Fay turned his attention to California's coastal redwood forests. He spent a year hiking 2,900 kilometers (1,900 miles) through this one-of-kind ecosystem, collecting ecological data with the goal of helping to protect it. This was one more example of how Fay, a modern-day explorer, uses his scientific knowledge and his passion for conservation to help us focus on the urgent need to preserve the earth's remaining wild places.

Background photo: Sharon Eisenzopf/Shutterstock.com

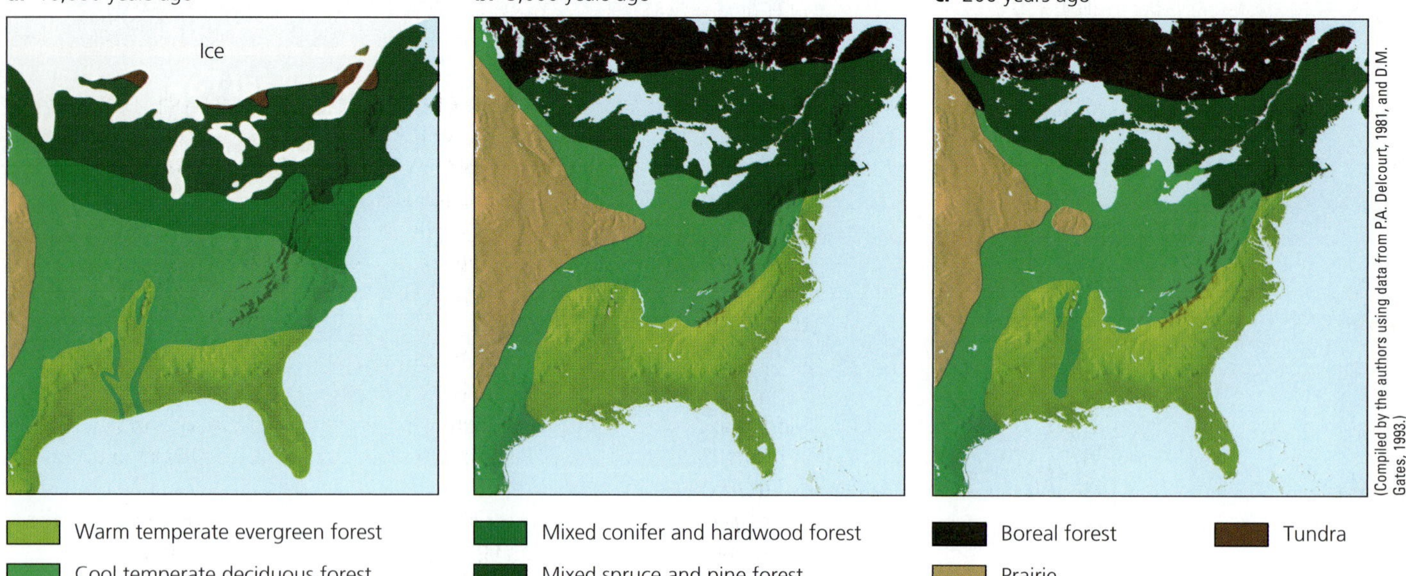

Figure 7-19 As temperatures rose during the last 18,000 years, huge quantities of ice melted and the global climate changed. This led to shifts in the sizes and locations of several major biomes in eastern North America. ***Question:*** Based on these maps, how might you expect the sizes and locations of these biomes to change if the earth's average atmospheric temperature rises by about 2–5 C° (4–9 F°) during this century?

Many environmental scientists call for a global effort to better understand the nature and state of the world's major ecosystems and to use such scientific data to protect the world's remaining wild areas from development (see Individuals Matter 7.1). In addition, they call for us to restore many of the land areas that have been degraded, especially in areas that are rich in biodiversity. However, such efforts are highly controversial because of the timber, mineral, fossil fuel, and other resources found on or under many of the earth's remaining wild land areas. These issues are discussed in Chapter 10.

Sizes and Locations of Biomes Can Change

The locations and sizes of the world's biomes shown in Figure 7-9 are not fixed and they change as the earth's climate changes. For example, the climate has changed drastically since the end of the last ice age, about 18,000 years ago (see Figure 4-10, p. 87). During the warmer interglacial period of the last 10,000 years, the earth's average atmospheric temperature rose by about 5 C° (9 F°). This changed the locations and sizes of biomes in several parts of the world, including North America (Figure 7-19).

Considerable scientific evidence indicates that human activities are likely to raise the average atmospheric temperature by 2–5 C° (4–9 F°) during this century, mostly as a result of human activities. This will likely change the sizes and locations of many biomes and alter the ecological map of the earth's land areas. These changes could also wipe out many species and degrade important ecosystem services. If these scientific projections are correct, such changes will take place within 100 years or so, instead of within thousands of years as they have in the past. This gives us and other species very little time to deal with such projected changes.

Big Ideas

- **Differences in climate, based mostly on long-term differences in average temperature and precipitation, largely determine the types and locations of the earth's deserts, grasslands, and forests.**
- **The earth's terrestrial ecosystems provide important ecosystem and economic services.**
- **Human activities are degrading and disrupting many of the ecosystem and economic services provided by the earth's terrestrial ecosystems.**

TYING IT ALL TOGETHER A Temperate Deciduous Forest and Sustainability

mihalec/Shutterstock.com

In this chapter we discussed the influence of climate on terrestrial biodiversity in the formation of biomes—deserts, grasslands, and forests. In the **Core Case Study**, we focused on the *temperate deciduous forest* biome, noting how the plants and animals that have evolved within such forests are adapted to climatic conditions there. Thus we saw that *climate*—the weather conditions in a given area averaged over periods of time ranging from three decades to thousands of years—plays a key role in determining the nature of terrestrial ecosystems, as well as the life-forms that live in those systems.

These relationships are in keeping with the three scientific **principles of sustainability** (see Figure 1-2, p. 6 or back cover). The earth's dynamic climate system helps to distribute heat from solar energy and to recycle the earth's nutrients. This in turn helps to generate and support the biodiversity found in the earth's various biomes.

Scientists have made some progress in understanding the ecology of the world's terrestrial systems, as well as how the vital ecosystem and economic services they provide are being degraded and disrupted. One of the major lessons from their research is: *in nature, everything is connected.* According to these scientists, we urgently need more research on the components and workings of the world's biomes, on how they are interconnected, and on which connections are in the greatest danger of being disrupted by human activities. With such information, we will have a clearer picture of how our activities affect the natural capital that supports the earth's life and of what we can do to help sustain that natural capital on which we and all other species depend.

Chapter Review

Core Case Study

1. Describe a temperate deciduous forest (**Core Case Study**) and explain why it serves as an example of how differences in climate lead to the formation of different types of ecosystems.

Section 7-1

2. What is the key concept for this section? Distinguish between **weather** and **climate**. Define **ocean currents**. Describe three major factors that determine how air circulates in the lower atmosphere. Explain how varying combinations of precipitation and temperature, along with global air circulation and ocean currents, lead to the formation of various types of forests, grasslands, and deserts.
3. Define and give four examples of a **greenhouse gas**. What is the **greenhouse effect** and why is it important to the earth's life and climate?
4. What is the **rain shadow effect** and how can it lead to the formation of deserts? Why do cities tend to have more haze and smog, higher temperatures, and lower wind speeds than the surrounding countryside?

Section 7-2

5. What is the key concept for this section? Describe how climate and vegetation vary with latitude and elevation. What is a **biome**? Explain why there are three major types of each of the major biomes (deserts, grasslands, and forests). Explain why biomes are not uniform. Describe how climate and vegetation vary with latitude and elevation.
6. Describe how the three major types of deserts differ in their climate and vegetation. Why are desert ecosystems fragile? How do desert plants and animals survive?
7. Explain how the three major types of grasslands differ in their climate and vegetation. What is a savanna? Why have many of the world's temperate grasslands disappeared? What is **permafrost**?

8. Explain how the three major types of forests differ in their climate and vegetation. Why is biodiversity so high in tropical rain forests? Why do most soils in tropical rain forests hold few plant nutrients? Why do temperate deciduous forests typically have a thick layer of decaying litter? How do most species of coniferous evergreen trees survive the cold winters in boreal forests? What are coastal coniferous or temperate rain forests? What important ecological roles do mountains play?

Section 7-3

9. What is the key concept for this section? About what percentage of the world's major terrestrial ecosystems are being degraded or used unsustainably? Summarize the ways in which human activities have affected the world's deserts, grasslands, forests, and mountains. How is a warming climate likely to change the earth's biomes?

10. What are this chapter's *three big ideas*? Summarize the connections between climate and terrestrial ecosystems, and explain how these connections are in keeping with the three scientific **principles of sustainability** (see Figure 1-2, p. 6 or back cover).

Note: Key terms are in bold type.

Critical Thinking

1. Why do you think temperate deciduous forests (**Core Case Study**) are among the biomes most extensively disturbed by human activities?

2. Describe the roles of temperature and precipitation in determining what parts of the earth's land are covered with **(a)** desert, **(b)** arctic tundra, **(c)** temperate grassland, **(d)** tropical rain forest, and **(e)** temperate deciduous forest (**Core Case Study**).

3. Why do deserts and arctic tundra support a much smaller number and variety of animals than do tropical forests? Why do most animals in a tropical rain forest live in its trees?

4. How might the distribution of the world's forests, grasslands, and deserts shown in Figure 7-9 differ if the prevailing winds shown in Figure 7-3 did not exist?

5. Which biomes are best suited for **(a)** raising crops and **(b)** grazing livestock? Use the three scientific **principles of sustainability** to come up with three guidelines for growing crops and grazing livestock more sustainably in these biomes.

6. What type of biome do you live in? (If you live in a developed area, what type of biome was the area before it was developed?) List three ways in which your lifestyle could be contributing to the degradation of this biome. What changes could you make in order to reduce your contribution, if any?

7. You are a defense attorney arguing in court for sparing a tropical rain forest from being cut down. Give your three best arguments for the defense of this ecosystem. Do the same for the case of temperate deciduous forest (**Core Case Study**).

8. Congratulations! You are in charge of the world. What are the three most important features of your plan for helping to sustain the earth's terrestrial biodiversity and ecosystem services?

Doing Environmental Science

Using Google Earth, find an undeveloped ecosystem somewhere on the planet that you can zoom in on to observe some features. For example, find a forest or grassland area that is undeveloped. Take careful notes on what you see. Then find such an example of a biome that has been developed, study it, and write a detailed comparison of your undeveloped and developed biome samples. If possible, check in on these systems once during your course term and again at the end of the term and describe any changes to the systems that you can observe. Suggest a hypothesis to explain any major changes.

Global Environment Watch Exercise

Search for *tropical rain forests* and use the topic portal to find information on **(a)** trends in the global rate of destruction of these forest; **(b)** what areas of the world are seeing rising rates of destruction and what areas are seeing falling rates; and **(c)** what is being done to protect them in various areas. Write a report on your findings.

Data Analysis

In this chapter, you learned how long-term variations in average temperatures and average precipitation play a major role in determining the types of deserts, forests, and grasslands found in different parts of the world. Below are typical annual climate graphs for a tropical grassland (savanna) in Africa and a temperate grassland in the Midwestern United States.

Tropical grassland (savanna)

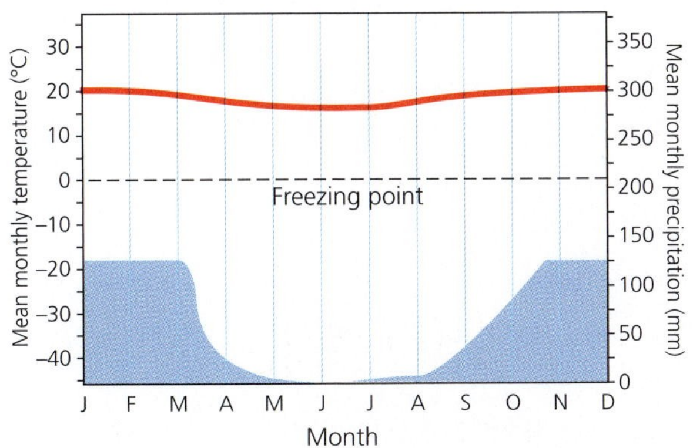

Temperate grassland (prairie)

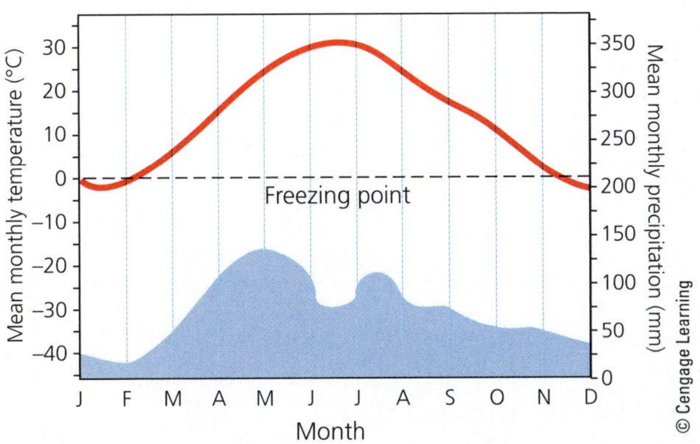

1. In what month (or months) does the most precipitation fall in each of these areas?
2. What are the driest months in each of these areas?
3. What is the coldest month in the tropical grassland?
4. What is the warmest month in the temperate grassland?

8 Aquatic Biodiversity

The sea, once it casts its spell, holds one in its net of wonders forever.

JACQUES-YVES COUSTEAU

Key Questions

8-1 What is the general nature of aquatic systems?

8-2 Why are marine aquatic systems important?

8-3 How have human activities affected marine ecosystems?

8-4 Why are freshwater ecosystems important?

8-5 How have human activities affected freshwater ecosystems?

Coral reef off the coast of Fiji in the South Pacific ocean.

CORE CASE STUDY Why Should We Care about Coral Reefs?

Coral reefs form in clear, warm coastal waters in tropical areas. These stunningly beautiful natural wonders (see chapter-opening photo) are among the world's oldest, most diverse, and most productive ecosystems.

Coral reefs are formed by massive colonies of tiny animals called *polyps* (close relatives of jellyfish). They slowly build reefs by secreting a protective crust of limestone (calcium carbonate) around their soft bodies. When the polyps die, their empty crusts remain behind as a platform for more reef growth. The resulting elaborate network of crevices, ledges, and holes serves as calcium carbonate "condominiums" for a variety of marine animals.

Coral reefs are the result of a mutually beneficial relationship between polyps and tiny single-celled algae called *zooxanthellae* ("zoh-ZAN-thel-ee") that live in the tissues of the polyps. The algae provide the polyps with food and oxygen through photosynthesis and help produce the coral's calcium carbonate skeleton. Algae also give the reefs their stunning coloration. The polyps, in turn, provide the algae with a well-protected home and some of their nutrients.

Rainer von Brandis/iStockphoto.com

Figure 8-1 This bleached coral has lost most of its algae because of changes in the environment such as warming of the waters and deposition of sediments.

Although shallow- and deep-water coral reefs occupy only about 0.2% of the ocean floor, they provide important ecosystem and economic services. They act as natural barriers that help to protect 15% of the world's coastlines from erosion caused by battering waves and storms. These reefs provide habitats for one-quarter of all marine organisms, and they produce about one-tenth of the global fish catch.

Studies by the Global Coral Reef Monitoring Network and other scientist groups estimate that since the 1950s, some 45–53% of the world's shallow coral reefs have been destroyed or degraded by pollution and other stressors, and another 25–33% could be lost within 20–40 years. Also, deep coral reefs that are thousands of years old are being destroyed by large numbers of trawler fishing boats that drag huge weighted nets across the ocean bottom. One result of such stresses is *coral bleaching* (Figure 8-1). Pollution or water that is too warm can cause the algae, on which corals depend for food, to die off. Without food, the coral polyps die, leaving behind a white skeleton of calcium carbonate.

Coral reefs are vulnerable to damage because they grow slowly and are disrupted easily. Runoff of soil and other materials from the land can cloud the water and block the sunlight that the algae in shallow reefs need for photosynthesis. Also, the water in which shallow reefs live must have a temperature of 18–30°C (64–86°F) and cannot be too acidic. This explains why the two major long-term threats to coral reefs are projected *climate change,* which could raise the water temperature above this limit in most reef areas, and *ocean acidification,* which could make it harder for polyps to build reefs and could even dissolve some of their calcium carbonate formations.

In this chapter, we explore ocean and freshwater ecosystems and the threats they face. We also consider some measures we can all take to help sustain these vital systems.

8-1 What Is the General Nature of Aquatic Systems?

CONCEPT 8-1A
Saltwater and freshwater aquatic life zones cover almost three-fourths of the earth's surface, with oceans dominating the planet.

CONCEPT 8-1B
The key factors determining biodiversity in aquatic systems are temperature, dissolved oxygen content, availability of food, and access to light and nutrients necessary for photosynthesis.

Most of the Earth Is Covered with Water

When viewed from outer space, the earth appears to be almost completely covered with water (Figure 8-2). Saltwater covers about 71% of the earth's surface, and freshwater occupies roughly another 2.2%.

Although the *global ocean* is a single and continuous body of water, geographers divide it into four large areas—the Atlantic, Pacific, Arctic, and Indian Oceans—separated by the continents. The largest ocean is the Pacific, which contains more than half of the earth's water and covers one-third of the earth's surface. Together, the oceans hold almost 98% of the earth's water. Each of us is connected to, and utterly dependent on, the earth's global ocean through the water cycle (see Figure 3-15, p. 63).

The aquatic equivalents of biomes are called **aquatic life zones**—saltwater and freshwater portions of the biosphere that can support life. The distribution of many aquatic organisms is determined largely by the water's *salinity*—the amounts of various salts such as sodium chloride (NaCl) dissolved in a given volume of water. As a result, aquatic life zones are classified into two major types: **saltwater** or **marine life zones** (oceans and their bays, estuaries, coastal wetlands, shorelines, coral reefs, and mangrove forests) and **freshwater life zones** (lakes, rivers, streams, and inland wetlands). Some systems such as estuaries are a mix of saltwater and freshwater, but scientists classify them as marine systems for purposes of discussion.

Ocean hemisphere

Land–ocean hemisphere

Figure 8-2 *The ocean planet:* The salty oceans cover 90% of the planet's ocean hemisphere (left) and nearly half of its land–ocean hemisphere (right) (**Concept 8-1A**).

Figure 8-3 Jellyfish are drifting zooplankton that use their long tentacles with stinging cells to stun or kill their prey. Jellyfish populations are expanding globally and taking over many marine ecosystems because excessive nutrient runoff from the land has spurred the growth of algae on which small jellyfish feed, and because many of their predators have been removed by overfishing.

Aquatic Species Drift, Swim, Crawl, and Cling

Saltwater and freshwater life zones contain several major types of organisms. One type consists of **plankton**, which can be divided into three groups. The first group consists of drifting organisms called *phytoplankton* ("FY-toe-plank-ton"), which includes many types of algae. These tiny aquatic plants and even smaller *ultraplankton*—the second group of plankton—are the producers that make up the base of most aquatic food chains and webs (see Figure 3-12, p. 60). Through photosynthesis, they produce about half of the earth's oxygen, on which we depend for survival.

The third group is made up of drifting animals called *zooplankton* ("ZOH-uh-plank-ton"), which feed on phytoplankton and on other zooplankton (see Figure 3-12, p. 60). The members of this group range in size from single-celled protozoa to large invertebrates such as jellyfish (Figure 8-3).

A second major type of aquatic organism is **nekton**, strongly swimming consumers such as fish, turtles, and whales. The third type, **benthos**, consists of bottom-dwellers such as oysters and sea stars (Figure 8-4), which anchor themselves to ocean-bottom structures; clams and worms, which burrow into the sand or mud; and lobsters

Figure 8-4 Bottom-dwelling starfish, four types of which appear here, use their tube feet to attach themselves to surfaces on the ocean bottom and move slowly in search of food.

and crabs, which walk about on the sea floor. A fourth major type is **decomposers** (mostly bacteria), which break down organic compounds in the dead bodies and wastes of aquatic organisms into nutrients that aquatic primary producers can use.

Key factors determining the types and numbers of organisms found in different areas of the ocean are *temperature, dissolved oxygen content, availability of food,* and *availability of light and nutrients required for photosynthesis,* such as carbon (as dissolved CO_2 gas), nitrogen (as NO_3^-), and phosphorus (mostly as PO_4^{3-}) (**Concept 8-1B**).

In deep aquatic systems, photosynthesis is largely confined to the upper layer—the *euphotic* or *photic zone,* through which sunlight can penetrate. The depth of the euphotic zone in oceans and deep lakes is reduced when the water is clouded by excessive growth of algae—called *algal blooms*—that results from nutrient overloads. This cloudiness, called **turbidity**, can occur naturally, such as from algal growth. It can also be caused by soil and other sediments being carried by rain and melting snow from cleared land into adjoining bodies of water. This is one of the problems plaguing shallow coral reefs (**Core Case Study**), as excessive turbidity due to silt runoff prevents photosynthesis and causes the corals to die.

In shallow systems such as small open streams, lake edges, and ocean shorelines, ample supplies of nutrients for primary producers are usually available, which tends to make these areas high in biodiversity. By contrast, in most areas of the open ocean, nitrates, phosphates, iron, and other nutrients are often in short supply, and this limits net primary productivity (NPP) (see Figure 3-14, p. 62).

8-2 Why Are Marine Aquatic Systems Important?

CONCEPT 8-2

Saltwater ecosystems provide major ecosystem and economic services and are irreplaceable reservoirs of biodiversity.

Oceans Provide Vital Ecosystem and Economic Services

Oceans provide enormously valuable ecosystem and economic services (Figure 8-5). One estimate of the combined value of these goods and services from all marine coastal ecosystems is over $12 trillion per year.

As land dwellers, we have a distorted and limited view of the aquatic wilderness that covers most of the earth's surface. We know more about the surface of the moon than we know about the oceans. According to aquatic scientists, the scientific investigation of poorly understood marine and freshwater aquatic systems could yield immense ecological and economic benefits.

Marine aquatic systems are enormous reservoirs of biodiversity. They include many different ecosystems, which host a great variety of species, genes, and biological and chemical processes, thus helping to sustain the four major components of the earth's biodiversity (see Figure 4-2, p. 79). Marine life is found in three major *life zones:* the coastal zone, the open sea, and the ocean bottom (Figure 8-6).

Natural Capital

Marine Ecosystems

Ecosystem Services	Economic Services
Oxygen supplied through photosynthesis	Food
Water purification	Energy from waves and tides
Climate moderation	Pharmaceuticals
CO_2 absorption	Harbors and transportation routes
Nutrient cycling	Recreation and tourism
Reduced storm damage (mangroves, barrier islands, coastal wetlands)	Employment
Biodiversity: species and habitats	Minerals

Figure 8-5 Marine systems provide a number of important ecosystem and economic services (**Concept 8-2**). ***Questions:*** Which two ecosystem services and which two economic services do you think are the most important? Why?

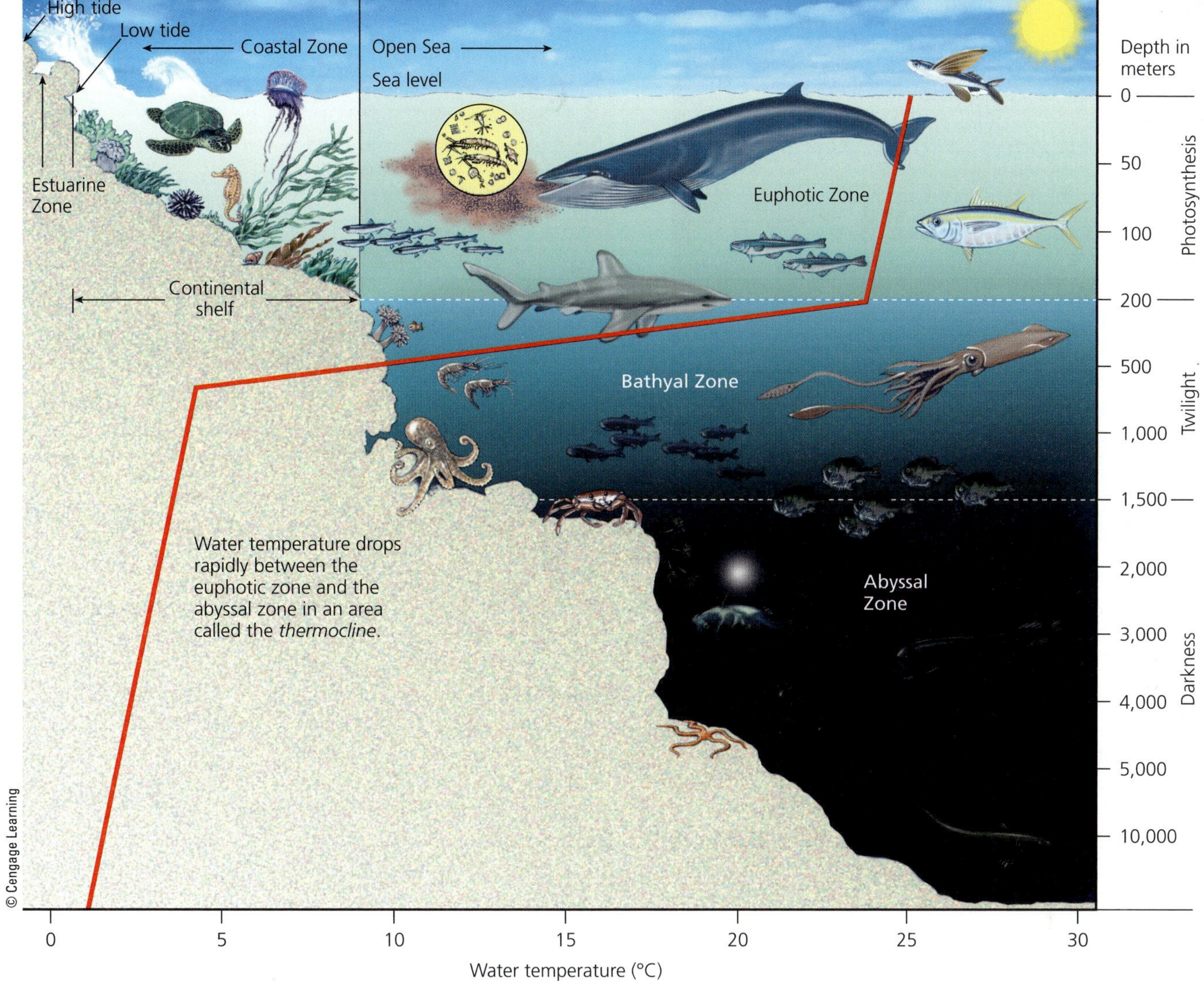

Figure 8-6 This diagram illustrates the major life zones and vertical zones (not drawn to scale) in an ocean. Actual depths of zones may vary. Available light determines the euphotic, bathyal, and abyssal zones. Temperature zones also vary with depth, shown here by the red line. ***Question:*** How is an ocean like a rain forest? (*Hint:* see Figure 7-14, p. 157.)

The **coastal zone** is the warm, nutrient-rich, shallow water that extends from the high-tide mark on land to the gently sloping, shallow edge of the *continental shelf* (the submerged part of the continents). It makes up less than 10% of the world's ocean area, but it contains 90% of all marine species and is the site of most large commercial marine fisheries. This zone's aquatic systems include estuaries, coastal marshes, mangrove forests, and coral reefs.

Estuaries and Coastal Wetlands Are Highly Productive

An **estuary** is where a river meets the sea (Figure 8-7). It is a partially enclosed body of water where seawater mixes with the river's freshwater, as well as nutrients and pollutants in runoff from the land.

Estuaries are associated with **coastal wetlands**—coastal land areas covered with water all or part of the year. These wetlands include *coastal marshes* (Figure 8-8) and *mangrove forests* (Figure 8-9). They are some of the earth's most productive ecosystems (see Figure 3-14, p. 62) because of high nutrient inputs from rivers and from adjoining land, rapid circulation of nutrients by tidal flows, and ample sunlight penetrating the shallow waters. For example, mangrove forests around the world host 69 different species of trees that can grow in salty water. They provide habitat, food, and nursery sites for a variety of fishes and other aquatic species.

Sea-grass beds are another component of coastal marine biodiversity (Figure 8-10). They consist of at least 60 species of plants that grow underwater in estuaries and shallow waters along most continental coastlines.

NASA

Figure 8-7 This satellite photo shows a view of an *estuary* taken from space. A sediment plume (turbidity caused by runoff) forms at the mouth of Madagascar's Betsiboka River as it flows through the estuary and into the Mozambique Channel.

These highly productive and physically complex systems support a variety of marine species. Like other coastal systems, they owe their high NPP (see Figure 3-14, p. 62) to ample supplies of sunlight and plant nutrients that flow from land and are distributed by wind and ocean currents.

These coastal aquatic systems provide important ecosystem and economic services. They help to maintain water quality in tropical coastal zones by filtering toxic pollutants, excess plant nutrients, and sediments and by absorbing other pollutants. They provide food, habitats, and nursery sites for a variety of aquatic and terrestrial species. They also reduce storm damage and coastal erosion by absorbing waves and storing excess water produced by storms and tsunamis.

Michael Melford/National Geographic Creative

Figure 8-8 This coastal marsh is located on Cape Cod, Massachusetts (USA).

Rocky and Sandy Shores Host Different Types of Organisms

The gravitational pull of the moon and sun causes *tides* to rise and fall about every 6 hours in most coastal areas. The area of shoreline between low and high tides is called the **intertidal zone**. Organisms living in this zone must be able to avoid being swept away or crushed by waves, and must deal with being immersed during high tides and left high and dry (and much hotter) at low tides. They must also survive changing levels of salinity when heavy rains dilute saltwater. To deal with such stresses, most intertidal organisms hold on to something, dig in, or hide in protective shells.

Theo Allofs/Visuals Unlimited

Figure 8-9 This is a mangrove forest on the coast of Queensland, Australia.

Figure 8-10 Sea-grass beds, such as this one near the coast of San Clemente Island, California, support a variety of marine species.

On some coasts, steep *rocky shores* are pounded by waves. The numerous pools and other habitats in these intertidal zones contain a great variety of species that occupy different niches in response to daily and seasonal changes in environmental conditions such as temperature, water flows, and salinity (Figure 8-11, top).

Other coasts have gently sloping *barrier beaches,* or *sandy shores,* that support other types of marine organisms (Figure 8-11, bottom), most of which keep hidden from view and survive by burrowing, digging, and tunneling in the sand. These sandy beaches and their adjoining coastal wetlands are also home to a variety of shorebirds that have evolved in specialized niches to feed on crustaceans, insects, and other organisms (see Figure 4-16, p. 92). Many of these same species also live on *barrier islands*—low, narrow, sandy islands that form offshore, parallel to nearby coastlines.

Undisturbed barrier beaches generally have one or more rows of natural sand dunes in which the sand is held in place by the roots of plants, usually grasses. These dunes are the first line of defense against the ravages of the sea. These areas are attractive to real estate developers, and such real estate is scarce and valuable. However, coastal developers frequently remove the protective dunes or build behind the first set of dunes, covering them with buildings and roads. Large storms can then flood and even sweep away seaside construction and severely erode the sandy beaches.

Coral Reefs Are Amazing Centers of Biodiversity

As we noted in the **Core Case Study**, coral reefs are among the world's oldest and most diverse and productive ecosystems (see chapter-opening photo). These amazing centers of aquatic biodiversity are the marine equivalents of tropical rain forests, with complex interactions among their diverse populations of species.

Worldwide, coral reefs are being damaged and destroyed at an alarming rate by a variety of human activities. A growing threat, as mentioned in this chapter's **Core Case Study**, is **ocean acidification**—the increasing levels of acid in the world's oceans. This occurs because the oceans absorb about a third of the CO_2 emitted into the atmosphere by human activities, especially the burning of carbon-containing fossil fuels. The CO_2 reacts with ocean water to form a weak acid and decreases the levels of carbonate ions (CO_3^{2-}) needed to form coral and the shells and skeletons of organisms such as crabs, oysters, and some phytoplankton.

The Open Sea and the Ocean Floor Host a Variety of Species

The sharp increase in water depth at the edge of the continental shelf separates the coastal zone from the vast volume of the ocean called the **open sea**. Primarily on the basis of the penetration of sunlight, this deep blue sea is divided into three *vertical zones* (Figure 8-6). Temperatures also change with depth (Figure 8-6, red line) and we can use them to define zones that help to determine species diversity in these zones.

The *euphotic zone* is the brightly lit upper zone, where drifting phytoplankton carry out about 40% of the world's photosynthetic activity. Nutrient levels are low and levels of dissolved oxygen are high in the euphotic zone. One exception to this is those areas called *upwelling zones,* where ocean currents driven by coastal winds or by differences in water temperature at different depths bring water up from the deepest waters (see Figure 7-2, p. 145). Upwellings carry nutrients from the ocean bottom to the surface for use by producers, and thus these zones contain high levels of nutrients. Large, fast-swimming predatory fishes such as swordfish, sharks, and bluefin tuna populate the euphotic zone. They feed on secondary and higher-level consumers, which are supported directly or indirectly by producers.

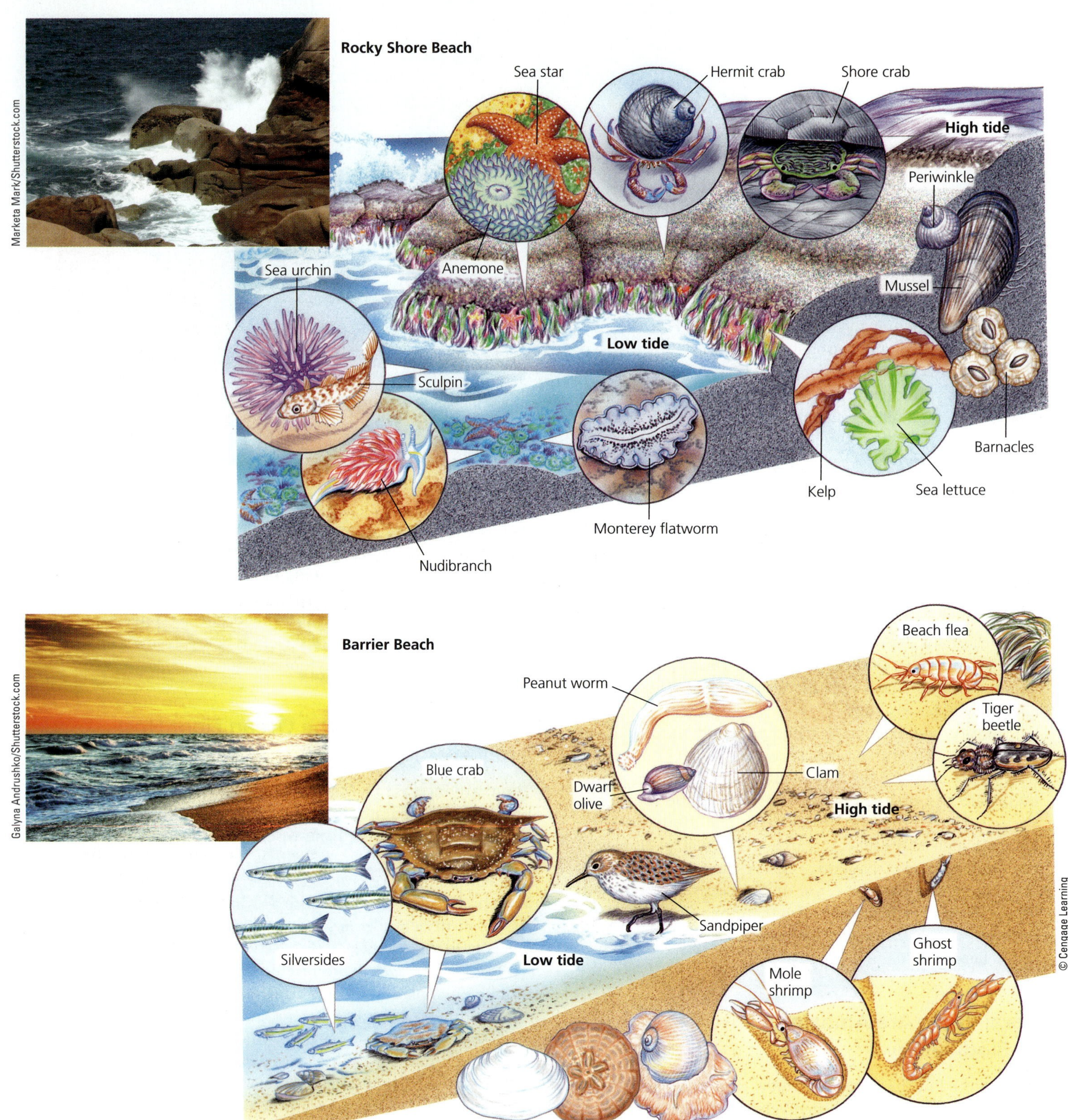

Figure 8-11 *Living between the tides:* Some organisms with specialized niches are found in various zones on rocky shore beaches (top) and barrier or sandy beaches (bottom). Organisms are not drawn to scale.

SCIENCE FOCUS 8.1

WE ARE STILL LEARNING ABOUT THE OCEAN'S BIODIVERSITY

Scientists have long assumed that open-ocean waters contained few microbial life forms. But a 2007 report challenged that assumption and drastically increased our knowledge of the ocean's genetic diversity.

A team of scientists led by J. Craig Venter (Figure 8-A) took 2 years to conduct a census (an estimated count based on sampling) of ocean microbes. They sailed around the world, stopping every 320 kilometers (200 miles) to pump seawater through extremely fine filters, from which they gathered data on bacteria, viruses, and other microbes. It was the most thorough of such censuses ever conducted.

Using a supercomputer, they counted genetic coding for 6 million new proteins—double the number that had previously been known. They also reported that they were discovering new genes and proteins at the same rate at the end of their voyage as they had at the start of it, meaning there is still much more of this biodiversity to discover.

This means that the ocean contains a much higher diversity of microbial life than had previously been thought. Ocean-water microbes play an important role in the absorption of carbon by the ocean, as well as in the ocean food web. Since the 2007 report, Venter has led more expeditions to continue the sampling in other areas, including the waters of the Baltic, Mediterranean, and Black Seas.

Critical Thinking

Why is it that the *rate* of discovery of new genes and proteins was important to Venter and his colleagues? Explain.

Figure 8-A J. Craig Venter

The *bathyal zone* is the dimly lit middle zone, which receives little sunlight and therefore does not contain photosynthesizing producers. Zooplankton and smaller fishes, many of which migrate to feed on the surface at night, populate this zone.

The lowest zone, called the *abyssal zone,* is dark and very cold. There is no sunlight to support photosynthesis, and this zone has little dissolved oxygen. Nevertheless, the deep ocean floor is teeming with life, because it contains enough nutrients to support a large number of species. Most organisms of the deep waters and ocean floor get their food from showers of dead and decaying organisms—called *marine snow*—drifting down from upper, lighted levels of the ocean.

Some abyssal-zone organisms, including many types of worms, are *deposit feeders,* which take mud into their guts and extract nutrients from it. Others such as oysters, clams, and sponges are *filter feeders,* which pass water through or over their bodies and extract nutrients from it.

NPP is quite low in the open sea, except in upwelling areas. However, because the open sea covers so much of the earth's surface, it makes the largest contribution to the earth's overall NPP. Also, scientists are learning that the open sea contains more biodiversity than it was thought to hold until a few years ago (see Science Focus 8.1, above).

8-3 How Have Human Activities Affected Marine Ecosystems?

CONCEPT 8-3
Human activities threaten aquatic biodiversity and disrupt ecosystem and economic services provided by saltwater systems.

Human Activities Are Disrupting and Degrading Marine Ecosystems

Human activities are disrupting and degrading some ecosystem and economic services provided by marine aquatic systems, especially coastal marshes, shorelines, mangrove forests, and coral reefs (Concept 8-3).

For example, according to a 2008 United Nations (UN) Food and Agriculture Organization report, at least one-fifth of the world's mangrove forests were eliminated between 1980 and 2005, mostly to make way for human coastal developments. Also, since 1980, about 29% of the world's sea-grass beds have been lost to pollution and other disturbances.

In 2008, the U.S. National Center for Ecological Analysis and Synthesis (NCEAS) used computer models to analyze and provide the first-ever comprehensive map of the effects of 17 different types of human activities on the world's oceans. In this 4-year study, an international team of scientists found that human activities have heavily affected 41% of the world's ocean area. No area of the oceans has been left completely untouched, according to the report.

In their desire to live near a coast, many people are unwittingly taking part in this degradation of natural capital. In 2012, about 45% of the world's population and more than half of the U.S. population lived along or near coasts and these percentages are rising rapidly.

Major threats to marine systems from human activities include:

- Coastal development, which destroys and pollutes coastal habitats
- Runoff of nonpoint sources of pollutants such as silt, fertilizers, pesticides, and livestock wastes
- Point-source pollution such as sewage from cruise ships and spills from oil tankers
- Pollution and degradation of coastal wetlands and estuaries (see Case Study at right)
- Overfishing, which depletes populations of commercial fish species
- Use of fishing trawlers that drag weighted nets across the ocean bottom, degrading and destroying its habitats
- Invasive species, introduced by humans, which can deplete populations of native aquatic species and cause economic damage
- Ocean warming
- Ocean acidification

Figure 8-12 Human activities are having major harmful impacts on all marine ecosystems (left) and particularly on coral reefs (right) (**Concept 8-3**). ***Questions:*** Which two of the threats to marine ecosystems do you think are the most serious? Why? Which two of the threats to coral reefs do you think are the most serious? Why?

Top left: Jorg Hackemann/Shutterstock.com. Top right: Rich Carey/Shutterstock.com. Bottom left: Piotr Marcinski/Shutterstock .com. Bottom right: Rostislav Ageev/Shutterstock.com.

A major threat that is very alarming to some marine scientists is that of projected climate change, enhanced by human activities, which is warming and acidifying the oceans and slowly melting land-based glaciers in Greenland and other parts of the world. This could cause a rise in sea levels during this century that would destroy shallow coral reefs and flood coastal marshes and many coastal cities.

A second threat, which some scientists view as more serious than the threat of climate change, is ocean acidification, known as the "other CO_2 problem." The level of acids in the world's oceans has increased by about 30% since the beginning of the Industrial Revolution and is projected to continue increasing during this century by 100–150%.

This trend is a threat to coral reefs (**Core Case Study**) and to phytoplankton and many shellfish that form their shells from calcium carbonate. It could kill off much of the oceans' phytoplankton that form the base of the entire marine food web. A 2010 study led by Boris Worm concluded that since 1950, the average global concentration of phytoplankton in the upper ocean had declined by 40% and that this decline was steadily continuing at about 1%

per year. Some scientists consider this decline a major threat to the entire global marine ecosystem. According to a 2007 study by Ove Hoegh-Guldberg and 16 other scientists, unless we take action soon to significantly reduce CO_2 emissions, the oceans may be too acidic and too warm for most of the world's coral reefs to survive this century, and the important ecosystem and economic services they provide will be lost.

Figure 8-12 shows some of the effects of these human impacts on marine systems in general (left) and on coral reefs in particular (right). We examine some of these impacts more closely in Chapters 11 and 19.

CONSIDER THIS. . .

THINKING ABOUT Coral Reef Destruction

How might the loss of most of the world's remaining tropical coral reefs (**Core Case Study**) affect your life and the lives of any children or grandchildren you might have? What are two things you could do to help reduce this loss?

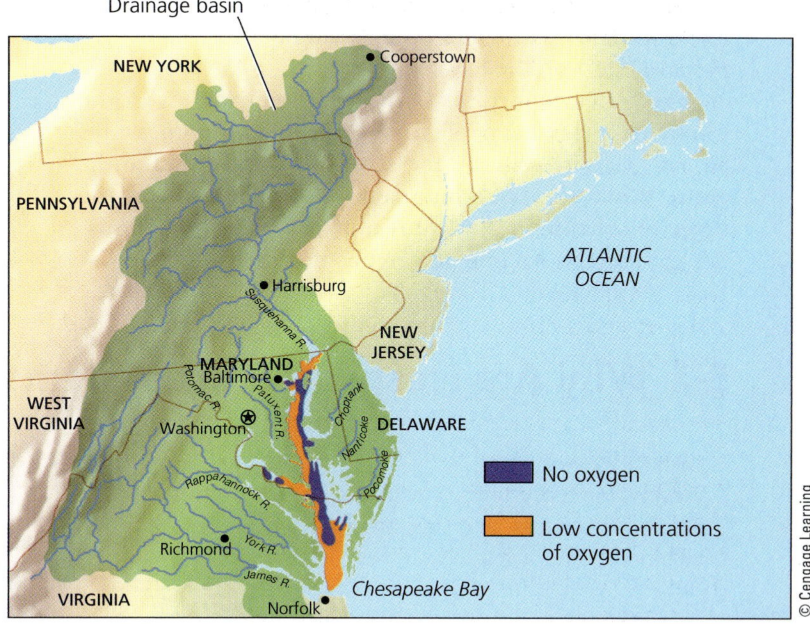

Figure 8-13 The Chesapeake Bay is severely degraded as a result of water pollution and atmospheric deposition of air pollutants.

CASE STUDY

The Chesapeake Bay—An Estuary in Trouble

Since 1960, the Chesapeake Bay (Figure 8-13)—the largest estuary in the United States—has been in serious trouble from water pollution, mostly because of human activities. One problem is population growth. Between 1940 and 2012, the number of people living in the Chesapeake Bay area grew from 3.7 million to more than 17 million, and could reach 18 million by 2020, according to estimates by the Chesapeake Bay Program.

The estuary receives wastes from point and nonpoint sources scattered throughout its huge drainage basin, which lies in parts of six states and the District of Columbia (Figure 8-13). The shallow bay has become a huge pollution sink because only 1% of the waste entering it is flushed into the Atlantic Ocean. Phosphate and nitrate levels have been high in many parts of the bay, causing algal blooms that deplete the oxygen dissolved in the waters, making them unsuitable for most forms of aquatic life. Commercial harvests of the bay's once-abundant oysters and crabs, as well as several important fish species, fell sharply after the 1960s because of a combination of pollution, overfishing, and disease.

Point sources, primarily sewage treatment plants and industrial plants, account for 60% by weight of the phosphates. Nonpoint sources—mostly runoff of fertilizer, animal wastes from urban, suburban, and agricultural land, and deposition of pollutants from the atmosphere—account for 60% by weight of the nitrates. In addition, runoff of sediment, mostly from soil erosion, harms the submerged grasses on which crabs and young fish depend. Runoff also increases when trees near the bay are cut down for development.

A century ago, oysters were so abundant that they filtered and cleaned the Chesapeake's entire volume of water every 3 days. This important form of natural capital provided by these *keystone species* (see Chapter 4, pp. 94–96) helped remove or reduce excess nutrients and algal blooms. Now the oyster population has been reduced to the point where this filtration process takes a year and the keystone role of this species has been severely weakened.

In 1983, the United States implemented the Chesapeake Bay Program. In this ambitious attempt at *integrated coastal management,* citizens' groups, communities, state legislatures, and the federal government worked together to reduce pollution inputs into the bay. One strategy of the program was to establish land-use regulations to reduce agricultural and urban runoff in the bay's drainage area. Other strategies included banning phosphate detergents, upgrading sewage treatment plants, and monitoring industrial discharges more closely. Some adjoining wetlands have been restored and large areas of the bay were replanted with sea grasses to help filter out excessive nutrients and other pollutants.

In 2008, after 25 years of effort costing almost $6 billion, the Chesapeake Bay Program had failed to meet its goals. This was because of increased population and development, a drop in state and federal funding, and a lack of cooperation and enforcement among local, state, and federal officials. That year, the Chesapeake Bay Foundation reported that the bay's water quality was "very poor" and only 21% of the established goals had been met.

GOOD NEWS

However, in 2011, a team of scientists led by Rebecca R. Murphy reported encouraging news. After analyzing 60 years of water-quality data, the scientists found that the sizes of oxygen-depleted zones that had been growing every summer in the bay had leveled off in the 1980s and had been declining since then. This meant that the integrated efforts to reduce inputs of fertilizers, animal wastes, and other pollutants were having a positive effect on the bay.

The other good news is that crab populations in the Chesapeake Bay have rebounded due to a set of measures put in place in 2008 by the states of Maryland and Virginia. While the bay's blue crab population was on the verge of collapse in 2003, it is now growing again to sustainable levels. As one researcher noted, efforts such as these must continue at the same or higher levels if the Chesapeake Bay Program is to meet its goals.

CONSIDER THIS. . .

THINKING ABOUT The Chesapeake Bay

What are three ways in which Chesapeake Bay area residents could apply the three scientific **principles of sustainability** (see Figure 1-2, p. 6 or back cover) to try to improve the environmental quality of the bay?

8-4 Why Are Freshwater Ecosystems Important?

CONCEPT 8-4
Freshwater ecosystems provide major ecosystem and economic services, and they are irreplaceable reservoirs of biodiversity.

Water Stands in Some Freshwater Systems and Flows in Others

Precipitation that does not sink into the ground or evaporate becomes **surface water**—freshwater that flows or is stored in bodies of water on the earth's surface. *Freshwater life zones* include *standing* (lentic) bodies of freshwater such as lakes, ponds, and inland wetlands, and *flowing* (lotic) systems such as streams and rivers. Surface water that flows into such bodies of water is called **runoff**. A **watershed**, or **drainage basin**, is the land area that delivers runoff, sediment, and dissolved substances to a stream, lake, or wetland.

Although freshwater systems cover less than 2.2% of the earth's surface, they provide a number of important ecosystem and economic services (Figure 8-14).

Lakes are large natural bodies of standing water formed when precipitation, runoff, streams, rivers, and groundwater seepage fill depressions in the earth's surface. Causes of such depressions include glaciation (as in Lake Louise in Alberta, Canada), displacement of the earth's crust (Lake Nyasa in East Africa), and volcanic activity (Crater Lake, Figure 8-15). Drainage areas on surrounding land supply lakes with water from rainfall, melting snow, and streams.

Freshwater lakes vary tremendously in size, depth, and nutrient content. Deep lakes normally consist of four distinct zones that are defined by their depth and distance from shore (Figure 8-16). The top layer, called the *littoral* ("LIT-tore-el") *zone*, is near the shore and consists of the shallow sunlit waters to the depth at which rooted plants stop growing. It has a high level of biodiversity because of ample sunlight and inputs of nutrients from the surrounding land. Species living in the littoral zone include many rooted plants; animals such as turtles, frogs, and crayfish; and fish such as bass, perch, and carp.

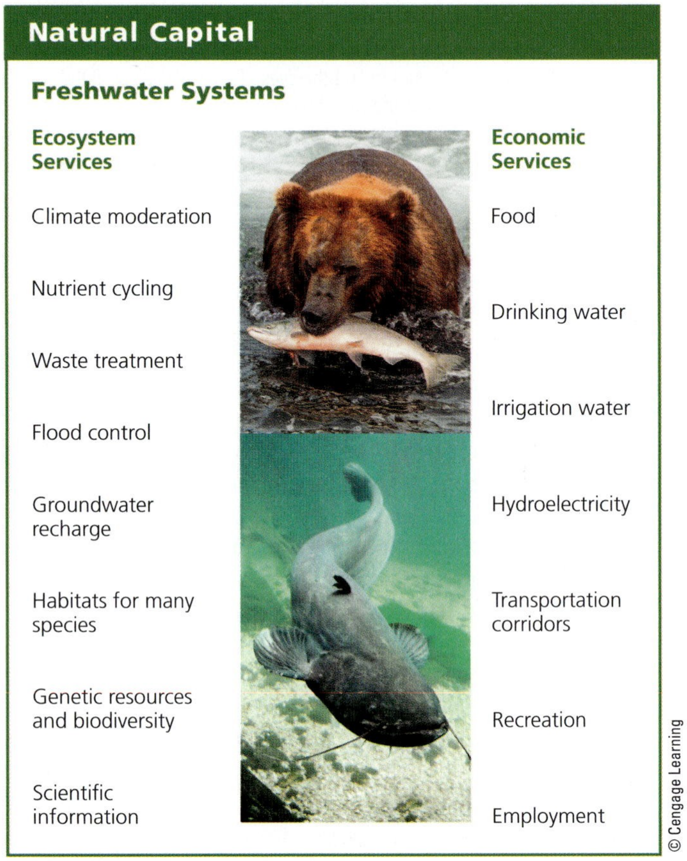

Natural Capital

Freshwater Systems

Ecosystem Services	Economic Services
Climate moderation	Food
Nutrient cycling	Drinking water
Waste treatment	Irrigation water
Flood control	Hydroelectricity
Groundwater recharge	Transportation corridors
Habitats for many species	Recreation
Genetic resources and biodiversity	Employment
Scientific information	

Figure 8-14 Freshwater systems provide many important ecosystem and economic services (Concept 8-4). ***Questions:*** Which two ecosystem services and which two economic services do you think are the most important? Why?

The next layer is the *limnetic* ("lim-NET-ic") *zone*, the open, sunlit surface layer away from the shore that extends to the depth penetrated by sunlight. This is the main photosynthetic zone of the lake, the layer that produces the food and oxygen that support most of the lake's consumers. Its most abundant organisms are microscopic phytoplankton and zooplankton. Some large species of fish spend most of their time in this zone, with occasional visits to the littoral zone to feed and reproduce.

Figure 8-15 Crater Lake, in the U.S. state of Oregon, formed in the crater of an extinct volcano.

Next comes the *profundal* ("pro-FUN-dahl") *zone,* a layer of deep, open water where it is too dark for photosynthesis. Without sunlight and plants, oxygen levels are often low here. Fishes adapted to the lake's cooler and darker water are found in this zone.

The bottom layer of the lake is called the *benthic* ("BEN-thic") *zone,* inhabited mostly by decomposers, detritus feeders, and some species of fish (benthos). The benthic zone is nourished mainly by dead matter that falls from the littoral and limnetic zones and by sediment washing into the lake.

Some Lakes Have More Nutrients Than Others

Ecologists classify lakes according to their nutrient content and primary productivity. Lakes that have a small supply of plant nutrients are called **oligotrophic lakes**. This type of lake (Figure 8-15) is often deep and has steep banks. Glaciers and mountain streams supply water to many such lakes, bringing little in the way of sediment or microscopic life to cloud the water. These lakes usually have crystal-clear water and small populations of phytoplankton and fish species (such as smallmouth bass and trout). Because of their low levels of nutrients, these lakes have a low NPP.

Over time, sediments, organic material, and inorganic nutrients wash into most oligotrophic lakes, and plants grow and decompose to form bottom sediments. A lake with a large supply of nutrients is called a **eutrophic lake** (Figure 8-17). Such lakes typically are shallow and have murky brown or green water with high turbidity. Because of their high levels of nutrients, these lakes have a high NPP.

Human inputs of nutrients through the atmosphere and from urban and agricultural areas within a lake's watershed can accelerate the eutrophication of the lake. This process, called **cultural eutrophication**, often puts excessive nutrients into lakes. Many lakes fall somewhere between the two extremes of nutrient enrichment. They are called **mesotrophic lakes**.

Freshwater Streams and Rivers Carry Large Volumes of Water

In drainage basins, water accumulates in small streams that join to form rivers. Some rivers join with other rivers and, altogether, the planet's streams and rivers carry huge amounts of water from highlands to lakes and oceans (Figure 8-18).

Animated Figure 8-16 This diagram illustrates the distinct zones of life in a fairly deep temperate-zone lake. ***Question:*** How are deep lakes like tropical rain forests? (*Hint:* See Figure 7-14, p. 157.)

Figure 8-17 This eutrophic lake, found in western New York State, has received large flows of plant nutrients. As a result, its surface is covered with mats of algae.

In many areas, streams begin in mountainous or hilly areas, which collect and release water falling to the earth's surface as rain or as snow that melts during warm seasons. The downward flow of surface water and groundwater from mountain highlands to the sea typically takes place in three aquatic life zones characterized by different environmental conditions: the *source zone,* the *transition zone,* and the *floodplain zone* (Figure 8-18). Rivers and streams can differ somewhat from this generalized model.

In the narrow *source zone* (Figure 8-18, left), headwater streams are usually shallow, cold, clear, and swiftly flowing (Figure 8-18, photo inset). As this turbulent water flows and tumbles downward over obstacles such as rocks, waterfalls, and rapids, it dissolves large amounts of oxygen from the air. Most of these streams are not very productive because of a lack of nutrients and primary producers. Their nutrients come primarily from organic matter, mostly leaves, branches, and the bodies of living and dead insects that fall into the stream from nearby land.

The source zone is populated by cold-water fish species (such as trout in some areas), which need lots of dissolved oxygen. Many fishes and other animals in fast-flowing headwater streams have compact and flattened bodies that allow them to live under stones. Others have streamlined and muscular bodies that allow them to swim in the rapid, strong currents. Most of the plants in this zone are algae and mosses attached to rocks and other surfaces under water.

In the *transition zone* (Figure 8-18, center), headwater streams merge to form wider, deeper, and warmer streams that flow down gentler slopes with fewer obstacles. They can be more turbid (containing suspended sediment) and slower flowing than headwater streams, and they tend to have less dissolved oxygen. The warmer water and other conditions in this zone support more producers, as well as cool-water and warm-water fish species (such as black bass) with slightly lower oxygen requirements.

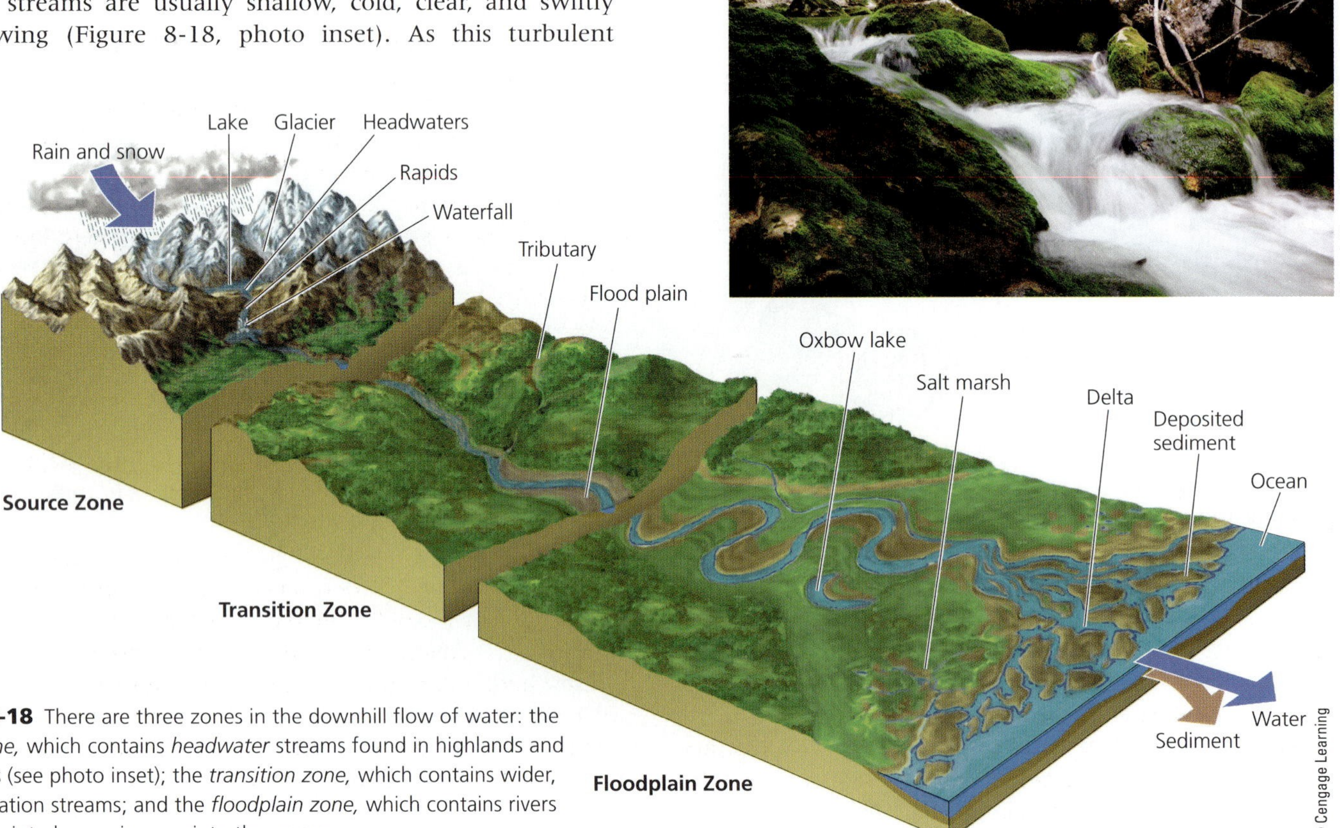

Figure 8-18 There are three zones in the downhill flow of water: the *source zone,* which contains *headwater* streams found in highlands and mountains (see photo inset); the *transition zone,* which contains wider, lower-elevation streams; and the *floodplain zone,* which contains rivers that empty into larger rivers or into the ocean.

As streams flow downhill, they shape the land through which they pass. Over millions of years, the friction of moving water has leveled mountains and cut deep canyons, and rocks and soil removed by the water have been deposited as sediment in low-lying areas. In these *floodplain zones* (Figure 8-18, right), streams join into wider and deeper rivers that flow across broad, flat valleys. Water in this zone usually has higher temperatures and less dissolved oxygen than water in the two higher zones. The slow-moving rivers sometimes support fairly large populations of producers such as algae and cyanobacteria, as well as rooted aquatic plants along the shores.

Because of increased erosion and runoff over a larger area, water in the floodplain zone often is muddy and contains high concentrations of suspended particulate matter (silt). The main channels of these slow-moving, wide, and murky rivers support distinctive varieties of fishes (such as carp and catfish), whereas their backwaters support species similar to those found in lakes. At its mouth, a river may divide into many channels as it flows through its *delta*—an area at the mouth of a river that was built up by deposited sediment and contains coastal wetlands and estuaries (Figure 8-7).

Coastal deltas and wetlands, as well as inland wetlands and floodplains, are important parts of the earth's natural capital (see Figure 1-3, p. 7). They absorb and slow the velocity of floodwaters from coastal storms, hurricanes, and tsunamis and provide habitats for a diversity of marine life (see the Case Study that follows).

CONSIDER THIS. . .

CONNECTIONS Stream Water Quality and Watershed Land

Streams receive most of their nutrients from bordering land ecosystems. Such nutrients come from falling leaves, animal feces, insects, and other forms of biomass washed into streams during heavy rainstorms or by melting snow. Chemicals and other substances flowing off the land can also pollute streams. Thus, the levels and types of nutrients and pollutants in a stream depend on what is happening in the stream's watershed.

CASE STUDY

River Deltas and Coastal Wetlands—Vital Components of Natural Capital Now in Jeopardy

Coastal deltas, mangrove forests, and coastal wetlands provide considerable natural protection against flood and wave damage from coastal storms, hurricanes, typhoons, and tsunamis.

When we remove or degrade these natural speed bumps and sponges, any damage from a natural disaster such as a hurricane or typhoon is intensified. As a result, flooding in places like New Orleans, Louisiana (USA), other parts of the U.S. Gulf Coast, and Venice, Italy, are largely self-inflicted unnatural disasters. For example, Louisiana, which contains about 40% of all coastal wetlands in the lower 48 states, has lost more than a fifth of such wetlands since 1950 to oil and gas wells and other forms of coastal development.

Humans have built dams, levees, and hydroelectric power plants on many of the world's rivers to control water flows and to generate electricity. This helps to reduce flooding along rivers, but it also reduces flood protection provided by the coastal deltas and wetlands. Because river sediments are deposited in the reservoirs behind dams, the river deltas do not get their normal inputs of sediment to build them back up, and they naturally sink into the sea.

As a result, 24 of the world's 33 major river deltas are sinking rather than rising and their protective coastal wetlands are flooding, according to a 2009 study by geologist James Syvitski and his colleagues. The study found that 85% of the world's sinking deltas have experienced severe flooding in recent years, and that global delta flooding is likely to increase by 50% by the end of this century. This is because of dams and other human-made structures that reduce the flow of silt and also the projected rise in sea levels resulting from climate change. It poses a serious threat to the roughly 500 million people in the world who live on river deltas.

For example, the Mississippi River once delivered huge amounts of sediments to its delta each year. But the multiple dams, levees, and canals built in this river system funnel much of this sediment load through the wetlands and out into the Gulf of Mexico. Instead of building up delta lands, this causes them to *subside,* or to sink. Other human activities that are leading to such subsidence include the extraction of groundwater, oil, and natural gas. As many of the river delta's freshwater wetlands have been lost to this subsidence, saltwater from the Gulf has intruded and killed many plants that depended on the river water, further degrading this coastal aquatic system.

This subsidence helps to explain why the city of New Orleans, Louisiana (Figure 8-19), has long been 3 meters (10 feet) below sea level. Dams and levees were built to help protect the city from flooding. However, in 2005, the powerful winds and waves from Hurricane Katrina overwhelmed these defenses. They are being rebuilt, but subsidence will probably put New Orleans 6 meters (20 feet) below sea level at some point in the future. Add to this the reduced protection from coastal and inland wetlands and barrier islands, and you have a recipe for a major and much more damaging unnatural disaster if the area is struck by another major hurricane.

To make matters worse, global sea levels have risen almost 0.3 meters (1 foot) since 1900 and are projected to rise another 0.3–0.9 meter (1–3 feet) by the end of this century. This is because the projected warming of the atmosphere will also warm the ocean, causing its waters to expand, and melt glaciers and other land-based ice, which will add to the ocean's volume. Such a rising sea level would put many of the world's coastal areas, including New Orleans and most of Louisiana's present-day coast, under water (Figure 8-20).

Figure 8-19 Much of the U.S. city of New Orleans, Louisiana, was flooded by the storm surge that accompanied Hurricane Katrina, which made landfall just east of the city on August 29, 2005.

The good news is that we now understand how our building of dams and other structures on rivers can affect their deltas and associated coastal wetlands. We know how the engineering of rivers can degrade or eliminate the ecosystem services provided by these freshwater systems and the marine coastal systems they feed. Environmental scientists argue that we must use such knowledge to sustain, rather than degrade or destroy, these components of natural capital and their vital ecosystem and economic services.

Freshwater Inland Wetlands Are Vital Sponges

Inland wetlands are lands located away from coastal areas that are covered with freshwater all or part of the time—excluding lakes, reservoirs, and streams. They include *marshes* (Figure 8-21, left), *swamps* (Figure 8-21, right), and *prairie potholes* (which are depressions carved out by ancient glaciers). Other examples of inland wetlands are *floodplains,* which receive excess water during heavy rains and floods.

Some wetlands are covered with water year-round. Others, called *seasonal wetlands,* remain under water or are soggy for only a short time each year. The latter include prairie potholes, floodplain wetlands, and arctic tundra (see Figure 7-11, bottom, p. 153). Some can stay dry for years before water covers them again. In such cases, scientists must use the composition of the soil or the presence of certain plants (such as cattails, bulrushes, or red maples) to determine that a particular area is a wetland. Wetland plants are highly productive because of an abundance of nutrients available to them. Many wetlands are important habitats for game fishes, muskrats, otters, beavers, migratory waterfowl, and other bird species.

Inland wetlands provide a number of free ecosystem and economic services, which include:

- filtering and degrading toxic wastes and pollutants;
- reducing flooding and erosion by absorbing storm water and releasing it slowly, and by absorbing overflows from streams and lakes;
- helping to sustain stream flows during dry periods;
- helping to recharge groundwater aquifers;
- helping to maintain biodiversity by providing habitats for a variety of species;
- supplying valuable products such as fishes and shellfish, blueberries, cranberries, wild rice, and timber; and
- providing recreation for birdwatchers, nature photographers, boaters, anglers, and waterfowl hunters.

CONSIDER THIS. . .

THINKING ABOUT Inland Wetlands

Which two ecosystem services and which two economic services provided by inland wetlands do you believe are the most important? Why? List two ways in which our daily activities directly or indirectly degrade inland wetlands.

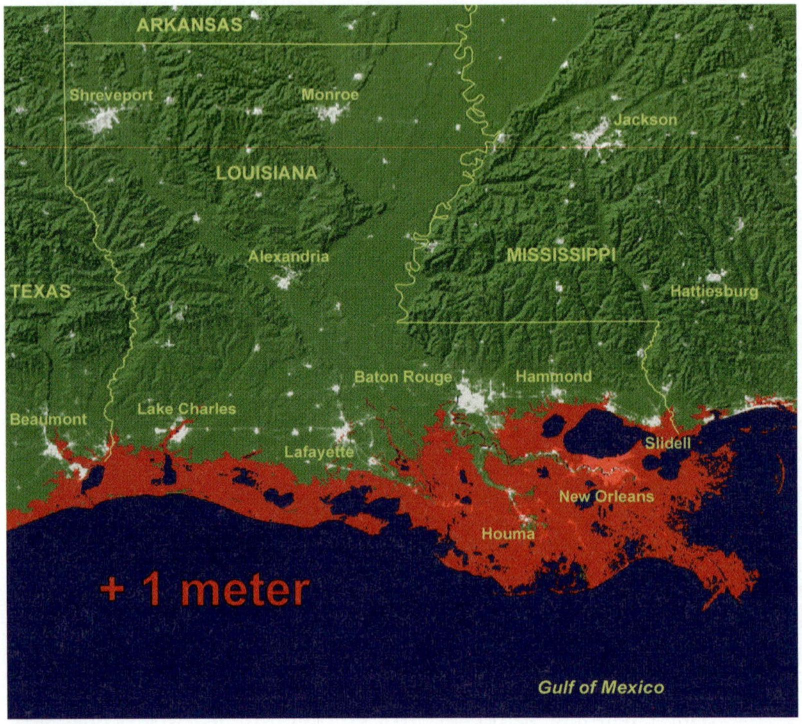

Used by permission from Jonathan Overpeck and Jeremy Weiss, University of Arizona

Figure 8-20 The areas in red represent projected coastal flooding that would result from a 1-meter (3-foot) rise in sea level due to projected climate change by the end of this century.

individuals matter 8.1

Alexandra Cousteau: Environmental Advocate, Filmmaker, and National Geographic Emerging Explorer

Alexandra Cousteau is proud of her heritage as granddaughter of Captain Jacques-Yves Cousteau and daughter of Philippe Cousteau. Her father and grandfather were legendary underwater explorers who brought the mysteries and wonders of the oceans into living rooms around the world with their films and books. She is also determined to build her own legacy as an advocate focused on water-related environmental issues.

The focus of Cousteau's life is to advocate the importance of conservation and sustainable management of water in order to preserve a healthy planet. Her global initiatives seek to inspire and empower individuals to protect not only the ocean and its inhabitants, but also the human communities that rely on freshwater resources. She seeks to make water one of the defining issues of this century, arguing: "We inhabit a water planet, and unless we protect, manage, and restore that resource, the future will be a very different place from the one we imagine today."

To that end, she is carrying on her family's proud tradition of storytelling. She says, "We evolved as a storytelling species, but the environmental community hasn't fully leveraged this approach. . . . That's why my grandfather was so successful. He was a master storyteller."

However, Cousteau is exploring a territory not even imagined by her grandfather—that of social networking and other emerging modes of communication. She is studying such trends along with video games and fantasy sports networks to see what makes them so popular and successful, and she believes that environmental advocates can use such new media tools to inform people about how their actions affect our water resources and the environments we inhabit. As an example, she cites the smart phone apps that help people to make sustainable choices in seafood when ordering from a restaurant menu or buying seafood at a grocery store.

One of Cousteau's major projects is Blue Legacy, an organization dedicated to "telling the story of our Water Planet" and shaping society's dialogue to include water as one of the defining issues of the 21st century by leading the conversation on the importance of "watershed-first thinking" and inspiring mainstream audiences to fully participate in the conservation and restoration of the water in their local communities. It is a source for articles and short films about a variety of issues related to water issues. Her website also features her many expeditions around the world, exploring and filming the intersection between water and our society. Led by Cousteau, these are explorations of ecosystems such as the Great Lakes and the Chesapeake Bay (background photo)—systems that are greatly stressed by various human activities.

Background photo: Lh8971/Dreamstime.com

FloridaStock/Shutterstock.com

Jayne Chapman/Shutterstock.com

a. b.

Figure 8-21 This great white egret lives in an inland marsh in the Florida Everglades (left). This cypress swamp (right) is located in South Carolina.

8-5 How Have Human Activities Affected Freshwater Ecosystems?

> **CONCEPT 8-5**
> Human activities threaten biodiversity and disrupt ecosystem and economic services provided by freshwater lakes, rivers, and wetlands.

Human Activities Are Disrupting and Degrading Freshwater Systems

Human activities are disrupting and degrading many of the ecosystem and economic services provided by freshwater rivers, lakes, and wetlands (**Concept 8-5**) in four major ways. *First,* dams and canals restrict the flows of about 40% of the world's 237 largest rivers. This alters or destroys terrestrial and aquatic wildlife habitats along these rivers and in their coastal deltas and estuaries. By reducing the flow of sediments to river deltas, these structures also lead to degraded coastal wetlands and greater damage from coastal storms (Case Study, p. 181). *Second,* flood-control levees and dikes built along rivers disconnect the rivers from their floodplains, destroy aquatic habitats, and alter or degrade the functions of adjoining wetlands.

A *third* major human impact on freshwater systems comes from cities and farms, which add pollutants and excess plant nutrients to streams, rivers, and lakes. For example, runoff of nutrients into a lake (cultural eutrophication, Figure 8-17) causes explosions in the populations of algae and cyanobacteria, which deplete the lake's dissolved oxygen. When these organisms die and sink to the lake bottom, decomposers go to work and further deplete the oxygen in deeper waters. Fishes and other species may then die off, which can mean a major loss in biodiversity.

Fourth, many inland wetlands have been drained or filled to grow crops or have been covered with concrete, asphalt, and buildings. More than half of the inland wetlands estimated to have existed in the continental United States during the 1600s no longer exist. About 80% of lost wetlands were drained to grow crops. The rest were lost to mining, logging, oil and gas extraction, highway construction, and urban development. The heavily farmed U.S. state of Iowa has lost about 99% of its original inland wetlands.

This loss of natural capital has been an important factor in greater and more frequent flood damage in parts of the United States. Many other countries have suffered similar losses. For example, 80% of all inland wetlands in Germany and France have been destroyed.

When we look further into human impacts on aquatic systems in Chapter 11, we will also explore possible solutions to environmental problems that result from these impacts, as well as ways to help sustain aquatic biodiversity. This is an area that will offer great opportunities to young scientists and other professionals in the years to come. (See Individuals Matter 8.1.)

Big Ideas

- Saltwater and freshwater aquatic life zones cover almost three-fourths of the earth's surface, and oceans dominate the planet.
- The earth's aquatic systems provide important ecosystem and economic services.
- Certain human activities threaten biodiversity and disrupt ecosystem and economic services provided by aquatic systems.

TYING IT ALL TOGETHER Coral Reefs and Sustainability

Vlad61/Shutterstock.com

This chapter's **Core Case Study** pointed out the ecological and economic importance of the world's incredibly diverse coral reefs. They are living examples of the three scientific **principles of sustainability** (see Figure 1-2, p. 6 or back cover) in action. They thrive on solar energy, participate in the cycling of carbon and other chemicals, and sustain a great deal of aquatic biodiversity.

SUSTAINABILITY

In this chapter, we have also seen that coral reefs and other aquatic systems are being severely stressed by a variety of human activities. Research shows that when such harmful human activities are reduced, some coral reefs and other endangered aquatic systems can recover fairly quickly.

As with terrestrial systems, scientists have made a start in understanding the ecology of the world's aquatic systems and how humans are degrading and disrupting the vital ecosystem and economic services they provide, but we still know far too little about the vital parts of the earth's life-support system. In studying these systems, scientists have again found that *in nature, everything is connected.* They argue that we urgently need more research on the components and workings of the world's aquatic life zones, on how they are interconnected, and on which systems are in the greatest danger of being disrupted by human activities.

We can take the lessons on how life in aquatic ecosystems has sustained itself for billions of years and use these scientific **principles of sustainability** to help sustain our own systems, as well as the ecosystems on which we depend. By relying more on solar energy and less on fossil fuels, we could drastically cut pollution of aquatic systems and CO_2 emissions that are causing ocean warming and ocean acidification. By reusing and recycling more of the materials and chemicals we use, we could further reduce pollution and disruption of the chemical cycling within aquatic systems. And by learning more about aquatic biodiversity and its importance, we could go a long way toward preserving it and sustaining its valuable ecosystem services.

Chapter Review

Core Case Study

1. What are **coral reefs** and why should we care about them? What is coral bleaching? What are the major threats to coral reefs?

Section 8-1

2. What are the two key concepts for this section? What percentage of the earth's surface is covered with water? What is an **aquatic life zone**? Distinguish between a **saltwater (marine) life zone** and a **freshwater life zone**, and give two examples of each. Define **plankton** and describe three types of plankton. Distinguish among **nekton**, **benthos**, and **decomposers** and give an example of each. List five factors that determine the types and numbers of organisms found in the three layers of aquatic life zones. What is **turbidity** and how does it occur? Describe one of its harmful impacts.

Section 8-2

3. What is the key concept for this section? What major ecosystem and economic services are provided by marine systems? What are the three major life zones in an ocean? Define and distinguish between the **coastal zone** and the **open sea**. Distinguish between an **estuary** and a **coastal wetland** and explain why each has high net primary productivity. Explain the ecological and economic importance of coastal marshes, mangrove forests, and sea-grass beds.

4. What is the **intertidal zone**? Distinguish between rocky and sandy shores and describe some of the organisms often found on each type of shoreline. Explain the importance of coral reefs. What is **ocean acidification** and why is it a threat to coral reefs? Describe the three major zones in the open sea. Why does the open sea have a low net primary productivity? What have scientists recently learned about the ocean's biodiversity?

Section 8-3

5. What is the key concept for this section? List five human activities that pose major threats to marine systems and eight human activities that threaten coral reefs. Explain why the Chesapeake Bay is an estuary in trouble. What is being done about some of its problems?

Section 8-4

6. What is the key concept for this section? Define **surface water**, **runoff**, and **watershed (drainage basin)**. What major ecosystem and economic services do freshwater systems provide? What is a **lake**? What four zones are found in deep lakes? Distinguish among **oligotrophic**, **eutrophic**, and **mesotrophic lakes**. What is **cultural eutrophication**?

7. Describe the three zones that a stream passes through as it flows from highlands to lower elevations. Explain how the building of dams and other structures on rivers can affect the river deltas and associated coastal wetlands. How do these effects in turn threaten human coastal communities?

8. Give three examples of **inland wetlands** and describe the ecological and economic importance of such wetlands.

Section 8-5

9. What is the key concept for this section? What are four ways in which human activities are disrupting and degrading freshwater systems? Describe losses of inland wetlands in the United States in terms of the area of wetlands lost and the resulting loss of ecosystem and economic services.

10. What are this chapter's three big ideas? How do coral reefs showcase the three scientific **principles of sustainability**? How can we use these three principles to help sustain the earth's vital aquatic ecosystems?

Note: Key terms are in **bold** type.

Critical Thinking

1. What are three steps that governments and private interests could take to protect the world's remaining coral reefs (**Core Case Study**)?

2. Can you think of any ways in which you might be contributing to the degradation of a nearby or distant aquatic ecosystem? Describe the system and how your actions might be affecting it. What are three things you could do to reduce your impact?

3. You are a defense attorney arguing in court for protecting a coral reef (**Core Case Study**) from harmful human activities. Give your three most important arguments for the defense of this ecosystem.

4. How would you respond to someone who argues that we should use the deep portions of the world's oceans to deposit our radioactive and other hazardous wastes because the deep oceans are vast and are located far away from human habitats? Give reasons for your response.

5. Why is increasing ocean acidification considered a very serious problem? If acid levels in the ocean rise sharply during your lifetime, how might this affect you? Can you think of ways in which you might be contributing to this problem? What could you do to reduce your impact?

6. Suppose a developer builds a housing complex overlooking a coastal marsh (Figure 8-8) and the result is pollution and degradation of the marsh. Describe the effects of such a development on the wildlife in the marsh, assuming at least one species is eliminated as a result.

7. Suppose you have a friend who owns property that includes a freshwater wetland and the friend tells you she is planning to fill the wetland to make more room for her lawn and garden. What would you say to this friend?

8. Congratulations! You are in charge of the world. What are the three most important features of your plan to help sustain the earth's aquatic biodiversity?

Doing Environmental Science

Using Google Earth, find a relatively undisturbed aquatic ecosystem somewhere on the planet that you can zoom in on to observe some features. For example, find a coastal wetland or a river system that seems to be undisturbed by human activities on adjoining land, such as farming or urban development. Write a description of what you see. Then find an example of a comparable system that has been disturbed by human activities and write a detailed comparison of these two areas. If possible, check in on these systems once during your course term and again at the end of the term and describe any changes to the systems that you can observe. Suggest a hypothesis to explain major changes.

Global Environment Watch Exercise

Search for *Coral reefs* and use the topic portal to find information on **(a)** trends in the global rate of coral destruction; **(b)** what areas of the world are seeing rising rates and what areas are seeing falling rates; and **(c)** what is being done to protect them in various areas. Write a report on your findings.

Data Analysis

Some 45–53% of the world's shallow coral reefs have been destroyed or severely damaged (**Core Case Study**). A number of factors have played a role in this serious loss of aquatic biodiversity, including ocean warming, sediment from coastal soil erosion, excessive algal growth from fertilizer runoff, coral bleaching, rising sea levels, ocean acidification, overfishing, and damage from hurricanes.

In 2005, scientists Nadia Bood, Melanie McField, and Rich Aronson conducted research to evaluate the recovery of coral reefs in Belize from the combined effects of mass bleaching and Hurricane Mitch in 1998. Some of these reefs are in protected waters where no fishing is allowed. The researchers speculated that reefs in waters where no fishing is allowed should recover faster than reefs in waters where fishing is allowed. The graph to the left shows some of the data they collected from three highly protected (unfished) sites and three unprotected (fished) sites to evaluate their hypothesis. Study this graph and answer the questions below.

This graph tracks the effects of restricting fishing on the recovery of unfished and fished coral reefs damaged by the combined effects of mass bleaching and Hurricane Mitch in 1998.

(Compiled by the authors with data from Melanie McField, et al., *Status of Caribbean Coral Reefs after Bleaching and Hurricanes in 2005*, NOAA, 2008. Report available at www.coris.noaa.gov/activities/caribbean_rpt/.)

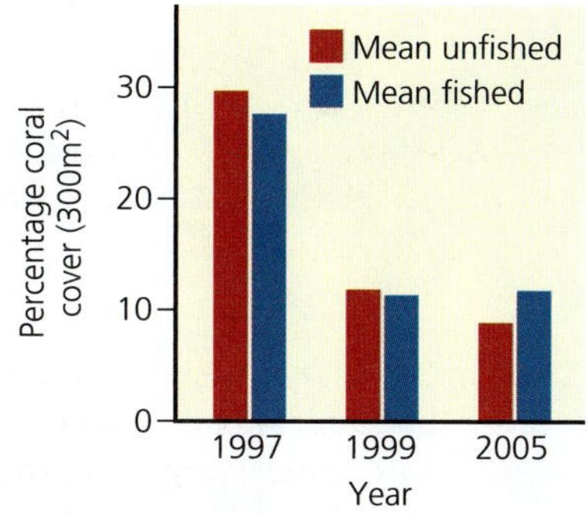

1. By about what percentage did the mean coral cover drop in the protected (unfished) reefs between 1997 and 1999?
2. By about what percentage did the mean coral cover drop in the protected (unfished) reefs between 1997 and 2005?
3. By about what percentage did the coral cover drop in the unprotected (fished) reefs between 1997 and 1999?
4. By about what percentage did the coral cover change in the unprotected (fished) reefs between 1997 and 2005?
5. Do these data support the hypothesis that coral reef recovery should occur faster in areas where fishing is prohibited? Explain.

9 Sustaining Biodiversity: Saving Species and Ecosystem Services

The last word in ignorance is the person who says of an animal or plant: "What good is it?" . . . If the land mechanism as a whole is good, then every part of it is good, whether we understand it or not. Harmony with land is like harmony with a friend; you cannot cherish his right hand and chop off his left.

ALDO LEOPOLD

Key Questions

9-1 What role do humans play in the loss of species and ecosystem services?

9-2 Why should we care about sustaining species and the ecosystem services they provide?

9-3 How do humans accelerate species extinction and degradation of ecosystem services?

9-4 How can we sustain wild species and their ecosystem services?

Endangered wild Siberian tiger.

Volodymyr Burdiak/Shutterstock.com

CORE CASE STUDY Where Have All the Honeybees Gone?

In meadows, forests, farm fields, and gardens around the world, industrious honeybees (Figure 9-1) flit from one flowering plant to another. They spend their days collecting nectar and pollen that they take back to their hives where the typical bee lives with 30,000 to 100,000 other bees. They feed their young the protein-rich pollen collected from the flowers, and the adults live on the honey made from the collected nectar and stored in the hive's wax honeycombs.

Honeybees play a key role in providing us with one of nature's most important ecosystem services: *pollination,* the transfer of pollen within and among flowering plants that enables them to produce seeds and fruit. Bees pollinate many flower species and some of our most important food crops, including many vegetables, fruits, and tree nuts such as almonds. Globally, about one-third of the human food supply comes from insect-pollinated plants, and European honeybees are responsible for 80% of that pollination.

Many U.S. growers rent European honeybees from commercial beekeepers who together truck about 2.7 million hives to farms across the country to pollinate different crops at bloom times. These growers have largely replaced the earth's free pollination service provided by a diversity of bees and other pollinators with reliance mostly on a single bee species.

Some producers believe we need this industrialized pollination system in order to grow enough food. Others see such heavy dependence on a single bee species as a potentially dangerous violation of the earth's biodiversity **principle of sustainability**. They argue that this dependence will put food supplies at risk if the European honeybees decline.

In fact, bee experts assembled by the U.S. National Academy of Sciences have reported a 30% drop in populations of European honeybee in the United States since the 1980s. Since 2006, this situation has worsened in the United States and in parts of Europe as massive numbers of European honeybees have been disappearing from their colonies during the winter and not returning as expected in the spring—a phenomenon called **colony collapse disorder (CCD)**. In each year since 2006, about 30–50% of the European honeybee colonies in the United States have suffered from CCD. Researchers are searching for the causes of this problem and for ways to deal with it.

Scientists project that during this century, human activities, especially those that contribute to habitat loss and climate change, are likely to play a key role in the extinction of one-fourth to one-half of the world's identified plant and animal species (Figure 9-2). Many scientists view this threat to the earth's vital ecosystem services as one of the most serious and long-lasting environmental and economic problems we face. In this chapter, we discuss the causes of this problem and possible ways to deal with it.

Figure 9-1 European honeybee drawing nectar from a flower.

9-1 What Role Do Humans Play in the Loss of Species and Ecosystem Services?

CONCEPT 9-1
Species are becoming extinct 100 to 1,000 times faster than they were before modern humans arrived on earth, and by the end of this century, the extinction rate is projected to be 10,000 times higher than that background rate.

Extinctions Are Natural but Sometimes They Increase Sharply

When a species can no longer be found anywhere on the earth, it has suffered **biological extinction**. The disappearance of any species, and especially those that play keystone roles (see Chapter 4, pp. 94–96), is irreversible. The loss of a keystone species or a major reduction in its populations can lead to population declines or extinctions of species with strong connections to such species and to a breakdown in ecosystem services that depend on those connections.

Ecologists refer to such a series of changes as a *trophic cascade*. For example, a 2011 report by ecologist James A. Estes and his colleagues pointed out that sharp declines in the populations of top predators such as sea otters (see Chapter 5 Core Case Study, p. 102), lions, tigers, wolves, and some shark species can affect populations of the species on which they feed. Such declines can also result in degradation of habitats and ecosystem services such as chemical cycling and energy flows (see Chapter 4 Case Study, p. 95).

The extinction of many species in a relatively short period of geologic time is called a **mass extinction**. Geologic, fossil, and other records indicate that the earth has experienced five mass extinctions, when 50–95% of the world's species appear to have become extinct. After each mass extinction, the earth's overall biodiversity eventually returned to equal or higher levels, but each recovery required millions of years. Such long-term recovery is an example of the biodiversity **principle of sustainability** in action.

The causes of past mass extinctions are poorly understood but probably involved global changes in environmental conditions. Examples are sustained and significant global warming or cooling, large changes in sea level, and catastrophes such as multiple large-scale volcanic eruptions. One hypothesis is that the last mass extinction, which took place about 65 million years ago, occurred after a large asteroid hit the planet and spewed huge amounts of dust and debris into the atmosphere. This could have reduced the input of solar energy and cooled the planet long enough to wipe out the dinosaurs and many other forms of life.

Some Human Activities Hasten Extinctions and Threaten Ecosystem Services

Extinction is a natural process. Scientists who study it base much of their work on an estimated **background extinction rate**—the rate that existed before modern humans evolved some 200,000 years ago—which scientists believe was about 1 species per year for every 1 million wild species living on the earth.

However, scientific evidence indicates that extinction rates have risen in some areas as human populations have spread over most of the globe, destroying and degrading habitats, consuming huge quantities of resources, and creating large and growing ecological footprints (see Figure 1-13, p. 14). In some areas, these trends are threatening the ecosystem services that sustain our lives and economies. In the words of biodiversity expert Edward O. Wilson (see Individuals Matter 4.1, p. 82), "The natural world is everywhere disappearing before our eyes—cut to pieces, mowed down, plowed under, gobbled up, replaced by human artifacts."

Evidence is piling up that human activities are causing losses in global, regional, and local biodiversity at an increasing rate. Each year, the World Wildlife Fund publishes its annual Living Planet Index (LPI) based on its monitoring of populations of over 2,500 vertebrate species. Between 1970 and 2008, the global LPI fell by 28% with a 31% drop in temperate areas and a 61% drop in tropical areas.

Scientists from around the world who conducted the 2005 Millennium Ecosystem Assessment estimated that the current annual rate of species extinction is at least 100 to 1,000 times the estimated background extinction rate (Science Focus 9.1). We have identified about 2 million species so far, but scientists estimate that there are many millions more to identify. For simplicity, let's assume there are 10 million species on earth. Then, at the background extinction rate of 1 species per million per year, about 10 species would disappear naturally each year. However, at today's estimated rate of 100 to 1,000 times the background rate, we are losing between 1,000 and 10,000 species per year, or between 2 and 27 species every day, on average.

Biodiversity researchers project that during this century, the extinction rate is likely to rise to at least 10,000 times the background rate—mostly because of habitat loss and degradation, climate change, and other environmentally harmful effects of human activities (**Concept 9-1**). At this rate, if there are 10 million species on the earth, then about 100,000 species would be expected to disappear each year—an average of about 274 species per day or about 11 every hour. By the end of this century, most of the big carnivorous cats, including tigers (see chapter-opening photo), cheetahs, and lions, will probably exist only in zoos and

SCIENCE FOCUS 9.1

ESTIMATING EXTINCTION RATES

Figure 9-A Painting of a pair of North American passenger pigeons, which once were one of the world's most abundant bird species. They became extinct in the wild in 1913 mostly because of habitat loss and overhunting.

Scientists who try to catalog extinctions, estimate past extinction rates, and project future extinction rates face three problems. *First,* because the natural extinction of a species typically takes a very long time, it is difficult to document. *Second,* we have identified only about 2 million of the world's estimated 7–10 million and perhaps as many as 100 million species. *Third,* scientists know little about the ecological roles of most of the species that have been identified, or about how vulnerable they are to extinction and how their extinction might affect other species and the ecosystem services in the ecosystems where they are found.

One approach to estimating future extinction rates is to study records documenting past rates at which easily observable mammals and birds (Figure 9-A) have become extinct. Most of these extinctions have occurred since humans began to dominate the planet about 10,000 years ago, when we began the shift from hunting and gathering food in the wild to growing our food. This information can be compared with fossil records of extinctions that occurred before that time.

Another approach is to observe how reductions in habitat area affect extinction rates. The *species–area relationship,* studied by Edward O. Wilson (see Individuals Matter 4.1, p. 82) and Robert MacArthur, suggests that, on average, a 90% loss of land habitat in a given area can cause the extinction of about 50% of the species living in that area. Scientists use this model to estimate the number of current and future extinctions in patches or "islands" of shrinking wild habitat that are surrounded by degraded habitats or by rapidly growing human developments.

Scientists also use mathematical models to estimate the risk of a particular species becoming endangered or extinct within a certain period of time. These *population viability analysis* (PVA) models include factors such as trends in population size, past and projected changes in habitat availability, interactions with other species, and genetic factors.

Researchers know that their estimates of extinction rates are based on incomplete data and sampling, and on imperfect models. Thus, they are continually striving to get more and better data and to improve the models they use in order to estimate extinction rates and to project the effects of such extinctions on vital ecosystem services such as pollination (**Core Case Study**).

At the same time, they point to considerable and growing evidence that human activities have accelerated the rate of species extinction and that this rate is increasing. According to these biologists, arguing over the numbers and waiting to get better data and models should not delay our acting now to help prevent extinctions, along with the resulting threats to key ecosystem services, that result mostly from human activities.

Critical Thinking

Does the fact that extinction rates can only be estimated make them unreliable? Why or why not? (*Hint:* See Chapter 2, pp. 34–35.)

---

small wildlife sanctuaries. And most elephants and rhinoceroses will likely disappear from the wild, along with gorillas, chimpanzees, and orangutans. According to conservation biologist Thomas Lovejoy, we are experiencing the beginning of a tsunami of species extinction and degradation of ecosystem services that is likely to spread over much of the world during this century.

So why is this a big deal? According to biodiversity researchers Edward O. Wilson and Stuart Pimm, at this extinction rate, at least 25% and as many as 50% of the world's current 2 million identified animal and plant species could vanish from the wild by the end of this century, along with many of the millions of unidentified species. This would amount to a sixth mass extinction caused primarily by human activities with much of it taking place within just one century.

These experts note that with the loss of such a huge portion of the planet's biodiversity, we would also likely lose whole ecosystems that depend on the vanishing species, along with the vital ecosystem services they provide, including air and water purification, natural pest control, and pollination (**Core Case Study**). According to the 2005 Millennium Ecosystem Assessment, 15 of 24 major ecosystem services are in decline. If such estimates are only half correct, we can see why many biologists warn that such a massive loss of biodiversity and ecosystem services within the span of a single human lifetime is one of the most important and long-lasting environmental and economic problems we face.

CONSIDER THIS. . .

THINKING ABOUT Extinction

How might your lifestyle change if human activities contribute to the extinction of up to half of the world's identified species during this century? How might this affect the lives of any children or grandchildren you might have? List two aspects of your lifestyle that contribute to this threat to the earth's natural capital.

In fact, Wilson, Pimm, and other extinction experts consider a projected extinction rate of 10,000 times the background extinction rate to be low, for several reasons. *First,* both the rate of extinction and the resulting threats to ecosystem services are likely to increase sharply during the next 50–100 years because of the harmful environmental impacts of the rapidly growing human population and its growing use of resources per person (see Figure 1-14, p. 15).

Second, the current and projected extinction rates in the world's *biodiversity hotspots*—areas that are highly endangered centers of biodiversity—are much higher than the global average. Biodiversity expert Norman Myers and several other researchers urge us to focus our efforts on lowering the rates of extinction in such areas as soon as possible. They see such emergency action as the best and quickest way to prevent much of the earth's biodiversity from being lost during this century. Other scientists urge us to identify and protect areas where vital ecosystem services such as pollination by honeybees (**Core Case Study**) and topsoil formation are being threatened.

Third, we are eliminating, degrading, fragmenting, and simplifying many biologically diverse environments—including tropical forests, coral reefs, wetlands, and estuaries—that serve as potential sites for the emergence of new species. Thus, in addition to greatly increasing the rate of extinction, we may be limiting the long-term recovery of biodiversity by eliminating these places where new species can evolve. In other words, we are also creating a *speciation crisis.* (See the online Guest Essay by Norman Myers on this topic.) Based on what scientists have learned about the recovery of biodiversity after past mass extinctions, it will take 5 million to 10 million years for the earth's processes to replace the projected number of species that are likely to go extinct during this century primarily because of human activities.

In addition, Philip Levin, Donald Levin, and other biologists argue that, while our activities are likely to reduce the speciation rates for some species, they might increase the speciation rates for other rapidly reproducing species such as weeds and rodents, as well as cockroaches and many other species of insects. Rapidly expanding populations of such species could crowd and compete with various other species, further accelerating their extinction and also threatening key ecosystem services.

Endangered and Threatened Species Are Ecological Smoke Alarms

Biologists classify species that are heading toward biological extinction as either *endangered* or *threatened.* An **endangered species** has so few individual survivors that the species could soon become extinct. A **threatened species** (also known as a *vulnerable species*) still has enough remaining individuals to survive in the short term, but because of declining numbers, it is likely to become endangered in the near future.

Figure 9-2 shows four of the 20,219 species listed in 2012 by the International Union for Conservation of Nature (IUCN) as critically endangered, endangered, or

Geoffrey Kuchera/Shutterstock.com

a. Mexican gray wolf: About 42 in the forests of Arizona and New Mexico

Ferenc Cegledi/Shutterstock.com

b. California condor: 226 in the southwestern United States (up from 9 in 1986)

Catcher of Light, Inc./Shutterstock.com

c. Whooping crane: 437 in North America

Tiago Jorge da Silva Estima/Shutterstock.com

d. Sumatran tiger: No more than 500 on the Indonesian island of Sumatra

Figure 9-2 *Endangered natural capital:* These four critically endangered species are threatened with extinction, largely because of human activities. The number below each photo indicates the estimated total number of individuals of that species remaining in the wild, as of 2012.

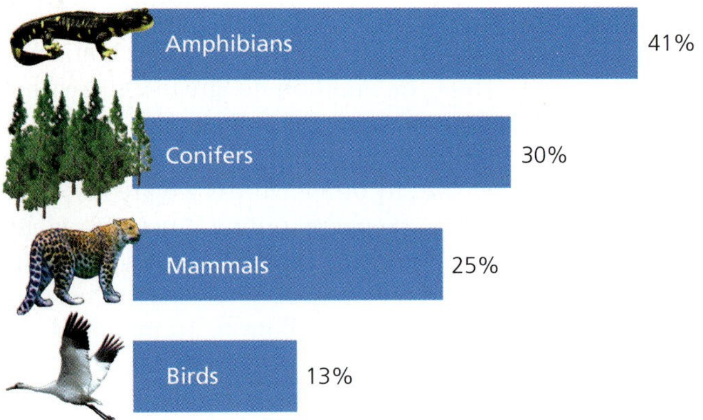

Figure 9-3 **Endangered natural capital:** Comparison of the percentages of various types of known species that are threatened with extinction hastened by human activities as of 2012 (**Concept 9-1**). ***Question:*** Why do you think so many of the world's amphibians are threatened with extinction? (See Chapter 4 Core Case Study, p. 78.)

(Compiled by the authors using data from 2012 IUCN Red List of Threatened Species.)

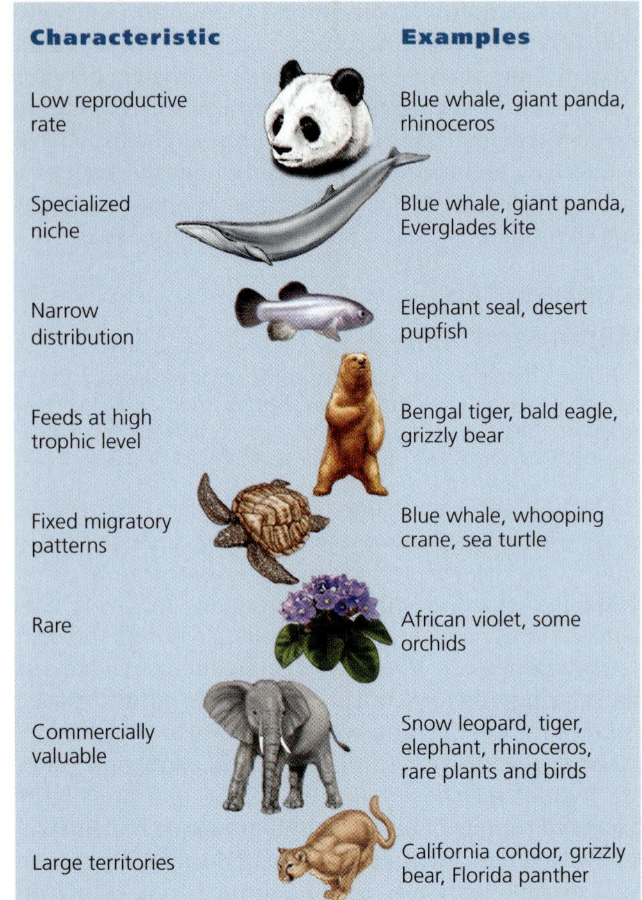

Figure 9-4 Certain characteristics can put a species in greater danger of becoming extinct.

threatened. This is an 92% increase over the number listed in 1996. The real number of species in trouble is very likely much higher. This annual assessment by the IUCN shows that some types of species are threatened with extinction hastened by human activities more than others are (Figure 9-3). In 2012 there were 1,143 U.S. species listed as endangered or threatened under the U.S. Endangered Species Act. Again, the real number is very likely much higher.

Some species have characteristics that increase their chances of becoming extinct (Figure 9-4). As biodiversity expert Edward O. Wilson puts it, "The first animal species to go are the big, the slow, the tasty, and those with valuable parts such as tusks and skins."

Species can also become *regionally extinct* in the areas where they are normally found, as well as *functionally extinct* when their populations crash to the point where they can no longer play their functional roles in an ecosystem. The latter is also called *extinction of ecological interactions,* because when a species' numbers drop to a certain point, its interactions with other species are lost or greatly diminished. Important ecosystem services that depend on these interactions might also then be lost or diminished, and this is often difficult to detect until it is too late.

For example, the American alligator is a keystone species in its marsh and swamp habitats of the southeastern United States. (See Case Study, Chapter 4, p. 95.) When its numbers dwindled in the 1960s, certain ecosystem services, such as the building of gator nests, were not performed, and bird species that depended on these nesting sites also declined. After the alligator was placed on the U.S. endangered species list, it made a strong comeback. The ecosystem services it performed were once again available, and its ecosystems recovered.

9-2 Why Should We Care about Sustaining Species and the Ecosystem Services They Provide?

CONCEPT 9-2
We should avoid speeding up the extinction of wild species because of the ecosystem and economic services they provide, because it can take millions of years for nature to recover from large-scale extinctions, and because many people believe that species have a right to exist regardless of their usefulness to us.

Species Are a Vital Part of the Earth's Natural Capital

According to the World Wildlife Fund, only 50,000–60,000 orangutans (Figure 9-5) remain in the wild, most of them in the tropical forests of Indonesia. These highly intelligent animals are disappearing at a rate of more than 1,000–2,000

SuperStock

Figure 9-5 **Natural capital degradation:** These endangered orangutans depend on a rapidly disappearing tropical forest habitat. ***Question:*** What difference will it make if human activities hasten the extinction of the orangutan?

per year because of illegal smuggling and the clearing of their tropical forest habitat to make way for plantations of oil palms that supply palm oil used in cosmetics, cooking, and the production of biodiesel fuel. An illegally smuggled, live orangutan sells for a street price of up to $10,000. Without urgent protective action, the endangered orangutan may disappear in the wild within the next two decades.

Does it matter that orangutans might soon become extinct mostly because of human activities? Does it matter if some unknown plant or insect in a tropical forest meets the same fate? If all species eventually become extinct, why should we worry about the rate of extinction? New species eventually evolve through speciation to take the places of those species lost through mass extinctions, so why should we care if we greatly speed up the extinction rate over the next 50–100 years?

According to biologists, there are three major reasons why we should work to prevent our activities from causing or hastening the extinction of other species. *First,* the world's species provide vital *ecosystem services* (see Figure 1-3, p. 7) that help to keep us alive and support our economies (**Concept 9-2**). For example, we depend on honeybees (**Core Case Study**) and other insects for pollination of many food crops and on certain bird species for natural pest control. When species disappear, we can lose such services, and at some point, such losses can threaten our own health and our economies in the affected areas.

Species also depend on other species within their food webs and, thus, by eliminating any species or sharply reducing its populations, especially a species that plays a keystone role (see Chapter 4, pp. 94–95), we can speed up the extinction of other species. This is another way in which, by hastening extinction, we can upset an ecosystem and degrade its important ecosystem services.

CONSIDER THIS. . .

CONNECTIONS Species and Ecosystem Services

Plant and animal species provide us with services in ways that you might not expect. For example, those that live in streams help to purify the flowing water. Trees and other forest plants produce oxygen, without which we could not survive. And earthworms aerate topsoil that we use for growing our food. These and other amazing ecosystem services are free and available around the clock for us as long as we don't degrade them through pollution or overuse.

Many species also contribute to *economic services* on which we depend (**Concept 9-2**). For example, various plant species provide economic value as food crops, and from trees, we produce fuelwood, lumber, and paper. *Bioprospectors* search tropical forests and other ecosystems to find plants and animals that scientists can use to make medicinal drugs (Figure 9-6). According to a United Nations University report, 62% of all cancer drugs were derived from the discoveries of bioprospectors. Despite their economic and medicinal potential, less than 0.5% of the world's known plant species have been examined for their medicinal properties.

Figure 9-6 **Natural capital:** These plant species are examples of *nature's pharmacy.* Once the active ingredients in the plants have been identified, scientists can usually produce them synthetically. The active ingredients in nine of the ten leading prescription drugs originally came from wild organisms.

In 2011, scientists estimated that as many as 10,000 *phytochemicals*—certain chemicals such as antioxidants that occur naturally in plants—have the potential to slow aging, reduce pain, and help us to reduce our weight, prevent various cancers, and control diseases such as diabetes. Many of these potentially beneficial chemicals are available as over-the-counter supplements that are far less expensive than prescription drugs. However, more research is needed to measure their benefits.

By preserving species and their habitats, we can also gain economic benefits from wildlife tourism, or *ecotourism,* which generates more than $1 million per minute in tourist expenditures, worldwide. Conservation biologist Michael Soulé estimates that a male lion living to age 7 generates about $515,000 from tourism in Kenya, but only about $1,000 if it is killed for its skin. Ecotourism is thriving because people enjoy wildlife (Figure 9-7). **GREEN CAREER:** Ecotourism guide

GOOD NEWS

A *second* reason for preventing extinctions caused or hastened by human activities is that analysis of past mass extinctions indicates it will take 5 million to 10 million years for natural speciation to rebuild the biodiversity that is likely to be lost during this century. As a result, any grandchildren we might have and many thousands of future generations are unlikely to be able to depend on the life-sustaining biodiversity and vital ecosystem services that currently support our well-being and our economies.

Third, many people believe that wild species have a right to exist, regardless of their usefulness to us (**Concept 9-2**). According to the stewardship view, we have a responsibility to protect the earth's species—our only known living companions in the universe—from becoming extinct as a result of human activities, and to prevent

ROY TOFT/National Geographic Creative

Figure 9-7 Many species of wildlife such as this endangered hyacinth macaw in Mato Grosso, Brazil, are sources of beauty and pleasure. It is endangered because of habitat loss and illegal capture in the wild by pet traders.

the degradation of the world's ecosystem services and thus the human life-support system.

This ethical viewpoint raises a number of challenging questions. Since we cannot save all species from the harmful consequences of our actions, we have to make choices about which ones to protect. Should we protect more animal species than plant species and, if so, which ones should we protect? Some people support protecting familiar and appealing species such as elephants, whales, tigers, and orangutans (Figure 9-5), but care much less about protecting plants or insects that serve as the base of the food supply for many other species (**Core Case Study**). Others might think little about getting rid of species that most people fear or hate, such as mosquitoes, cockroaches, disease-causing bacteria, snakes, sharks, and bats.

Some scientists argue for halting activities that hasten extinctions because no one has ever seen or studied most of the life-sustaining biodiversity that is being lost. To biologist Edward O. Wilson, carelessly eliminating species is like burning millions of books that we have never read.

9-3 How Do Humans Accelerate Species Extinction and Degradation of Ecosystem Services?

CONCEPT 9-3

The greatest threats to species and ecosystem services are (in order) loss or degradation of habitat, harmful invasive species, human population growth, pollution, climate change, and overexploitation.

Loss of Habitat Is the Single Greatest Threat to Species: Remember HIPPCO

Biodiversity researchers summarize the most important direct causes of extinction and threats to ecosystem services using the acronym **HIPPCO**: **H**abitat destruction, degradation, and fragmentation; **I**nvasive (nonnative) species; **P**opulation growth and increasing use of resources; **P**ollution; **C**limate change; and **O**verexploitation (**Concept 9-3**).

According to biodiversity researchers, the greatest threat to wild species is habitat loss (Figure 9-8), degradation, and fragmentation. Specifically, deforestation in tropical areas (see Chapter 3 opening photo and Figure 3-1, p. 52) is the greatest threat to species and to the ecosystem services they provide, followed by the destruction and degradation of coastal wetlands and coral reefs (see Chapter 8 opening photo, pp. 166–167), the plowing of grasslands (see Figure 7-12, p. 154), and the pollution of streams, lakes, and oceans.

Island species—many of them found nowhere else on earth—are especially vulnerable to extinction when their habitats are destroyed, degraded, or fragmented, because they have nowhere else to go. This is why the collection of islands that make up the U.S. state of Hawaii is America's "extinction capital"—with 63% of its species at risk.

Habitat fragmentation occurs when a large, intact area of habitat such as a forest or natural grassland is divided, typically by roads, logging operations, crop fields, and urban development, into smaller, isolated patches or *habitat islands.* This process can decrease tree cover in forests and block animal migration routes. It can also divide populations of a species into smaller, increasingly isolated groups that are more vulnerable to predators, competitor species, disease, and catastrophic events such as storms and fires. In addition, habitat fragmentation creates barriers that limit the abilities of some species to disperse and colonize new areas, to locate adequate food supplies, and to find mates.

Most national parks and other nature reserves are habitat islands, many of them surrounded by potentially damaging logging and mining operations, coal-burning power plants, industrial activities, and human settlements. Freshwater lakes are also habitat islands that are especially vulnerable to the introduction of harmful invasive species and pollution from human activities. Can you think of other examples of habitat islands?

We Have Moved Disruptive Species into Some Ecosystems

After habitat loss and degradation, the next biggest threat to animal and plant extinctions and the ecosystem services they provide is the deliberate or accidental introduction of harmful species into ecosystems (**Concept 9-3**).

GOOD NEWS

Many introductions of nonnative species have been beneficial to us. According to a study by ecologist David Pimentel, nonnative species such as corn, wheat, rice, and other food crops, as well as some species of cattle, poultry, and other livestock, provide more than 98% of the U.S. food supply. Similarly, nonnative tree species are grown in about 85% of the world's tree plantations. Some deliberately introduced species have helped to control pests. And highly beneficial European honeybees (**Core Case Study**) were brought to North America in the 1600s by English settlers who harvested the bees' honey and used the wax from their hives to make candles.

The problem is that, in their new habitats, some introduced species do not face the natural predators, competi-

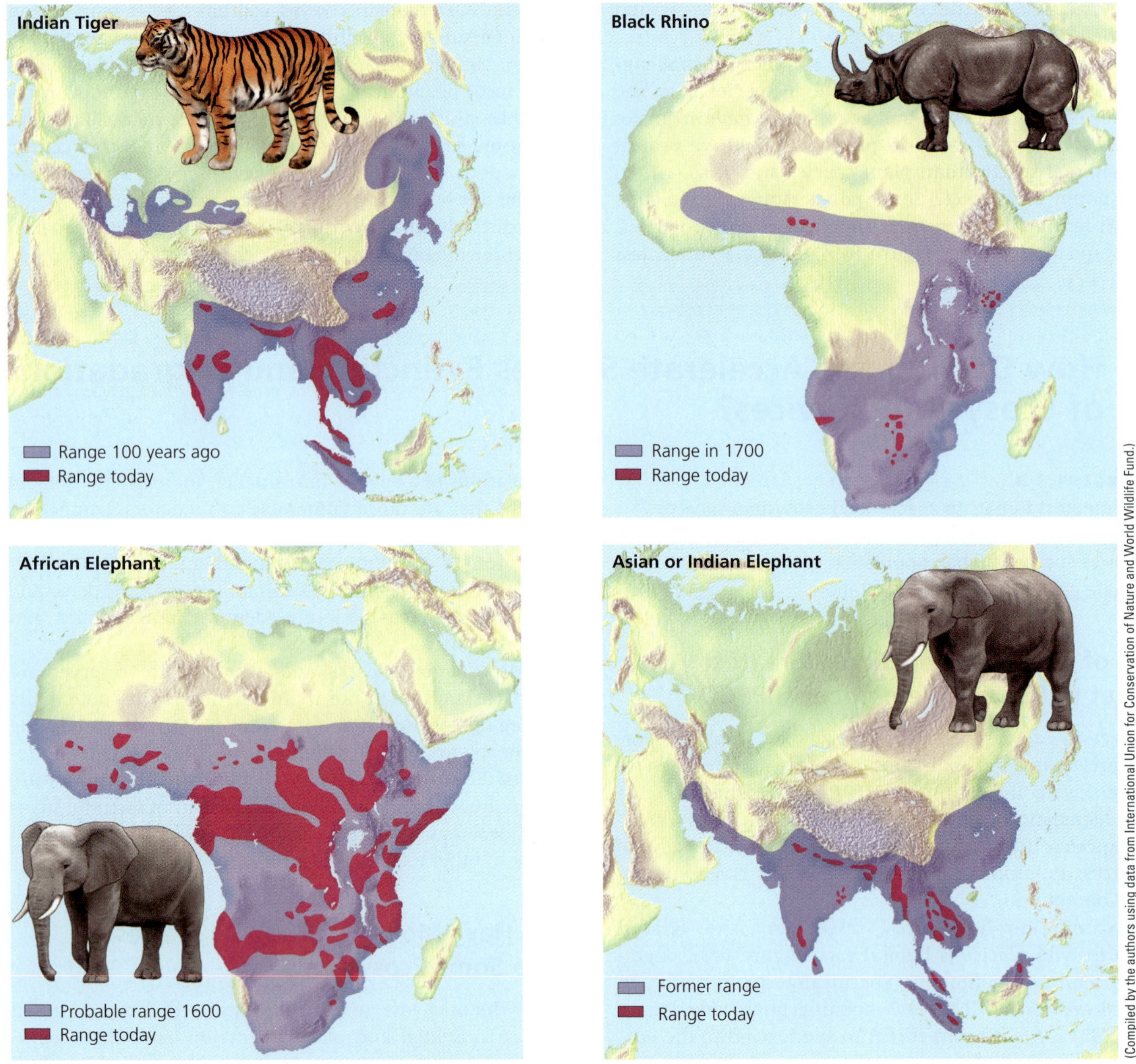

Animated Figure 9-8 Natural capital degradation: These maps reveal the reductions in the ranges of four wildlife species, mostly as the result of severe habitat loss and fragmentation and illegal hunting for some of their valuable body parts. ***Question:*** Would you support expanding these ranges even though this would reduce the land available for human habitation and farming? Explain.

tors, parasites, viruses, bacteria, or fungi that had helped to control their numbers in their original habitats. Such nonnative species can thus crowd out populations of many native species, disrupt ecosystem services, cause human health problems, and lead to economic losses. When this happens the nonnative species are viewed as harmful *invasive species.* Invasive species rarely cause the global extinction of other species, but they can cause population declines and local and regional extinctions of some native species.

CONSIDER THIS. . .

CONNECTIONS Giant Snails and Meningitis

In 1988, the giant East African land snail was imported to Brazil from East Africa as a cheap substitute for conventional escargot (snails), used as a source of food. It is the world's largest land snail, growing to the size of a human fist and weighing 1 kilogram (2.2 pounds) or more, and it can feed on at least 500 different types of plants. Mating adults lay about 1,200 eggs a year. When export prices for escargot fell, breeders dumped the imported snails into forests and other natural systems. Since then, they have spread widely, devouring many native plants and food crops such as lettuce. They also can carry rat lungworm, a parasite that burrows into the human brain and causes potentially lethal meningitis. Authorities eventually banned imports of the snail, but so far, this invasive species has been unstoppable.

Figure 9-9 shows some of the 7,100 or more invasive species that, after being deliberately or accidentally introduced into the United States, have caused ecological and economic harm. According to the U.S. Fish and Wildlife Service, about 40% of the species listed as endangered in the United States and 95% of those in the U.S. state of Hawaii are on the list because of threats from invasive species.

In 2009, Achim Steiner, head of the UN Environment Program (UNEP), and environmental scientist David Pimentel estimated that, globally, invader species cause $1.4 trillion a year in economic and ecological damages—an average of $2.7 million a minute—and the damages are rising rapidly. An example of such damage can be found in the story of the deliberately introduced kudzu vine (see the following Case Study).

CASE STUDY

The Kudzu Vine and Kudzu Bugs

An example of a deliberately introduced plant species is the *kudzu* ("CUD-zoo") *vine,* which in the 1930s was imported from Japan and planted in the southeastern United States to help control soil erosion. It worked too well. Kudzu does control erosion, but it grows so rapidly that it engulfs hillsides, gardens, trees, stream banks, cars (Figure 9-10), and anything else in its path. Dig it up or burn it, and it still keeps spreading. It is very difficult to kill, even with the use of grazing goats and herbicides, which can also damage other plants and contaminate water supplies. Scientists have found a common fungus that can kill kudzu within a few hours, but they need to investigate any harmful side effects it may have.

This plant—sometimes called "the vine that ate the South"—has spread throughout much of the southeastern United States. It could spread to the north if the climate gets warmer as scientists project.

Kudzu is considered a menace in the United States, but Asians use a powdered kudzu starch in beverages, gourmet confections, and herbal remedies for a range of diseases. Almost every part of the kudzu plant is edible. Its leaves are delicious when deep-fried and contain high levels of vitamins A and C. Also, ingesting small amounts of kudzu powder can lessen one's desire for alcohol, and thus it could be used to reduce alcoholism and binge

Deliberately Introduced Species

Accidentally Introduced Species

Figure 9-9 These are some of the estimated 7,100 harmful invasive species that have been deliberately or accidentally introduced into the United States.

© MELISSA FARLOW/National Geographic Creative

Figure 9-10 Kudzu has grown over this car in the U.S. state of Georgia.

drinking. And although kudzu can engulf and kill trees, it might eventually help to save some of them. Researchers at the Georgia Institute of Technology have found that kudzu could be used in place of trees as a source of fiber for making paper.

GOOD NEWS

The brown, pea-sized Kudzu bug is another invasive species that was imported from Japan. It breeds in and feeds on patches of kudzu, and it can help to reduce the spread of the vine. However, it spreads even more rapidly than the kudzu vine, and it also feeds on soybeans and thus could pose a major threat to soy crops. When it is disturbed, the bug emits a chemical that can irritate human skin and stain it yellow. It smells like industrial-strength drain cleaner, which is why it is also called a stinkbug.

The very adaptable Kudzu bug is a strong flier and an excellent hitchhiker, catching rides on vehicles that help it to spread to new areas. Since its arrival in 2009 in Georgia, it has spread rapidly throughout much of North Carolina and South Carolina, is moving north and west, and probably will soon be found in areas where soybeans are grown. Some pesticides can kill this bug, but might end up boosting their numbers by promoting genetic resistance to the pesticides (see Figure 4-6, p. 84). Researchers hope to change this bug through genetic engineering in such a way that it will stop eating soybeans. They are also evaluating the use of a wasp whose larvae attack kudzu bug embryos. However, scientists see no immediate way to eradicate this rapidly expanding invader species.

Some Accidentally Introduced Species Can Disrupt Ecosystems

Many unwanted nonnative invaders arrive from other continents as stowaways on aircraft, in the ballast water of tankers and cargo ships, and as hitchhikers on imported products such as wooden packing crates. Cars and trucks can also spread the seeds of nonnative plant species embedded in their tire treads. Many tourists return home with living plants that can multiply and become invasive. Some of these plants might also contain insects that can invade new areas, multiply rapidly, and threaten crops.

In the 1930s, the extremely aggressive Argentina fire ant (Figure 9-9) was accidentally introduced into the United States in Mobile, Alabama. The ants may have arrived on shiploads of lumber or coffee imported from South America. They can float on water and have no natural predators in the southern United States where they have spread rapidly. Now, the insect has stowed away on exported goods in shipping containers and has invaded other countries, including China, Taiwan, Malaysia, and Australia.

When these ants invade an area, they can wipe out as much as 90% of native ant populations. Spreading mounds containing fire ant colonies are found in many fields and yards in the southeastern United States. Walk on one of these mounds, and as many as 100,000 ants may swarm out of their nest to simultaneously attack you with painful, burning stings. They have killed deer fawns, ground-nesting birds, baby sea turtles, newborn calves, pets, and at least 80 people who were allergic to their venom—some of them infants and elderly people who could not escape the ants.

Widespread pesticide spraying in the 1950s and 1960s temporarily reduced fire ant populations. But this chemical warfare actually hastened the advance of the rapidly multiplying fire ants by reducing populations of many native ant species. Even worse, it promoted development of genetic resistance to pesticides in the fire ants through natural selection (see Figure 4-6, p. 84).

In 2009, pest management scientist Scott Ludwig reported some success in using tiny parasitic flies to reduce fire ant populations. The flies dive-bomb the fire ants and inject eggs inside them, from which their larvae hatch and, within 6 months, eat away the brains of the ants. Then the parasitic fly emerges looking for more fire ants to attack and kill. The researchers say that the flies do not attack native ant species. But more research is needed to see how well this approach will work.

GOOD NEWS

In recent years, several U.S. southern states have been invaded by biting, flea-sized *hairy crazy ants,* which are very resistant to most pesticides. They could help to wipe out fire ants but they are harder to control than fire ants. They also invade beehives and thus could be playing a role in honeybee declines in the United States (**Core Case Study**).

CASE STUDY

Burmese Pythons Are Eating Their Way through the Florida Everglades

Burmese pythons, along with African pythons and several species of boa constrictors, have been accidentally introduced in Everglades National Park in the U.S. state of Florida. About a million of these snakes, imported from

AP Photo/Michael R. Rochford/University of Florida

Figure 9-11 University of Florida researchers hold a 4.6-meter-long (15-foot-long), 74-kilogram (162-pound) Burmese python captured in Everglades National Park shortly after it had eaten a 1.8-meter-long (6-foot-long) American alligator.

Africa and Asia, have been sold as pets. After learning that these reptiles do not make good pets, some owners have dumped them into the wetlands of the Everglades.

The Burmese python (Figure 9-11) can live 20–25 years, growing as long as 5 meters (16 feet). It can weigh as much 77 kilograms (170 pounds) and be as big around as a telephone pole. Pythons are hard to find and kill and they reproduce rapidly. They have huge appetites and feed at night, eating a variety of birds and mammals and occasionally other reptiles, including the American alligator. They seize their prey with their sharp teeth, wrap themselves around the prey, and squeeze them to death before feeding on them. They have also been known to eat pet cats and dogs, small farm animals, and geese.

According to a 2012 National Academy of Sciences peer-reviewed study, since 1993, the rapidly growing population of the Burmese pythons in the Everglades has greatly depleted populations of marsh and eastern cottontail rabbits, red and gray foxes, raccoons, Virginia opossums, and white-tailed deer. The pythons also eat a variety of bird species (some of them endangered) and American alligators—a keystone species and top predator in the Everglades ecosystem (see Chapter 4, Case Study, pp. 95–96).

Researchers say that the Burmese python population in Florida's wetlands cannot be controlled. Some fear that the species could spread to other swampy wetlands in the southern half of the United States by finding natural routes to, or by being released to other wetlands by pet owners. In 2012, the U.S. Department of the Interior made it illegal to import Burmese pythons, North and South African pythons, and yellow anacondas, as well as to move them across state lines. In 2013, more than 1,300 people took part in a 1-month contest to find and remove Burmese pythons in Florida's wetlands.

Wildlife officials estimate that there are many thousands of pythons and constrictors living in the Everglades, and their numbers are increasing rapidly. Research indicates that predation by these snakes is altering the complex food web and ecosystem services of the Everglades. This is an excellent example of what can happen if an invasive top predator ends up in an ecosystem where it has no natural enemies.

Prevention Is the Best Way to Reduce Threats from Invasive Species

Once a harmful nonnative species becomes established in an ecosystem, its removal is almost impossible—somewhat like trying to collect smoke after it has come out of a chimney. Americans are paying more than $160 billion a year to eradicate or control an increasing number of invading species, with not much success. Clearly, the best way to limit the harmful impacts of nonnative species is to prevent them from being introduced into ecosystems.

Scientists suggest several ways to do this, including:

- Funding a massive research program to identify the major characteristics of successful invaders, the types of ecosystems that are vulnerable to invaders, and the natural predators, parasites, bacteria, and viruses that could be used to control populations of established invaders.
- Greatly increasing ground surveys and satellite observations to track invasive plant and animal species, and developing better models for predicting how they will spread and what harmful effects they might have.
- Identifying major harmful invader species and establishing international treaties banning their transfer from one country to another, as is now done for endangered species, while stepping up inspection of imported goods to enforce such bans.
- Requiring cargo ships to discharge their ballast water and to replace it with saltwater at sea before entering ports, or to sterilize such water or to pump nitrogen into the water to displace dissolved oxygen and kill most invader organisms.
- Educating the public about the effects of releasing exotic plants and pets into the environment near where they live.

What Can You Do?

Controlling Invasive Species

- Do not buy wild plants and animals or remove them from natural areas.
- Do not release wild pets in natural areas.
- Do not dump aquarium contents or unused fishing bait into waterways or storm drains.
- When camping, use only local firewood.
- Brush or clean pet dogs, hiking boots, mountain bikes, canoes, boats, motors, fishing tackle, and other gear before entering or leaving wild areas.

Figure 9-12 Individuals matter: Here are some ways to prevent or slow the spread of harmful invasive species. ***Questions:*** Which two of these actions do you think are the most important to take? Why? Which of these actions do you plan to take?

Figure 9-12 shows some of the things you can do to help prevent or slow the spread of harmful invasive species.

Population Growth, High Rates of Resource Use, Pollution, and Climate Change Can Cause Species Extinctions

Past and projected *human population growth* and rising rates of *resource use per person* have greatly expanded the human ecological footprint (see Figure 1-13, p. 14). This has eliminated, degraded, and fragmented vast areas of wildlife habitat (Figure 9-8). Acting together, these two growth factors have caused the extinction of many species (**Concept 9-3**).

Pollution also threatens some species with extinction (**Concept 9-3**), as has been shown by the unintended effects of certain pesticides. According to the U.S. Fish and Wildlife Service, each year, pesticides kill about one-fifth of the European honeybee colonies that pollinate almost a third of U.S. food crops (**Core Case Study** and Science Focus 9.2). According to the U.S. Fish and Wildlife Service, pesticides also kill more than 67 million birds and 6–14 million fish each year, and they threaten about 20% of the country's endangered and threatened species.

During the 1950s and 1960s, populations of fish-eating birds such as ospreys, brown pelicans, and bald eagles plummeted. A chemical derived from the pesticide DDT was magnified as it moved up through their food web through processes called *bioaccumulation* and *biomagnification* (Figure 9-13). The chemical made these top predator birds' eggshells so fragile that they could not reproduce successfully. Also hard hit in those years, in other ecosystems, were such predatory birds as the prairie falcon, sparrow hawk, and peregrine falcon, which help to control populations of rabbits, ground squirrels, and other crop eaters. Since the U.S. ban on DDT in 1972, most of these bird species have made a comeback. GOOD NEWS

According to a study by Conservation International, projected *climate change* could help to drive a quarter to half of all land animals and plants to extinction by the end of this century. Scientific studies indicate that the polar bear (see Case Study that follows) is threatened because of higher temperatures and melting sea ice in its polar habitat. Similarly, other species, especially some marine species such as coral polyps, are threatened by a warming environment that is making their habitats unfit for their survival.

However, a 2013 research article by ecologists Roland Jánsson and Christer Nilsson indicated that as global warming expands temperate climate zones, 43 of the 61 species they studied are likely to expand their habitats. On the other hand, populations of some cold-weather species such as the lemming and Artic fox are likely to decrease but are unlikely to become extinct.

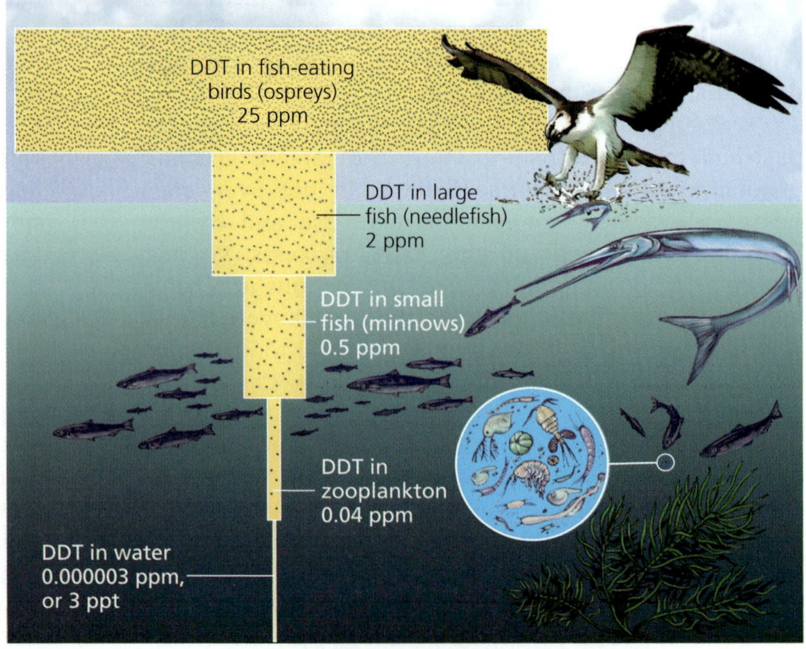

Figure 9-13 *Bioaccumulation and biomagnification:* DDT is a fat-soluble chemical that can accumulate in the fatty tissues of animals. In a food chain or web, the accumulated DDT is biologically magnified in the bodies of animals at each higher trophic level, as it was in the case of a food chain in the U.S. state of New York, illustrated here. (Dots in this figure represent DDT.) ***Question:*** How does this story demonstrate the value of pollution prevention?

SCIENCE FOCUS 9.2

HONEYBEE LOSSES: A SEARCH FOR CAUSES

U.S. Department of Agriculture

Figure 9-B The parasitic varroa mite, shown here on a honeybee host, can weaken and kill honeybees.

Over the past 50 years the European honeybee population in the United States has been cut in half. Since 2006, at least one-third of the U.S. population of this species has disappeared because of colony collapse disorder (**Core Case Study**). This problem also occurs in parts of Europe, China, and India.

Scientific research has found a number of possible reasons for this decline, including:

- The *varroa mite* (Figure 9-B), a parasitic insect, can weaken and kill honeybee adults by feeding on their blood and larvae.
- *Harmful interactions between viruses and fungi* found in European honeybees and in almost all colonies suffering from colony collapse disorder. For example, the *Israeli acute paralysis virus,* which can be spread to hives by the varroa mite, and the *nosema fungus* can interact to kill bees by weakening their immune systems.
- *Pesticides.* As honeybees forage for nectar they can come into contact with harmful pesticides that they can carry back to the hives. Some research indicates that *neonictinoids*—among the world's most widely use pesticides—can disrupt the nervous systems of bees and can decrease their ability to find their way back to their hives. Neonictinoids have been found in the pollen and nectar of corn plants serviced by bees and in high-fructose corn syrup that many beekeepers feed their bees after harvesting their honey. Pesticides in the air can also end up in the honeycomb wax and stored pollen in hives. A U.S. researcher at a USDA lab in North Carolina found more than 170 different pesticides in samples of bees, honeycomb wax, and stored pollen. Exposure to such a cocktail of pesticides can weaken bees' immune systems and make them vulnerable to deadly parasites, viruses, and fungi.
- *Stress and poor nutrition* from being transported long distances around the United States as insect workers for the industrial pollination business (Figure 9-C). Overworking and overstressing honeybees by moving them around the country can weaken their immune systems and make them more vulnerable to death from parasites, viruses, fungi, and pesticides. In natural ecosystems, honeybees gather nectar and pollen from a variety of flowering plants, but industrial worker honeybees feed mostly on pollen or nectar from one crop or a small number of crops that may lack the nutrients they need.

The growing consensus among bee researchers is that one or more combinations of the factors listed here are probably responsible for colony collapse disorder.

Critical Thinking

Can you think of some ways in which commercial beekeepers could lessen one or more of the threats described here? Explain.

© Cristi111 | Dreamstime.com

Figure 9-C European honeybee hive boxes in an acacia orchard. Each year, commercial beekeepers rent and deliver several million hives by truck to farmers throughout the United States.

CASE STUDY

Polar Bears and Climate Change

The world's 20,000–25,000 polar bears are found in 19 subpopulations scattered across the frozen Arctic Circle. About 60% of them are in Canada, and the rest live in arctic areas of Greenland, Norway, Russia, and the U.S. state of Alaska.

Throughout the winter, polar bears hunt for ringed seals on floating sea ice (Figure 9-14) that expands each winter and contracts as the temperature rises during summer. By eating the seals, the bears build up their body fat. In the summer and fall, they live off this fat until hunting resumes when the ice expands again during winter.

©age footstock/SuperStock

Figure 9-14 On floating ice in Svalbard, Norway, a polar bear feeds on its ringed seal prey. ***Question:*** Do you think it matters that the polar bear may become extinct during this century primarily because of human activities? Explain.

Scientific measurements reveal that the earth's atmosphere has been getting warmer since 1975 and that this warming is occurring twice as fast in the Arctic as in the rest of the world. Thus, arctic ice is melting faster and the average annual area of floating sea ice in the Arctic during the summer is decreasing. The floating winter ice is also breaking up earlier each year, shortening the polar bears' hunting season. In addition, much of the remaining ice is getting too thin to support the weight of an adult polar bear.

These changes mean that polar bears must swim longer distances to find prey and have less time to feed and store fat. As a result, they must fast longer, which weakens them. As females become weaker, their ability to reproduce and keep their young cubs alive declines. In 2008, the U.S. Fish and Wildlife Service placed the Alaskan polar bear on its list of threatened species.

According to the International Union for Conservation of Nature, the world's total polar bear population is likely to decline by 30–35% by 2050. By the end of this century, polar bears might be found only in zoos. However, the evidence for such population projections varies with the different polar bear sub-populations, their locations, and what they eat. Of the 19 sub-populations, in 2011, some were shrinking and some were steady or growing.

The Illegal Killing, Capturing, and Selling of Wild Species Threaten Biodiversity

Some protected species are illegally killed (poached) for their valuable parts or are sold live to collectors. Globally, this illegal trade in wildlife brings in an average of at least $600,000 an hour. Organized crime has moved into illegal wildlife smuggling because of the huge profits involved. Few of the smugglers are caught or punished.

To poachers, a highly endangered, live mountain gorilla (of which there are about 880 left in the wild) is worth $150,000, and the pelt of an endangered giant panda (1,600 to 3,000 left in the wild in China) can bring $100,000. A poached rhinoceros horn (Figure 9-15) can be worth as much as $99,000 per kilogram ($45,000 per pound) on the black market. Their horns are used to make dagger handles in Yemen and Oman. Powdered rhino horn has long been used in traditional medicines for a variety of ailments and as an alleged male aphrodisiac in many Asian countries including China, India, and Vietnam. There is no verifiable evidence for these benefits and the horn is largely made of the same material in fingernails and animal hooves.

Mostly because of poaching, three of the five rhino species—the black rhino (4,880 left in the wild, see Figure 9-8), the Sumatran (2,700 left), and the Javan rhino (about 30 left)—are critically endangered. They have survived for about 40 million years, but poachers very likely could wipe them out within the next 50–80 years. However, during the last century, the number of southern white rhinos in South Africa rose from 50 to 17,400 because people worked to protect their habitat and reduce poaching. GOOD NEWS

Rhinos can be raised on farms and if their horns are carefully cut to no more than 8 centimeters (3 inches) from the base, new horns will grow back in about 2 years. Such sustainable harvesting of rhino horns could help to reduce poaching. Also, park rangers in some rhino areas are using motion-sensitive cameras to photograph and identify poachers.

In 1900, there were an estimated 100,000 tigers in the wilds of Asia in a rapidly shrinking range (Figure 9-8, top left). Today there are about 3,200, half of them in India. Three of the nine subspecies of tigers have become extinct and the other six, including the Siberian (see chapter-opening photo) and Sumatran tigers (Figure 9-2d), are endangered.

This is mostly due to a 90% loss of habitat, caused by rapid human population growth, and poaching, much of which is motivated by poverty. The Indian, or Bengal, tiger is at risk because a coat made from its fur can sell for as much as $100,000 in Tokyo, and tiger skins sell for up to $35,000. The bones and penis of a single tiger can fetch as much as $70,000 in China, the world's biggest market for such illegal items. According to the World Wildlife Fund, without emergency action to curtail poaching and preserve tiger habitat, few if any tigers will be left in the wild by 2022.

Around the globe, the legal and illegal trade in wild species for use as pets is a huge and very profitable business. Many owners of wild pets do not know that, for every live animal captured and sold in the pet market, many others are killed or die in transit. According to the IUCN, more than 60 bird species, mostly parrots

Photoshot Holdings Ltd/Alamy

Figure 9-15 A poacher in South Africa killed this critically endangered northern white rhinoceros for its two horns. This species is now extinct in the wild. With a rhino horn worth up to $300,000 on the Asian black market, thieves have been stealing the horns from museums, antique stores, and even private collections. ***Question:*** What would you say if you could talk to the poacher who killed this animal for its horns?

(Figure 9-7), are endangered or threatened because of the wild bird trade (see the Case Study that follows).

CONSIDER THIS. . .

CONNECTIONS Exotic Pets and Human Diseases

Most people are unaware that some imported exotic animals carry diseases such as hantavirus, Ebola virus, Asian bird flu, herpes B virus (carried by most adult macaques), and salmonella (from pets such as hamsters, turtles, and iguanas). These diseases can spread quite easily from pets to their owners and then to other people.

Other wild species whose populations are depleted because of the pet trade include many amphibians (see Chapter 4 Core Case Study, p. 78), various reptiles, and tropical fishes taken mostly from the coral reefs of Indonesia and the Philippines. Divers catch tropical fish by using plastic squeeze bottles of poisonous cyanide to stun them. For each fish caught alive, many more die. In addition, the cyanide solution kills the coral animals (polyps) that create the reef.

Some exotic plants, especially orchids and cacti (see Figure 7-10, center, p. 151), are endangered because they are gathered, often illegally, and sold to collectors to decorate houses, offices, and landscapes. A mature crested saguaro cactus can earn a cactus rustler as much as $15,000, and an orchid collector might pay $5,000 for a single rare orchid.

According to the U.S. Fish and Wildlife Service, collectors of exotic birds might pay $10,000 for an endangered hyacinth macaw parrot (Figure 9-7) smuggled out of Brazil. However, during its lifetime, a single hyacinth macaw left in the wild could account for as much as $165,000 in ecotourist revenues.

CONSIDER THIS. . .

THINKING ABOUT Collecting Wild Species

Some people believe it is unethical to collect wild animals and plants for display and personal pleasure and that such plants and animals should be left to live in the wild. Explain why you agree or disagree with this view.

A Rising Demand for Bushmeat Threatens Some African Species

For centuries, indigenous people in much of West and Central Africa have sustainably hunted wildlife for *bushmeat* as a source of food. But in the last two decades, bushmeat hunting in some areas has skyrocketed as hunters have tried to provide food for rapidly growing populations or to make a living by supplying restaurants in major cities with exotic meats from gorillas (Figure 9-16) and other species. Logging roads in once-inaccessible forests have made such hunting much easier. As a result, some forests in areas such as Africa's Congo basin are being stripped of many of their monkeys, apes, antelope (the most commonly hunted bushmeat animal), elephants, hippos, and other wild animals.

Bushmeat hunting has driven at least one species—Miss Waldron's red colobus monkey—to complete extinction. It is also a factor in reducing some populations of orangutans (Figure 9-5). Another problem is that the butchering and eating of some forms of bushmeat have helped to spread fatal diseases such as HIV/AIDS and the Ebola virus from animals to humans.

The U.S. Agency for International Development (USAID) is trying to reduce unsustainable hunting for bushmeat in some areas of Africa by introducing alternative sources of food, including farmed fish. They are also showing villagers how to breed large rodents such as cane rats as a source of food. GOOD NEWS

CASE STUDY

A Disturbing Message from the Birds

Approximately 70% of the world's more than 10,000 known bird species are declining in numbers, and much of this decline is related to human activities, summarized by HIPPCO. According to the IUCN 2012 Red List of Endangered Species, roughly one of every eight (13%) of all bird species is threatened with extinction mostly by habitat loss, degradation, and fragmentation (the H in HIPPCO). About three-fourths of the threatened bird species live in forests, many of which are being cleared at a rapid rate, especially in many tropical areas of Asia, Latin America, and Africa.

According to a 2011 study, *State of the Birds,* almost one-third of the more than 800 bird species in the United States are endangered (Figure 9-17), threatened, or in decline, mostly because of habitat loss and degradation, invasive species, and climate change. About one-third of all endangered and threatened bird species in the United States live in Hawaii.

Sharp declines in bird populations have occurred among songbird species that migrate long distances, such as tanagers, orioles, thrushes, vireos, and warblers. These birds nest deep in North American woods in the summer and spend their winters in Central or South America or on the Caribbean Islands. The primary causes of these

Photoshot Holdings Ltd/Alamy

Figure 9-16 *Bushmeat* such as this severed head of an endangered lowland gorilla in the Congo is consumed as a source of protein by local people in parts of West and Central Africa and is sold in national and international marketplaces and served in some restaurants where wealthy patrons regard gorilla meat as a source of status and power. ***Question:*** How, if at all, is this different from killing a cow for food?

population declines appear to be habitat loss and fragmentation of the birds' breeding habitats. In North America, road construction and housing developments result in the breaking up or clearing of woodlands. In Central and South America, tropical forest habitats, mangroves, and wetland forests are suffering the same fate. In addition, the populations of 40% of the world's water birds are in decline because of the global loss of wetlands.

After habitat loss, the intentional or accidental introduction of nonnative species such as bird-eating rats is the second greatest danger, affecting about 28% of the world's threatened birds. Other such invasive species (the I in HIPPCO) include snakes (such as the brown tree snake), and mongooses, which kill hundreds of millions of birds each year.

Population growth, the first P in HIPPCO, also threatens some bird species, as more people spread out over the landscape and increase their use of timber, food, and other resources, destroying or disturbing some bird habitats. The second P in HIPPCO is for pollution, another major threat to birds. Countless birds are exposed to oil spills, pesticides, herbicides, and toxic lead from shotgun pellets that fall into wetlands and from lead sinkers left by anglers.

CONSIDER THIS. . .

CONNECTIONS Vultures, Wild Dogs, and Rabies

The IUCN placed four species of carcass-eating vultures found in India on the endangered list after their populations fell by more than 95% during the early 1990s. Scientists discovered that they were dying from kidney failure when they fed on the carcasses of cows that had been given an anti-inflammatory drug to increase milk production. As the vultures died off, huge numbers of cow carcasses, normally a source of food for the vultures, were consumed by wild dogs and rats whose populations the vultures had helped to control by reducing their food supply. As wild dog populations exploded, the number of dogs with rabies also increased. About 48,000 people bitten by the rabid dogs died of rabies. The drug that caused this problem was banned, and by 2012, the vultures were making a slow comeback.

Overexploitation (the O in HIPPCO) of birds and other species is also a major threat to bird populations. Fifty-two of the world's 388 parrot species (Figure 9-7) are threatened partly because so many parrots are captured for the pet trade (often illegally) for sale usually to buyers in Europe and the United States.

Industrialized fishing fleets also pose a threat to birds. At least 23 species of seabirds, including albatrosses, face extinction. Many of these diving birds drown after becoming hooked on baited lines or trapped in huge nets that are set out by fishing boat crews.

Biodiversity scientists view this decline of bird species with alarm. One reason is that birds are excellent *indicator species* because they live in every climate and biome, respond quickly to environmental changes in their habitats, and are relatively easy to track and count. What their decline is indicating to these scientists is widespread environmental degradation.

A second reason is that birds perform critically important economic and ecosystem services in ecosystems throughout

BATES LITTLEHALES/National Geographic Creative

Figure 9-17 **Endangered natural capital:** This endangered Attwater's prairie chicken lives in a wildlife refuge in the U.S. state of Texas.

individuals matter 9.1

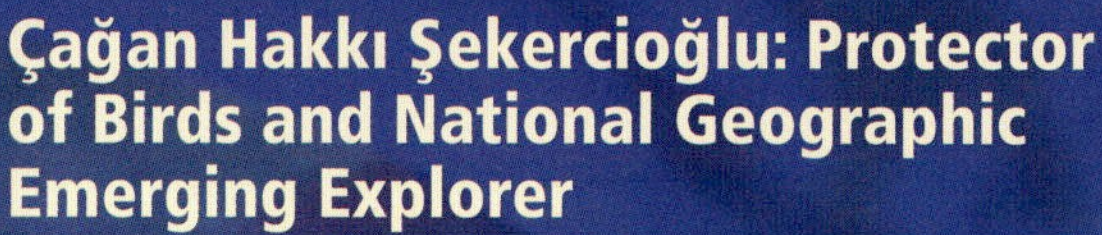

Çağan Hakkı Şekercioğlu: Protector of Birds and National Geographic Emerging Explorer

©Courtesy of Dr. Sekercioglu

Çağan Şekercioğlu, assistant professor at the University of Utah Department of Biology, is a bird expert, conservation ecologist, tropical biologist, and accomplished wildlife photographer. He has seen over 60% of the planet's known bird species in 70 countries, developed a global database on bird ecology, and become an expert on the causes and consequences of bird extinctions around the world. One example of an endangered species he has studied is the great green macaw of South and Central America (shown here in the background photo).

In 2010, this scientist with an undergraduate degree from Harvard and a PhD in biology from Stanford University was listed in the top 1% of world's scientists as measured by the number of times his research was cited in ecology and environmental science reports between 2000 and 2010. In 2011, he was selected as a National Geographic Emerging Explorer and named as Turkey's Scientist of the Year.

In 2007 Şekercioğlu founded KuzeyDoğa, an award-winning ecological research and community-based conservation organization to help conserve and protect the wildlife of northeastern Turkey. One of its goals is to promote ecotourism, especially bird watching, as a way to protect threatened bird species and their habitats and improve economic conditions in the participating Turkish communities. He also developed Turkey's first protected wildlife corridor, which would stretch across the eastern half of the country.

Based on his extensive research Şekercioğlu estimates that the percentage of the world's known bird species that are endangered could nearly double from 13% in 2012 to 25% by the end of this century. According to this champion for the world's biodiversity, "My ultimate goal is to prevent extinctions and consequent collapses of critical ecosystem processes while making sure that human communities benefit from conservation as much as the wildlife they help conserve. . . . I don't see conservation as people versus nature, I see it as a collaboration."

Background photo: ©Courtesy of Dr. Sekercioglu

the world. For example, many birds play specialized roles in pollination and seed dispersal, especially in tropical areas, so extinctions of these bird species could lead to extinctions of plants that depend on the birds for pollination. Then, some specialized animals that feed mostly on these plants might also become extinct. Such a cascade of extinctions, in turn, could affect our own food supplies and well-being. Biodiversity scientists (Individuals Matter 9.1) urge us to listen more carefully to what birds are telling us about the state of the environment, for their sake, as well as for ours.

9-4 How Can We Sustain Wild Species and Their Ecosystem Services?

CONCEPT 9-4

We can reduce species extinction and sustain ecosystem services by establishing and enforcing national environmental laws and international treaties, creating protected wildlife sanctuaries, and taking precautionary measures to prevent such harm.

International Treaties and National Laws Can Help to Protect Species

Several international treaties and conventions help to protect endangered and threatened wild species (**Concept 9-4**). One of the most far reaching is the 1975 *Convention on International Trade in Endangered Species (CITES).* This treaty, signed by 178 countries, bans the hunting, capturing, and selling of threatened or endangered species. It lists 926 species that are in danger of extinction and cannot be commercially traded as live specimens or for their parts or products. It also restricts international trade of roughly 5,000 species of animals and 29,000 species of plants that are at risk of becoming threatened.

CITES has helped to reduce the international trade of many threatened animals, including elephants, crocodiles, cheetahs, and chimpanzees. But the effects of this treaty are limited because enforcement varies from country to country (Figure 9-18), and convicted violators often pay only small fines. Also, member countries can exempt themselves from protecting any listed species, and much of the highly profitable illegal trade in wildlife and wildlife products goes on in countries that have not signed the treaty.

Another important treaty is the *Convention on Biological Diversity (CBD),* ratified or accepted by 193 countries (but as of 2012, not by the United States). It legally commits participating governments to reducing the global rate of biodiversity loss and to equitably sharing the benefits from use of the world's genetic resources. This includes efforts to prevent or control the spread of ecologically harmful invasive species.

This convention is a landmark in international law because it focuses on ecosystems rather than on individual species, and it links biodiversity protection to issues such as the traditional rights of indigenous peoples. However, because some key countries, including the United States, have not ratified it, implementation has been slow. Also, the law contains no severe penalties or other enforcement mechanisms.

National laws are also important for sustaining wild species and ecosystem services. Several countries have strong laws (see the following Case Study), but scientists call for more, stronger, and more widespread national legal efforts.

CASE STUDY

The U.S. Endangered Species Act

The *Endangered Species Act of 1973* (ESA; amended in 1982, 1985, and 1988) was designed to identify and protect endangered species in the United States and abroad (**Concept 9-4**). This act is probably the most far-reaching environmental law ever adopted by any nation, which has made it controversial.

Under the ESA, the National Marine Fisheries Service (NMFS) is responsible for identifying and listing endangered and threatened ocean species, while the U.S. Fish and Wildlife Service (USFWS) is to identify and list all other endangered and threatened species. Any decision by either agency to list or delist a species must be based on biological factors alone, without consideration of economic or political factors. However, the two agencies can use economic factors in deciding whether and how to protect endangered habitat and in developing recovery plans for listed species. The ESA also forbids federal agencies (except the Defense Department) to carry out, fund, or authorize projects that would jeopardize any endangered or threatened species or destroy or modify its critical habitat.

Figure 9-18 These poachers were caught while trying to sell a tiger skin in Madhya Pradesh State, India.

For offenses committed on private lands, fines as high as $100,000 and 1 year in prison can be imposed to ensure protection of the habitats of endangered species, although this provision has rarely been used. This part of the act has been controversial because at least 90% of the listed species live totally or partially on private land. Since 1982, however, the ESA has been amended to give private landowners various economic incentives to help save endangered species living on their lands. The ESA also makes it illegal for Americans to sell or buy any product made from an endangered or threatened species or to hunt, kill, collect, or injure such species in the United States.

Between 1973 and 2012, the number of U.S. species on the official endangered and threatened species lists increased from 92 to 1,476. According to a study by the Nature Conservancy, about 33% of the country's species are at risk of extinction, and 15% of all species are at high risk—far more than the current number listed. In 2012, 77% of the protected species were covered by active recovery plans. Successful recovery plans include those for the American alligator (see Chapter 4, Case Study, pp. 95–96), the gray wolf, the peregrine falcon, the bald eagle, and the brown pelican. GOOD NEWS

The ESA also requires that all commercial shipments of wildlife and wildlife products enter or leave the country through one of 17 designated airports and ocean ports. The 120 full-time USFWS inspectors can inspect only a small fraction of the more than 200 million wild animals brought legally into the United States annually. Each year, tens of millions of wild animals are also brought in illegally, but few illegal shipments of endangered or threatened animals or plants are confiscated. Even when they are caught, many violators are not prosecuted, and convicted violators often pay only a small fine.

Since 1995, there have been numerous efforts to weaken the ESA and to reduce its already meager annual budget. Opponents of the act contend that it puts the rights and welfare of endangered plants and animals above those of people. Some critics would do away with this act. They call it an expensive failure because only 26 species have recovered enough to be removed from the endangered species list (Figure 9-19), and because 10 became extinct while on the list. Most biologists insist that it has not been a failure, for three main reasons.

First, species are listed only when they face serious danger of extinction. ESA supporters argue that this is similar to a hospital emergency room set up to take only the most desperate cases, often with little hope for recovery. Such a facility could not be expected to save all of its patients.

Second, according to federal data, the conditions of more than half of the listed species are stable or improving, 90% are recovering at rates specified by their recovery plans, and 99% of the protected species are still surviving. A hospital emergency room taking only the most desperate cases and then stabilizing or improving the conditions of more than half of those patients while keeping 99% of them alive and 90% recovering would be considered to be highly effective.

Third, the 2012 budget for protecting endangered species amounted to an average expenditure of about 86 cents per U.S. citizen. To its supporters, it is amazing that the federal agencies, on such a small budget, have managed to stabilize or improve the conditions of more than half of the listed species.

ESA supporters agree that the act can be improved and that federal regulators have sometimes been too heavy handed in enforcing it. They cite a study by the U.S. National Academy of Sciences that recommended three major changes in the way the law is being implemented, in order to make it more scientifically sound and effective:

1. Greatly increase the meager funding for implementing the act.
2. Put more emphasis on developing recovery plans more quickly, as scientists have found that species with recovery plans have a better chance of getting off the endangered list.
3. When a species is first listed, establish the core of its habitat as critical for its survival and give that area the maximum protection.

Most biologists and wildlife conservationists believe that the United States also needs a new law that emphasizes protecting and sustaining biological diversity and ecosystem services rather than focusing mostly on saving individual species. (We examine this idea further in Chapter 10.)

Figure 9-19 The American bald eagle has been removed from the U.S. endangered species list. Here, an eagle is about to catch a fish in its powerful talons.

CONSIDER THIS. . .

CAMPUS SUSTAINABILITY

The Endangered Species Lab at Washington State University

©Tania Thomson/Shutterstock.com

How do we save endangered species and prevent extinction processes? That is one of the questions that drives researchers at Washington State University's Endangered Species Lab. Students there conduct field studies in both natural and managed landscapes to explore the roles played by evolution and species behavior in extinction processes. They also study how landscape factors affect the process of reintroduction of threatened and endangered species, seeking to apply such knowledge to efforts to restore native habitats. The researchers have focused on a number of species, including Columbian sharp-tailed grouse, burrowing owls (see photo), northern leopard frogs, a variety of bat species, and the Columbia Basin pygmy rabbit.

We Can Establish Wildlife Refuges and Other Protected Areas

In 1903, President Theodore Roosevelt established the first U.S. federal wildlife refuge at Pelican Island, Florida (Figure 9-20), to help protect birds such as the brown pelican (Figure 9-20, inset photo) from extinction. It took more than a century but this protection worked. In 2009, the brown pelican was removed from the U.S. Endangered Species list. GOOD NEWS By 2012, there were more than 560 refuges in the National Wildlife Refuge System. Each year, more than 47 million Americans visit these refuges to hunt, fish, hike, and watch birds and other wildlife.

More than three-fourths of the refuges serve as wetland sanctuaries that are vital for protecting migratory waterfowl. More than one-fourth of all U.S. endangered and threatened species have habitats in the refuge system, and some refuges have been set aside specifically for certain endangered species (**Concept 9-4**). Such areas have helped Florida's key deer, the brown pelican, and the trumpeter swan to recover.

There is also bad news about refuges. According to a General Accounting Office study, activities considered harmful to wildlife, such as mining, oil drilling, and use of off-road vehicles, occur in nearly 60% of the nation's wildlife refuges. Also, a 2008 study prepared for Congress found that, for years, the country's wildlife refuges have received so little funding that a third of them have no staff, and boardwalks, public buildings, and other structures in some refuges are in disrepair.

Biodiversity scientists are urging the U.S. government to set aside more refuges for endangered plants and to significantly increase the long-underfunded budget for the refuge system. They are also calling on the Congress and state legislatures to allow abandoned military lands that contain significant wildlife habitat to be designated as wildlife refuges.

©Chuck Wagner/Shutterstock.com

George Gentry/U.S. Fish and Wildlife Service

Figure 9-20 The Pelican Island National Wildlife Refuge in Florida was America's first National Wildlife Refuge.

Seed Banks, Botanical Gardens, and Wildlife Farms Can Help to Protect Species

Seed banks preserve genetic information and endangered plant species by storing their seeds in refrigerated, low-humidity environments. More than 1,000 seed banks around the world collectively hold about 3 million samples.

Some species cannot be preserved in seed banks. The banks are of varying quality and expensive to operate, and some can be destroyed by fires and other mishaps. However, the Svalbard Global Seed Vault, a new underground facility on a remote island in the Arctic, will eventually contain 100 million of the world's seeds. It is not vulnerable to power losses, fires, storms, or war.

The world's 1,600 *botanical gardens* and *arboreta* contain living plants that represent almost one-third of the world's known plant species. But they contain only about 3% of the world's rare and threatened plant species and have too little space and funding to preserve most of those species.

We can take pressure off some endangered or threatened species by raising individuals of these species on *farms* for commercial sale. In Florida, for example, alligators are raised on farms for their meat and hides. Butterfly farms established to raise and protect endangered species flourish in Papua New Guinea, where many butterfly species are threatened by development activities. These farms are also used to educate visitors about the need to protect butterfly species.

Zoos and Aquariums Can Protect Some Species

Zoos, aquariums, game parks, and animal research centers are being used to preserve some individuals of critically endangered animal species, with the long-term goal of reintroducing the species into protected wild habitats.

Two techniques for preserving endangered terrestrial species are egg pulling and captive breeding. *Egg pulling* involves collecting wild eggs laid by critically endangered bird species and then hatching them in zoos or research centers. In *captive breeding,* some or all of the wild individuals of a critically endangered species are collected for breeding in captivity, with the aim of reintroducing the offspring into the wild. Captive breeding has been used to save the peregrine falcon and the California condor (Figure 9-2b).

Other techniques for increasing the populations of captive species include artificial insemination, embryo transfer (surgical implantation of eggs of one species into a surrogate mother of another species), use of incubators, and cross fostering (in which the young of a rare species are raised by parents of a similar species). Scientists also use computer databases that hold information on family lineages of endangered zoo animals along with DNA analysis to match individuals for mating—a computer dating service for zoo animals.

The ultimate goal of captive breeding programs is to build up populations to a level where they can be reintroduced into the wild. Successes include the black-footed ferret, the golden lion tamarin (a highly endangered monkey species), the Arabian oryx, and the California condor (Figure 9-2b). However, most reintroductions fail because of lack of suitable habitat, inability of individuals bred in captivity to survive in the wild, and renewed overhunting or poaching.

One problem for captive breeding programs is that a captive population of an endangered species must typically number 100–500 individuals in order for it to avoid extinction through accident, disease, or loss of genetic diversity through inbreeding. Recent genetic research indicates that 10,000 or more individuals are needed for an endangered species to maintain its capacity for biological evolution. Zoos and research centers do not have the funding or space to house such large populations.

Public aquariums (Figure 9-21) that exhibit unusual and attractive species of fish and marine animals such as seals and dolphins help to educate the public about the need to protect such species. However, mostly because of limited funds, public aquariums have not served as effective gene banks for endangered marine species, especially marine mammals that need large volumes of water.

While zoos, aquariums, and botanical gardens perform valuable services, they cannot by themselves solve the growing problem of species extinction. Figure 9-22 lists some things you can do to help deal with this problem. GOOD NEWS

The Precautionary Principle

Biodiversity scientists call for us to take precautionary action to avoid hastening species extinction and disrupting essential ecosystem services. This approach is based on the **precautionary principle**: When substantial preliminary evidence indicates that an activity can harm human health or the environment, we should take precautionary measures to prevent or reduce such harm even if some of the cause-and-effect relationships have not been fully established scientifically. It is based on the commonsense idea behind many adages, including "Better safe than sorry" and "Look before you leap."

SISSE BRIMBERG/National Geographic Creative

Figure 9-21 The Monterey Bay Aquarium in Monterey, California (USA), contains this tidewater pool, which is used to train rescued sea otter pups to survive in the wild.

Scientists use the precautionary principle to argue for both the preservation of species and protections of entire ecosystems and their ecosystem services, which is the focus of the next chapter. The precautionary principle is also used as a strategy for dealing with other challenges such as preventing exposure to harmful chemicals in the air we breathe, the water we drink, and the food we eat.

Using limited financial and human resources to protect biodiversity based on the precautionary principle involves dealing with three important questions:

1. Should we focus mostly on protecting species or on protecting ecosystems and the ecosystem services they provide for a variety of species, and how do we allocate limited resources between these two priorities?
2. How do we decide which species should get the most attention in our efforts to protect as many species as possible? For example, should we focus on protecting the most threatened species or on protecting keystone species? Protecting species that are appealing to humans, such as tigers (see Figure 9-2d) and orangutans (Figure 9-5), can increase public awareness of the need for wildlife conservation. But some argue that we should instead protect more ecologically important endangered species (see the following Case

What Can You Do?

Protecting Species

- Do not buy furs, ivory products, or other items made from endangered or threatened animal species.
- Do not buy wood or wood products from tropical or old-growth forests.
- Do not buy pet animals or plants taken from the wild.
- Tell friends and relatives what you're doing about this problem.

© Cengage Learning

Figure 9-22 Individuals matter: You can help to prevent the extinction of species. ***Questions:*** Which two of these actions do you believe are the most important? Why?

©Martha Cooper/National Geographic Creative

Figure 9-23 This woman is hand-pollinating pear tree blossoms on Honshu Island, Japan, where overuse of pesticides decimated populations of natural pollinators such as honeybees. This is a time-consuming and expensive alternative to natural pollination.

Study), especially in severely threatened biodiversity hotspots.

3. How do we determine which habitat areas are the most critical to protect?

A 2012 study by an international team of conservation scientists, led by environmental economist Donald McCarthy, estimated that we could sharply reduce the threat of species extinction and protect key conservation sites by spending about $80 billion a year. This investment in sustaining biodiversity is less than 20% of the amount spent globally each year on soft drinks.

CASE STUDY

Protecting Honeybees and Other Pollinators

Failure to protect honeybees could lead to a loss in some areas of the ecosystem service of pollination that they provide. For example, in some parts of Japan and China, overuse of pesticides and other factors have caused farmers to have to pollinate some crops by hand (Figure 9-23)—a time-consuming and expensive undertaking.

Researchers, farmers, and beekeepers are searching for ways to protect European honeybees and other pollinators and to sustain their populations (**Core Case Study**). They are taking several steps, some of which involve applying the precautionary principle.

Some beekeepers are breeding bees that are not as vulnerable to harmful parasitic mites and fungi. They are also improving nutrition for bees that are transported from place to place, for example, by not feeding their bees sugar syrups or pollen substitutes. Some farmers are raising their own honeybee colonies to avoid having to bring in stressed, unhealthy honeybees for pollination.

We can all work to preserve healthy populations of honeybees that help to supply us with at least a third of our food. Ways to do this include buying food from local, USDA-certified organic farmers who do not use pesticides and buying local, organic honey to support organic beekeepers. Some people are setting up their own beehives and producing their own organic honey. Others, by growing their own food organically, are supporting native, wild honeybee populations that benefit from such crops. Many people are replacing their lawns with a variety of native plants that make good sources of nectar for bees, including foxglove, red clover, and bee balm.

Big Ideas

- **We are hastening the extinction of wild species and degrading the ecosystem services they provide by destroying and degrading their habitats, introducing harmful invasive species, and increasing human population growth, pollution, climate change, and overexploitation.**
- **We should avoid causing the extinction of wild species because of the ecosystem and economic services they provide and because their existence should not depend primarily on their usefulness to us.**
- **We can work to prevent the extinction of species and to protect overall biodiversity and ecosystem services by using laws and treaties, protecting wildlife sanctuaries, and making greater use of the precautionary principle.**

TYING IT ALL TOGETHER Honeybees and Sustainability

Malwina Szweda/Shutterstock.com

In this chapter, we have looked at the human activities that are hastening the extinction of many species and at how we might curtail those activities. We learned that there is considerable evidence that as many as half of the world's wild species could go extinct during this century, largely as a result of human activities that threaten many species and some of the vital ecosystem services they provide. For example, populations of honeybees, vital for pollinating crops that supply much of our food, have been declining for a variety of reasons (**Core Case Study**), many of them related to human activities. One of the key reasons for such problems is that most people are simply unaware of the highly valuable ecosystem and economic services provided by the earth's species.

In keeping with two of the three scientific **principles of sustainability** (see Figure 1-2, p. 6 or back cover), acting to prevent the extinction of species as a result of human activities helps to preserve not only the earth's biodiversity, but also the vital ecosystem services that sustain us, including chemical cycling. Thus, it is not only for these species that we ought to act, but also for the long-term health and well-being of our own species in keeping with the win-win and future generations' **principles of sustainability** (see Figure 1-5, p. 9 or back cover).

SUSTAINABILITY

Chapter Review

Core Case Study

1. What economic and ecological roles do honeybees play? What is *pollination*? How are human activities contributing to the decline of many populations of European honeybees, and why should we care? What is **colony collapse disorder (CCD)**?

Section 9-1

2. What is the key concept for this section? What is **biological extinction**? Explain what a trophic cascade is and give an example. What is a **mass extinction**? What is the **background extinction rate**, and how do estimated current and projected extinction rates compare with it? What percentage of the earth's land and what percentage of the earth's oceans have been disturbed by human activities? Explain how scientists estimate extinction rates and describe three problems they face. Give three reasons why many extinction experts believe that projected extinction rates are probably on the low side. What percentage of the world's species are likely to go extinct, largely as a result of human activities, during this century? Distinguish between **endangered species** and **threatened species** and give an example of each. List four characteristics that make some species especially vulnerable to extinction.

Section 9-2

3. What is the key concept for this section? What are three reasons for trying to avoid hastening the extinction of wild species? Describe two economic and two ecological benefits of species diversity. Explain how saving other species and the ecosystem services they provide can help us to save our own species and our cultures and economies.

Section 9-3

4. What is the key concept for this section? What is **HIPPCO**? What is the greatest threat to wild species? What is **habitat fragmentation**? Describe the major effects of habitat loss and fragmentation. Why are island species especially vulnerable to extinction? What are habitat islands?

5. Give two examples of the benefits that have been gained by the introduction of nonnative species. Give two examples of the harmful effects of nonnative species that have been introduced deliberately. Describe the harmful and beneficial effects of introducing the kudzu vine. Give two examples of harmful results of accidental introductions of nonnative species. List four ways to limit the harmful impacts of nonnative species. Explain why prevention is the best way to reduce threats from invasive species and list five ways to implement this strategy. Summarize the roles of population growth, overconsumption, pollution, and climate change in the extinction of wild species. Explain how pesticides such as DDT can accumulate and be biomagnified in food chains and webs.

6. Summarize the roles of population growth, overconsumption, pollution, and climate change in the extinction of wild species. List possible causes of the decline of European honeybee populations in the United States. Describe how human activities threaten polar bears in the Arctic. Why does poaching occur? Give three examples of species that are threatened by this illegal activity. Why are wild tigers likely to disappear within a few decades? What is the connection between infectious diseases in humans and the pet trade? Describe the threat to some forms of wildlife from the increased hunting for bush meat.

7. List the major threats to the world's bird populations and give two reasons for protecting bird species from extinction. Describe environmental explorer Çağan Şekercioğlu's contributions to our understanding of the ecological importance of birds and threats to their extinction.

Section 9-4

8. What is the key concept for this section? Name two international treaties that are used to help protect species. What is the U.S. Endangered Species Act? How successful has it been, and why is it controversial?

9. Summarize the roles and limitations of wildlife refuges, gene banks, botanical gardens, wildlife farms, zoos, and aquariums in protecting some species. Describe the role of captive breeding in efforts to prevent species extinction and give an example of success in returning a nearly extinct species to the wild. What is the **precautionary principle** and how can we use it to help protect wild species and overall biodiversity? What are three important questions related to the use of this principle? List five ways to lessen the environmental threats to European honeybees and other pollinators.

10. What are this chapter's *three big ideas*? How do two of the three scientific **principles of sustainability** apply to protecting honeybees and other wild species from extinction along with protecting the ecosystem services provided by species? What two social science **principles of sustainability** are involved in protecting species from extinction due to human activities?

Note: Key terms are in bold type.

Critical Thinking

1. What are three aspects of your lifestyle that might directly or indirectly contribute to declines in European honeybee populations and the endangerment of other pollinator species (**Core Case Study**)?

2. Give your response to the following statement: "Eventually, all species become extinct. So it does not really matter that the world's remaining tiger species or a tropical forest plant is endangered mostly because of human activities." Be honest about your reaction, and give arguments to support your position.

3. Do you accept the ethical position that each species has the inherent right to survive without human interference, regardless of whether it serves any useful purpose for humans? Explain. Would you extend this right to the *Anopheles* mosquito, which transmits malaria, and to infectious bacteria? Explain.

4. Wildlife ecologist and environmental philosopher Aldo Leopold wrote this with respect to preventing the extinction of wild species: "To keep every cog and wheel is the first precaution of intelligent tinkering." Explain how this statement relates to the material in this chapter.

5. What would you do if fire ants invaded your yard and house? Explain your reasoning behind your course of action. How might your actions affect other species or the ecosystem you are dealing with?

6. How do you think your lifestyle might contribute directly or indirectly to the extinction of some bird species? What are two things that you think should be done to reduce the rate of extinction of bird species?

7. Which of the following statements best describes your feelings toward wildlife?
   a. As long as it stays in its space, wildlife is okay.
   b. As long as I do not need its space, wildlife is okay.
   c. I have the right to use wildlife habitat to meet my own needs.
   d. When you have seen one redwood tree, elephant, or some other form of wildlife, you have seen them all, so lock up a few of each species in a zoo or wildlife park and do not worry about protecting the rest.
   e. Wildlife should be protected in its current ranges.

8. Environmental groups in a heavily forested state want to restrict logging in some areas to save the habitat of an endangered squirrel. Timber company officials argue that the well-being of one type of squirrel is not as important as the well-being of the many families who would be affected if the restriction were to cause the company to lay off hundreds of workers. If you had the power to decide this issue, what would you do and why? Describe any trade-offs included in your solution.

Doing Environmental Science

Identify examples of habitat destruction or degradation in the area in which you live or go to school. Try to determine and record any harmful effects that these activities have had on the populations of one wild plant and one animal species. (Name each of these species and describe how they have been affected.) Do some research on the Internet and/or in a school library on management plans, and then develop a management plan for restoring the habitats and species you have studied. Try to determine whether trade-offs are necessary with regard to the human activities you have observed, and account for these trade-offs in your management plan. Compare your plan with those of your classmates.

Global Environment Watch Exercise

Search for *Extinction,* and scroll to statistics on the portal's page. Click on "Known Causes of Animal Extinction since 1600." You will find four general categories of causes. Thinking about history from 1600 through today, how do you think humans have changed their impact on species in each of these categories? Has the impact increased or decreased over this time period? Give specific examples of changes in this timeframe to support your answers.

Data Analysis

Examine the following data released by the World Resources Institute and answer the questions that follow the table.

Country	Total Land Area in Square Kilometers (Square Miles)	Protected Area as Percent of Total Land Area (2003)	Total Number of Known Breeding Bird Species (1992–2002)	Number of Threatened Breeding Bird Species (2002)	Threatened Breeding Bird Species as Percent of Total Number of Known Breeding Bird Species
Afghanistan	647,668 (250,000)	0.3	181	11	
Cambodia	181,088 (69,900)	23.7	183	19	
China	9,599,445 (3,705,386)	7.8	218	74	
Costa Rica	51,114 (19,730)	23.4	279	13	
Haiti	27,756 (10,714)	0.3	62	14	
India	3,288,570 (1,269,388)	5.2	458	72	
Rwanda	26,344 (10,169)	7.7	200	9	
United States	9,633,915 (3,718,691)	15.8	508	55	

(Compiled by the authors using data from World Resources Institute, *Earth Trends, Biodiversity and Protected Areas, Country Profiles.*)

1. Complete the table by filling in the last column. For example, to calculate this value for Costa Rica, divide the number of threatened breeding bird species by the total number of known breeding bird species and multiply the answer by 100 to get the percentage.
2. Arrange the countries from largest to smallest according to total land area. Does there appear to be any correlation between the size of country and the percentage of threatened breeding bird species? Explain your reasoning.

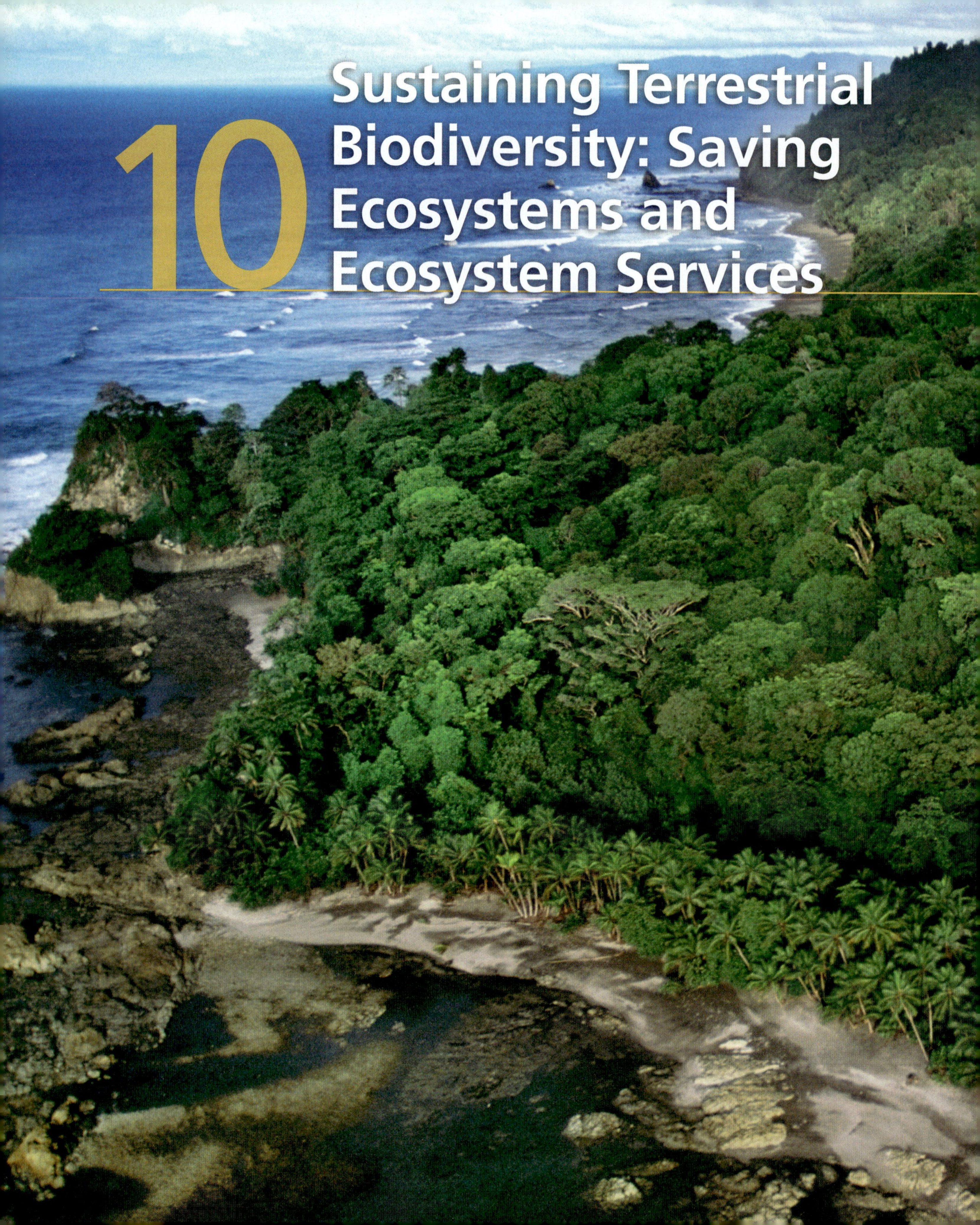

10 Sustaining Terrestrial Biodiversity: Saving Ecosystems and Ecosystem Services

There is no solution, I assure you, to save Earth's biodiversity other than preservation of natural environments in reserves large enough to maintain wild populations sustainably.

EDWARD O. WILSON

Key Questions

10-1 What are the major threats to forest ecosystems?

10-2 How should we manage and sustain forests?

10-3 How should we manage and sustain grasslands?

10-4 How should we manage and sustain parks and nature reserves?

10-5 What is the ecosystem approach to sustaining biodiversity and ecosystem services?

Coastal rain forest in Costa Rica's Corcovado National Park.

FRANS LANTING/National Geographic Creative

CORE CASE STUDY Costa Rica—A Global Conservation Leader

Tropical forests once completely covered Central America's Costa Rica, which is smaller in area than the U.S. state of West Virginia and about one-tenth the size of France. Between 1963 and 1983, politically powerful ranching families cleared much of the country's forests to graze cattle.

Despite such widespread forest loss, tiny Costa Rica is a superpower of biodiversity, with an estimated 500,000 plant and animal species. A single park in Costa Rica is home to more bird species (Figure 10-1, left) than are found in all of North America. This oasis of biodiversity is also home to an amazing variety of other exotic wildlife, including monkeys (Figure 10-1, right), jaguars, lizards, snakes, spiders, and frogs.

This biodiversity results mostly from two factors. One is the country's tropical geographic location, lying between two oceans and having both coastal (see chapter-opening photo) and mountainous regions that provide a variety of microclimates and habitats for wildlife. The other factor is the government's strong conservation efforts.

In the mid-1970s, Costa Rica established a system of nature reserves and national parks that, by 2012, included more than 25% of its land—6% of it reserved for indigenous peoples. Costa Rica now devotes a larger proportion of its land to biodiversity conservation than does any other country, in keeping with the biodiversity **principle of sustainability** (see Figure 1-2, p. 6 or back cover).

SUSTAINABILITY

To reduce *deforestation,* or the widespread removal of forests, the government has eliminated subsidies for converting forestland to rangeland. Instead, it pays landowners to maintain or restore tree cover. The strategy has worked: Costa Rica has gone from having one of the world's highest deforestation rates to having one of the lowest.

GOOD NEWS

Since 1980, *biodiversity* (see Figure 4-2, p. 79) has emerged as a key concept of biology. Ecologists warn that human population growth, economic development, and poverty are exerting increasing pressure on the earth's ecosystems and on the ecosystem services they provide that help to sustain biodiversity. In 2010, a report by two United Nations environmental bodies warned that unless radical and creative action is taken now to conserve the earth's biodiversity, many local and regional ecosystems that support human lives and livelihoods are at risk of collapsing.

This chapter is devoted to helping us understand the threats to the earth's forests, grasslands, and other storehouses of terrestrial biodiversity, and to seeking ways to help sustain these vital ecosystems. To many scientists, one of our most important challenges is to sustain the world's vital biodiversity and the ecosystem services it provides.

Figure 10-1 Costa Rica is one of the world's most biologically rich places. Two of its half-million species are the scarlet macaw parrot (left) and the white-faced capuchin monkey (right).

Left: Vladimir Melnik/Shutterstock.com. Right: © Vilainecre... | Dreamstime.com.

10-1 What Are the Major Threats to Forest Ecosystems?

CONCEPT 10-1A
Forest ecosystems provide ecosystem services far greater in value than the value of raw materials obtained from forests.

CONCEPT 10-1B
Unsustainable cutting and burning of forests, along with diseases and insects, all made worse by projected climate change, are the chief threats to forest ecosystems.

Forests Vary in Their Age, Makeup, and Origins

Natural and planted forests occupy about 31% of the earth's land surface (excluding Greenland and Antarctica). Figure 7-9 (p. 150) shows the distribution of the world's northern coniferous, temperate, and tropical forests.

Forest managers and ecologists classify natural forests into two major types based on their age and structure: old-growth and second-growth forests. An **old-growth forest**, or **primary forest**, is an uncut or regenerated forest that has not been seriously disturbed by human activities or natural disasters for 200 years or more (Figure 10-2 and Chapter 1 opening photo). Old-growth forests are reservoirs of biodiversity because they provide ecological niches for a multitude of wildlife species (see Figure 7-14, p. 157). According to the United Nations Environment Programme (UNEP), they make up about 36% of the world's forests.

A **second-growth forest** is a stand of trees resulting from secondary ecological succession (see Figure 5-12, p. 110). These forests develop after the trees in an area have been removed by human activities, such as clear-cutting for timber or conversion to cropland, or by natural forces such as fire, hurricanes, or volcanic eruptions.

A **tree plantation**, also called a **tree farm** or **commercial forest** (Figure 10-3), is a managed forest containing only one or two species of trees that are all of the same age. They are usually harvested by clear-cutting as soon as they become commercially valuable. The land is then replanted and clear-cut again in a regular cycle. When managed carefully, such plantations can produce wood at a rapid rate and thus increase their owners' profits. Some analysts project that eventually, tree plantations could supply most of the wood used for industrial purposes such as papermaking. This would help to protect the world's remaining old-growth and second-growth forests, as long as they are not cleared to make room for tree plantations.

The downside of tree plantations is that, with only one or two tree species, they are much less biologically diverse

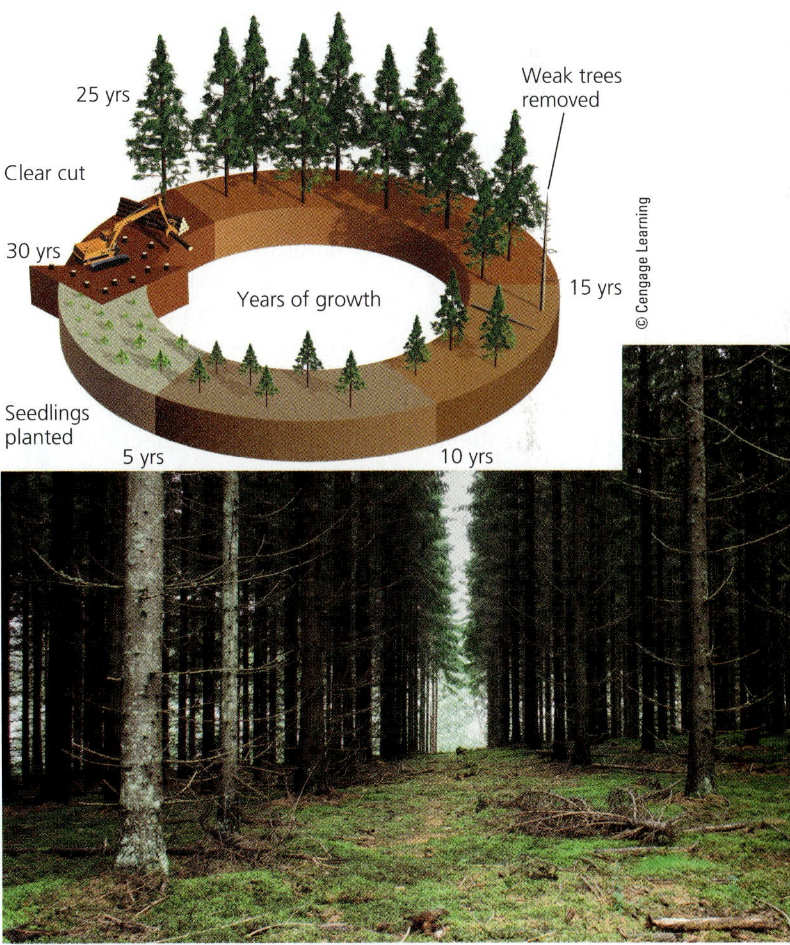

Figure 10-3 The rotation cycle of cutting and regrowth of monoculture tree plantations is short, usually 20–30 years. In tropical countries, where trees can grow more rapidly year-round, the rotation cycle can be 6–10 years. Like most tree plantations, the pine tree plantation in this photo was grown on land that had been cleared of an old-growth or second-growth forest.

©FRANS LANTING/National Geographic Creative

Figure 10-2 This protected old-growth rain forest at a high altitude in Monteverde, Costa Rica (**Core Case Study**), is home for a rich diversity of plant and animal species.

Natural Capital

Forests

Ecosystem Services	Economic Services
Support energy flow and chemical cycling	Fuelwood
Reduce soil erosion	Lumber
Absorb and release water	Pulp to make paper
Purify water and air	Mining
Influence local and regional climate	Livestock grazing
Store atmospheric carbon	Recreation
Provide numerous wildlife habitats	Jobs

Figure 10-4 Forests provide many important ecosystem and economic services (Concept 10-1A). ***Question:*** Which two ecosystem services and which two economic services do you think are the most important?

Photo: Val Thoermer/Shutterstock.com

and less sustainable than old-growth and second-growth forests because they violate nature's biodiversity **principle of sustainability**. Tree plantations do not provide the wildlife habitats and ecosystems services such as water storage and purification that diverse natural forests do. Also, repeated cycles of cutting and replanting can eventually deplete the topsoil of nutrients and hinder the regrowth of any type of forest on such land.

Forests Provide Important Economic and Ecosystem Services

We should care about sustaining forests because they provide highly valuable economic and ecosystem services (Figure 10-4 and Concept 10-1A). For example, through photosynthesis, forests remove CO_2 from the atmosphere and store it in organic compounds (biomass). By performing this ecosystem service as a part of the global carbon cycle (see Figure 3-17, p. 66), forests help to stabilize average atmospheric temperatures and the earth's climate. Forests also provide habitats for about two-thirds of the earth's terrestrial species. In addition, according to the UN Food and Agriculture Organization (FAO) and the UNEP, forests are home to more than 300 million people, and about 1 billion people living in extreme poverty depend on forests for their survival.

Along with highly valuable ecosystem services, forests provide us with raw materials, especially wood. More than half of the wood removed from the earth's forests is used as *biofuel* for cooking and heating. The remainder of the harvest, called *industrial wood,* is used primarily to make lumber and paper.

Forests also provide us with important health benefits. For example, traditional medicines, used by 80% of the world's people, are derived mostly from plant species that are native to forests, and chemicals found in tropical forest plants are used as blueprints for making most of the world's prescription drugs (see Figure 9-6, p. 196). Scientists and economists have various ways of estimating the economic value of major ecosystem services provided by the world's forests and other ecosystems (Science Focus 10.1).

There Are Several Ways to Harvest Trees

Because of the immense economic value of forests, the harvesting of wood is one of the world's major industries. The first step in harvesting trees is to build roads for access and timber removal. Even carefully designed logging roads have a number of harmful effects (Figure 10-5)—namely, increased topsoil erosion and sediment runoff into waterways, habitat fragmentation, and loss of biodiversity. Logging roads also expose forests to invasion by nonnative pests, diseases, and wildlife species. And they open once-inaccessible forests to miners, ranchers, farmers, hunters, and off-road vehicles.

Once loggers reach a forest area, they use a variety of methods to harvest the trees (Figure 10-6). With *selective cutting,* intermediate-aged or mature trees in a forest are cut singly or in small groups (Figure 10-6a). Loggers

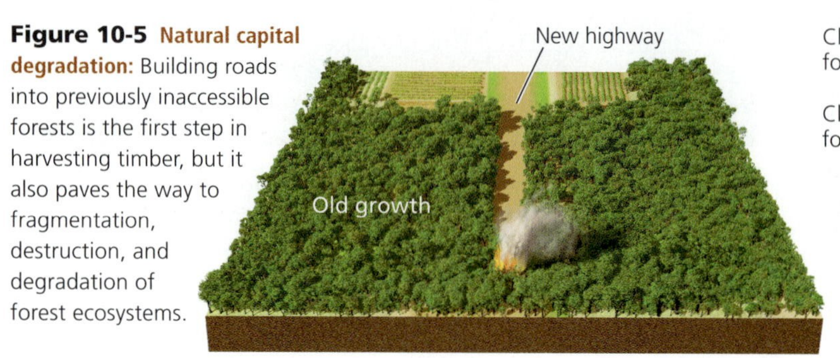

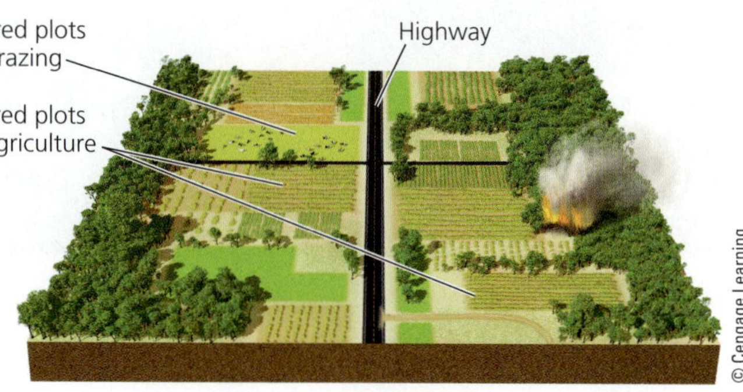

Figure 10-5 Natural capital degradation: Building roads into previously inaccessible forests is the first step in harvesting timber, but it also paves the way to fragmentation, destruction, and degradation of forest ecosystems.

SCIENCE FOCUS 10.1

PUTTING A PRICE TAG ON NATURE'S ECOSYSTEM SERVICES

Currently, forests and other ecosystems are valued mostly for their economic services (Figure 10-4, right). Ecologists and ecological economists call for us to also calculate the monetary value of the ecosystem services provided by forests (Figure 10-4, left), as a way to implement the full-cost pricing **principle of sustainability** (see Figure 1-5, p. 9 or back cover).

In 1997, a team of ecologists, economists, and geographers, led by ecological economist Robert Costanza of the University of Vermont, estimated the monetary worth of the earth's ecosystem services, which can be thought of as *ecological income*, somewhat like interest income earned from a savings account. We can think of the earth's stock of natural resources that provide the ecosystem services as the savings account from which the ecological income flows.

Costanza's team estimated the monetary value of 17 ecosystem services provided by all ecosystems (including, for example, pollination and regulation of atmospheric temperatures) to be at least $33.2 trillion per year—equal to about 39% of the value of all of the goods and services produced in 2012 throughout the world. The researchers also estimated that the amount of money we would need to put into a savings account in order to earn that amount of interest income would be at least $500 trillion—an average of about $70,400 for each person on earth in 2012.

According to Costanza's study, the world's forest ecosystems alone provide us with ecosystem services worth at least $4.7 trillion per year—hundreds of times more than their economic value in terms of lumber, paper, and other wood products (**Concept 10-1A**). Some of these comparative value estimates are shown in Figure 10-A. The researchers pointed out that their estimates were very conservative.

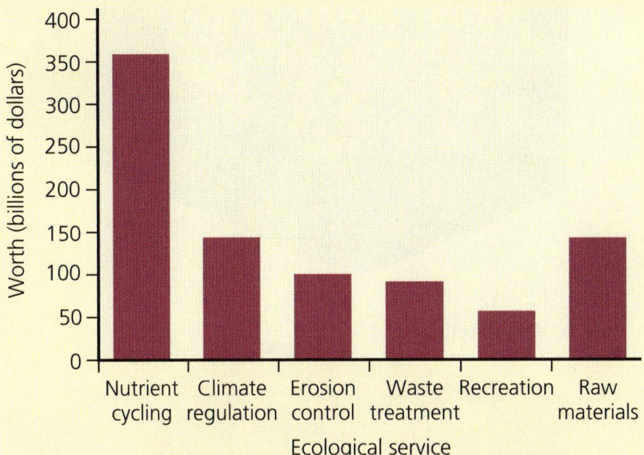

Figure 10-A Estimates of the annual global economic values of some ecosystem services provided by forests, compared to the value of the raw materials they produce (in billions of dollars). (Compiled by the authors using data from Robert Costanza.)

Costanza's team had examined more than 100 studies and a variety of methods used to estimate the values of ecosystems. For example, one method that had been used was to estimate the costs of replacing ecosystems' services such as water purification with technological versions of the same services.

In 2002, Costanza and other researchers reported on a similar analysis comparing the value of preserving natural ecosystems with the values that could be obtained by using such systems to create farmland, to harvest timber, and to create aquaculture ponds, among other uses. The researchers estimated that preserving ecosystems in a global network of nature reserves occupying 15% of the earth's land surface and 30% of the ocean would provide $4.4 trillion to $5.2 trillion worth of ecosystem services—about 100 times the economic value of converting those systems to human uses.

If the estimates from these studies are reasonable, we can draw three important conclusions: **(1)** the earth's ecosystem services are essential for all humans and their economies; **(2)** the economic value of these services is huge; and **(3)** ecosystem services are an ongoing source of ecological income, as long as they are used sustainably.

Critical Thinking

Some analysts believe that we should not try to put economic values on the world's irreplaceable ecosystem services because their value is infinite. Do you agree with this view? Explain. What is the alternative?

often remove all the trees from an area in what is called a *clear-cut* (Figure 10-6b and Figure 10-7). Clear-cutting is the most efficient and often the least costly way to harvest trees, but it can do considerable harm to an ecosystem. Figure 10-8 summarizes some advantages and disadvantages of clear-cutting.

A variation of clear-cutting that allows a more sustainable timber yield without widespread destruction is *strip cutting* (Figure 10-6c). It involves clear-cutting a strip of trees along the contour of the land within a corridor narrow enough to allow natural forest regeneration within a few years. After regeneration, loggers cut another strip next to the first, and so on.

One of the major threats to forests is the unsustainable harvesting of trees (**Concept 10-B**). We examine this growing problem later in this chapter.

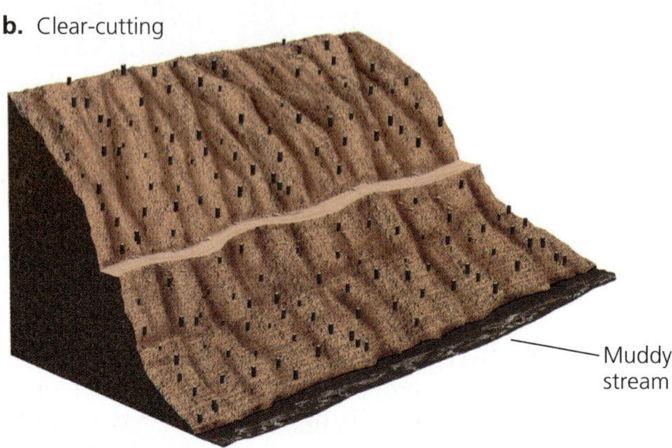

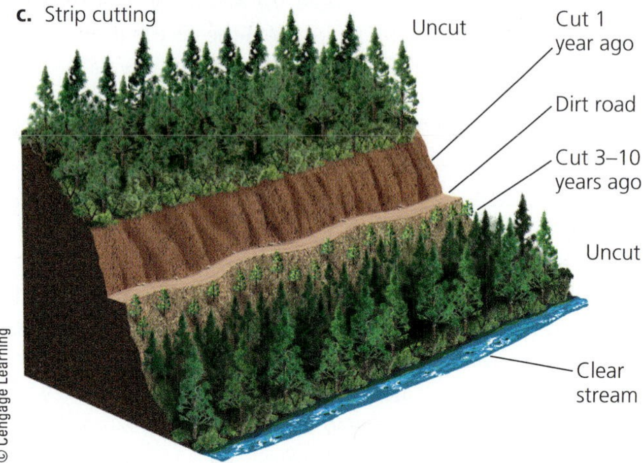

Figure 10-6 There are three major ways to harvest trees. ***Question:*** If you were cutting trees in a forest you owned, which method would you choose and why?

Figure 10-7 Clear-cut forest.

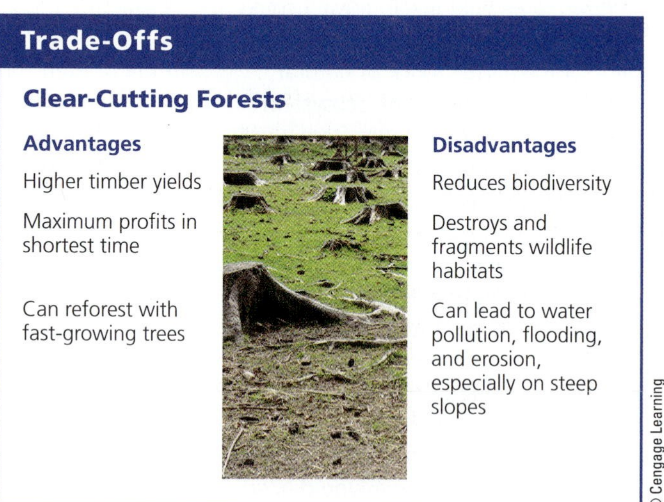

Figure 10-8 There are advantages and disadvantages to clear-cutting a forest. ***Questions:*** Which single advantage and which single disadvantage do you think are the most important? Why?

Photo: © Kirill Livshitskiy/Shutterstock.com

Fire, Insects, and Climate Change Can Threaten Forest Ecosystems

Two types of fires can affect forest ecosystems. *Surface fires* (Figure 10-9, left) usually burn only undergrowth and leaf litter on the forest floor. They may kill seedlings and small trees, but they spare most mature trees and allow most wild animals to escape.

Occasional surface fires have a number of ecological benefits. They:

- burn away flammable ground material such as dry brush and help to prevent more destructive fires;
- free valuable mineral nutrients tied up in slowly decomposing litter and undergrowth;
- release seeds from the cones of tree species such as lodgepole pines;
- stimulate the germination of certain tree seeds such as those of the giant sequoia and jack pine; and
- help to control destructive insects and tree diseases.

Figure 10-9 Surface fires (left) usually burn only undergrowth and leaf litter on a forest floor. They can help to prevent more destructive crown fires (right) by removing flammable ground material.

Left: David J. Moorhead/The University of Georgia. Right: age fotostock/SuperStock.

Another type of fire, called a *crown fire* (Figure 10-9, right), is an extremely hot fire that leaps from treetop to treetop, burning whole trees. Crown fires usually occur in forests that have not experienced surface fires for several decades, a situation that allows dead wood, leaves, and other flammable ground litter to accumulate. These rapidly burning fires can destroy most vegetation, kill wildlife, increase topsoil erosion, and burn or damage human structures in their paths. As part of a natural cycle, forest fires are not a major threat to forest ecosystems, except in the parts of the world where people intentionally burn forests to clear the land. Another threat to forests is the accidental or deliberate introduction of disease-causing organisms and destructive insects.

On top of these threats, projected climate change could harm many forests. Rising temperatures and increased drought influenced by a warmer atmosphere will likely make many forest areas more suitable for insect pests, which would then multiply and kill more trees. The resulting combination of drier forests and more dead trees could also increase the number and intensity of forest fires (**Concept 10-1B**).

Almost Half of the World's Forests Have Been Cut Down

Deforestation is the temporary or permanent removal of large expanses of forest for agriculture, settlements, or other uses. Surveys by the World Resources Institute (WRI) indicate that during the past 8,000 years, human activities have reduced the earth's virgin or frontier forest cover by about 47%, with most of this loss occurring in the last 60 years (Figure 10-10).

Worldwide, deforestation leads to a net loss of about 52,000 square kilometers (20,000 square miles) of forest per year, according to surveys by the FAO and the WRI. Some good news is that this is down from an annual loss of about 83,000 square kilometers (32,000 square miles) in the 1990s. However, the deforestation rate is still quite high and each year amounts to the clearing of forest areas totaling roughly the area of Costa Rica.

These forest losses are concentrated in less-developed countries, especially those in the tropical areas of Latin America, Indonesia, and Africa. However, scientists are also concerned about the increased clearing of the northern boreal forests of Alaska, Canada, Scandinavia, and Russia, which together make up about one-fourth of the world's forested area.

According to the WRI, if current deforestation rates continue, about 40% of the world's remaining intact forests will have been logged or converted to other uses within two decades if not sooner. Clearing large areas of forests, especially old-growth forests, has important short-

©National Geographic Maps/National Geographic Stock

Figure 10-10 Global deforestation over the past 8,000 years.

Natural Capital Degradation

Deforestation

- Water pollution and soil degradation from erosion
- Acceleration of flooding
- Local extinction of specialist species
- Habitat loss for native and migrating species
- Release of CO_2 and loss of CO_2 absorption

Figure 10-11 Deforestation has some harmful environmental effects that can reduce biodiversity and degrade the ecosystem services provided by forests (Figure 10-4, left).

term economic benefits (Figure 10-4, right column), but it also has a number of harmful environmental effects (Figure 10-11), including severe erosion and loss of topsoil (Figure 10-12) that was once renewed largely by forest ecosystems at no cost to us.

In 2011, the FAO reported that the net total forest cover in several countries, including the United States (see the Case Study that follows), changed very little or even increased between 2000 and 2010. Some of the increases resulted from natural reforestation by secondary ecological succession on cleared forest areas and abandoned croplands (see Figure 5-12, p. 110). Other increases in forest cover were due to the spread of commercial tree plantations (Figure 10-3, right) and to a global program, sponsored by the UNEP, to plant billions of trees throughout much of the world—many of them in tree plantations. China now leads the world in new forest cover, mostly due to its plantations of fast-growing trees.

GOOD NEWS

CASE STUDY

Many Cleared Forests in the United States Have Grown Back

Forests cover about 30% of the U.S. land area, providing habitats for more than 80% of the country's wildlife species and containing about two-thirds of the nation's surface water. Today, forests in the United States (including tree plantations) cover more area than they did in 1920. The primary reason is that many of the old-growth forests that were cleared or partially cleared between 1620 and 1920 have grown back naturally through secondary ecological succession (Figure 10-13).

GOOD NEWS

There are now fairly diverse second-growth (and in some cases third-growth) forests in every region of the United States except in much of the West. In 1995, environmental writer Bill McKibben cited forest regrowth in the United States—especially in the East—as "the great environmental success story of the United States, and in some ways, the whole world." Protected forests make up about 40% of the country's total forest area, mostly in the

Figure 10-12 Severe soil erosion occurred in this area of Brazil after a large area of the tropical Atlantic forest was removed.

National Forest System, which consists of 155 national forests managed by the U.S. Forest Service (USFS) (Figure 10-14).

On the other hand, since the mid-1960s, a large area of the nation's remaining old-growth and fairly diverse second-growth forests has been cut down and replaced with biologically simplified tree plantations. According to biodiversity researchers, this reduces overall forest biodiversity and it can disrupt important ecosystem services.

Tropical Forests Are Disappearing Rapidly

Tropical forests (see Figure 7-13, top, p. 156) cover about 6% of the earth's land area—roughly the area of the continental United States. Climatic and biological data suggest that mature tropical forests once covered at least twice as much area as they do today. Most of this loss of half of the world's tropical forests has taken place since 1950 (see Chapter 3, Core Case Study).

Satellite scans and ground-level surveys indicate that large areas of tropical rain forests and tropical dry forests are being cut rapidly in parts of Africa, Southeast Asia, and South America (see Figure 10-15 and Figure 3-1, p. 52)—especially in Brazil's vast Amazon Basin, which has more than 40% of the world's remaining tropical forests. Such clearing of trees, which absorb carbon dioxide as they grow, helps to hasten climate change because tropical forests absorb and store about one-third of the terrestrial carbon on the planet as part of the carbon cycle (see Figure 3-17, p. 66).

The drier climate increases the risk of more and bigger natural forest fires, which add more climate-changing

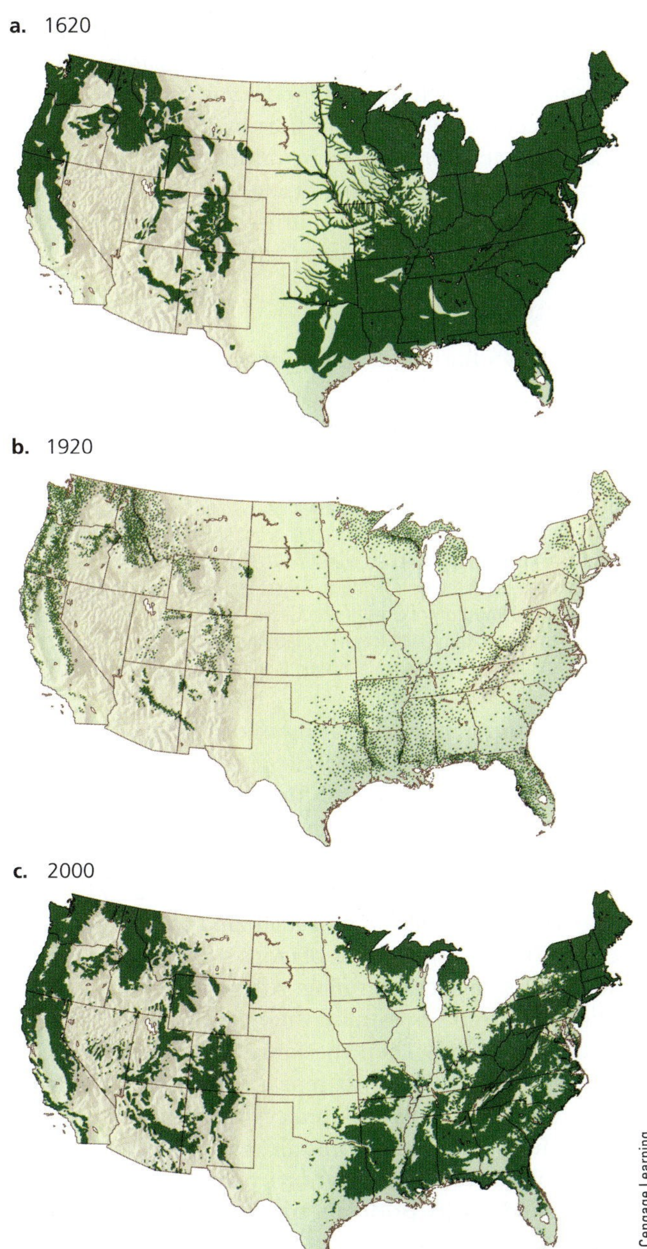

Figure 10-13 In 1620, **(a)** when European settlers were moving to North America, forests covered more than half of the current land area of the continental United States. By 1920, **(b)** most of these forests had been decimated. Since then, a combination of secondary ecological succession and the expansion of commercial forests has resulted in greatly expanded forest cover. In 2000, **(c)** secondary and commercial forests covered about a third of U.S. land in the lower 48 states.

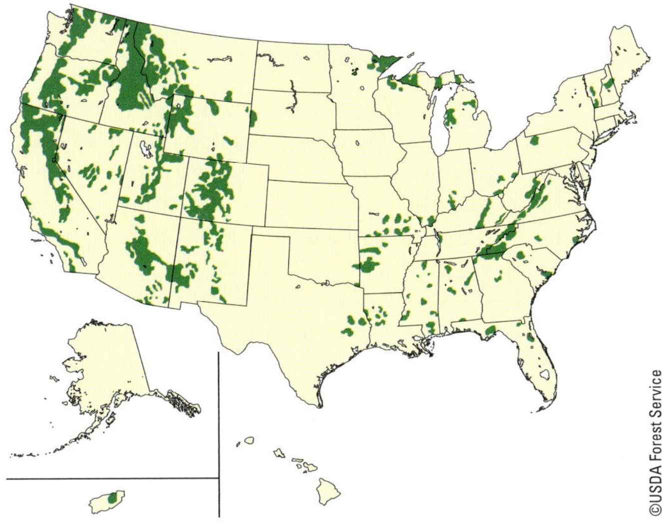

Figure 10-14 U.S. National Forests. ***Question:*** Why do think most national forests are in the western half of the United states?

Figure 10-15 **Natural capital degradation:** Conversion of tropical rain forest into agricultural settlements near Santa Cruz la Sierra, Bolivia, between 1970 and 2000.

carbon dioxide to the atmosphere and further reduce the number of trees that remove carbon dioxide from the atmosphere. It also dehydrates the topsoil by exposing it to sunlight. The dry topsoil can then be blown away, making it difficult for new plants to become established. This can lead to an irreversible *ecological tipping point* (p. 45), beyond which a forest cannot grow back in the area and is then often replaced by tropical grassland, or savanna. Scientists project that if current burning and deforestation rates continue, 20–30% of the Amazon Basin will be turned into savanna in the next 50 years, and most of it could become savanna by 2080.

Studies indicate that at least half of the world's known species of terrestrial plants, animals, and insects live in tropical forests. Because of their specialized niches (see Figure 7-14, p. 157) many of these species are highly vulnerable to extinction when their forest habitats are destroyed or degraded. The FAO warns that, at the current rate of global tropical deforestation, as much as 50% of the world's remaining old-growth tropical forests will be gone or severely degraded by the end of this century (**Concept 10-1B**).

Causes of Tropical Deforestation Are Varied and Complex

Tropical deforestation results from a number of underlying and direct causes. Underlying causes, such as pressures from population growth and poverty, push subsistence farmers and the landless poor into tropical forests, where they cut or burn trees for firewood or to try to grow enough food to survive (see Figure 1-17, p. 18). Government subsidies can accelerate other direct causes such as large-scale logging and ranching by reducing the costs of timber harvesting, cattle grazing, and farming on vast plantations of crops such as soybeans and oil palms.

The major direct causes of deforestation vary in different tropical areas. Tropical forests in the Amazon and other South American countries are cleared (see Figure 1-4, p. 8) or burned primarily for cattle grazing and large soybean plantations. In Indonesia, Malaysia, and other areas of Southeast Asia, tropical forests are being replaced with major plantations of oil palm, which produces an oil used in cooking, cosmetics, and biodiesel fuel for motor vehicles (especially in Europe). In Africa, the primary direct cause of deforestation is people clearing plots for small-scale farming and harvesting wood for fuel.

Global trade has also furthered tropical forest degradation. In 2011, tropical forest researcher William Laurance reported that China was by far the biggest consumer of tropical timber. It is the destination for more than half of the world's timber shipments and much of the paper pulp shipped internationally. This includes large volumes of illegally harvested timber, worth about $15 billion a year, according to a 2011 report by Interpol and the World Bank. As a result, huge amounts of wood and paper products are produced in China and shipped all over the world through a process that includes the unsustainable cutting of tropical forests.

The degradation of a tropical forest usually begins when a road is cut deep into the forest interior for logging and settlement (Figure 10-5). Loggers then use selective cutting (Figure 10-6a) to remove the largest and best trees. When these big trees fall, many other trees often fall with them because of their shallow roots and the network of vines connecting the trees in the forest's canopy.

Burning is widely used to clear forest areas for agriculture, settlement, and other purposes. Healthy rain forests do not burn naturally, but roads, settlements, and other developments fragment them. The resulting patches of forest dry out and readily ignite. According to a 2005 study by forest scientists, widespread fires in the Amazon basin are changing weather patterns by raising temperatures and reducing rainfall. The resulting droughts dry out the forests and make them more likely to burn—an example of a runaway positive feedback loop (see Figure 2-16, p. 45).

CONSIDER THIS. . .

CONNECTIONS Burning Tropical Forests and Climate Change

The burning of tropical forests releases CO_2 into the atmosphere. Rising concentrations of this gas are warming the atmosphere and causing changes in the global climate. Scientists estimate that tropical forest fires account for at least 17% of all human-created greenhouse gas emissions, and that each year, they emit twice as much CO_2 as all of the world's cars and trucks emit. The large-scale burning of the Amazon rain forest accounts for 75% of Brazil's greenhouse gas emissions, making Brazil the world's fourth largest emitter of such gases, according to the National Inventory of Greenhouse Gases. And with these forests gone, even if savanna or second-growth forests replace them, far less CO_2 will be absorbed for photosynthesis, resulting in even more atmospheric warming.

Foreign corporations operating under government concession contracts do much of this logging. After they remove the best timber, they often sell the land to ranchers who burn the remaining timber to clear the land for cattle grazing. Within a few years, their cattle typically overgraze the land and the ranchers move their operations to another forest area. Then they sell the degraded land to farmers, who plow it up for large plantations of crops such as soybeans, or to settlers for small-scale farming. After a few years of crop growing and erosion from rain, the nutrient-poor topsoil is depleted of nutrients. Then the farmers and settlers move on to newly cleared land to repeat this environmentally destructive process. Scientists project that if current burning and deforestation rates continue, much of the Amazon forest will be converted to savanna by this failure to follow the biodiversity **principle of sustainability**.

SUSTAINABILITY

CONSIDER THIS. . .

THINKING ABOUT Tropical Forests

Why should you care if most of the world's remaining tropical forests are burned or cleared or converted to savanna within your lifetime? What are three ways in which this might affect your life or the lives of any children and grandchildren that you might have?

10-2 How Should We Manage and Sustain Forests?

CONCEPT 10-2

We can sustain forests by emphasizing the economic value of their ecosystem services, removing government subsidies that hasten their destruction, protecting old-growth forests, harvesting trees no faster than they are replenished, and planting trees.

We Can Manage Forests More Sustainably

Biodiversity researchers and a growing number of foresters have called for more sustainable forest management. Figure 10-16 lists ways to achieve this goal (**Concept 10-2**). Certification of sustainably grown timber and of sustainably produced forest products can help con-

GOOD NEWS

Solutions

More Sustainable Forestry

- Include ecosystem services of forests in estimates of their economic value
- Identify and protect highly diverse forest areas
- Stop logging in old-growth forests
- Stop clear-cutting on steep slopes
- Reduce road-building in forests and rely more on selective and strip cutting
- Leave most standing dead trees and larger fallen trees for wildlife habitat and nutrient cycling
- Put tree plantations only on deforested and degraded land
- Certify timber grown by sustainable methods

Figure 10-16 There are a number of ways to grow and harvest trees more sustainably (Concept 10-2). ***Questions:*** Which three of these methods of more sustainable forestry do you think are the best methods? Why?

sumers to play their part in reaching this goal (Science Focus 10.2).

Loggers could use more sustainable practices in tropical forests. For example, they can use sustainable selective cutting (Figure 10-6a) and strip cutting (Figure 10-6c) to harvest tropical trees for lumber instead of clear-cutting the forests (Figure 10-6b). They could also be more careful when cutting individual trees, taking care to cut canopy vines (lianas) before felling a tree and using the least obstructed paths to remove the logs. These practices would sharply reduce damage to neighboring trees.

Many economists are urging governments to begin making a shift to more sustainable forest management strategies by phasing out government subsidies and tax breaks that encourage forest degradation and deforestation and replacing them with forest-sustaining economic rewards. This would be an application of the full-cost pricing **principle of sustainability** (see Figure 1-5, p. 9 or back cover) because it would have the effect of raising prices on unsustainably produced timber and wood products. Costa Rica (**Core Case Study**) is taking a lead in using this approach. Governments can also encourage tree planting programs to help restore degraded forests. **GREEN CAREER:** Sustainable forestry

We Can Improve the Management of Forest Fires

In the United States, the Smokey Bear educational campaign undertaken by the Forest Service and the National Advertising Council has likely prevented many forest fires, saved many lives, and prevented billions of dollars in losses of trees, wildlife, and human structures. At the same time, this educational program has convinced much of the public that all forest fires are bad and should be prevented or put out. Ecologists warn that trying to prevent all forest fires can make matters worse by increasing the likelihood of destructive crown fires (Figure 10-9, right) due to the accumulation of highly flammable underbrush and smaller trees in some forests.

Ecologists and forest fire experts have proposed several strategies for reducing fire-related harm to forests and to people who use or live in the forests. One approach is to set small, contained surface fires to remove flammable small trees and underbrush in the highest-risk forest areas. Such *prescribed burns* require careful planning and monitoring to keep them from getting out of control. However, in 2012, research by ecologists William Baker and Mark Williams, which included studies of descriptions of past fires and tree ring data, indicated that natural low-level fires were not as widespread as other studies had indicated and that they rarely prevented more severe fires in much of the western United States. These researchers contend that expensive prescribed burns and thinning should not be used to prevent large fires in most western forests except to protect human structures.

A second strategy is to allow some fires on public lands to burn, thereby removing flammable underbrush and smaller trees, as long as the fires do not threaten human structures and life.

A third approach is to protect houses and other buildings in fire-prone areas by thinning trees and other vegetation in a zone of about 60 meters (200 feet) around them, and eliminating the use of highly flammable construction materials such as wood shingles.

A fourth approach is to thin forest areas that are vulnerable to fire by clearing away small fire-prone trees and underbrush under careful environmental controls. Many forest fire scientists warn that such thinning operations should not remove economically valuable medium-size and large trees for two reasons. *First,* these are usually the most fire-resistant trees. *Second,* their removal encourages dense growth of more flammable young trees and underbrush and leaves behind highly flammable *slash,* the debris left behind by a logging operation. Many of the worst fires in U.S. history burned through cleared forest areas that contained slash. A 2006 study by Forest Service researchers found that thinning forests without using prescribed burning to remove the slash can greatly increase rather than decrease the risk of heavy fire damage. The previously cited research in 2012 by ecologists Baker and Williams also questioned the value of thinning in most western forests.

We Can Reduce the Demand for Harvested Trees

According to the Worldwatch Institute and to forestry analysts, *up to 60% of the wood consumed in the United States is wasted unnecessarily.* This results from inefficient use of construction materials, excessive packaging, overuse of

SCIENCE FOCUS 10.2

CERTIFYING SUSTAINABLY GROWN TIMBER AND PRODUCTS SUCH AS THE PAPER USED IN THIS BOOK

Collins Pine is a forest products company that owns and manages a large area of productive timberland in the northeastern part of the U.S. state of California. Since 1940, the company has used selective cutting to help maintain the ecological and economic sustainability of its timberland.

Since 1993, Scientific Certification Systems (SCS) of Oakland, California, has evaluated the company's timber production. SCS, which is part of the nonprofit Forest Stewardship Council (FSC), was formed to develop environmentally sound and sustainable practices for use in certifying timber and timber products.

Each year, SCS evaluates Collins Pine's landholdings and has consistently found that their cutting of trees has not exceeded long-term forest regeneration; roads and harvesting systems have not caused unreasonable ecological damage; topsoil has not been damaged; and downed wood (boles) and standing dead trees (snags) are left to provide wildlife habitat. As a result, SCS judges the company to be a good employer and a good steward of its land and water resources.

The FSC reported that, by 2012, about 5% of the world's forest area in 80 countries had been certified according to FSC standards. The FSC also certifies 5,400 manufacturers and distributors of wood products. The paper used in this book was produced with the use of sustainably grown timber, as certified by the FSC symbol, and contains recycled paper fibers. Figure 10-B shows the FSC certification and recycling symbol used for this textbook.

Figure 10-B This Forest Stewardship Council (FSC) symbol certifies that the paper used in this textbook was produced from environmentally responsible sources with the use of recycled fibers. It also appears on the back cover of this book.

Critical Thinking

Should governments provide subsidies or tax breaks for sustainably grown timber to encourage this practice? Explain.

junk mail, inadequate paper recycling, and failure to reuse or find substitutes for wooden shipping containers.

One reason for cutting trees is to provide pulp for making paper, but paper can be made by using fiber from sources other than trees. China uses rice straw and other agricultural residues to make much of its paper. Most of the small amount of tree-free paper produced in the United States is made from the fibers of a rapidly growing woody annual plant called *kenaf* (pronounced "kuh-NAHF," Figure 10-17). Kenaf and other nontree fibers such as hemp yield more paper pulp per area of land than tree farms do and require less use of pesticides and herbicides.

According to the USDA, kenaf is "the best option for tree-free papermaking in the United States" and could replace wood-based paper within 20–30 years. GOOD NEWS However, while timber companies successfully lobby for government subsidies to grow and harvest trees

Figure 10-17 **Solutions:** The pressure to cut trees to make paper could be greatly reduced by planting and harvesting a fast-growing plant known as kenaf.

James P. Blair/National Geographic Creative

Figure 10-18 **Natural capital degradation:** Haiti's deforested brown landscape (left) contrasts sharply with the heavily forested green landscape of its neighboring country, the Dominican Republic.

to make paper, there are no major lobbying efforts or subsidies for producing paper from kenaf and other alternatives to trees.

Another way to reduce the demand for tree cutting is to sharply reduce the use of throwaway paper products made from trees. We can instead choose reusable plates, cups, and cloth napkins and handkerchiefs. Recycled paper products—including soft toilet paper—are becoming more available every year.

Humans have always used trees for fuel, but now the demand for wood as fuel is becoming unsustainable in many areas (see the following Case Study).

CASE STUDY

Deforestation and the Fuelwood Crisis

About 50% of the wood harvested globally each year, and 75% of that harvested in less-developed countries, is burned directly for fuel or converted to charcoal fuel. More than 2 billion people in less-developed countries use fuelwood (see Figure 6-17, p. 135) and charcoal made from wood for heating and cooking.

However, most of these countries are suffering from fuelwood shortages because people are cutting trees for fuelwood and forest products 10–20 times faster than new trees are being planted. As the demand for fuelwood in urban areas of some less-developed countries exceeds the sustainable yield of nearby forests, expanding rings of deforested land encircle such cities. The FAO warns that, by 2050, the demand for fuelwood could easily be 50% greater than the amount that can be sustainably supplied.

For example, Haiti, a country with 10.3 million people, was once a tropical paradise, 60% of it covered with forests. Now it is an ecological disaster. Largely because its trees were cut for fuelwood and to make charcoal, less than 2% of its land is now covered with trees (Figure 10-18). With the trees gone, soils have eroded away in many areas, making it much more difficult to grow crops.

The U.S. Agency for International Development funded the planting of 60 million trees over more than two decades in Haiti. However, the local people cut most of them down for firewood and to make charcoal before they could grow into mature trees. This unsustainable use of natural capital and failure to follow the **principles of sustainability** have led to a downward spiral of environmental degradation, poverty, disease, social injustice, crime, and violence in Haiti.

One way to reduce the severity of the fuelwood crisis in less-developed countries is to establish small plantations of fast-growing fuelwood trees and shrubs around farms and in community woodlots. Another approach to this problem is to encourage more efficient use of wood by providing villagers with more fuel-efficient wood stoves. Villagers could stop using wood if they could get access to solar ovens and electric hotplates powered by solar- or wind-generated electricity. Another option is stoves that burn renewable biomass, such as sun-dried roots of various gourds and squash plants, or methane produced from crop and animal wastes—technologies that could be provided inexpensively. All of these are also low-pollution options that would greatly reduce the number of deaths caused by indoor air pollution from open fires and poorly designed stoves.

In addition, scientists are looking for ways to produce charcoal for heating and cooking without cutting down trees. For example, Professor Amy Smith of the Massachusetts Institute of Technology is developing a way to make charcoal from the fibers in a waste product called bagasse, which is left over from sugar cane processing in Haiti, Brazil, and many other countries.

Countries such as South Korea, China, Nepal, and Senegal have used such methods to reduce fuelwood shortages, while sustaining biodiversity through reforestation and efforts to reduce topsoil erosion. Indeed, the mountainous country of South Korea is a global model for its successful reforestation following severe deforestation during the war between North and South Korea, which ended in 1953. Today, forests cover almost two-thirds of the country, and tree plantations near villages supply fuelwood on a sustainable basis.

Solutions

Sustaining Tropical Forests

Prevention

Protect the most diverse and endangered areas

Educate settlers about sustainable agriculture and forestry

Subsidize only sustainable forest use

Protect forests through debt-for-nature swaps and conservation concessions

Certify sustainably grown timber

Reduce poverty and slow population growth

Restoration

Encourage regrowth through secondary succession

Rehabilitate degraded areas

Concentrate farming and ranching in already-cleared areas

Sustainably grown timber

Figure 10-19 These are some effective ways to protect tropical forests and to use them more sustainably (Concept 10-2). ***Questions:*** Which three of these solutions do you think are the best ones? Why?

Top: STILLFX/Shutterstock.com. Center: Manfred Mielke/USDA Forest Service Bugwood.org.

There Are Several Ways to Reduce Tropical Deforestation

Analysts have suggested various ways to protect tropical forests and use them more sustainably (Figure 10-19).

At the international level, *debt-for-nature swaps* can make it financially attractive for countries to protect their tropical forests. In such swaps, participating countries act as custodians of protected forest reserves in return for foreign aid or debt relief. In a similar strategy, called *conservation concessions,* governments or private conservation organizations pay nations for agreeing to preserve their natural resources.

National governments can also take important steps to reduce deforestation. Between 2005 and 2011, Brazil cut its deforestation rate by 80% by cracking down on illegal logging and setting aside a conservation reserve in the Amazon Basin that is roughly equal to the size of France. In 2011, the UNEP reported that 75% of the protected areas created around the world between 2003 and 2011 are located in Brazil. Governments can also end subsidies that fund the construction of logging roads and instead subsidize more sustainable forestry (Figure 10-16) and the planting of trees in *reforestation programs.*

Consumers can reduce the demand for unsustainable and illegal logging in tropical forests by buying only wood and wood products that have been FSC-certified (see Science Focus 10.2). For building projects, using recy-

United Nations Environment Programme

Figure 10-20 In 1977, Wangari Maathai (1940–2011) founded the internationally acclaimed Green Belt Movement.

Figure 10-21 These women joined Africa's Green Belt Movement and are planting trees.

©Green Belt Movement

cled waste lumber is an option for some. People are also choosing wood substitutes such as recycled plastic building materials and bamboo.

Around the world, and especially in tropical forest areas, tree planting has become a high priority. The late Wangari Maathai (Figure 10-20), a Nobel Peace Prize winner, promoted and inspired tree planting in her native country of Kenya and throughout the world in what became the Green Belt Movement (Figure 10-21). Her efforts inspired the UNEP to implement a global effort to plant at least 1 billion trees a year beginning in 2006. By 2011, the year Maathai died, about 12.6 billion trees had been planted in 193 countries.

GOOD NEWS

10-3 How Should We Manage and Sustain Grasslands?

CONCEPT 10-3
We can sustain the productivity of grasslands by controlling the numbers and distribution of grazing livestock and by restoring degraded grasslands.

Some Rangelands Are Overgrazed

Grasslands provide many important ecosystem services, including soil formation, erosion control, chemical cycling, storage of atmospheric carbon dioxide in biomass, and maintenance of biodiversity.

After forests, grasslands are the ecosystems most widely used and altered by human activities. **Rangelands** are unfenced grasslands in temperate and tropical climates that supply *forage,* or vegetation for grazing (grass-eating) and browsing (shrub-eating) animals. Cattle, sheep, and goats graze on about 42% of the world's grassland. The 2005 UN Millennium Ecosystem Assessment—a 4-year study by 1,360 experts from 95 countries—estimated that this could increase to 70% by 2050. Livestock also graze in **pastures,** which are managed grasslands or fenced meadows often planted with domesticated grasses or other forage crops such as alfalfa and clover.

Blades of rangeland grass grow from the base, not at the tip as broadleaf plants do. Thus, as long as only the upper half of the blade is eaten and its lower half remains, rangeland grass is a renewable resource that can be grazed again and again. Moderate levels of grazing are healthy for grasslands, because removal of mature vegetation stimulates rapid regrowth and encourages greater plant diversity.

Overgrazing occurs when too many animals graze for too long, damaging the grasses and their roots, and exceeding the carrying capacity of a rangeland area (Figure 10-22, left). Overgrazing reduces grass cover, exposes the topsoil to erosion by water and wind, and compacts the soil, which diminishes its capacity to hold water. Overgrazing also encourages the invasion of once-productive rangeland by species such as sagebrush, mesquite, cactus, and cheatgrass, which cattle will not eat.

Figure 10-22 **Natural capital degradation:** To the left of the fence is overgrazed rangeland. The land to the right of the fence is lightly grazed.

USDA, Natural Resources Conservation Service

Limited data from FAO surveys in various countries indicate that overgrazing by livestock has caused a loss in productivity in as much as 20% of the world's rangeland.

We Can Manage Rangelands More Sustainably

The most widely used way to manage rangelands more sustainably is to control the number of grazing animals and the duration of their grazing in a given area so the carrying capacity of the area is not exceeded (**Concept 10-3**). One method for doing this is called *rotational grazing,* in which cattle are confined by portable fencing to one area for a short time (often only 1–2 days) and then moved to a new location.

Cattle tend to aggregate around natural water sources, especially along streams or rivers lined by thin strips of lush vegetation known as *riparian zones,* and around ponds created to provide water for livestock. Overgrazing by cattle can destroy the vegetation in such areas (Figure 10-23, left). Ranchers can protect overgrazed land from further grazing by moving their livestock around and by fencing off these damaged areas, which eventually leads to their natural restoration by ecological succession (Figure 10-23, right). Ranchers can also move cattle around by providing supplemental feed at selected sites and by strategically locating watering ponds and tanks and salt blocks.

A more expensive and less widely used method of rangeland management is to suppress the growth of unwanted invader plants by the use of herbicides, mechanical removal, or controlled burning. A cheaper way to discourage unwanted vegetation in some areas is through controlled, short-term trampling by large numbers of livestock such as sheep, goats, and cattle that destroy the invasive plants' root systems. The least expensive way to deal with degradation of rangelands is to prevent it, by using methods described above and in the following Case Study.

CASE STUDY

Grazing and Urban Development in the American West—Cows or Condos?

The landscape is changing in U.S. ranch country. Since 1980, millions of people have moved to parts of the southwestern United States, and a growing number of ranchers are selling their land to developers. Housing developments, condos, and small "ranchettes" are creeping out from the edges of many southwestern cities and towns. Most people moving to the southwestern states value the landscape for its scenery and recreational opportunities, but uncontrolled urban development can degrade these very qualities.

For decades, some environmental scientists and environmentalists have sought with limited success to reduce overgrazing on these lands and, in particular, to reduce or eliminate livestock grazing permits on public lands. Now, because of the population surge in the Southwest and the resulting development of land, ranchers, ecologists, and environmentalists are joining together to try to preserve a number of cattle ranches as the best hope for sustaining the key remaining grasslands and the habitats they provide for native species.

One preservation strategy involves land trust groups, which pay ranchers for *conservation easements*—deed restrictions that bar future owners from developing the land. These groups are also pressuring local governments to zone the land in order to prevent large-scale development in ecologically fragile rangeland areas.

The main goal of groups that are trying to preserve western lands is to identify areas that are sustainable for grazing, areas that are best for sustainable urban development, and areas that should be neither grazed nor developed. This is in keeping with the win-win **principle of sustainability** (see Figure 1-5, p. 9 or back cover).

Figure 10-23 **Natural capital restoration:** In the mid-1980s, cattle had degraded the vegetation and soil on this stream bank along the San Pedro River in the U.S. state of Arizona (left). Within 10 years, the area was restored through secondary ecological succession (right) after grazing and off-road vehicle use were banned (**Concept 10-3**).

Left: U.S. Bureau of Land Management. Right: U.S. Bureau of Land Management.

10-4 How Should We Manage and Sustain Parks and Nature Reserves?

CONCEPT 10-4
Sustaining biodiversity will require more effective protection of existing parks and nature reserves, as well as the protection of much more of the earth's remaining undisturbed land area.

National Parks Face Many Environmental Threats

According to the International Union for the Conservation of Nature (IUCN), there are now more than 6,600 major national parks (see chapter-opening photo) located in more than 120 countries. However, most of these parks are too small to sustain many large animal species. And many parks suffer from invasions by harmful nonnative species that compete with and reduce the populations of native species. Some national parks are so popular that large numbers of visitors are degrading the natural features that make them attractive (see the Case Study that follows).

Parks in less-developed countries have the greatest biodiversity of all the world's parks, but only about 1% of these parklands are protected. Local people in many of these countries enter the parks illegally in search of wood, game animals, and other natural products that they need for their daily survival. Loggers and miners operate illegally in many of these parks, as do wildlife poachers who kill animals to obtain and sell items such as rhino horns (see Figure 9-15, p. 205), elephant tusks, and furs. Park services in most of the less-developed countries have too little money and too few personnel to fight these invasions, either by force or through education.

CASE STUDY

Stresses on U.S. Public Parks

The U.S. National Park System, established in 1912, includes 59 major national parks, sometimes called the country's crown jewels (Figure 10-24), along with 339 monuments and historic sites. States, counties, and cities also operate public parks.

Popularity is one of the biggest problems for many parks. Between 1960 and 2011, the number of recreational visitors to U.S. national parks more than tripled, reaching about 279 million. In some U.S. parks and other public lands, noisy and polluting dirt bikes, dune buggies, jet skis, snowmobiles, and other off-road vehicles destroy or damage fragile vegetation, disturb wildlife, and degrade the aesthetic experience for many visitors. Some visitors expect parks to have grocery stores, laundries, bars, and other such conveniences.

A number of parks also suffer damage from the migration or deliberate introduction of nonnative species. European wild boars, imported into the state of North Carolina in 1912 for hunting, threaten vegetation in parts of the very popular Great Smoky Mountains National Park. Nonnative mountain goats in Washington State's Olympic National Park trample and destroy the root systems of native vegetation and accelerate soil erosion.

At the same time, native species—some of them threatened or endangered—are killed in, or illegally removed from, almost half of all U.S. national parks. This is what happened to the gray wolf in Yellowstone National Park until it was successfully reintroduced there after a 50-year absence (Science Focus 10.3).

Many U.S. national parks have become threatened islands of biodiversity surrounded by seas of commercial development. Nearby human activities that threaten wildlife and recreational values in many national parks include mining, logging, livestock grazing, coal-fired power plants, water diversion, and urban development. According to the National Park Service, air pollution, mostly from coal-fired power plants and dense vehicle traffic, degrades scenic views in many U.S. national parks more than 90% of the time.

The National Park Service estimated that in 2011, the national parks had at least an $8 billion backlog for long overdue maintenance and repairs to trails, buildings, and other park facilities. Some analysts say more of these funds could come from private concessionaires who provide campgrounds, restaurants, hotels, and other services for park visitors. They pay the government franchise fees averaging only about 6–7% of their gross receipts, and many large concessionaires with long-term contracts pay as little as 0.75%. Analysts say these percentages could reasonably be increased to around 20%.

Nature Reserves Occupy Only a Small Part of the Earth's Land

Most ecologists and conservation biologists believe the best way to preserve biodiversity is to create a worldwide network of protected areas. A map of the earth's remaining wildlands can be found in Supplement 6, Figure 16, p. S41, and Figure 17, pp. S42–S43, contains a map of the world's protected areas.

As of 2012, less than 13% of the earth's land area (not including Antarctica) was protected either strictly or partially in about 150,000 wildlife refuges, nature reserves, parks, and wilderness areas. This 13% figure is misleading because no more than 6% of the earth's land is strictly protected from potentially harmful human activities. In other words, *we have reserved 94% of the earth's land for human use.*

Conservation biologists call for full protection of at least 20% of the earth's land area in a global system of biodiversity reserves that would include multiple exam-

Figure 10-24 Grand Teton National Park in the U.S. state of Wyoming.

ples of all the earth's biomes (**Concept 10-4**). In 2012, the IUCN estimated that making this investment in protecting a vital part of our life-support system would cost roughly $23 billion a year—more than 4 times the current expenditure.

Most developers and resource extractors oppose protecting even 13% of the earth's remaining undisturbed ecosystems. They contend that these areas might contain valuable resources that would add to current economic growth. Ecologists and conservation biologists disagree. They view protected areas as islands of biodiversity and ecosystem services that help to sustain all life and economies indefinitely and that serve as centers of future evolution. In other words, they serve as an "ecological insurance policy" for us and other species. (See the online Guest Essay on this topic by Norman Myers.)

In establishing nature reserves, the size of the reserve is important, as is the design. Whenever possible, conservation biologists call for using the *buffer zone concept* to design and manage nature reserves. This means strictly protecting an inner core of a reserve, usually by establishing two buffer zones in which local people can extract resources sustainably without harming the inner core (see the Case Study that follows). Instead of shutting people out of the protected areas and likely creating enemies, this approach enlists local people as partners in protecting a reserve from unsustainable uses such as illegal logging and poaching. It is an application of the biodiversity and win-win **principles of sustainability**.

Another important concept in nature reserve design is the *habitat corridor*—a strip of protected land connecting two reserves that allows animals to migrate from one area to another as needed. By 2012, the United Nations had used these design concepts to create a global network of 621 *biosphere reserves* in 117 countries. However, most biosphere reserves fall short of these design ideals and receive too little funding for their protection and management.

CASE STUDY

Identifying and Protecting Biodiversity in Costa Rica

For several decades, Costa Rica (**Core Case Study**) has been using government and private research agencies to identify the plants and animals that make it one of the world's most biologically diverse countries. The government has consolidated the country's parks and reserves into several large conservation areas, or *megareserves,* distributed throughout the country (Figure 10-25). Each reserve contains a protected inner core surrounded by two buffer zones that local and indigenous people can use for sustainable logging, crop farming, cattle grazing, hunting, fishing, and ecotourism. They were designed with the goal of sustaining about 80% of the country's rich biodiversity.

In addition to its ecological benefits, Costa Rica's biodiversity conservation strategy has paid off financially. Today, the country's largest source of income is its $1-billion-a-year tourism industry, almost two-thirds of which involves ecotourism (Figure 10-26).

SCIENCE FOCUS 10.3

REINTRODUCING THE GRAY WOLF TO YELLOWSTONE NATIONAL PARK

Around 1800, at least 350,000 gray wolves (Figure 10-C) roamed over about three-quarters of America's lower 48 states, especially in the West. They survived mostly by preying on abundant bison, elk, caribou, and deer. However, between 1850 and 1900, most of them were shot, trapped, or poisoned by ranchers, hunters, and government employees.

Ecologists recognize the important role that this keystone predator species once played in parts of the West, especially in the six northern Rocky Mountain states of Montana, Idaho, Wyoming (where Yellowstone National Park is located), Utah, Oregon, and Washington. The wolves culled herds of bison, elk, moose, and mule deer, and kept down coyote populations. By leaving some of their kills partially uneaten, they provided meat for scavengers such as ravens, bald eagles, ermines, grizzly bears, and foxes.

When wolves declined, herds of plant-browsing elk, moose, and mule deer expanded and devastated vegetation such as willow and aspen trees growing near streams and rivers. This led to increased soil erosion and to declining populations of other wildlife species such as beaver, which eat willow and aspen. This in turn affected species that depended on wetlands created by the beavers.

When Congress passed the U.S. Endangered Species Act in 1973, only a few hundred gray wolves remained outside of Alaska, primarily in Minnesota and Michigan. In 1974, the gray wolf was listed as an endangered species in the lower 48 states.

In 1987, the U.S. Fish and Wildlife Service (USFWS) proposed reintroducing gray wolves into the Yellowstone National Park ecosystem in Wyoming to help restore and sustain biodiversity and to prevent further environmental degradation of the ecosystem. The proposal brought angry protests, some from area ranchers who feared the wolves would leave the park and attack their cattle and sheep. Other objections came from hunters who feared the wolves would kill too many big-game animals, and from mining and logging companies that feared the government would halt their operations on wolf-populated federal lands.

In 1995 and 1996, federal wildlife officials caught gray wolves in Canada and relocated 31 of them in Yellowstone National Park. Scientists estimate that the long-term carrying capacity of the park is 110 to 150 gray wolves. By 2012, the park had about 97 gray wolves.

For more than a decade, wildlife ecologist Robert Crabtree and a number of other scientists have been studying the effects of reintroducing the gray wolf into Yellowstone National Park. They have put radio collars on most of the wolves to gather data and track their movements. They have also studied changes in vegetation and in the populations of various plant and animal species.

This research has suggested that the return of this keystone predator species has sent ecological ripples through the park's ecosystem. It has contributed to a decline in populations of elk, the wolves' primary food source, which had grown too large for the carrying capacity of much of the park. The leftovers of elk killed by wolves have been an important food source for scavengers such as bald eagles and ravens. Also, wary elk are gathering less near streams and rivers, which has helped to spur the regrowth of trees in these areas. This, in turn, has helped to stabilize and shade stream banks, lowering the water temperature and making better habitat for trout. Beavers seeking willow and aspen for food and for dam building materials have returned, and the dams they build establish additional wetlands and create more favorable habitat for aspens.

The wolves have also cut in half the population of coyotes—the top predators in the absence of wolves. This has reduced coyote attacks on cattle from surrounding ranches and has led to larger populations of small animals such as ground squirrels, mice, and gophers, which are hunted by coyotes, eagles, and hawks. Overall, this experiment in ecosystem restoration has helped to reestablish and sustain some of the biodiversity that the Yellowstone ecosystem once had.

Between 1974 and 2012, the wolf population in the six Northern Rocky Mountain states and the Great Lakes grew from around 100 to around 6,000 individuals. In 2013, the USFWS proposed removing the gray wolf from the U.S. Endangered Species list in the lower 48 states. The agency considers the return of the gray wolf "one of the world's great conservation successes." If this rule is implemented, protection of this species would be left up to state wildlife agencies. Conservation groups plan to fight such a ruling in court, arguing that it could lead to greatly increased killing of the gray wolf.

Figure 10-C After becoming almost extinct in much of the western United States, the *gray wolf* was listed and protected as an endangered species in 1974.

Tom Kitchin/Tom Stack & Associates

Critical Thinking

Do you favor removing federal protection of the gray wolf as an endangered species? Explain.

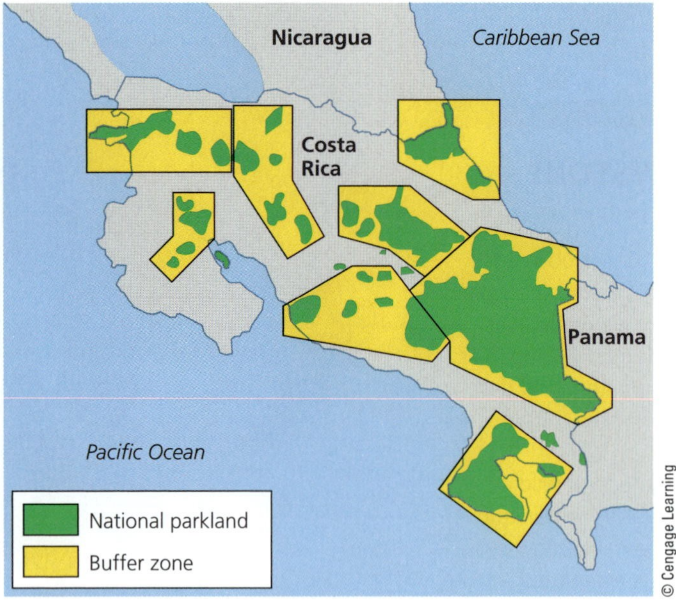

Figure 10-25 **Solutions:** Costa Rica has created several *mega-reserves.* Green areas are protected natural parklands and yellow areas are the surrounding buffer zones.

Figure 10-26 These tourists are exploring a forest canopy via a walkway in Costa Rica's Monteverde Cloud Forest Reserve.

CONSIDER THIS. . .

CAMPUS SUSTAINABILITY

Protecting Green Space at Emory University

Emory University is located in Atlanta, Georgia (USA), where, as in many urban areas, green space is being lost to development every year. At Emory, however, more than half of the campus area is designated as green space, off-limits to development. In 2003, the university put in place a "no net loss of forest canopy" policy. It requires that for every tree removed to make way for a building or other development, another tree must be planted. For these and other efforts, Emory has received many notices in publications such as *Sierra Magazine* as being in the forefront of campus sustainability.

Protecting Wilderness Is an Important Way to Preserve Biodiversity

One way to protect undeveloped lands from human exploitation is to set them aside as **wilderness**—land officially designated as an area where natural communities have not been seriously disturbed by humans and where harmful human activities are limited by law (**Concept 10-4**). Theodore Roosevelt, the first U.S. president to set aside protected areas, summarized what we should do with wilderness: "Leave it as it is. You cannot improve it."

Some critics oppose protecting large areas for their scenic and recreational value for a relatively small number of people. They believe this keeps some areas of the planet from being economically useful to large numbers of people who need it now. Most conservation biologists disagree. To them, the most important reasons for protecting wilderness and other areas from exploitation and degradation involve the long-term needs of all species—to *preserve biodiversity* as a vital part of the earth's natural capital and to *protect wilderness areas as centers for evolution* in response to mostly unpredictable changes in environmental conditions. In other words, protecting wilderness areas is equivalent to investing in a long-term biodiversity insurance policy for the earth's life and ecosystems.

In the United States, conservationists have been trying to save wild areas from development since 1900. Overall, they have fought a losing battle. Not until 1964 did Congress pass the Wilderness Act. It allowed the government to protect undeveloped tracts of public land from development as part of the National Wilderness Preservation System (Figure 10-27).

The area of protected wilderness in the United States grew by nearly 12-fold between 1964 and 2012. Even so, only about 5% of all U.S. land is protected as wilderness—more than 54% of it in Alaska. Only about 2.7% of the land area of the lower 48 states is protected, most of it in the West.

One problem is that only a small number of the 709 wilderness areas in the lower 48 states are large enough to sustain all of the species they contain. Also, the system includes only 81 of the country's 233 distinct ecosystems. Most wilderness areas in the lower 48 states are also threatened habitat islands in a sea of development.

According to a 2008 study by conservation biologists Oliver Pergans and Patricia Zaradic, wilderness protection is being eroded because since the 1980s, fewer people are taking part in outdoor activities such as hiking, camping, fishing, and hunting. Instead, people are more often opting for indoor activities involving video games, the Internet, social networking, and viewing rented movies. Some scientists and other analysts warn that, as activities in the natural world become less popular, there could be less citizen pressure for protecting wilderness and for setting aside nature reserves.

Figure 10-27 The John Muir Wilderness in the U.S. state of California was among the first areas to be designated as wilderness.

©Photofurl.com

10-5 What Is the Ecosystem Approach to Sustaining Biodiversity and Ecosystem Services?

CONCEPT 10-5

We can help to sustain terrestrial biodiversity by identifying and protecting severely threatened areas (biodiversity hotspots), sustaining ecosystem services, restoring damaged ecosystems (using restoration ecology), and sharing with other species much of the land we dominate (using reconciliation ecology).

The Ecosystems Approach: A Five-Point Strategy

Most biologists and wildlife conservationists believe that the best way to keep from hastening the extinction of wild species through human activities is to protect threatened habitats and ecosystem services. This *ecosystems approach* would generally employ the following five-point plan:

1. Map the world's terrestrial ecosystems and create an inventory of the species contained in each of them and the ecosystem services these ecosystems provide.
2. Identify terrestrial ecosystems that are resilient and that can recover if not overwhelmed by harmful human activities, along with ecosystems that are fragile and that need protection.
3. Locate and protect the most endangered terrestrial ecosystems and species, with emphasis on protecting plant biodiversity and ecosystem services.
4. Seek to restore as many degraded ecosystems as possible.
5. Make development *biodiversity-friendly* by providing significant financial incentives (such as tax breaks and write-offs) and technical help to private landowners who agree to help protect endangered ecosystems.

These steps have been applied in varying combinations in several locations around the world, as we discuss further throughout the rest of this chapter.

Protecting Global Biodiversity Hotspots Is an Urgent Priority

To protect as much of the earth's remaining biodiversity as possible, some biodiversity scientists urge the adoption of an *emergency action* strategy to identify and quickly protect **biodiversity hotspots**—areas especially rich in plant species that are found nowhere else and are in great danger of extinction (Concept 10-5). These areas have suffered serious ecological disruption, mostly because of rapid human population growth and the resulting pressure on natural resources and ecosystem services.

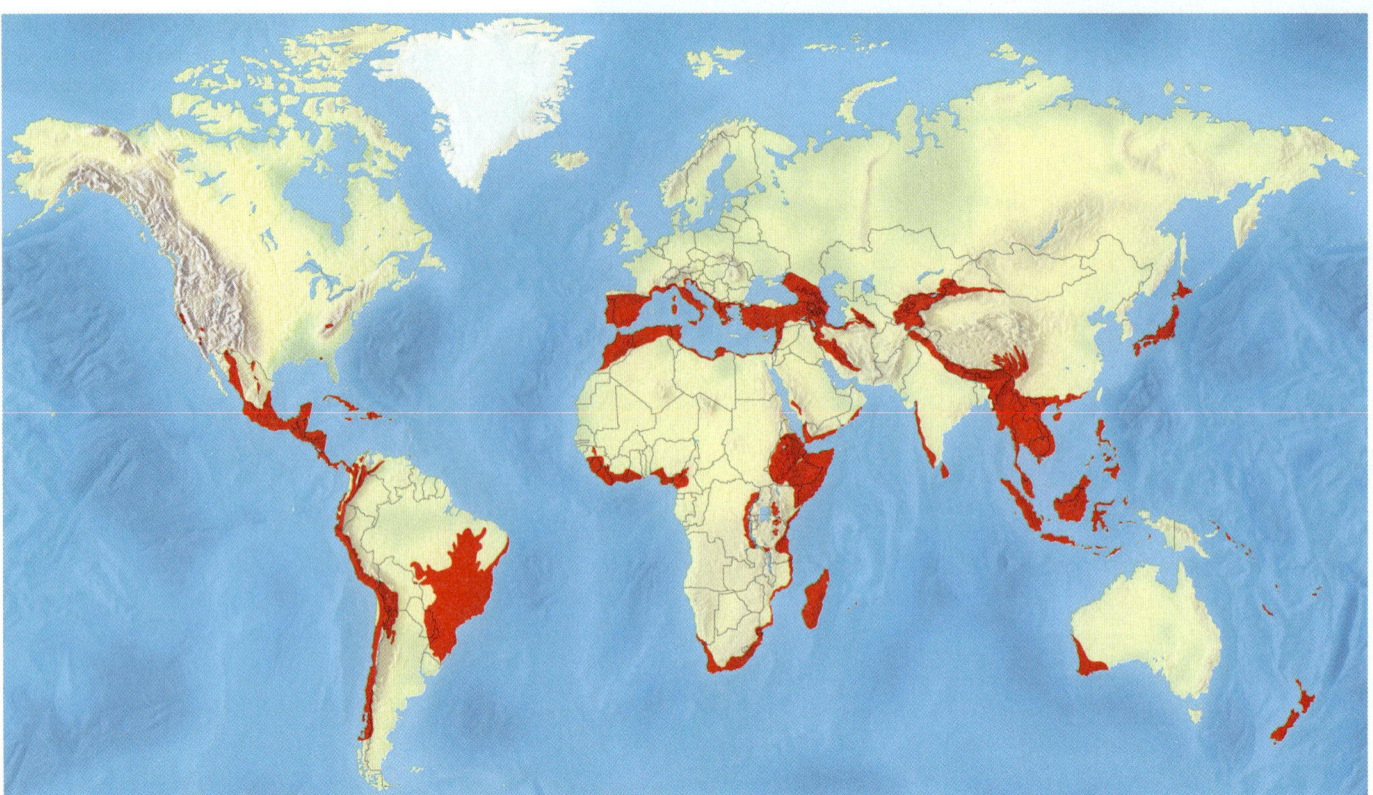

Animated Figure 10-28 **Endangered natural capital:** Biologists have identified these 34 biodiversity hotspots. Compare this map with the global map of the human ecological footprint, shown in Figure 1-12, p. 13. ***Question:*** Why do you think so many hotspots are located near coastal areas?

(Compiled by the authors using data from Center for Applied Biodiversity Science at Conservation International.)

Environmental scientist Norman Myers first proposed this idea in 1988 (see his online Guest Essay on this topic). Myers and his colleagues at Conservation International relied primarily on diversity of plant species as a way to identify biodiversity hotspots because data on plant diversity were the most readily available. They also viewed plant diversity as an indicator of animal diversity.

Figure 10-28 shows 34 terrestrial biodiversity hotspots identified by biologists. (For a map of hotspots in the United States, see Figure 9, p. S34, in Supplement 6.) Identified biodiversity hotspots cover only a little more than 2% of the earth's land surface, but they contain an estimated 50% of the world's flowering plant species and 42% of all terrestrial vertebrates (mammals, birds, reptiles, and amphibians), according to Conservation International. Yet, on average, only about 5% of the hotspots is truly protected with government funding and law enforcement.

Biodiversity hotspots are also home to a large majority of the world's endangered or critically endangered species, as well as to 1.2 billion people—one-fifth of the world's population. Says Norman Myers, "I can think of no other biodiversity initiative that could achieve so much at a comparatively small cost, as the hotspots strategy." In 2012, the IUCN began publishing its Red List of Ecosystems that are vulnerable, endangered, or critically endangered as a companion to its Red List of Threatened Species.

CASE STUDY

Madagascar: An Endangered Center of Biodiversity

Madagascar, the world's fourth largest island, lies in the Indian Ocean off the east coast of Africa. Most of its numerous species have evolved in near isolation from mainland Africa and all other land areas for at least 40 million years. As a result, roughly 90% of the more than 200,000 plant and animal species (Figure 10-29) found in this Texas-size biodiversity hotspot are found nowhere else on the earth.

Many of Madagascar's plant and animal species are among the world's most endangered, primarily because of habitat loss. People have cut down or burned more than 90% of Madagascar's original forests to get firewood and lumber and to make way for small farms, large terraced rice plantations, or cattle grazing. Only about 17% of the island's original vegetation remains, and as a result, Madagascar is one of the world's most eroded countries. Huge quantities of its precious topsoil have run off its hills, flowing as sediment in its rivers and emptying into its coastal waters (Figure 8-7, p. 172). All of this explains why Madagascar is one of the world's most threatened biodiversity hotspots.

Since 1984, the government, conservation organizations, and scientists (see Individuals Matter 10.1) world-

Figure 10-29 Madagascar is the only home for six of the world's eight baobab tree species (left), old-growth trees that are disappearing. This tree survives the desertlike conditions on part of the island by storing water in its large bottle-shaped trunk. The island is also the only home of more than 70 species of lemurs, including the threatened Verreaux's sifaka, or dancing lemur (right).

Left: David Thyberg/Shutterstock.com. Right: © Richlindie | Dreamstime.com.

wide have mounted efforts to slow the country's rapid loss of biodiversity. However, such efforts are hampered by Madagascar's rapid population growth. Between 1994 and 2012, its population grew from 12 million to 22 million and is projected to more than double, growing to about 54 million by 2050. The country is also very poor with 90% of its population struggling to survive on the equivalent of less than $2.25 per day. This puts pressure on its dwindling forest resources.

Despite the efforts to preserve Madagascar's biodiversity, less than 3% of its land area is officially protected. To reduce the rapid losses of biodiversity, the country will need to slow its population growth drastically and teach many of its people how to make a living from reforestation, ecotourism, and more sustainable uses of its forests, wildlife, and soil resources.

Protecting Ecosystem Services Is Also an Urgent Priority

Another way to help sustain the earth's biodiversity is to identify and protect areas where vital ecosystem services (see the orange boxed labels in Figure 1-3, p. 7) are being impaired enough to reduce biodiversity and harm local residents.

This approach has received more attention since the release of the 2005 UN Millennium Ecosystem Assessment. It identified key ecosystem services that provide numerous ecological and economic benefits, including those provided by forests (Figure 10-4). The study pointed out that humans are degrading or overusing about 60% of the ecosystem services provided by various ecosystems around the world, and it outlined ways to help sustain these vital services.

The ecosystem services approach recognizes that most of the world's ecosystems are already dominated or influenced by human activities and that such pressures are increasing as the human population, urbanization, resource use, and the human ecological footprint all expand (see Figures 1-12, p. 13, and 1-13, p. 14). Proponents of this approach recognize that setting aside and protecting reserves and wilderness areas, especially highly endangered biodiversity hotspots (Figure 10-28) and ecosystems, are vital. However, they contend that such efforts by themselves will not significantly slow the steady erosion of the earth's biodiversity and ecosystem services, nor will they help to reduce the poverty that plays a major role in ecosystem degradation.

Proponents of this strategy would also identify highly stressed *liferaft ecosystems*. These would be areas where poverty levels are high and where a large part of the economy depends on various ecosystem services that are being degraded severely enough to threaten the well-being of people and other forms of life. In such areas, residents, public officials, and conservation scientists would work together to develop strategies to help protect human communities along with natural biodiversity and the ecosystem services that support all life and economies. Thus, instead of pitting nature against people, this approach applies the win-win **principle of sustainability** (see Figure 1-5, p. 9 or back cover).

SUSTAINABILITY

We Can Rehabilitate and Partially Restore Ecosystems That We Have Damaged

Almost every natural place on the earth has been affected or degraded to some degree by human activities. We can at least partially reverse much of this harm through **ecological restoration**: the process of repairing damage caused by humans to the biodiversity and ecosystem services pro-

individuals matter 10.1

Luke Dollar: A National Geographic Emerging Explorer Working to Save Biodiversity in Madagascar

Courtesy of Luke Dollar

Luke Dollar, with a PhD in ecology from Duke University, is Associate Professor of Biology at Pfeiffer University. He is also a research associate with the Duke University Primate Center and founder of the Carnivore Conservation and Research Trust.

For 18 years, this boots-on-the-ground ecologist has been conducting research in Madagascar to help preserve its threatened biodiversity. He is an expert on the fosa (also spelled "fossa") shown here in his arms and in the background photo. This top predator and keystone species is found nowhere else in the world. It is threatened by the loss of about 97% of its habitat, mostly due to deforestation.

Each year, Dollar conducts Earthwatch Institute expeditions, during which students and citizens participate in his research and learn about Madagascar's amazing but endangered biodiversity. He also works to help locals find alternatives to traditional slash-and-burn agriculture in order to help reduce the pressure on Madagascar's remaining forests and wildlife. Dollar says, "I wake up every morning knowing I am one of the luckiest guys on Earth because I am doing exactly what I want to do and it's going to make a difference."

Background photo: Hotshotsworldwide/Dreamstime.com

vided by ecosystems. Examples include replanting forests (Science Focus 10.4), reintroducing native species (Science Focus 10.3), removing invasive species, freeing river flows by removing dams, and restoring grasslands, coral reefs, wetlands, and stream banks (Figure 10-23, right).

Ecological restoration takes time, which is not a problem for the earth, but it could be for our species. During the past 12,000 years and especially during the past 100 years, we have dramatically altered the land and life that took 3.5 billion years to develop on much of the planet. Through secondary succession (Figure 5-12, p. 110) many disturbed areas of land can return to forest or to other natural ecosystems, but this can take hundreds of years. Some scientists are concerned that much of the damage we have done will not be mended by nature in time for us to be able to depend on the damaged ecosystems for maintaining our life-support system. When we degrade natural capital (see Figure 1-3, p. 7), including the earth's biodiversity and ecosystem services that are crucial parts of our life-support system, in the long run, we threaten ourselves.

SCIENCE FOCUS 10.4

ECOLOGICAL RESTORATION OF A TROPICAL DRY FOREST IN COSTA RICA

Costa Rica (**Core Case Study**) is the site of one of the world's largest ecological restoration projects. In the lowlands of its Guanacaste National Park, a small, tropical dry forest was burned, degraded, and fragmented for large-scale conversion of the area to cattle ranches and farms. Now it is being restored and reconnected to a rain forest on nearby mountain slopes. The goal is to eliminate damaging nonnative grasses and reestablish a tropical dry-forest ecosystem during the next 100–300 years.

Daniel Janzen (Figure 10-D), professor of conservation biology at the University of Pennsylvania and a leader in the field of restoration ecology, helped to galvanize international support for this restoration project. He used his own MacArthur Foundation grant money to purchase this Costa Rican land for designation as a national park. He also raised more than $10 million for restoring the park.

Janzen recognizes that ecological restoration and protection of the park will fail unless the people in the surrounding area believe they will benefit from such efforts. His vision is to see that the nearly 40,000 people who live near the park play an essential role in the restoration of the degraded forest, a concept he calls *biocultural restoration.*

By actively participating in the project, local residents reap educational, economic, and environmental benefits. Local farmers are paid to remove nonnative species, to sow large areas with tree seeds, and to plant tree seedlings started in Janzen's lab. Local grade school, high school, and university students and citizens' groups study the park's ecology during field trips. The park's location near the Pan American Highway makes it an ideal area for ecotourism, which stimulates the local economy. GOOD NEWS

The project also serves as a training ground in tropical forest restoration for scientists from all over the world. Research scientists working on the project give guest classroom lectures and lead field trips.

In a few decades, today's Costa Rican children will be running the park and the local political system. If they understand the ecological importance of their local environment, they will be more likely to protect and sustain its biological resources. Janzen believes that education, awareness, and involvement—not guards and fences—are the best ways to restore degraded ecosystems and to protect largely intact ecosystems from unsustainable use. This is an application of the biodiversity and win-win **principles of sustainability**. SUSTAINABILITY

Figure 10-D Daniel Janzen, Professor of Conservation Biology at the University of Pennsylvania.

Critical Thinking

Would such an ecological restoration project be possible in the area where you live? Explain.

By studying how natural ecosystems recover, scientists are learning how to speed up ecological succession processes using a variety of approaches, including the following four:

1. *Restoration:* returning a degraded habitat or ecosystem to a condition as similar as possible to its natural state. The problem is that in some cases, we do not know what the original natural state of an area was, and restoring it to its estimated natural state often is no longer feasible because environmental conditions have changed.
2. *Rehabilitation:* turning a degraded ecosystem into a functional or useful ecosystem without trying to restore it to its original condition. Examples include removing pollutants from abandoned mining or industrial sites and replanting trees to reduce soil erosion in clear-cut forests.
3. *Replacement:* replacing a degraded ecosystem with another type of ecosystem. For example, a degraded forest could be replaced by a productive pasture or tree plantation.
4. *Creating artificial ecosystems:* for example, artificial wetlands have been created in some areas to help reduce flooding and to treat sewage.

Researchers have suggested a science-based, four-step strategy for carrying out most forms of ecological restoration and rehabilitation.

- *First,* identify the causes of the degradation (such as pollution, farming, overgrazing, mining, or invasive species).
- *Second,* stop the abuse by eliminating or sharply reducing these factors. This would include removing toxic soil pollutants, improving depleted soil by adding nutrients and new topsoil, preventing fires, and controlling or eliminating disruptive nonnative species.
- *Third,* if necessary, reintroduce key species to help restore natural ecological processes, as was done with gray wolves in the Yellowstone ecosystem (Science Focus 10.3).

- *Fourth,* protect the area from further degradation and allow secondary ecological succession to occur (Figure 10-23, right).

Some analysts worry that ecological restoration could encourage continuing environmental destruction and degradation by suggesting that any ecological harm we do can be undone. Restoration ecologists disagree. They point out that so far, we have been able to protect only about 5% of the earth's land from the effects of human activities, so ecological restoration is badly needed for many of the world's ecosystems.

We Can Share Areas We Dominate with Other Species

Ecologist Michael L. Rosenzweig suggests that we develop a form of conservation biology called **reconciliation ecology**. This science focuses on inventing, establishing, and maintaining new habitats to conserve species diversity in places where people live, work, or play. In other words, the focus is on learning how to share with other species some of the spaces we dominate.

For example, people can learn how protecting local wildlife and ecosystems can provide economic resources for their communities by encouraging sustainable forms of ecotourism. GOOD NEWS In the Central American country of Belize, for instance, conservation biologist Robert Horwich has helped to establish a local sanctuary for the black howler monkey. He convinced local farmers to set aside strips of forest to serve as habitats and corridors through which these monkeys can travel. The reserve, run by a local women's cooperative, has attracted ecotourists and biologists. The community has built a black howler museum, and local residents receive income by housing and guiding ecotourists and biological researchers.

However, without proper controls popular ecotourism sites can be degraded when they are overrun with visitors. In addition, developers have erected hotels and other tourist facilities that have added to the degradation, often without adding many jobs for local residents.

In many areas of the world, people are learning how to protect vital insect pollinators, such as native butterflies and bees (see Chapter 9, Core Case Study, p. 190), which are vulnerable to insecticides and habitat loss. Neighborhoods and municipal governments are doing this by agreeing to reduce or eliminate the use of pesticides on their lawns, fields, golf courses, and parks. Neighbors also work together to plant gardens of flowering plants as a source of food for bees and other pollinating insect species. Some neighborhoods and farmers have built devices made of wood and plastic straws that serve as beehives.

People have also worked together to protect bluebirds within human-dominated habitats where most of the bluebirds' nesting trees have been cut down and bluebird populations have declined. Special boxes were designed to accommodate nesting bluebirds, and the North American Bluebird Society has encouraged Canadians and Americans to use these boxes on their properties and to keep house cats away from nesting bluebirds. Now bluebird numbers are rising again.

There are many other examples of individuals and groups working together on projects to restore grasslands, wetlands, streams, and other degraded areas. They involve applying the biodiversity and win-win **principles of sustainability** (see Figure 1-2, p. 6, and Figure 1-5, p. 9, or back cover).

Figure 10-30 lists some ways in which you can help to sustain the earth's terrestrial biodiversity. GOOD NEWS

What Can You Do?

Sustaining Terrestrial Biodiversity

- Plant trees and take care of them
- Recycle paper and buy recycled paper products
- Buy sustainably produced wood and wood products and wood substitutes such as recycled plastic furniture and decking
- Help restore a degraded forest or grassland
- Landscape your yard with a diversity of native plants

Figure 10-30 Individuals matter: These are some ways in which you can help to sustain terrestrial biodiversity. ***Questions:*** Which two of these actions do you think are the most important? Why? Which of these things do you already do?

Big Ideas

- The economic values of the important ecosystem services provided by the world's ecosystems are far greater than the value of raw materials obtained from those systems.
- We can manage forests, grasslands, and nature reserves more effectively by protecting more land and by preventing overuse and degradation of these areas and the renewable resources they contain.
- We can sustain terrestrial biodiversity and ecosystem services by protecting biodiversity hotspots and ecosystem services, restoring damaged ecosystems, and sharing with other species much of the land we dominate.

TYING IT ALL TOGETHER Sustaining Costa Rica's Biodiversity

Eduardo Rivero/Shutterstock.com

In this chapter, we looked at how humans are destroying or degrading terrestrial biodiversity in a variety of ecosystems. We also saw how we can reduce this destruction and degradation by using the earth's resources more sustainably. The **Core Case Study** introduced us to what Costa Rica is doing to protect and restore its precious biodiversity—primarily by protecting a larger percentage of its land from unsustainable development (see chapter-opening photo) than any other country. This strategy has also paid off financially by promoting the country's various parks and reserves as centers for ecotourism.

We also discussed the importance of preserving what remains of richly diverse and highly endangered ecosystems (biodiversity hotspots) and of sustaining the earth's ecosystem services. We examined the key strategy of restoring or rehabilitating some of the ecosystems we have degraded (restoration ecology). In addition, we explored ways in which people can share with other species some of the land they occupy in order to help sustain biodiversity (reconciliation ecology).

Preserving biodiversity involves applying the three scientific **principles of sustainability** (see Figure 1-2, p. 6 or back cover, top). First, it means respecting biodiversity and understanding the value of sustaining it. Then, in helping to sustain biodiversity by planting trees, for example, we also help to restore and preserve the flows of energy from the sun through food webs and the cycling of nutrients within ecosystems. Also, if we rely less on fossil fuels and more on direct solar energy and its indirect forms, such as wind and flowing water, we will generate less pollution and interfere less with chemical cycling and other forms of natural capital that sustain biodiversity and our own lives and societies.

We can also apply the three social science **principles of sustainability** (see Figure 1-5, p. 9 or back cover, bottom) to help preserve biodiversity. First, placing economic value on nature's ecosystem services tends to make the preservation of those services more important. Second, we know that people can successfully work together to find win-win solutions to problems of environmental degradation. And third, all of these efforts are driven by our ethical responsibility to sustain biodiversity for current and future generations.

Chapter Review

Core Case Study

1. Summarize the story of Costa Rica's efforts to preserve its rich biodiversity.

Section 10-1

2. What are the two key concepts for this section? Distinguish among **old-growth (primary) forests**, **second-growth forests**, and **tree plantations** (**tree farms** or **commercial forests**). What major ecological and economic benefits do forests provide? Describe the efforts of scientists and economists to put a price tag on the major ecosystem services provided by forests and other ecosystems.

3. Explain how building roads into previously inaccessible forests can harm the forests. Distinguish among selective cutting, clear-cutting, and strip cutting in the harvesting of trees. What are the major advantages and disadvantages of clear-cutting forests? What are two types of forest fires? What are some possible ecological benefits of occasional surface fires? How can the introduction of insects and disease-causing organisms harm forests? Explain some possible effects of climate change on forests.

4. What is **deforestation** and what parts of the world are experiencing the greatest forest losses? List some major harmful environmental effects of deforestation. Describe the encouraging news about reforestation in the United States. Explain how increased reliance on tree plantations can reduce overall forest biodiversity and degrade forest topsoil. How serious is tropical deforestation? Describe some major causes of tropical deforestation. Explain how widespread tropical deforestation can irreversibly convert a tropical forest to a tropical grassland.

Section 10-2

5. What is the key concept for this section? Describe four ways to manage forests more sustainably. What is certified sustainably grown timber? What are four ways to reduce the harm caused by forest fires to forests and to people? What is a prescribed fire? What are three ways to reduce the need to harvest trees? Describe the global fuelwood crisis. What are five ways to protect tropical forests and use them more sustainably? Describe the efforts by Wangari Maathai and the Green Belt Movement to restore forests.

Section 10-3

6. What is the key concept for this section? Distinguish between **rangelands** and **pastures**. What is **overgrazing** and what are its harmful environmental effects? What are three ways to reduce overgrazing and use rangelands more sustainably? Explain why there is a growing conflict between grazing and urban development in the American southwest.

Section 10-4

7. What is the key concept for this section? What are the major environmental threats to national parks in the world and in the United States? Why are many U.S. national parks considered to be threatened islands of biodiversity? Describe some of the ecological effects of reintroducing the gray wolf to Yellowstone National Park. What percentage of the world's land has been set aside and protected as nature reserves, and what percentage do conservation biologists believe should be protected?

8. How should nature reserves be designed and managed? What is the buffer zone concept? What is a biosphere reserve? Describe what Costa Rica has done to identify and protect much of its biodiversity. What is **wilderness** and why is it important? Summarize the controversy over protecting wilderness in the United States.

Section 10-5

9. What is the key concept for this section? Summarize the five-point strategy recommended by biologists for protecting ecosystems. What is a **biodiversity hotspot** and why is it important to protect such areas? About how much of the earth's land surface is occupied by hotspots and what percentages of the world's flowering plants and terrestrial vertebrates live in these areas? Explain the importance of protecting ecosystem services and list three ways to do this.

10. Define **ecological restoration**. What are four approaches to restoration? Summarize the science-based, four-point strategy for carrying out ecological restoration and rehabilitation. Describe the ecological restoration of the Guanacaste National Park forest in Costa Rica. Define and give three examples of **reconciliation ecology**. What are this chapter's *three big ideas*? Explain the relationship between preserving biodiversity as it is done in Costa Rica and the scientific and social science **principles of sustainability**.

Note: Key terms are in bold type.

Critical Thinking

1. Why do you think Costa Rica has set aside a larger percentage of its land for biodiversity conservation than the United States has? Should the United States reserve more of its land for this purpose? Explain.

2. If we fail to protect a much larger percentage of the world's remaining old-growth forests and tropical rain forests, what are three harmful effects that this failure is likely to have on any children and grandchildren you might have?

3. In the early 1990s, Miguel Sanchez, a subsistence farmer in Costa Rica, was offered $600,000 by a hotel developer for a piece of land that he and his family had been using sustainably for many years. The land, which contained an old-growth rain forest and a black sand beach, was surrounded by an area under rapid development. Sanchez refused the offer. Explain how Sanchez's decision was an application of one of the social science **principles of sustainability**(see Figure 1-5, p. 9 or back cover). What would you have done if you were in Sanchez's position? Explain your thinking.

4. There is controversy over whether Yellowstone National Park should be accessible by snowmobile during winter. Conservationists and backpackers who use cross-country skis and snowshoes for winter excursions in the park are opposed to this. They contend that snowmobiles are noisy and that they pollute the air, destroy vegetation, and disrupt some of the park's wildlife. Proponents say that snowmobiles should be allowed so that snowmobilers can enjoy the park during winter when cars are mostly banned. They point out that new snowmobiles are made to cut pollution and noise. As of 2013, this issue was yet to be resolved on a long-term basis. What is your view on this issue? Explain.

5. In 2009, environmental analyst Lester R. Brown estimated that reforesting the earth and restoring its degraded rangelands would cost about $15 billion a year. Suppose the United States, the world's most afflu-

ent country, agreed to put up half of this money, which would amount to about $23 per American citizen per year. Would you support doing this? Explain. If your answer was yes, what part or parts of the federal budget would you decrease to come up with these funds?

6. Should more-developed countries provide at least half of the money needed to help preserve remaining tropical forests in less-developed countries? Explain. Do you think that the long-term economic and ecological benefits of doing this would outweigh the short-term economic costs? Explain.

7. Are you in favor of establishing more wilderness areas in the United States, especially in the lower 48 states (or in the country where you live)? Explain. What might be some drawbacks of doing this?

8. You are a defense attorney arguing in court for preserving an old-growth forest that developers want to clear for cropland and other uses. Give your three strongest arguments for preserving this ecosystem. How would you counter the argument that preserving the forest would harm the local economy by causing a loss of jobs in the timber industry?

Doing Environmental Science

Pick an area near where you live or go to school that hosts plants and animals. It could be a yard, an abandoned lot, a park, a forest, or some part of your campus. Visit this area at least 3 times and make a survey of the plants and animals that you find there, including any trees, shrubs, groundcover plants, insects, reptiles, amphibians, birds, and mammals. Also, take a small sample of the topsoil and find out what organisms are living there. (Be careful to get permission from whomever owns or manages the land before doing any digging.) Using guidebooks and other resources to help identify different species, record your findings and categorize them into the general types of organisms listed above. Then think of, and record, five ecosystem services that you think some or all of these organisms provide.

Global Environment Watch Exercise

Go to the *Forests and Deforestation* portal and next to the Statistics heading click "View All." On this page, click on "Share of Tropical Deforestation, 2000–2005." Choose one of these countries and research the deforestation in this country further (tip: use the World Map feature). Write a report on your findings and include possible solutions for this problem. Solutions may include those legislated by governments, as well as those being tried by private individuals or companies.

Ecological Footprint Analysis

Use the table below to answer the questions that follow.

Country	Area of Tropical Rain Forest (square kilometers)	Area of Deforestation per Year (square kilometers)	Annual Rate of Tropical Forest Loss
A	1,800,000	50,000	
B	55,000	3,000	
C	22,000	6,000	
D	530,000	12,000	
E	80,000	700	

1. What is the annual rate of tropical rain forest loss, as a percentage of total forest area, in each of the five countries? Answer by filling in the blank column in the table.

2. What is the annual rate of tropical deforestation collectively in all of the countries represented in the table?

3. According to the table, and assuming the rates of deforestation remain constant, which country's tropical rain forest will be completely destroyed first?

4. Assuming the rate of deforestation in country C remains constant, how many years will it take for all of its tropical rain forests to be destroyed?

5. Assuming that a hectare (1.0 hectare = 0.01 square kilometer) of tropical rain forest absorbs 0.85 metric tons (1 metric ton = 2,200 pounds) of carbon dioxide per year, what would be the total annual growth in the carbon footprint (carbon emitted but not absorbed by vegetation because of deforestation) in metric tons of carbon dioxide per year for each of the five countries in the table?

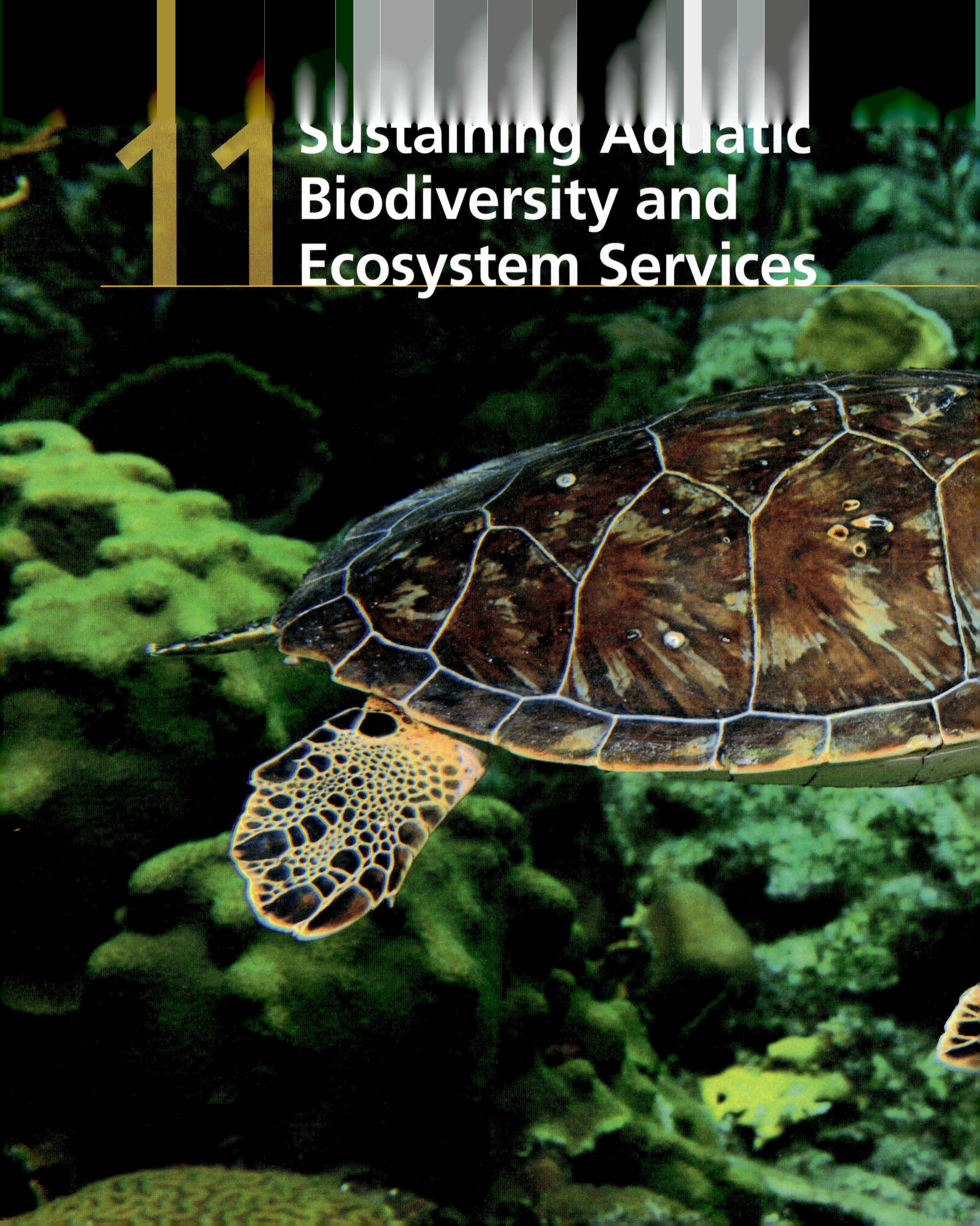

11 Sustaining Aquatic Biodiversity and Ecosystem Services

The coastal zone may be the single most important portion of our planet. The loss of its biodiversity may have repercussions far beyond our worst fears.

G. CARLETON RAY

Key Questions

11-1 What are the major threats to aquatic biodiversity and ecosystem services?

11-2 How can we protect and sustain marine biodiversity?

11-3 How should we manage and sustain marine fisheries?

11-4 How should we protect and sustain wetlands?

11-5 How should we protect and sustain freshwater lakes, rivers, and fisheries?

11-6 What should be our priorities for sustaining aquatic biodiversity?

Endangered green sea turtle swimming over a coral reef.

George Grall/National Geographic Creative

Sea turtles have been roaming the oceans for more than 100 million years—about 500 times longer than our species has been around. Today all seven species of sea turtles (Figure 11-1) are in danger of becoming extinct, mostly because of human activities taking place during the last 100 years.

Sea turtles spend most of their lives traveling throughout the world's oceans, but adult females almost always return to the beaches where they were born to lay their eggs. They come ashore at night and use their back flippers to dig nests on sand beaches and coastal dunes. Each female lays a clutch of around 100–110 eggs, then buries them and returns to the ocean. After the baby turtles hatch, they dig their way out of the nest, often at night, to scamper toward the water. During this dangerous trip, birds and other predators eat many of them. Only about one of every thousand hatchlings survives to adulthood.

The leatherback sea turtle (Figure 11-1) is one of the most endangered of the seven sea turtle species. Named for its leathery shell, it is the largest of all sea turtles. Adult leatherbacks are 1.8–2 meters (3–6 feet) long and weigh 250–700 kilograms (550–1,500 pounds). They feed mostly on jellyfish (Figure 8-3, p. 169) and help control jellyfish populations. They also swim great distances, migrating across the Atlantic and Pacific Oceans, as scientists have learned by using satellites to track their movements.

While leatherbacks survived the impact of the giant asteroid that probably wiped out the dinosaurs more than 60 million years ago, they and the other sea turtle species may not survive the growing human impact on their environment. A method of fishing called *trawler fishing* has destroyed many of the coral gardens that have served as their feeding grounds. The turtles are hunted for leather, and their eggs are taken for food. They often drown after becoming entangled in fishing nets and lines, as well as in lobster and crab traps.

Pollution of ocean water is another threat. Leatherbacks and some other sea turtles can mistake discarded plastic bags for jellyfish and choke to death trying to eat them. Beachgoers and motor vehicles sometimes crush their nests. Artificial lights can disorient newly hatched baby turtles, which try to find their way to the ocean by moving toward moonlight reflected from the ocean's surface. Going in the wrong direction increases their chances of ending up as food for predators. Add to this the threat of rising sea levels from projected climate change during this century. This will flood many nesting habitats and change ocean currents, which could disrupt the turtles' migration routes.

In this chapter, we will examine the effects of human activities on sea turtles, as well as on many other ocean and freshwater species. We will also explore ways to prevent or lessen these effects in order to help sustain aquatic life—a major component of the biodiversity that sustains our lives and economies.

Figure 11-1 There are seven species of sea turtles, all of them endangered. All but the flatback are found in U.S. waters.

11-1 What Are the Major Threats to Aquatic Biodiversity and Ecosystem Services?

CONCEPT 11-1
Aquatic species and the ecosystem and economic services they provide are threatened by habitat loss, invasive species, pollution, climate change, and overexploitation, all made worse by the growth of the human population and resource use.

We Have Much to Learn about Aquatic Biodiversity

Although we live on a water planet, we have explored only about 5% of the earth's interconnected oceans (see Figure 8-2, p. 169) and know relatively little about marine biodiversity and its many functions. Marine biologist Chris Bowler has estimated that only 1% of the life-forms in the sea have been properly identified and studied. We also have limited knowledge about freshwater biodiversity.

Scientists have observed three general patterns related to marine biodiversity. *First,* the greatest marine biodiversity occurs around coral reefs, in estuaries, and on the deep-ocean floor. *Second,* biodiversity is greater near the coasts than in the open sea because of the larger variety of producers and habitats in coastal areas. *Third,* biodiversity is generally greater in the bottom region of the ocean than in the surface region because of the larger variety of habitats and food sources on the ocean bottom.

The deepest part of ocean, where sunlight does not penetrate (Figure 8-6, p. 171), is the planet's least explored environment. But this is changing. More than 2,400 scientists from 80 countries are working on a 10-year project to catalog the species in this region of the ocean. So far, they have used remotely operated deep-sea vehicles to identify more than 17,000 species living in this ocean zone and they are adding a few thousand new species every year.

Marine ecosystems provide important ecosystem and economic services (see Figure 8-5, p. 170). For example, a 2010 study, *The Economics of Ecosystems and Biodiversity,* done by an international team of economists and scientists, estimated that an area of coral reef roughly equal to the size of a city block provides economic and ecosystem services worth more than $1.2 million a year. Thus, scientific investigation of marine aquatic systems is an exciting research frontier that could lead to immense ecological and economic benefits. Freshwater systems, which occupy only 1% of the earth's surface, also provide important ecosystem and economic services (see Figure 8-14, p. 178).

Human Activities Are Destroying and Degrading Aquatic Habitat

Human domination of nature has now reached the world's vast oceans with an accelerating impact. As with terrestrial biodiversity, the greatest threats to aquatic biodiversity and ecosystem services (**Concept 11-1**) can be remembered with the aid of the acronym HIPPCO, with H standing for *habitat loss and degradation.* Some 90% of the fish living in the ocean spawn on coral reefs (see Chapter 8 opening photo, p. 166), in coastal wetlands and marshes (see Figure 8-8, p. 172), in mangrove forests (see Figure 8-9, p. 172), or in rivers that empty into the sea.

All of these ecosystems are under intense pressure from human activities (see Figure 8-12, p. 176). For example, United Nations Environment Programme (UNEP) scientists reported in 2010 that a fifth of the world's mangroves have been lost since 1980. They continue to be destroyed for firewood, coastal construction, and shrimp farming. In addition, a 2009 study revealed that 58% of the world's coastal sea-grass beds (see Figure 8-10, p. 173) have been degraded or destroyed, mostly by dredging and coastal development.

Sea-bottom habitats are faring no better, being threatened by dredging operations and trawler fishing boats. Like giant submerged bulldozers, trawlers drag huge nets weighted down with chains and steel plates over the ocean floor to harvest a few species of bottom fish and shellfish (Figure 11-2). Each year, thousands of trawlers scrape and disturb an area of ocean floor many times larger than the annual global total area of forests that are clear-cut. These ocean-floor communities could take decades or centuries to recover. According to marine scientist Elliot Norse, "Bottom trawling is probably the largest human-caused disturbance to the biosphere."

Coral reefs serve as habitat for many hundreds of marine species (see Chapter 8, Core Case Study, p. 168). Coastal development, pollution, ocean warming, and **ocean acidification**—the rising levels of acid in ocean waters due to their absorption of carbon dioxide from the atmosphere—threaten them. The Coral Reef Alliance reported in 2010 that more than a quarter of the world's shallow coral reefs had been destroyed or severely damaged. A 2011 study by the World Resources Institute estimated that currently 75% of the world's shallow reefs are threatened by climate change, overfishing, pollution, and ocean acidification, and that by 2050, some 90% will be threatened. In 2013, coral reef ecologist Daniel Barshis and his colleagues found genes in some coral species that might help them adapt to higher water temperatures projected to result from climate change.

Photos: © Peter J. Auster/National Undersea Research Center

Figure 11-2 **Natural capital degradation:** These photos show an area of ocean bottom before (left) and after (right) a trawler net scraped it like a gigantic bulldozer. ***Question:*** What land activities are comparable to this?

Habitat disruption is also a problem in freshwater aquatic zones. The main causes are the damming of rivers and excessive withdrawal of river water for irrigation and urban water supplies. These activities destroy aquatic habitats, degrade water flows, and disrupt freshwater biodiversity.

Invasive Species Are Degrading Aquatic Biodiversity

Another problem that threatens aquatic biodiversity is the deliberate or accidental introduction of hundreds of harmful invasive species (see Figure 9-9, p. 199)—the I in HIPPCO—into coastal waters, wetlands, and lakes throughout the world (Concept 11-1). These *bioinvaders* are displacing or causing the extinction of native species and disrupting ecosystem services and human economies. According to the U.S. Fish and Wildlife Service, bioinvaders are blamed for about two-thirds of all fish extinctions in the United States since 1900 and have caused huge economic losses.

Many of these invaders arrive in the ballast water that is stored in tanks in large cargo ships to keep them stable. The ships take in ballast water from one harbor, along with whatever microorganisms and tiny fish species it contains, and dump it into another—an environmentally and economically harmful effect of globalized trade. Even when ballast water is flushed from an oceangoing ship's tank before it enters a harbor—a measure now required in many ports—the ship can still bring invaders that are stuck to its hull.

One invader that worries scientists and fishers on the east coast of North America is a species of *lionfish* native to the western Pacific Ocean (Figure 11-3). Scientists believe it escaped from outdoor aquariums in Miami, Florida, that were damaged by Hurricane Andrew in 1992. Lionfish populations have exploded at the highest rate of any species ever recorded by scientists in this part of the world. One scientist described the lionfish as "an almost perfectly designed invasive species." It reaches sexual maturity rapidly, has large numbers of offspring, and is protected by venomous spines. It competes with popular reef fish species such as grouper and snapper, taking their food and eating their young. One ray of hope for controlling this population is the fact that the lionfish tastes good. Scientists are hoping to see a growing market for lionfish as seafood.

CONSIDER THIS. . .

CONNECTIONS Lionfish and Coral Reef Destruction

Researchers have found that lionfish eat at least 50 species of prey fish, including parrotfish, that normally consume enough algae around coral reefs to keep the algae from overgrowing and killing the corals. Scientists warn that, where lionfish are now the dominant species such as in the Bahamas, unchecked algae could overwhelm and destroy some reefs.

In addition to threatening native species, invasive species can disrupt and degrade whole ecosystems and their ecosystem services. This is the focus of study for a growing number of researchers (see Science Focus 11.1).

Population Growth and Pollution Can Reduce Aquatic Biodiversity

In 2010, according to UNEP, about 80% of the world's people were living along or near seacoasts, mostly in large coastal cities. This coastal population growth—the first P in HIPPCO—has added to the already intense pressure on the world's coastal zones, resulting in more pollution, habitat destruction, and other problems (Concept 11-1).

The UNEP estimates that about 80% of all ocean pollution—the second P in HIPPCO—comes from land-based coastal activities. Growing inputs of nitrogen, mostly from nitrate fertilizers, into marine and freshwater systems is causing algal blooms (see Figure 8-17, p. 180), lower levels of dissolved oxygen, fish die-offs, and degradation of ecosystem services in many parts of the world.

SCIENCE FOCUS 11.1

HOW INVASIVE CARP HAVE MUDDIED SOME WATERS

Lake Wingra lies within the city of Madison, Wisconsin, surrounded mostly by a forest preserve. The lake contains a number of invasive plant and fish species, including purple loosestrife (see Figure 9-9, p. 199) and common carp. The carp were introduced in the late 1800s and since then have made up as much as half of the fish biomass in the lake. They devour algae called *chara,* which would normally cover the lake bottom and stabilize its sediments. Consequently, fish movements and winds stir these sediments, which accounts for much of the water's excessive *turbidity,* or cloudiness.

Knowing this, Dr. Richard Lathrup, a limnologist (lake scientist) who worked with Wisconsin's Department of Natural Resources, hypothesized that removing the carp would help to restore the natural ecosystem of Lake Wingra. Lathrop speculated that if the carp were removed, the bottom sediments would settle and become stabilized, allowing the water to clear. Clearer water would in turn allow native plants to receive more sunlight and become reestablished on the lake bottom, replacing purple loosestrife and other invasive plants that now dominate its shallow shoreline waters.

Lathrop and his colleagues installed a thick, heavy vinyl curtain around a 1-hectare (2.5-acre), square-shaped perimeter that extended out from the shore. This barrier hung from buoys on the surface to the bottom of the lake, isolating the volume of water within it. The researchers then removed all of the carp from this study area and began observing results. Within 1 month, the waters within the barrier were noticeably clearer, and within a year, the difference in clarity was dramatic (Figure 11-A) and native plants once again grew in the shallow shoreline waters.

Lathrop notes that removing and keeping carp out of Lake Wingra would be a daunting task, perhaps impossible, but his controlled scientific experiment clearly shows the effects that an invasive species can have on an aquatic ecosystem. And it reminds us that preventing the introduction of invasive species in the first place is the best and least expensive way to avoid such effects.

©Mike Kakuska

Figure 11-A *Lake Wingra* in Madison, Wisconsin (USA) became clouded with sediment partly because of the introduction of invasive species such as the common carp. Removal of carp in the experimental area shown here resulted in a dramatic improvement in the clarity of the water.

Critical Thinking

What are two other results of this controlled experiment that you might expect? (*Hint:* Think food webs.)

©Cigdem Cooper/Shutterstock.com

Figure 11-3 The *common lionfish* has invaded the eastern coastal waters of North America, where it has few, if any, predators.

Toxic pollutants from industrial and urban areas can kill some forms of aquatic life by poisoning them. Ocean pollution from plastic items dumped from ships and garbage barges, and left as litter on beaches, kills up to 1 million seabirds and 100,000 mammals (Figure 11-4) and sea turtles each year (**Core Case Study**).

Climate Change Is a Growing Threat

Projected climate change—the C in HIPPCO—threatens aquatic biodiversity (**Concept 11-1**) and ecosystem services, partly by contributing to rising sea levels. During the past 120 years, average sea levels have risen by about 20 centimeters (8 inches). Computer models developed by climate scientists estimate they will rise another 18–60 centimeters (0.6–2 feet) and perhaps as high as

SCIENCE FOCUS 11.2

OCEAN ACIDIFICATION: THE OTHER CO_2 PROBLEM

Our greatly increased burning of fossil fuels, especially during the last 60 years, has added carbon dioxide (CO_2) to the atmosphere faster than it can be removed by the carbon cycle (see Figure 3-17, p. 66). Research indicates that this has played a key role in the observed increase in the atmosphere's average temperature, especially since 1975. Most climate scientists recognize this temperature increase as a serious environmental problem that will very likely change the earth's climate, especially during the latter half of this century.

However, there is another problem related to CO_2 emissions—that of *ocean acidification,* a term that can be somewhat misleading. Ocean water is *basic,* not *acidic.* This means that it has more hydroxide ions (OH^-) than hydrogen ions (H^+) (see p. S14 and Figure 4, p. S15, in Supplement 4). When atmospheric CO_2 combines with ocean water, it forms carbonic acid (H_2CO_3), a weak acid also found in carbonated drinks. The resulting higher levels of hydrogen ions (H^+) in the water makes the water less basic. Thus, *increasing acidity* of the oceans could also be described as *decreasing basicity.*

The problem is that, as ocean water becomes less basic, the level of carbonate ions (CO_3^{2-}) in the water drops because they react with hydrogen ions (H^+) to form bicarbonate ions (HCO_3^-). Carbonate ions are vital to many aquatic species, including phytoplankton, corals, sea snails, crabs, and oysters, that use them to produce calcium carbonate ($CaCO_3$), the main component of their shells and bones. As carbonate ion levels drop, these species are hindered in forming their shells and skeletons, as revealed in a 2012 study by Lloyd Peck and other ocean scientists. As a result, shell-building species and coral reefs will grow more slowly. And when the hydrogen ion concentration of seawater gets high enough, their calcium carbonate components will begin to dissolve. Colder waters with a greater capacity for removing CO_2 from the atmosphere will be affected first (Figure 11-B). Some scientists refer to the resulting thinner and weaker shells and bones of sea creatures as "osteoporosis of the sea."

This problem is already affecting some areas of our economy. On the coast of the U.S. state of Oregon, for example, oyster larvae have died off in large numbers in recent years, making it hard for some oyster growers to stay in business. Preliminary research indicates that these larvae deaths are occurring because carbonate ion levels are too low for oyster larvae to build shells during the first days of their lives.

Such harmful effects are now under intensive study. Previous studies have suggested that higher levels of hydrogen ions (H^+) would be diluted as they circulated evenly throughout the ocean. But to make matters worse, researchers have discovered that higher levels of H^+ and lower levels

---

1.5 meters (5 feet) by 2100 from a combination of thermal expansion as ocean water warms and the partial melting of land-based ice in glaciers and ice sheets.

Such a rise in sea level would destroy more shallow coral reefs, swamp some low-lying islands, drown many highly productive coastal wetlands, and put many coastal areas such as a large part of the U.S. Gulf and East Coasts underwater (Figure 8-20, p. 182). In addition, some Pacific island nations could lose more than half of their protective coastal mangrove forests by 2100, according to a 2006 study by UNEP. Coral reefs and a number of ocean organisms are also threatened by ocean acidification, a process that is closely related to climate change (Science Focus 11.2).

CONSIDER THIS. . .

CONNECTIONS Protecting Mangroves and Dealing with Climate Change

Protecting mangrove forests and restoring them in areas where they have been destroyed are important ways to reduce the impacts of rising sea levels and storm surges, because mangroves forests can slow storm-driven waves. These ecosystem services will become more important if tropical storms become more intense as a result of projected climate change. Protecting and restoring these natural coastal barriers is much cheaper and more effective than building concrete sea walls or moving threatened coastal towns and cities inland.

Warmer and more acidic ocean water is also stressing phytoplankton, the foundation of the marine food web (Figure 3-12, p. 60). These tiny life-forms produce half of the earth's oxygen and absorb a great deal of the carbon dioxide that we are adding to the atmosphere through the burning of fossil fuels and other activities. A team of scientists led by Boris Worm reported in 2010 that global phytoplankton populations have declined 40% since the 1950s, probably because warmer, more acidic ocean waters make it harder for these plankton to absorb their critical nutrients such as iron.

Overfishing and Overharvesting: Gone Fishing, Fish Gone

The human demand for seafood has been met historically through fishing. A **fishery** is a concentration of a particular wild aquatic species suitable for commercial harvesting in a given ocean area or inland body of water. However, scientists warn that the human demand for seafood has now become unsustainable. Just to keep consuming seafood at our current rate, we will need 2.5 times the area of the earth's oceans, according to the *Fishprint of Nations 2006,* a

of CO_3^{2-} are concentrated in the shallow waters where most coral reefs and shellfish are found (Figure 11-B). A major part of this problem is that the basicity of ocean waters is decreasing so rapidly that most organisms likely will not have time to adapt to them.

Scientists estimate that the oceans have soaked up roughly a third of the CO_2 emitted by human activities, and this has played a key role in the decreasing basicity of the world's oceans. Since we began burning fossil fuels in large quantities during the Industrial Revolution, there has been a 30% rise in the average acidity (actually a 30% decrease in average basicity) of surface ocean water, according to a 2010 UNEP summary of research on this problem. Scientists project that, by 2100, we will see a 150% drop in the average basicity of surface ocean water because of increasing acid levels.

These scientists call for a crash research program to monitor the decreases in basicity and carbonate ion levels of ocean waters around the world. They also want to learn more about which species and ocean habitats are the most vulnerable and which species will be better able to adapt to these changing environmental conditions.

According to most marine scientists, the only way to slow these changes is through a quick and sharp reduction in the use of fossil fuels around the world, which would lessen the massive inputs of CO_2 into the air and from there into the ocean. We can also slow the rise of acid levels in ocean waters by protecting and restoring mangrove forests, sea-grass beds, and coastal wetlands, because these aquatic systems take up and store some of the atmospheric CO_2 that is at the heart of this problem.

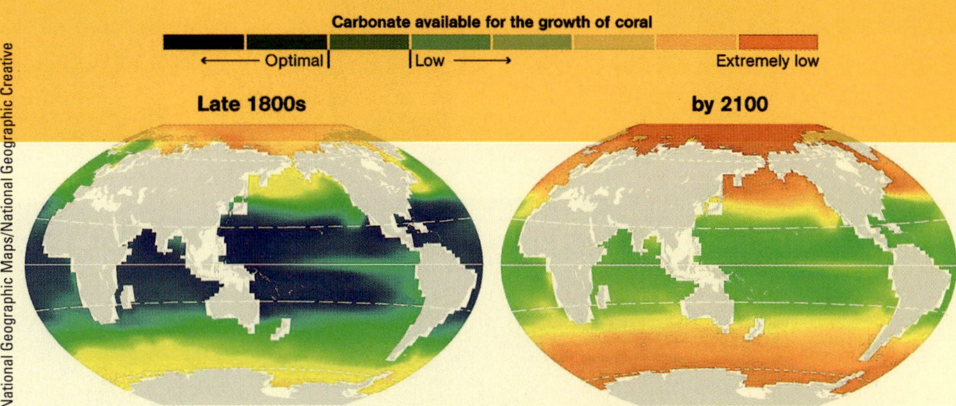

Figure 11-B Calcium carbonate levels in ocean waters, calculated from historical data (left), and projected for 2100 (right). Colors shifting from blue to red indicate where waters are becoming less basic. In the late 1800s, when CO_2 began to pile up rapidly in the atmosphere, tropical corals weren't yet affected by ocean acidification. But today carbonate levels have dropped substantially near the Poles, and by 2100, they may be too low even in the tropics for coral reefs to survive.

(Sources: Andrew G. Dickson, Scripps Institution of Oceanography, U.C. San Diego, and Sarah Cooley, Woods Hole Oceanographic Institution. Used by permission from National Geographic.)

Critical Thinking

How might a massive loss of life in the oceans affect life on land? How might it affect your life?

study based on the concept of the human ecological footprint (see **Concept 1-2**, p. 10, and Figure 1-12, p. 13).

Overfishing—the O in HIPPCO—is not new. Archaeological evidence indicates that for thousands of years, humans living in some coastal areas have overharvested fishes, shellfish, seals, turtles, whales, and other marine mammals (**Concept 11-1**). But today, fish are hunted throughout the world's oceans by a global fleet of about 4.4 million fishing boats using advanced technology to locate and harvest fish and shellfish in massive numbers. Research indicates that modern industrial fishing has been a key factor in the depletion of up to 80% of the populations of some wild fish species in only 10–15 years.

Industrial fishing fleets use global satellite positioning equipment, sonar fish-finding devices, huge nets and long fishing lines, spotter planes, and refrigerated factory ships that can process and freeze their enormous catches. These highly efficient fleets have helped to supply the growing demand for seafood (Figure 11-5), but critics say that they are vacuuming the seas, decreasing marine biodiversity, and degrading important marine ecosystem services.

For example, trawlers catch fishes and shellfish—especially cod, flounder, shrimp, and scallops—that live on or near the ocean floor. They have destroyed vast areas of ocean-bottom habitat (Figure 11-2). Also, these nets often capture endangered sea turtles (Figure 11-6), causing them to drown (**Core Case Study**).

Figure 11-4 This Hawaiian monk seal was slowly starving to death before a discarded piece of plastic was removed from its snout.

Another fishing method, *purse-seine fishing,* is used to catch surface-dwelling species such as tuna, mackerel,

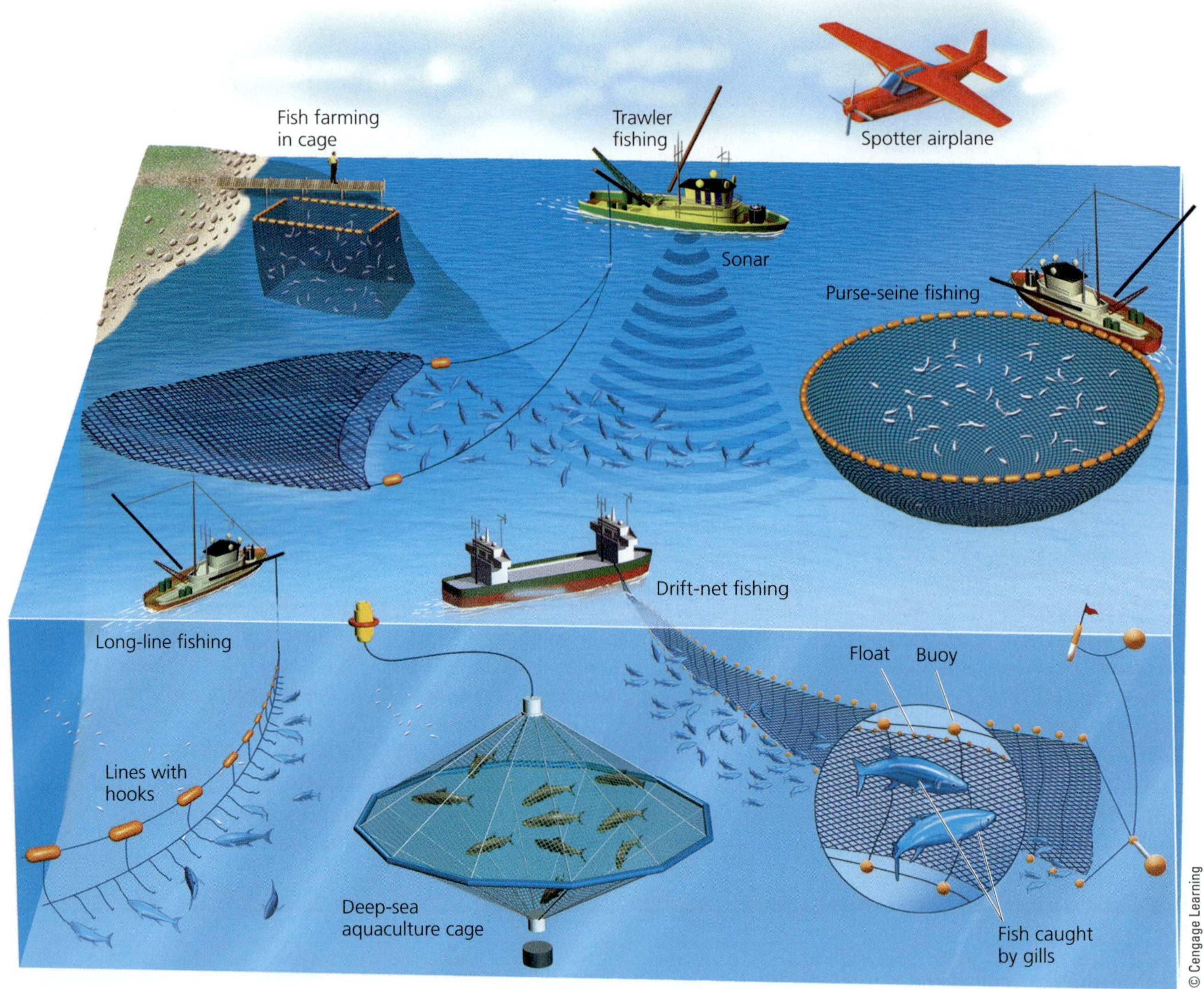

Figure 11-5 Major commercial fishing methods used to harvest various marine species, along with some methods used to raise fish through aquaculture.

anchovies, and herring (Figure 11-7), which tend to feed in schools near the surface or in shallow areas. After a spotter plane locates a school, the fishing vessel encloses it with a large purse-seine net. Some of these nets have killed large numbers of dolphins that swim on the surface above schools of tuna.

Some fishing vessels also use *long-lining,* which involves putting out lines up to 100 kilometers (60 miles) long, hung with thousands of baited hooks to catch swordfish, tuna, sharks, and ocean-bottom species such as halibut and cod. Long lines also hook and kill large numbers of sea turtles (**Core Case Study**), dolphins, and seabirds.

With *drift-net fishing,* fish are caught by drifting nets that can hang as deep as 15 meters (50 feet) below the surface and extend to 64 kilometers (40 miles) long. These nets trap and kill large quantities of unwanted fish, called *bycatch,* along with marine mammals, seabirds, and sea turtles (**Core Case Study**). Nearly one-third of the world's annual fish catch by weight consists of bycatch species that are mostly thrown overboard dead or dying. This adds to the depletion of these species and it puts stress on some of the species that feed on them. Also, according to a 2009 UNEP report, abandoned and lost nets known as *ghost nets* float beneath the surface in many areas, trapping and drowning aquatic animals for years, before they finally sink or are recovered.

In the Fishprint of Nations study, scientists used the term **fishprint**, defined as the area of ocean needed to sustain the fish consumption of an average person, a nation, or the world. They estimated that all nations together are taking 57% more wild fish than these species' populations can sustain in the long run and that 87% of oceanic commercial fisheries are being fished at or beyond their sustainable yields.

In most cases, overfishing leads to *commercial extinction,* which occurs when it is no longer profitable to continue harvesting the affected species. Overfishing can result in only a temporary depletion of fish stocks, as long as depleted areas and fisheries are allowed to recover. But

©Mark Foley/State Archives of Florida, Florida Memory

Figure 11-6 This green sea turtle died after being caught in a fishing net.

as industrialized fishing fleets take more and more of the world's available fish and shellfish, recovery times for severely depleted populations are increasing and can be two decades or more. Some depleted fisheries may not recover as jellyfish and other invasive species move in and take over their food webs (see the following Case Study).

For example, in the late 1950s, fishing fleets began overfishing the 500-year-old Atlantic cod fishery off the coast of Newfoundland, Canada. They used bottom trawlers to capture larger shares of the stock, reflected in the sharp rise in the graph in Figure 11-7. This extreme overexploitation of the fishery led to a steady decline in the fish catch throughout the 1970s. After a slight recovery in the 1980s, the fishery collapsed and in 1992, it was shut down, putting at least 35,000 fishers and fish processors out of work in more than 500 coastal communities. It was reopened on a limited basis in 1998 but then closed indefinitely in 2003. The Newfoundland fishing industry is now recovering by harvesting snow crab and northern shrimp.

One result of the increasingly efficient global hunt for fish is that larger individuals of commercially valuable wild species—including cod, marlin, swordfish, and tuna—are becoming scarce. Between 1950 and 2006, according to a study led by marine ecologist Boris Worm, 90% or more of these and other large, predatory, open-ocean fishes had disappeared. This resulted in a massive change to ocean food webs and the ecosystem services they provide, taking place in only a few decades.

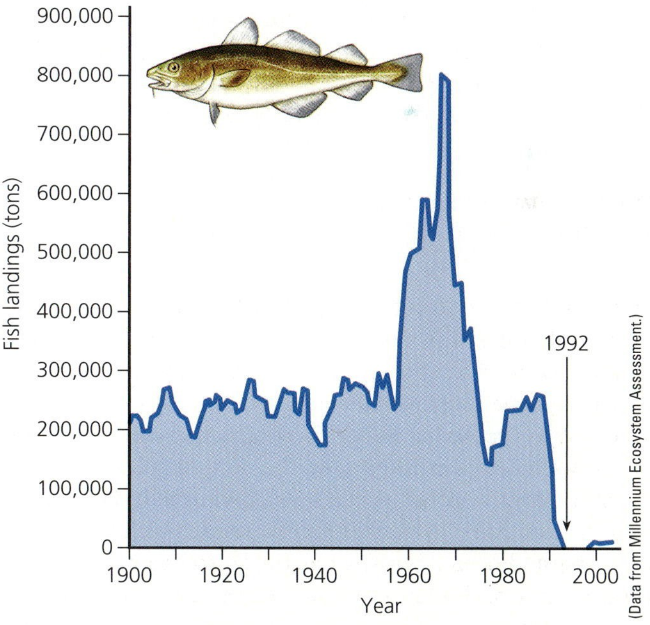

Figure 11-7 Natural capital degradation: This graph illustrates the collapse of Newfoundland's Atlantic cod fishery.

For example, the *Atlantic,* or *Northern bluefin tuna,* is among the largest of ocean fishes. It is highly prized for making sushi and sashimi, and the demand for those products has resulted in the severe depletion of this species. Japan consumes about 80% of the annual catch of Atlantic blue-

Brian J. Skerry/National Geographic Creative

Figure 11-8 *Bluefin tuna ranching:* These bluefin tunas have been corralled in an underwater pen and will be fattened for slaughter to make sushi.

fins. In 2001, one giant Atlantic bluefin sold for $175,000 in a Tokyo fish market. Such high prices have helped to decrease the populations of this species by 75–80% since the 1970s. As a result, the International Union for Conservation of Nature (IUCN) classifies the Atlantic bluefin tuna as critically endangered. So far, measures to regulate the catch for this species have largely failed.

A practice that some marine scientists strongly criticize is that of *tuna ranching,* in which schools of half-grown Atlantic bluefin and other tuna are herded by the thousands into underwater pens and towed to areas where they are held and fattened for slaughter (Figure 11-8). Richard Ellis, a leading U.S. marine conservationist, described it this way: "If you had to design a way to guarantee the decimation of a breeding population, this would be it: catch the fish before they are old enough to breed and keep them penned up until they are killed."

A number of marine scientists have called for a ban on the fishing and ranching of Atlantic bluefins until the species can recover. However, so far, strong opposition from Japan and from the fishing industry has prevented such a ban.

As commercially valuable species have become scarce, fishers have turned to other species such as sharks, which are now being overfished (see the second following Case Study). Also, as large species are overfished, the fishing industry is shifting to smaller marine species such as herring, sardines, and krill. About 90% of this catch is converted to fishmeal and fish oil, most of which is fed to farmed fish. One researcher referred to this as "stealing the ocean's food supply," because these smaller species make up much of the diet of larger predator fish, seabirds, and toothed whales.

Some marine mammals have also been threatened with extinction by overharvesting. The most prominent examples are whales, including the highly endangered blue whale. Whale harvesting in international waters has followed the classic pattern of a tragedy of the commons (see Chapter 1, p. 12), with whalers killing an estimated 1.5 million whales of all species between 1925 and 1975. This overharvesting drove 8 of the 11 major whale species to commercial extinction before commercial whaling was banned internationally in 1986.

After the ban, the estimated number of whales killed by commercial whale hunts worldwide dropped from 42,480 in 1970 to about 2,800 in 2011. Most of the whales killed now are taken by Norway, Japan (allegedly for scientific purposes), the Faroe Islands, and Iceland, despite the ban. Japan, Norway, and Iceland have worked to overthrow the international moratorium on commercial whaling and to reverse the separate international ban on the buying and selling of whale products.

CASE STUDY

The Great Jellyfish Invasion

Jellyfish, which existed before the dinosaurs, are not fish. A jellyfish (see Figure 8-3, p. 169) has no brain, head, heart, eyes, or bones and is basically a cluster of nerves that can detect warmth, food, odors, and vibration. Its generally bell-shaped body is filled with a jelly-like substance. It uses its tentacles dangling from its body to sting or stun prey, to draw food into its mouth, and to defend itself against predators such as other jellyfish, tuna, sharks, swordfish, and sea turtles. Most jellyfish feed on zooplankton, fish eggs, small fish, and other jellyfish.

Jellyfish sizes vary widely. Some are as small as a mosquito. The largest is the Arctic lion's mane jellyfish (Figure 11-9) with a diameter of up to 2.5 meters (8 feet) and tentacles as long as 37 meters (120 feet). It can weigh as much as 150 kilograms (330 pounds). The sting of a jellyfish can cause an itch like that of a bug bite or a more serious burning sensation lasting for several days. However, a sting from the Portuguese man-of-war, the Australian box jelly, or the tiny Irukandji jellyfish can kill a human within minutes.

Jellyfish are often found in large swarms or *blooms* of thousands, even millions of individuals. In recent years, the numbers of these blooms observed by scientists and fishers have been rising and more frequent jellyfish stings have had a harmful economic effect on a number of popular tourist beach areas. Each year, the stings of lethal jellyfish kill dozens of people—far more than the annual average number of people killed by sharks. In addition, jellyfish blooms often cause beach closings, disrupt commercial fishing operations by clogging nets, and close down coal-burning and nuclear power plants by clogging their cooling water intakes. Fish in some ocean aquacul-

Figure 11-9 The lion's mane is the largest of the roughly 1,500 known jellyfish species. It can be as wide as a full-size school bus and about 3 times as long.

ture pens (Figure 11-5) have been wiped out by swarms of jellyfish attracted to the pens by the fish feed and fish waste found there.

Several factors play a role in the rapid rise of jellyfish populations. One is the overfishing of species that eat small jellyfish, including endangered sea turtles (**Core Case Study**). The jellyfish, in turn, feed on fish eggs and larvae, making it harder for the depleted fish stocks to recover. Another factor is inputs of excessive plant nutrients from fertilizer runoff and sewage due to coastal urban development. This spurs the growth of phytoplankton, on which jellyfish feed, which helps them to increase their numbers. This results in oxygen-depleted zones, which jellyfish can tolerate better than most other species.

Warmer waters are also associated with jellyfish blooms, so continuing climate change may result in jellyfish becoming the dominant species in many marine ecosystems around the world. A 2010 UNEP report linked rising jellyfish populations to increasing ocean acidification, which harms some marine species but not jellyfish.

According to Chinese oceanographer Wei Hao and other marine scientists, the startling growth of jellyfish populations threatens to upset marine food webs and ecosystem services and turn some of the most productive seas into jellyfish empires. Once jellyfish take over a marine ecosystem they might dominate it indefinitely.

CASE STUDY

Why Should We Protect Sharks?

Sharks have roamed the world's oceans for more than 400 million years. As *keystone species,* some shark species play crucial roles in helping to keep their ecosystems functioning. Some shark species that feed at or near the tops of food webs remove injured and sick animals from the ocean. Without this ecosystem service provided by some shark species, the oceans would be teeming with dead and dying fish and marine mammals.

More than 400 known species of sharks inhabit the world's oceans. They vary widely in size and behavior, from the goldfish-sized dwarf dog shark to the whale shark (Figure 11-10, left), which can grow to the length of a city bus and weigh as much as two full-grown African elephants.

Many people, influenced by movies, popular novels, and widespread media coverage of shark attacks, think of sharks as people-eating monsters. In reality, the three largest species—the whale shark (Figure 11-10, left),

Figure 11-10 The threatened whale shark (left), which feeds on plankton, is the largest fish in the ocean and is quite friendly to humans. The scalloped hammerhead shark (right) is endangered.

Left: Colin Parker/National Geographic My Shot/National Geographic Creative. Right: Westend61/SuperStock.

©NOAA

Figure 11-11 This endangered loggerhead turtle is escaping a fishing net equipped with a turtle excluder device (TED).

basking shark, and megamouth shark—are gentle giants. These plant-eating sharks swim through the water with their mouths open, filtering out and swallowing huge quantities of phytoplankton.

Media coverage of shark attacks greatly exaggerates the danger from sharks. Every year, members of a few species, including the great white, bull, tiger, oceanic whitetip, and hammerhead sharks, injure 60–75 people and typically kill 6–10 people worldwide. Some of these sharks feed on sea lions and other marine mammals and sometimes mistake swimmers and surfers for their usual prey.

For every shark that injures or kills a person, people kill about 1.2 million sharks. As many as 73 million sharks are caught each year for their valuable fins and then thrown back alive into the water, fins removed, to bleed to death or drown because they can no longer swim. Sharks are also killed for their livers, meat, hides, and jaws, and because we fear them. Each year an estimated 50 million sharks die when fishing lines and nets trap them.

Harvested shark fins are worth as much as $1.2 billion a year. They are widely used in Asia as an ingredient in expensive soup (up to $100 a bowl) and as a pharmaceutical cure-all. According to the wildlife conservation group WildAid, there is no reliable evidence that the fins provide flavor or have any nutritional or medicinal value. The group also warns that consuming shark fins and shark meat can be harmful to human health because they contain very high levels of mercury and other toxins. In 2012, neurologist Deborah Mash found neurotoxins in the fins of sharks of seven different species with links to Parkinson's, Alzheimer's, and Lou Gehrig's diseases.

According to a 2009 IUCN study, 32% of the world's open-ocean shark species are threatened with extinction, including the scalloped hammerhead shark (Figure 11-10, right). Sharks are especially vulnerable to population declines because they grow slowly, mature late, and have only a few offspring per generation. Today, they are among the earth's most vulnerable and least protected animals. And because some sharks are keystone species, their ecosystems and the ecosystem services they provide are also threatened.

It is encouraging that in 2012, French Polynesia created the world's largest shark sanctuary, although protecting sharks in such a large area will be difficult. Also, in 2013, the Convention on International Trade in Endangered Species (CITES) placed trade limits on catches of oceanic whitetip, porbeagle, and three species of hammerhead sharks.

CONSIDER THIS. . .

CONNECTIONS Shark Declines and Scallop Fishing

In 2007, scientists reported that the decline in certain shark populations along the U.S. Atlantic coast led to an explosion in populations of rays and skates, which sharks normally feed on. As a result, the rays and skates were feasting on bay scallops, which resulted in a sharp decline in their population and in the area's bay scallop fishing business.

Extinction of Aquatic Species Is a Growing Threat

Beyond the *commercial extinction* of a number of fisheries, many fish species are also threatened with *biological extinction*, mostly from overfishing, water pollution, wetlands destruction, and excessive removal of water from rivers and lakes. According to the 2011 IUCN Red List of Threatened Species, of all species evaluated, 34% of the world's marine species and 71% of the world's freshwater fish species face extinction within the next 6 to 7 decades. A 2012 analysis by researchers at BirdLife International found that seabirds are more endangered than other types of birds. Some 27% of seabird species are threatened, compared to 12% of all bird species combined.

In 2011, an international panel of 27 marine scientists assessed how human impacts are affecting the world's oceans. Their startling conclusion was that because of a combination of habitat loss, overfishing, pollution, and ocean acidification, marine life is poised to enter a new period of mass extinction, which would eventually affect the world's terrestrial and marine ecosystems and thus the world's economies.

Among the most threatened of all marine species are sea turtles (**Core Case Study**). This is due to a number of factors, including people taking their eggs and loss or degradation of their beach nesting habitat. Female turtles are often blocked from reaching their nesting sites by sea-walls, sandbags, and other devices designed to slow beach erosion in developed coastal areas. Also, many adult sea turtles die when commercial and recreational fishers accidentally catch them in nets (Figure 11-6) or when they become entangled in ocean debris such as plastic bags and lost or abandoned fishing nets (ghost nets).

In U.S. waters, the National Marine Fisheries Service requires the use of turtle excluder devices (TEDs) on commercial fishing nets that allow captured sea turtles to escape (Figure 11-11). Since 1990, fishing regulations have reduced accidental sea turtle deaths in U.S. waters by 90%. GOOD NEWS In 2012, marine biologist Helen Bailer and her colleagues began using satellites to track highly endangered leatherback turtles and the routes of trawlers that trap and kill many of these turtles. Such data could help regulatory agencies in setting times and places where they might limit trawler fishing to protect the leatherbacks.

11-2 How Can We Protect and Sustain Marine Biodiversity?

CONCEPT 11-2

We can help to sustain marine biodiversity by using laws and economic incentives to protect species, setting aside marine reserves to protect ecosystems and ecosystem services, and using community-based integrated coastal management.

Laws and Treaties and Economic Incentives Can Help to Sustain Aquatic Biodiversity

Protecting marine biodiversity is difficult for several reasons. *First,* the human ecological footprint (see Figure 1-11, p. 13) and the fishprint are expanding so rapidly that it is difficult to monitor their impacts. *Second,* much of the damage to the oceans and other bodies of water is not visible to most people. *Third,* many people incorrectly view the seas as an inexhaustible resource that can absorb an almost infinite amount of waste and pollution and still produce all the seafood we want. *Fourth,* most of the world's ocean area lies outside the legal jurisdiction of any country. Thus, much of it is an open-access resource, subject to overexploitation—a classic case of the tragedy of the commons (see Chapter 1, p. 12).

Nevertheless, there are several ways to protect and sustain marine biodiversity, one of which is the regulatory approach (**Concept 11-2**). For example, in the United States and a number of other countries, laws have helped to protect sea turtle nesting sites (**Core Case Study**) from egg poachers and destruction by vehicles (Figure 11-12). GOOD NEWS

National and international laws and treaties to help protect marine species include the 1975 Convention on International Trade in Endangered Species (CITES), the 1979 Global Treaty on Migratory Species, the U.S. Marine Mammal Protection Act of 1972, the U.S. Endangered Species Act of 1973 (ESA; see Chapter 9, pp. 208–209), the U.S. Whale Conservation and Protection Act of 1976, and the 1995 International Convention on Biological Diversity. The ESA and several international agreements have been used to identify and protect endangered and threatened marine species, including whales, seals, sea lions, and sea turtles. The problem is that with some international agreements, it is hard to get all nations to comply, which can weaken the effectiveness of such agreements. Even when agreements and regulations are enforced, the resulting fines and punishments for violators are often inadequate.

Another way to protect endangered and threatened aquatic species is to use economic incentives (**Concept 11-2**). For example, according to a World Wildlife Fund study, sea turtles are worth more to coastal communities alive than dead (**Core Case Study**). The report estimates that sea turtle tourism typically brings in almost 3 times more money than the sale of turtle products such as meat, leather, and eggs brings in. In Brazil, a sea turtle program hires ex-poachers to protect rather than exploit the turtle populations. Educating citizens about this issue could inspire communities to protect more turtles.

Marine Sanctuaries Protect Ecosystems and Species

By international law, a country's offshore fishing zone extends to 370 kilometers (200 nautical miles) from its shores. Foreign fishing vessels can take certain quotas of fish within such zones, called *exclusive economic zones,* but only with a government's permission. Ocean areas beyond

Figure 11-12 Loggerhead sea turtle nests are protected by law in some areas, including Myrtle Beach, South Carolina.

the legal jurisdiction of any country are known as the *high seas,* and laws and treaties pertaining to them are difficult to monitor and enforce.

The United Nations Law of the Sea treaty, which went into effect in 1984, has been signed by 164 countries (but not by the United States). Under this treaty, the world's coastal nations have jurisdiction over 36% of the ocean surface and 90% of the world's fish stocks. Instead of using this treaty to protect their fishing grounds, many governments have promoted overfishing by subsidizing fishing fleets and failing to establish and enforce stricter regulation of fish catches in their coastal waters.

However, some countries are attempting to protect marine biodiversity and to sustain fisheries by establishing marine sanctuaries in their coastal waters. Since 1986, the IUCN has helped such nations to establish a global system of *marine protected areas* (MPAs)—areas of ocean partially protected from human activities. In 2010, the IUCN reported that there were 5,880 MPAs worldwide (355 in U.S. waters), covering about 1.6% of the world's ocean surface.

The number of MPAs is growing, but most of them are only partially protected. Nearly all allow dredging, trawler fishing, and other ecologically harmful resource–extraction activities. And many of them are too small to be effective in protecting larger species. However, since 2007 the U.S. state of California has been establishing the nation's most extensive network of MPAs in which fishing will be banned or strictly limited. In 2011, Costa Rica (see Core Case Study, Chapter 10, p. 218) expanded one of its MPAs to help protect a number of marine species, including the critically endangered leatherback sea turtle (**Core Case Study**) and the endangered scalloped hammerhead shark (Figure 11-10, right). And in 2012, Australia announced that it would create the world's largest MPA consisting of the Great Barrier Reef Marine Park and the neighboring Coral Sea Reserve. GOOD NEWS

Establishing a Global Network of Marine Reserves: An Ecosystem Approach to Marine Sustainability

Many scientists and policy makers call for the widespread use of an *ecosystem approach* focused on protecting and sustaining entire marine ecosystems and their ecosystem services for current and future generations rather than relying mostly on protecting individual species. The goal of this ecological approach is to establish a global network of fully protected *marine reserves,* areas that are declared off-limits to destructive human activities in order to enable their ecosystems to flourish and recover.

This global network would include large reserves on the high seas, especially near extremely productive nutrient upwelling areas (see Figure 7-2, p. 145), and a mixture of smaller reserves in coastal zones that are adjacent to well-managed, sustainable commercial fishing areas. Such protected "underwater wilderness" areas would be closed to activities such as commercial fishing, dredging, and mining, as well as to waste disposal. Most reserves in the proposed global network would permit less-harmful activities such as recreational boating, shipping, and in some cases, small-scale, nondestructive fishing. However, most reserves would also contain core zones where no human activity would be allowed.

Marine reserves work and they work quickly. Scientific studies show that within fully protected marine reserves, within 2–4 years after strict protection begins, commercially valuable fish populations can double, average fish size can grow by almost a third, fish reproduction can triple, and species diversity can increase by almost one-fourth. Furthermore, these improvements can last for decades (**Concept 11-2**). Research also shows that reserves benefit nearby fisheries, because fish move into and out of the reserves, and currents carry fish larvae produced inside reserves to adjacent fishing grounds, thus bolstering the populations there. GOOD NEWS

In 2012, the Australian government announced that it would establish the world's largest network of fully protected marine reserves around its coasts. Two other giant marine reserves have been created by the United States around the Northwestern Hawaiian Islands and by the United Kingdom around the Chagos Islands in the Indian Ocean. About 3% of U.S. territorial waters have been set aside in 223 marine reserves.

The IUCN and other scientific groups have identified *marine hotspots*—threatened areas in need of full protection because of their importance to marine biodiversity and ecosystem services (Figure 11-13). Despite the importance of such protection, only 0.8% of the world's oceans are fully protected—closed to fishing and other potentially harmful human activities—compared with about 5% of the world's land. In other words, 99.2% of the world's oceans are not effectively protected from harmful human activities. Furthermore, many marine reserves are too small to protect most of the species within them and do not provide adequate protection against illegal fishing, garbage dumping, or pollution that flows from the land into coastal waters.

The trend toward creating huge megareserves is encouraging. However, researchers are struggling to design ways to monitor changes over vast areas and to determine the effectiveness of such reserves in sustaining and rebuilding marine populations. In addition, resource managers are trying to find affordable ways to enforce fishing bans in such vast ocean areas.

Many marine scientists, including National Geographic's Explorer-in-Residence Sylvia Earle (Individuals Matter 11.1), call for protecting at least 30% of the world's oceans as fully protected marine reserves. They also urge that protected corridors be established to connect the global network of marine reserves, especially those in coastal waters. This would also help species to move to different habitats in the process of adapting to the effects of ocean warming, acidification, and many forms of ocean pollution.

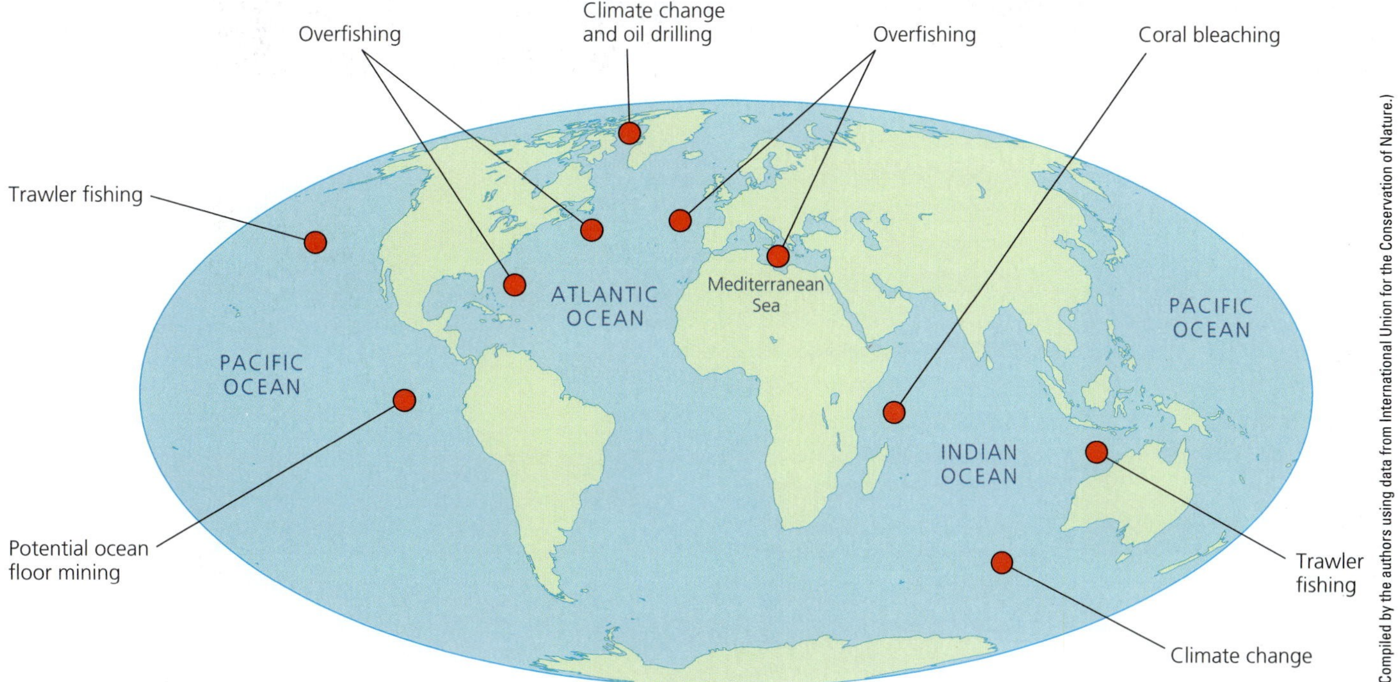

Figure 11-13 **Natural capital:** Ten *marine hotspots*—threatened areas, including coral reefs, and threatened fisheries that provide vital ecosystem services and that marine scientists say should be fully protected from harmful human activities.

A team of U.K. scientists led by Andrew Balmford studied 83 well-managed marine reserves and concluded that it would cost $12 billion to $14 billion a year to manage reserves covering 30% of the world's oceans. This would roughly equal to the annual subsidies that promote overfishing, which governments currently provide to the global fishing industry.

CONSIDER THIS. . .

THINKING ABOUT Marine Reserves

Do you support setting aside at least 30% of the world's oceans as fully protected marine reserves? Explain. How would this affect your life?

Restoration Helps to Protect Marine Biodiversity but Prevention Is the Key

A dramatic example of marine system restoration is Japan's attempt to restore its largest coral reef—90% of which has died—by seeding it with new corals. Divers drill holes into the dead reefs and insert ceramic discs holding sprigs of fledgling coral. Survival rates of the young corals were only at about 33% in 2009 but were rising, according to the Japanese government. Scientists see this experiment as possibly leading to more global efforts to save reefs by transplanting corals. Figure 11-14 also GOOD NEWS shows how protection has helped to restore coral reefs near Kanton Island, an atoll located in the South Pacific roughly halfway between Fiji and Hawaii.

While many scientists applaud such efforts to restore aquatic systems, they also note that these projects could fail if the problems that caused the degradation of the systems are not addressed (see Chapter 8, Core Case Study, p. 168). That is why they also argue that we must shift to taking a *prevention approach* toward aquatic ecosystem degradation, which is far less expensive and risky than restoration efforts. And they say we must make this shift soon.

For example, a study by IUCN and scientists from the Nature Conservancy concluded that the world's shallow coral reefs and mangrove forests could survive currently projected climate change if we relieve other threats such as overfishing and pollution. However, while some shallow coral species may be able to adapt to warmer temperatures, they may not have enough time to do this unless we act now to slow down the projected rapid rate of climate change, especially during the latter half of this century. In addition, ocean acidification can slow the growth of coral reefs.

To deal with problems of pollution and overfishing, communities must closely monitor and regulate fishing and coastal land development and greatly reduce pollution from land-based activities. Coastal residents must also think carefully about the chemicals they put on their lawns and the kinds of waste they generate because many of these chemicals will end up in coastal waters.

More important, each of us can make careful choices in purchasing only sustainably harvested or sustainably farmed seafood. People could also reduce their carbon footprints in order to slow ocean acidification along with other

individuals matter 11.1

Sylvia Earle—Ocean Protector and National Geographic Explorer-in-Residence

James A. Sugar/National Geographic Creative

Sylvia Earle is one of the world's most respected oceanographers and is a National Geographic Society Explorer-in-Residence. She has taken a leading role in helping us to understand the world's oceans and to protect them. *Time* magazine named her the first Hero for the Planet and the U.S. Library of Congress calls her "a living legend."

Earle has led more than 100 ocean research expeditions and has spent more than 7,000 hours underwater, either diving or descending in research submarines to study ocean life. She has focused her research on the ecology and conservation of marine ecosystems, with an emphasis on developing deep-sea exploration technology.

She is the author of more than 175 publications and has been a participant in numerous radio and television productions. During her long career, Earle has also been the Chief Scientist of the U.S. National Oceanic and Atmospheric Administration (NOAA) and she has founded three companies devoted to developing submarines and other devices for deep-sea exploration and research. She has received more than 100 major international and national honors, including a place in the National Women's Hall of Fame.

These days, Earle is leading a campaign called *Mission Blue* to finance research and to ignite public support for a global network of marine protected areas, which she dubs "hope spots." Her goal is to help save and restore the oceans, which she calls "the blue heart of the planet." She says, "There is still time, but not a lot, to turn things around."

Background photo: ©Igor Kovalchuk/Shutterstock.com

harmful effects of projected climate change on marine biodiversity, as discussed in more detail in Chapter 19.

One strategy emerging in some coastal communities is *integrated coastal management*—a community-based effort to develop and use coastal resources more sustainably (**Concept 11-2**). The overall aim of such programs is for fishers, business owners, developers, scientists, citizens, and politicians to identify shared problems and goals in their use of marine resources. The idea is to develop workable, cost-effective, and adaptable solutions that will help to preserve biodiversity, ecosystem services, and environmental quality, while also meeting economic and social goals.

This requires all participants to seek reasonable short-term trade-offs that can lead to long-term ecological and economic benefits—an example of applying the win-win **principle of sustainability** (see Figure 1-5, p. 9 or back cover). For example, fishers might have to give up harvesting various fish species in certain

Brian J. Skerry/National Geographic Creative

Brian J. Skerry/National Geographic Creative

Figure 11-14 Severe bleaching of coral near Kanton Island in the South Pacific (left), and an example of the recovery of coral in a protected area near the island (right).

areas until stocks recover enough to restore biodiversity in those areas. This might help to provide fishers with a more sustainable future for their businesses.

Australia manages its huge Great Barrier Reef Marine Park in this way, and more than 100 integrated coastal management programs are being developed throughout the world. GOOD NEWS Another example of such management in the United States is the Chesapeake Bay Program (see Chapter 8, Case Study, p. 168).

11-3 How Should We Manage and Sustain Marine Fisheries?

> **CONCEPT 11-3**
> Sustaining marine fisheries will require improved monitoring of fish and shellfish populations, cooperative fisheries management among communities and nations, reduction of fishing subsidies, and careful consumer choices in buying seafood.

Estimating and Monitoring Fishery Populations Is the First Step

The first step in protecting and sustaining the world's marine fisheries is to make the best possible estimates of their fish and shellfish populations (**Concept 11-3**). The traditional approach has used a *maximum sustained yield (MSY)* model to project the maximum number of individuals that can be harvested annually from fish or shellfish stocks without causing a population drop. However, the MSY concept has not worked very well because of the difficulty in estimating the populations and growth rates of fish and shellfish stocks. Also, harvesting a particular species at its estimated maximum sustainable level can affect the populations of other target and nontarget marine species.

In recent years, some fishery biologists and managers have begun using the *optimum sustained yield (OSY)* concept. They have attempted to take into account interactions among species and to provide more room for error. Similarly, another approach is *multispecies management* of a number of interacting species, which takes into account their competitive and predator–prey interactions. An even more ambitious approach is to develop complex computer models for managing multispecies fisheries in *large marine systems*. However, it is a political challenge to get groups of nations to cooperate in planning and managing such large systems.

There are uncertainties built into using any of these approaches because the biology of marine species and their interactions is not completely understood, and because of limited data on changing ocean conditions. As a result, many fishery and environmental scientists are increasingly interested in using the *precautionary principle* for managing fisheries and large marine systems. This means sharply reducing fish harvests and closing some overfished areas until they recover and until we have more information about what levels of fishing they can sustain.

Some Communities Cooperate to Regulate Fish Harvests

An obvious step to take in protecting marine biodiversity—and therefore fisheries—is to regulate fishing. Traditionally, many coastal fishing communities have developed allotment and enforcement systems for controlling fish catches in which each fisher gets a share of the total allowable catch. Such *catch-share systems* have sustained fisheries and jobs in many communities for hundreds and sometimes thousands of years.

An example of a successful catch-share system is Norway's Lofoten fishery, one of the world's largest cod fisheries. For 100 years, it has been self-regulated, with no participation by the Norwegian government (Figure 11-15). GOOD NEWS The key is to guarantee each participant a share of

Figure 11-15 For hundreds of years, fishing villages such as this one on one of the Lofoten Islands in northern Norway have used self-regulated cooperative *catch-share systems* to sustain their fisheries.

the total allowable catch. Another example is the Alaskan halibut fishery, which grew dramatically more profitable after a catch-share system was put in place in 1995.

However, the influx of large state-of-the-art fishing boats and international fishing fleets has weakened the ability of many coastal communities to regulate and sustain local fisheries. Community management systems have often been replaced by *comanagement,* in which coastal communities and the government work together to manage fisheries. Currently, more than 210 of the world's fisheries are comanaged.

With this approach, a central government typically sets quotas for various species and divides the quotas among communities. The government may also limit fishing seasons and regulate the types of fishing gear that can be used to harvest a particular species. Each community then allocates and enforces its quota among its members based on its own rules. Often, communities focus on managing inshore fisheries, and the central government manages the offshore fisheries. When it works, community-based comanagement illustrates that overfishing and the tragedy of the commons (see Chapter 1, p. 12) are not inevitable. GOOD NEWS

Government Subsidies Can Encourage Overfishing

Governments around the world give a total of more than $30 billion per year in subsidies to fishers to help them keep their businesses running, according to a 2006 study by fishery experts U. R. Sumaila and Daniel Pauly. Some marine scientists estimate that, each year, $10 billion to $14 billion of these subsidies is spent to encourage overfishing and expansion of the fishing industry. The result is too many boats chasing too few fish. Some argue that such subsidies are not a wise investment because they promote overfishing of targeted fish stocks, which causes economic losses of about $50 billion a year, according to a 2008 World Bank report.

CONSIDER THIS. . .

THINKING ABOUT Fishing Subsidies

Do you think that government fishing subsidies that promote unsustainable fishing should be eliminated or drastically reduced? Explain. Would your answer be different if your livelihood depended on commercial fishing?

Consumer Choices Can Help Sustain Fisheries and Aquatic Biodiversity

An important component of sustaining aquatic biodiversity and ecosystem services is bottom-up pressure from consumers demanding *sustainable seafood,* which will encourage more responsible fishing practices (**Concept 11-3**). One way to enable this is by labeling fresh and frozen seafood to inform consumers about how and where the fish and shellfish were caught. Another important component is the certification of sustainably caught seafood. GOOD NEWS The London-based Marine Stewardship Council (MSC) was created in 1999 to support sustainable fishing and to certify sustainably produced seafood. Only certified

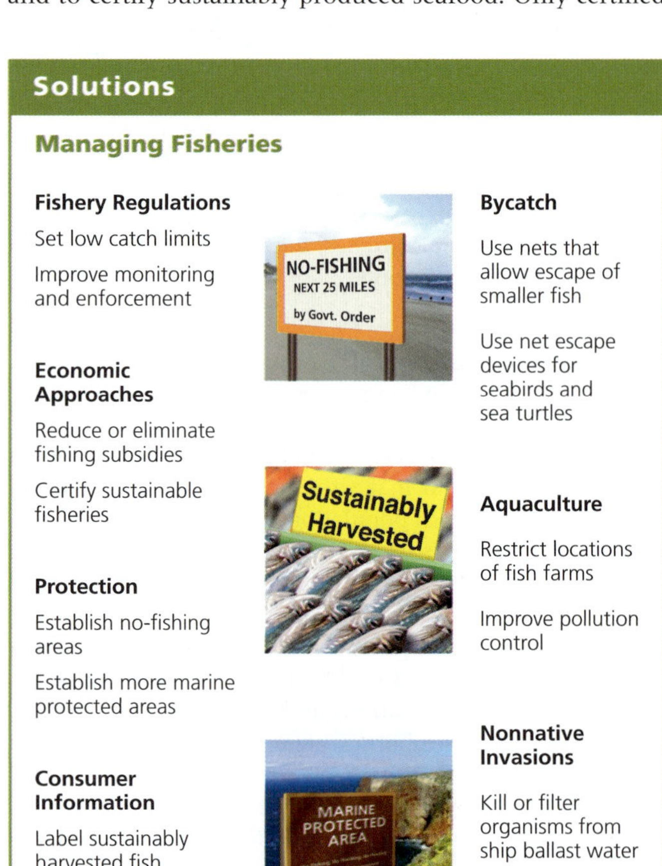

Figure 11-16 There are a number of ways to manage fisheries more sustainably and protect marine biodiversity. ***Questions:*** Which four of these solutions do you think are the best ones? Why?

fisheries are allowed to use the MSC's "Fish Forever" eco-label, which certifies that the fish were caught by fishers who used environmentally sound and socially responsible practices. Another approach is to certify and label products of sustainable *aquaculture,* or fish farming operations.

An important way for seafood consumers to help sustain aquatic biodiversity is to choose plant-eating species of fish, such as tilapia, raised through aquaculture. Carnivorous fish raised through aquaculture are fed fishmeal made from wild-caught fish. Some ocean fish species used to make fishmeal are being overfished at higher rates every year. Thus, by choosing herbivorous species, consumers can lessen the demand for this overfishing.

Figure 11-16 summarizes actions that individuals, organizations, and governments can take to help sustain global fisheries, marine biodiversity, and ecosystem services. History shows that most attempts to improve environmental quality and promote environmental sustainability require bottom-up political and economic pressure by concerned citizens. Individuals matter.

11-4 How Should We Protect and Sustain Wetlands?

CONCEPT 11-4

We can maintain the ecosystem and economic services of wetlands by protecting remaining wetlands and restoring degraded wetlands.

Coastal and Inland Wetlands Are Disappearing around the World

Coastal wetlands and marshes (see Figure 8-8, p. 172) and inland wetlands are important reservoirs of aquatic biodiversity that provide vital economic and ecosystem services. Despite their ecological value, the United States has lost more than half of its coastal and inland wetlands since 1900, and other countries have lost even more. New Zealand, for example, has lost 92% of its original coastal wetlands, and Italy has lost 95%.

For centuries, people have drained, filled in, or covered over swamps, marshes, and other wetlands to create rice fields or other cropland, to accommodate expanding cities and suburbs, and to build roads. Wetlands have also been destroyed in order to extract minerals, oil, and natural gas, and to eliminate breeding grounds for insects that cause diseases such as malaria.

To make matters worse, coastal wetlands in many parts of the world will probably be under water before the end of this century because of rising sea levels. If the atmosphere warms as projected, ocean waters will also warm and expand, and some of the world's land-based glaciers and ice sheets will melt, adding water to the ocean. This could seriously degrade the aquatic biodiversity and the ecosystem services provided by coastal wetlands, as well as the commercially important fish and shellfish species and millions of migratory waterfowl and other birds that depend on these wetlands.

We Can Preserve and Restore Wetlands

Scientists, land managers, landowners, and environmental groups are involved in intensive efforts to preserve existing wetlands and restore degraded ones (**Concept 11-4**), and laws have been passed to protect wetlands.

In the United States, zoning laws have been used to steer development away from wetlands. The U.S. government requires a federal permit to fill in wetlands occupying more than 1.2 hectares (3.0 acres) or to deposit dredged material in wetlands. According to the U.S. Fish and Wildlife Service, this law has helped to cut the average annual wetland loss by 80% since 1969. GOOD NEWS However, there are continuing attempts by land developers to weaken such wetlands protection. Only about 6% of the country's remaining inland wetlands are federally protected, and state and local wetland protection is inconsistent and generally weak.

The stated goal of current U.S. federal policy is *zero net loss* in the functioning and value of coastal and inland wetlands. A policy known as *mitigation banking* allows destruction of existing wetlands as long as an equal or greater area of the same type of wetland is created or restored. However, a study by the National Academy of Sciences found that at least half of the attempts to create new wetlands failed to replace lost ones, and most of the created wetlands did not provide the ecosystem services of natural wetlands. The study also found that wetland creation projects often fail to meet the standards set for them and are not adequately monitored.

Creating and restoring wetlands has become a profitable business. Private investment bankers make money by buying wetland areas and restoring or upgrading them by working with the U.S. Army Corps of Engineers and the U.S. Environmental Protection Agency. This creates wetlands banks or credits that the bankers can then sell to developers.

It is difficult to restore or create wetlands (see the following Case Study). Thus, most U.S. wetland banking systems require replacing each hectare of destroyed wetland with two or more hectares of restored or created wetland (Figure 11-17) as a built-in ecological insurance policy.

Ecologists urge that mitigation banking should be used only as a last resort. They also call for making sure that new replacement wetlands are created and evaluated *before* existing wetlands are destroyed. This example of applying the precautionary principle is often the reverse of what is actually done.

Figure 11-17 This human-created wetland is located near Orlando, Florida (USA).

CONSIDER THIS. . .

THINKING ABOUT Wetlands Mitigation

Is it good policy to require that a new wetland be created and evaluated before anyone is allowed to destroy the wetland it is supposed to replace? Explain.

CASE STUDY

Can We Restore the Florida Everglades?

South Florida's Everglades was once a 100-kilometer-wide (62-mile-wide), knee-deep sheet of water flowing slowly south from Lake Okeechobee to Florida Bay (Figure 11-18, red dashed lines). As this shallow body of water—known as the "River of Grass"—trickled south, it created a vast network of wetlands (Figure 11-18, photo) with a variety of wildlife habitats.

To help preserve the wilderness in the lower end of the Everglades system, in 1947, the U.S. government established Everglades National Park, which contains a portion of the remaining Everglades. But this protection effort did not work—as conservationists had predicted—because of a massive water distribution and land development project to the north. Between 1962 and 1971, the U.S. Army Corps of Engineers transformed the wandering 166-kilometer-long (103-mile-long) Kissimmee River into a mostly straight 84-kilometer (52-mile) canal flowing into Lake Okeechobee (Figure 11-18). The canal provided flood control by speeding the flow of water, but it drained large wetlands north of Lake Okeechobee, which farmers then converted to grazing land.

This and other projects have provided south Florida's rapidly growing population with a reliable water supply and flood protection. But as a result, much of the original Everglades has been drained, paved over, polluted by agricultural runoff, and invaded by a number of plant and animal species. The Everglades is now less than half its original size and much of it has dried out, leaving large areas vulnerable to summer wildfires. About 90% of the wading birds in Everglades National Park have vanished and populations of many remaining wading bird species have dropped sharply (Figure 11-19). Populations of other vertebrates, from deer to turtles, are down 75–95%.

By the 1970s, state and federal officials recognized that this huge plumbing project was reducing populations of native plants and wildlife—a major source of tourism revenues for Florida. It was also cutting the water supply for the more than 5 million residents of south Florida. In

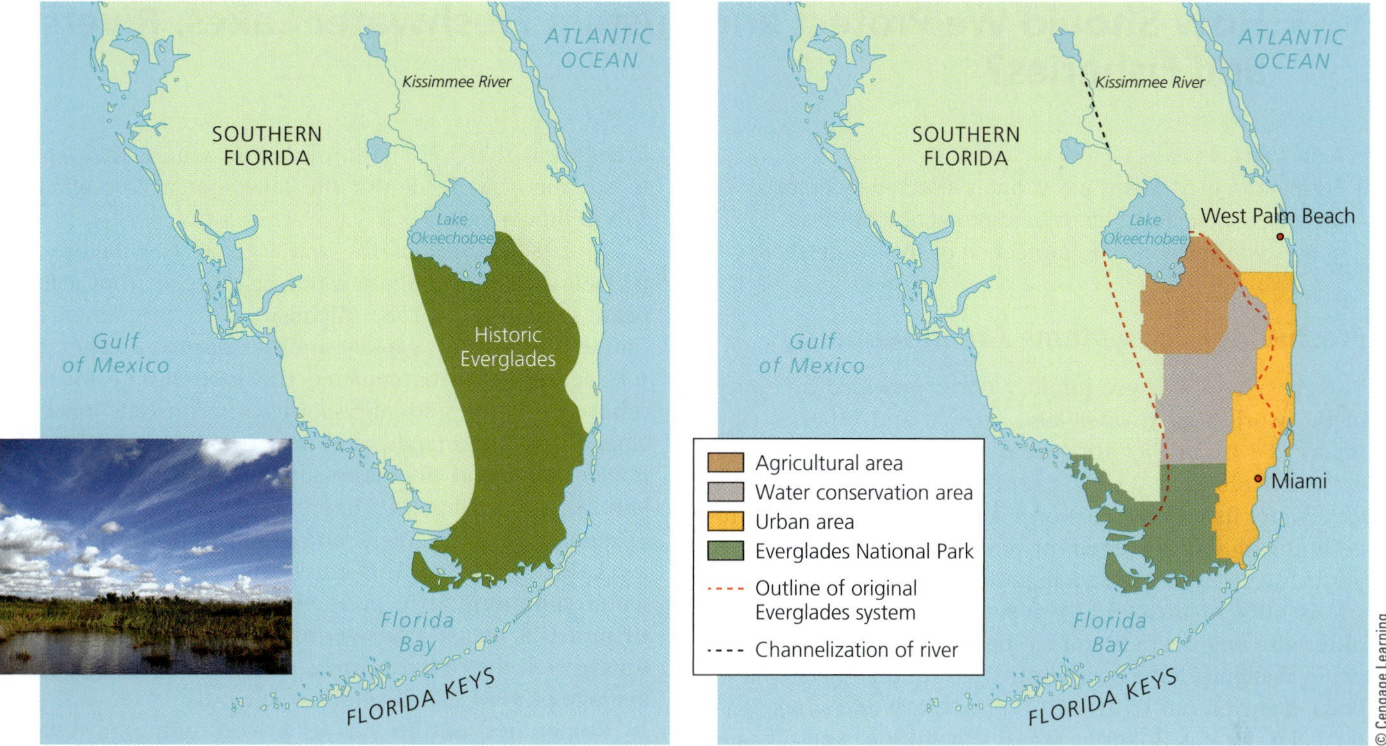

Figure 11-18 Florida's Everglades is the site of the world's largest ecological restoration project—an attempt to undo and redo an engineering project that has been destroying this vast wetland and threatening water supplies for south Florida's rapidly growing population.

Photo: ©slowfish/Shutterstock.com

1990, Florida's state government and the U.S. government agreed on a plan for the world's largest ecological restoration project, known as the Comprehensive Everglades Restoration Plan. The U.S. Army Corps of Engineers is supposed to carry out this joint federal and state plan to partially restore the Everglades.

The project has several ambitious goals, including restoration of the curving flow of more than half of the Kissimmee River; removal of 400 kilometers (248 miles) of canals and levees that block natural water flows south of Lake Okeechobee; conversion of large areas of farmland to marshes; the creation of 18 large reservoirs and underground water storage areas to store water for the lower Everglades and for south Florida's population; and building a canal–reservoir system for catching the water now flowing out to sea and pumping it back into the Everglades.

Will this huge ecological restoration project work? It depends not only on the abilities of scientists and engineers but also on prolonged political and economic support from citizens, the state's powerful sugarcane and agricultural industries, and elected state and federal officials. Already, some restrictions on phosphorus discharges from sugarcane plantations have been relaxed, which could worsen pollution problems. The project has had cost overruns and funding shortages and is considerably behind schedule.

CONSIDER THIS. . .

THINKING ABOUT Everglades Restoration

Do you support carrying out the proposed plan for partially restoring the Florida Everglades, including having the federal government provide half of the funding? Explain.

Figure 11-19 Since 1978, the population of the roseate spoonbill has dropped by more than 50%.

11-5 How Should We Protect and Sustain Freshwater Lakes, Rivers, and Fisheries?

CONCEPT 11-5
Freshwater ecosystems are strongly affected by human activities on adjacent lands, and protection of these ecosystems must include protection of their watersheds.

Freshwater Ecosystems Are in Jeopardy

The ecosystem and economic services provided by many of the world's freshwater lakes, rivers, and fisheries (see Figure 8-14, p. 178) are severely threatened by human activities (**Concept 11-5**). Currently, four of every ten freshwater fish species in North America are considered to be vulnerable, threatened, or endangered, according to a joint study by U.S., Canadian, and Mexican scientists.

As 40% of the world's rivers have been dammed or otherwise engineered, and as many of the world's freshwater wetlands have been destroyed, aquatic species have been crowded out of at least half of the world's freshwater habitat areas. Invasive species, pollution, and climate change threaten the ecosystems of many lakes, rivers, and wetlands. Many freshwater fish stocks are overharvested. And increasing human population pressure and projected climate change during this century will elevate these threats.

Sustaining and restoring the biodiversity and ecosystem services provided by freshwater lakes and rivers is a complex and challenging task, as shown by the following Case Study.

CASE STUDY

Can the Great Lakes Survive Repeated Invasions by Alien Species?

Invasions by nonnative species are a major threat to the biodiversity and ecological functioning of many lakes, as illustrated by what has happened to the five Great Lakes, located on the border between the United States and Canada.

Collectively, the Great Lakes are the world's largest body of freshwater. Since the 1920s, these lakes have been invaded by at least 180 nonnative species and the number keeps rising. Many of the alien invaders arrive on the hulls of, or in bilge-water discharges of, oceangoing ships that have been entering the Great Lakes through the St. Lawrence Seaway since 1959.

One of the biggest threats, the *sea lamprey,* reached the westernmost Great Lakes as early as 1920. This parasite attaches itself to almost any kind of fish and kills the victim by sucking out its blood (see Figure 5-8, p. 107). Over the years, it has depleted populations of many important sport fish species such as lake trout. The United States and Canada keep the lamprey population down by applying a chemical that kills lamprey larvae where they spawn in streams that feed into the lakes—at a cost of about $15 million a year.

In 1986, larvae of the *zebra mussel* (see Figure 9-9, p. 199) arrived in ballast water discharged from a European ship near Detroit, Michigan. This thumbnail-sized mollusk reproduces rapidly and has displaced other mussel species and thus depleted the food supply for some other Great Lakes aquatic species. The mussels have also clogged irrigation pipes, shut down water intake pipes for power plants and city water supplies, fouled beaches, and jammed ships' rudders. They have grown in thick masses on many boat hulls, piers, and other exposed aquatic surfaces (Figure 11-20). This mussel has also spread to freshwater communities in parts of southern Canada and 18 eastern U.S. states. Damages and attempts to control this mussel cost the two countries about $1 billion a year—an average of $114,000 per hour.

Sometimes, nature aids us in controlling an invasive alien species. For example, populations of zebra mussels are declining in some parts of the Great Lakes because a

Figure 11-20 These *zebra mussels* are attached to a water current meter in Lake Michigan.

native sponge growing on their shells is preventing them from opening up their shells to breathe. However, it is not clear whether the sponges will effectively control the invasive mussels in the long run and what harmful ecological effects the sponges might have.

In 1989, a larger and potentially more destructive species, the *quagga mussel,* invaded the Great Lakes, probably discharged in the ballast water of a Russian freighter. It can survive at greater depths and tolerate more extreme temperatures than the zebra mussel can. In 2009, scientists reported that quagga mussels had rapidly replaced many other bottom-dwellers in Lake Michigan. This has reduced the food supply for many fish and other species, thus leading to a major disruption of the lake's food web. There is concern that quagga mussels may spread by river transport and eventually colonize eastern U.S. ecosystems such as the Chesapeake Bay (see Chapter 8, Case Study, p. 177) and waterways in parts of Florida. In 2007, it was found to have crossed the United States, probably hitching a ride on a boat or trailer being hauled cross-country. It now resides in the Colorado River and its reservoir system.

The *Asian carp* (Figure 11-21) is the most recent threat to the Great Lakes system. In the 1970s, catfish farmers in the southern United States imported two species of Asian carp to help remove suspended matter and algae from their aquaculture farm ponds. Heavy flooding during the 1990s caused many of these ponds to overflow, which resulted in the release of some of the carp into the Mississippi River. After working their way up the Mississippi River system, these invaders are now close to entering Lake Michigan, if they have not done so already.

These highly prolific fish can quickly grow as long as 1.2 meters (4 feet) and weigh up to 50 kilograms (110 pounds), and they can eat as much as 20% of their own body weight in plankton every day. They can therefore rapidly disrupt lake food webs. They also have the ability to jump clear of the water, and several boaters have been injured after being hit by jumping carp.

These fish have no natural predators in the rivers they have invaded or in the Great Lakes. Joel Brammeler, president of the Alliance for the Great Lakes, warned that "if Asian carp get into Lake Michigan, there is no stopping them."

Figure 11-21 Asian carp may be the next major invasive species to threaten the Great Lakes.

Managing River Basins Is Complex and Controversial

Rivers and streams provide important ecosystem and economic services (Figure 11-22), but overfishing, pollution, dams, and water withdrawal for irrigation are disrupting these services. Currently, the services are given little or no monetary value when the costs and benefits of dam and reservoir projects are assessed. According to environmental economists, attaching even crudely estimated monetary values to these ecosystem services—an application of the full-cost pricing **principle of sustainability** (see Figure 1-5, p. 9 or back cover)—would help to sustain them.

An example of such disruption—one that especially illustrates biodiversity loss—is what happened in the Columbia River, which runs through parts of southwestern Canada and the northwestern United States. It has 119 dams, 19 of which are major generators of inexpensive hydroelectric power. It also supplies water for major urban areas and large irrigation projects.

The Columbia River dam system has benefited many people, but it has sharply reduced populations of wild salmon. These migratory fish hatch in the upper reaches of the streams and rivers that form the headwaters of the Columbia River, migrate to the ocean where they spend most of their adult lives, and then swim upstream to return to the place where they were hatched to spawn and die. Dams interrupt their life cycle by interfering with the migration of young fish downstream, and blocking the return of mature fish attempting to swim upstream to their spawning grounds.

Natural Capital

Ecosystem Services of Rivers

- Deliver nutrients to sea to help sustain coastal fisheries
- Deposit silt that maintains deltas
- Purify water
- Renew and renourish wetlands
- Provide habitats for wildlife

© Cengage Learning

Figure 11-22 Rivers provide some important ecosystem services. ***Questions:*** Which two of these services do you believe are the most important? Why?

Since the dams were built, the Columbia River's wild Pacific salmon population has dropped by 94% and nine of the Pacific Northwest salmon species are listed as endangered or threatened. Since 1980, the U.S. federal government has spent more than $3 billion in efforts to save the salmon, but none have been effective.

We Can Protect Freshwater Ecosystems by Protecting Watersheds

Sustaining freshwater aquatic systems begins with understanding that land and water are connected. For example, lakes and streams receive many of their nutrients from the ecosystems of bordering land. Such nutrient inputs come from falling leaves, animal feces, and pollutants generated by people, all of which are washed into bodies of water by rainstorms and melting snow. Therefore, to protect a stream or lake from excessive inputs of nutrients and pollutants, we must protect its watershed (**Concept 11-5**).

As with marine systems, freshwater ecosystems can be protected through laws, economic incentives, and restoration efforts. For example, restoring and sustaining the ecosystem and economic services of rivers will probably require taking down some dams and restoring river flows. In addition, some scientists and politicians have argued for protecting all remaining free-flowing rivers by prohibiting the construction of dams.

With that in mind, in 1968, the U.S. Congress passed the National Wild and Scenic Rivers Act to establish protection of rivers with outstanding wildlife, geological, scenic, recreational, historical, or cultural values. The law classified *wild rivers* as those that are relatively inaccessible (except by trail), and *scenic rivers* as rivers of great scenic value that are free of dams, mostly undeveloped, and accessible in only a few places by roads. These rivers are now protected from widening, straightening, dredging, filling, and damming (Figure 11-23).

Figure 11-23 The upper St. Croix River, flowing on the border between Minnesota and Wisconsin, was one of the first eight rivers to be designated a national wild and scenic river. It hosts a rich diversity of plant and animal species that are protected from riverside development and other harmful human activities.

In 2009, the U.S. Congress passed a law increasing the total length of wild and scenic rivers by half. But the Wild and Scenic Rivers System keeps only 3% of U.S. rivers free-flowing and protects less than 1% of the country's total river length.

CONSIDER THIS. . .

CAMPUS SUSTAINABILITY

Maine's College of the Atlantic

Courtesy College of the Atlantic

The College of the Atlantic (COA) in Bar Harbor, Maine, near Acadia National Park is well known for its student-led projects aimed at helping people to live more sustainably. COA students are working with people from the local community to preserve the Union River watershed near their campus. Like any river ecosystem, the Union River system depends on the environmental quality of its watershed—the land from which water drains to form the river. In working on this project, students are helping to map the watershed and prevent or control pollution within this aquatic system, which in turn helps to preserve the area's aquatic biodiversity.

Freshwater Fisheries Need Better Protection

Sustainable management of freshwater fisheries involves supporting populations of commercial and sport fish species, preventing such species from being overfished, and reducing or eliminating populations of harmful invasive species. The traditional way of managing freshwater fish species is to regulate the time and length of fishing seasons and the number and size of fish that can be taken.

Other techniques include building reservoirs and ponds and stocking them with fish, and protecting and creating fish spawning sites. In addition, some fishery managers try to protect fish habitats from sediment buildup and other forms of pollution. They also work to prevent or reduce large human inputs of plant nutrients that spur the excessive growth of aquatic plants.

Some fishery managers seek to control predators, parasites, and diseases by improving habitats, breeding genetically resistant fish varieties, and using antibiotics and dis-

National Park Service

infectants. Hatcheries can be used to restock ponds, lakes, and streams with prized species such as trout, and entire river basins can be managed to protect valued species such as salmon. However, all of these practices should be based on continuing studies of their effects on aquatic biodiversity and ecosystem services. **GREEN CAREERS:** Limnology and fishery management

11-6 What Should Be Our Priorities for Sustaining Aquatic Biodiversity?

CONCEPT 11-6
Sustaining the world's aquatic biodiversity requires mapping it, protecting aquatic hotspots, creating large and fully protected marine reserves, protecting freshwater ecosystems, and restoring degraded coastal and inland wetlands.

We Can Use an Ecosystem Approach to Sustain Aquatic Biodiversity and Ecosystem Services

Edward O. Wilson and other biodiversity experts have proposed the following priorities for an ecosystem approach to sustaining aquatic biodiversity and ecosystem services (Concept 11-6):

- Complete the mapping of the world's aquatic biodiversity, identifying and locating as many plant and animal species as possible in order to make conservation efforts more precise and cost-effective.
- Identify and preserve the world's aquatic biodiversity hotspots and areas where deteriorating ecosystem services threaten people and many other forms of life.
- Create large and fully protected marine reserves to allow damaged marine ecosystems to recover and to allow fish stocks to be replenished.
- Protect and restore the world's lakes and river systems, which are among the most threatened ecosystems of all.
- Initiate ecological restoration projects worldwide in systems such as coral reefs and inland and coastal wetlands to help preserve the biodiversity and important ecosystem services in those systems.
- Find ways to raise the incomes of people who live on or near protected lands and waters so that they can become partners in the protection and sustainable use of ecosystems.

There is growing evidence that the current harmful effects of human activities on aquatic biodiversity and ecosystem services could be reversed over the next two decades. Doing this will require implementing an ecosystem approach to sustaining both terrestrial and aquatic ecosystems. According to Edward O. Wilson (see Individuals Matter 4.1, p. 82), such a conservation strategy would cost about $30 billion per year—an amount that could be provided by a tax of one penny per cup of coffee consumed in the world each year.

GOOD NEWS

This strategy for protecting the earth's vital biodiversity and ecosystem services will not be implemented without bottom-up political pressure on elected officials from individual citizens and groups. It will also require cooperation among scientists, engineers, and key people in government and the private sector.

A key part of this strategy will be for individuals to "vote with their wallets" by trying to buy only products and services that do not have harmful impacts on terrestrial and aquatic biodiversity. For example, we can eat lower on the food chain by choosing plant-eating species such as tilapia, carp, and catfish instead of carnivorous species such as salmon and sea bass, and especially, instead of farm-raised species that are typically fed fishmeal made of wild-caught fish. Guides to sustainably produced seafood are available on websites such as that of the Monterrey Bay Aquarium.

Big Ideas

- **The world's aquatic systems provide important economic and ecosystem services, and scientific investigation of these poorly understood ecosystems could lead to immense ecological and economic benefits.**
- **Aquatic ecosystems and fisheries are being severely degraded by human activities that lead to aquatic habitat disruption and loss of biodiversity.**
- **We can sustain aquatic biodiversity by establishing protected sanctuaries, managing coastal development, reducing water pollution, and preventing overfishing.**

TYING IT ALL TOGETHER Sea Turtles and Sustainability

Rich Carey/Shutterstock.com

This chapter began with a look at the threatened and endangered species of sea turtles and how human activities are threatening their populations in several ways (**Core Case Study**), including habitat destruction and degradation of vital marine systems such as coral reefs (photo at left). Throughout the chapter, we also examined how these threats along with introductions of invasive species, population pressures, climate change, and overexploitation are harming sea turtles and many other marine and freshwater aquatic species. We looked at how many of the world's fisheries are being depleted.

We also explored possible solutions to these problems. Importantly, we know that when areas of the oceans are left undisturbed, marine ecosystems tend to recover their natural functions, and fish populations rebound fairly quickly. Similarly, it seems that the best approach to sustaining freshwater biodiversity is to use an ecosystem approach in order to use the world's wetlands, lakes, and rivers more sustainably.

We can achieve greater success in sustaining aquatic biodiversity by applying the three scientific **principles of sustainability** (see Figure 1-2, p. 6 or back cover). This means reducing inputs of sediments and excess nutrients, which cloud water, lessen the input of solar energy, and upset aquatic food webs and the natural cycling of nutrients in aquatic systems. It also means valuing aquatic biodiversity and putting a high priority on preserving the biodiversity and ecosystem services of aquatic systems. Applying the social science **principles of sustainability** (see Figure 1-5, p. 9 or back cover) can help us to achieve these goals.

Chapter Review

Core Case Study

1. Describe the status of sea turtles and explain how human activities are threatening their existence (**Core Case Study**).

Section 11-1

2. What is the key concept for this section? How much do we know about the habitats and species that make up the earth's aquatic biodiversity? What are three general patterns of marine biodiversity? Describe the threat to marine biodiversity from bottom trawling. How have coral reefs been threatened? What is **ocean acidification**? What are two causes of disruption of freshwater habitats?

3. Give two examples of threats to marine aquatic systems from invasive species and two examples of the same for freshwater systems. What are two harmful effects on aquatic systems resulting from the growth of the human population in coastal areas? Give two examples of how pollution is affecting aquatic systems. What are three ways in which projected climate change could threaten aquatic biodiversity? Explain how ocean acidification occurs and why it is a serious problem.

4. Define **fishery**. What are three major harmful effects of overfishing? Describe the effects of trawler fishing, purse-seine fishing, long-lining, and drift-net fishing. What is *bycatch*? What is a **fishprint**? Summarize the story of the collapse of the Atlantic cod fishery. Explain why the Atlantic bluefin tuna is seriously endangered. Explain how marine mammals are threatened by overfishing and give an example. Summarize the story of jellyfish invasions. Summarize the arguments for protecting sharks. About what percentage of marine species and what percentage of freshwater species are in danger of extinction?

Section 11-2

5. What is the key concept for this section? How have laws and treaties been used to help sustain aquatic species? What is the main problem that interferes with enforcing international agreements? How can economic incentives help to sustain aquatic biodiversity? Give an example of how this can happen.

6. Explain how marine protected areas and marine reserves can be used to help sustain aquatic biodiversity and ecosystem services. What are *marine hotspots*? What percentage of the world's oceans is strictly protected from harmful human activities in marine reserves? Summarize the contributions of Sylvia Earle to the protection of aquatic biodiversity. Give two examples of how marine systems can be restored. Describe the roles of fishing communities and individual consumers in regulating fishing and coastal development. What is *integrated coastal management*?

Section 11-3

7. What is the key concept for this section? Describe three ways of estimating the sizes of fish populations and list their limitations. How can the precautionary principle be applied in managing fisheries and large marine systems? What are catch-share and co-management systems and how can they help to sustain fisheries? How can government subsidies encourage overfishing? Explain how consumer choices can help to sustain fisheries and aquatic biodiversity and ecosystem services. List five ways to manage global fisheries more sustainably.

Section 11-4

8. What is the key concept for this section? What percentage of the U.S. coastal and inland wetlands has been destroyed since 1900? What are the major threats to wetlands and their ecosystem services? How does the United States attempt to reduce wetland losses? Summarize the story of efforts to restore the Florida Everglades.

Section 11-5

9. What is the key concept for this section? Describe the major threats to the world's rivers and other freshwater systems. Explain how invasions by nonnative species are threatening the Great Lakes. What are some ways to help sustain river systems? What are three ways to protect freshwater habitats and fisheries?

Section 11-6

10. What is the key concept for this section? List six priorities for applying the ecosystem approach to sustaining aquatic biodiversity. What are the three big ideas of this chapter? How can we apply the three scientific **principles of sustainability** in efforts to protect sea turtles from extinction (**Core Case Study**) and in helping to sustain aquatic biodiversity and ecosystem services?

SUSTAINABILITY

Note: Key terms are in bold type.

Critical Thinking

1. Why could sea turtles (**Core Case Study**) be considered indicator species? What does the plight of the world's sea turtles indicate about the ways in which we treat marine ecosystems?

2. Write a short essay describing how each of the six factors summarized by HIPPCO has affected sea turtles (**Core Case Study**) or could affect them further. Look for and describe connections among these factors. For example, how does one factor enhance the effects of one or more other factors? Suggest ways in which each of the factors could be reduced.

3. What do you think are the three greatest threats to aquatic biodiversity and ecosystem services? For each of them, explain your thinking. Overall, why are aquatic species more vulnerable to extinction hastened by human activities than terrestrial species are? Why is it more difficult to identify and protect endangered marine species than to identify and protect endangered terrestrial species?

4. Why should you be concerned about jellyfish populations taking over large areas of the ocean? Why can jellyfish be considered an indicator species, and what does the explosion of jellyfish populations indicate about marine ecosystems? What are three things you would do to try to slow the changes in the world's aquatic ecosystems that have likely caused the rise of jellyfish populations?

5. Why do you think no-fishing marine reserves recover their biodiversity faster and more effectively than do protected areas where fishing is allowed but restricted? Explain.

6. How might continued overfishing of marine species affect your lifestyle? How could it affect the lives of any children or grandchildren you might have? What are three things you could do to help prevent overfishing?

7. Should fishers who harvest fish from a country's publicly owned waters be required to pay the government fees for the fish they catch? Explain. If your livelihood depended on commercial fishing, would you be for or against such fees?

8. Some scientists consider ocean acidification to be one of the most serious environmental and economic threats that the world faces. How do you contribute to ocean acidification in your daily life? What are three things you could do to help reduce the threat of ocean acidification?

Doing Environmental Science

Pick a coastal area, river, stream, lake, or wetland near where you live and research and write a brief account of its history. Then survey and take notes on its present condition. Has its condition improved or deteriorated during the last 10 years? What governmental or private efforts are being used to protect this aquatic system? Write a report summarizing your findings. Based on your report along with your ecological knowledge of this system, write up some recommendations to policy makers for protecting it. Try presenting your recommendations to one or more local policy makers.

Global Environment Watch Exercise

Search for *Coral Reefs* and use the topic portal to do the following: **(a)** identify two cases of coral reef degradation—one where the reef continues to be degraded and one where a degraded reef is being restored; **(b)** for each case, list the main causes for this degradation; **(c)** for each case, describe any efforts to protect or restore the reef; **(d)** for each case, give your best estimation of whether the reef will survive and explain your reasoning.

Ecological Footprint Analysis

A fishprint provides a measure of a country's fish harvest in terms of area. The unit of area used in fishprint analysis is the global hectare (gha), a unit weighted to reflect the relative ecological productivity of the area fished. When compared with the fishing area's *sustainable biocapacity* (its ability to provide a stable supply of fish year after year, expressed in terms of yield per area), its fishprint indicates whether the country's annual fishing harvest is sustainable.

The fishprint and biocapacity are calculated using the following formulas:

Fishprint in (gha) = metric tons of fish harvested per year/productivity in metric tons per hectare × weighting factor

Biocapacity in (gha) = sustained yield of fish in metric tons per year/productivity in metric tons per hectare × weighting factor

The following graph shows the earth's total fishprint and biocapacity. Study it and answer the questions that follow.

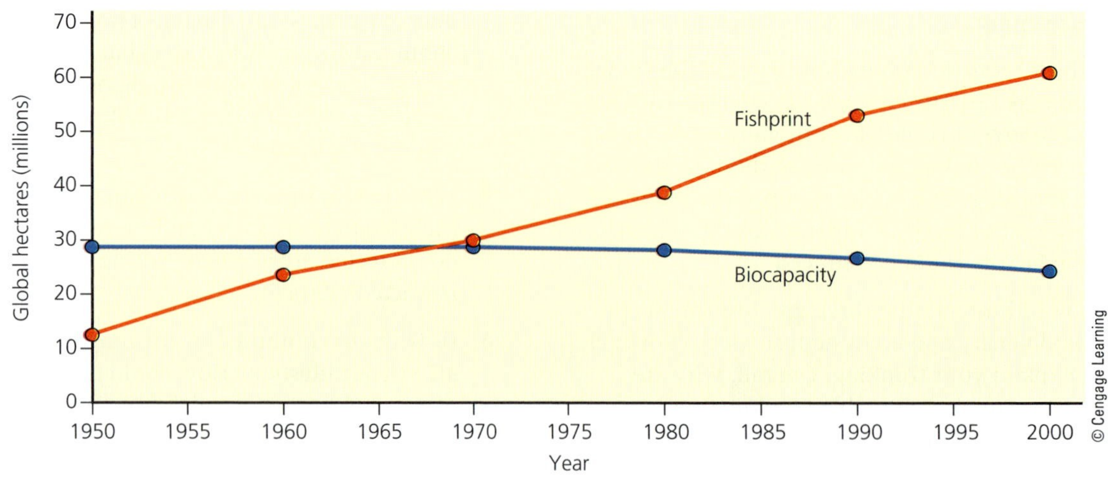

1. Based on the graph,
   a. What is the current status of the global fisheries with respect to sustainability?
   b. In what year did the global fishprint begin to exceed the biological capacity of the world's oceans?
   c. By how much did the global fishprint exceed the biological capacity of the world's oceans in 2000?
2. Assume a country harvests 18 million metric tons of fish annually from an ocean area with an average productivity of 1.3 metric tons per hectare and a weighting factor of 2.68. What is the annual fishprint of that country?
3. If biologists determine that this country's sustained yield of fish is 17 million metric tons per year,
   a. What is the country's sustainable biological capacity?
   b. Is the county's annual fishing harvest sustainable?
   c. To what extent, as a percentage, is the country undershooting or overshooting its biological capacity?

SUPPLEMENT 1 MEASUREMENT UNITS

Length

Metric

1 kilometer (km) = 1,000 meters (m)
1 meter (m) = 100 centimeters (cm)
1 meter (m) = 1,000 millimeters (mm)
1 centimeter (cm) = 0.01 meter (m)
1 millimeter (mm) = 0.001 meter (m)

English

1 foot (ft) = 12 inches (in)
1 yard (yd) = 3 feet (ft)
1 mile (mi) = 5,280 feet (ft)
1 nautical mile = 1.15 miles

Metric–English

1 kilometer (km) = 0.621 mile (mi)
1 meter (m) = 39.4 inches (in) or 3.28 feet (ft)
1 inch (in) = 2.54 centimeters (cm)
1 foot (ft) = 0.305 meter (m)
1 yard (yd) = 0.914 meter (m)
1 nautical mile = 1.85 kilometers (km)

Area

Metric

1 square kilometer (km^2) = 1,000,000 square meters (m^2)
1 square meter (m^2) = 1,000,000 square millimeters (mm^2)
1 square meter (m^2) = 10,000 square centimeters (cm^2)
1 hectare (ha) = 10,000 square meters (m^2)
1 hectare (ha) = 0.01 square kilometer (km^2)

English

1 square foot (ft^2) = 144 square inches (in^2)
1 square yard (yd^2) = 9 square feet (ft^2)
1 square mile (mi^2) = 27,880,000 square feet (ft^2)
1 acre (ac) = 43,560 square feet (ft^2)

Metric–English

1 hectare (ha) = 2.471 acres (ac)
1 square kilometer (km^2) = 0.386 square mile (mi^2)
1 square meter (m^2) = 1.196 square yards (yd^2)
1 square meter (m^2) = 10.76 square feet (ft^2)
1 square centimeter (cm^2) = 0.155 square inch (in^2)

Volume

Metric

1 cubic kilometer (km^3) = 1,000,000,000 cubic meters (m^3)
1 cubic meter (m^3) = 1,000,000 cubic centimeters (cm^3)
1 cubic meter (m^3) = 1,000 liters (L)
1 liter (L) = 1,000 milliliters (mL) = 1,000 cubic centimeters (cm^3)
1 cubic meter (m^3) = 1,000 liters (L)
1 milliliter (mL) = 0.001 liter (L)
1 milliliter (mL) = 1 cubic centimeter (cm^3)

English

1 gallon (gal) = 4 quarts (qt)
1 quart (qt) = 2 pints (pt)

Metric–English

1 liter (L) = 0.265 gallon (gal)
1 liter (L) = 1.06 quarts (qt)
1 liter (L) = 0.0353 cubic foot (ft^3)
1 cubic meter (m^3) = 35.3 cubic feet (ft^3)
1 cubic meter (m^3) = 1.30 cubic yards (yd^3)
1 cubic kilometer (km^3) = 0.24 cubic mile (mi^3)
1 barrel (bbl) = 159 liters (L)
1 barrel (bbl) = 42 U.S. gallons (gal)

Mass

Metric

1 kilogram (kg) = 1,000 grams (g)
1 gram (g) = 1,000 milligrams (mg)
1 gram (g) = 1,000,000 micrograms (μg)
1 milligram (mg) = 0.001 gram (g)
1 microgram (μg) = 0.000001 gram (g)
1 metric ton (mt) = 1,000 kilograms (kg)

English

1 ton (t) = 2,000 pounds (lb)
1 pound (lb) = 16 ounces (oz)

Metric–English

1 metric ton (mt) = 2,200 pounds (lb) = 1.1 tons (t)
1 kilogram (kg) = 2.20 pounds (lb)
1 pound (lb) = 454 grams (g)
1 gram (g) = 0.035 ounce (oz)

Energy and Power

Metric

1 kilojoule (kJ) = 1,000 joules (J)
1 kilocalorie (kcal) = 1,000 calories (cal)
1 calorie (cal) = 4.184 joules (J)

Metric–English

1 kilojoule (kJ) = 0.949 British thermal unit (Btu)
1 kilojoule (kJ) = 0.000278 kilowatt-hour (kW-h)
1 kilocalorie (kcal) = 3.97 British thermal units (Btu)
1 kilocalorie (kcal) = 0.00116 kilowatt-hour (kW-h)
1 kilowatt-hour (kW-h) = 860 kilocalories (kcal)
1 kilowatt-hour (kW-h) = 3,400 British thermal units (Btu)
1 quad (Q) = 1,050,000,000,000,000 kilojoules (kJ)
1 quad (Q) = 293,000,000,000 kilowatt-hours (kW-h)

Temperature Conversions

Fahrenheit (°F) to Celsius (°C): °C = (°F − 32.0) ÷ 1.80
Celsius (°C) to Fahrenheit (°F): °F = (°C × 1.80) + 32.0

SUPPLEMENT 2 READING GRAPHS AND MAPS

Graphs and Maps Are Important Visual Tools

A **graph** is a tool for conveying information that we can summarize numerically by illustrating that information in a visual format. This information, called *data,* is collected in experiments, surveys, and other information-gathering activities. Graphing can be a powerful tool for summarizing and conveying complex information.

In this textbook, we use three major types of graphs: *line graphs, bar graphs,* and *pie graphs.* Here, you will explore each of these types of graphs and learn how to read them.

An important visual tool used to summarize data that vary over small or large areas is a **map**. We discuss some aspects of reading maps relating to environmental science at the end of this supplement.

Line Graphs

Line graphs usually represent data that fall in some sort of sequence such as a series of measurements over time or distance. In most such cases, units of time or distance lie on the horizontal *x-axis.* The possible measurements of some quantity or variable such as temperature that changes over time or distance usually lie on the vertical *y-axis.*

In Figure 1, the x-axis shows the years between 1950 and 2009, and the y-axis displays the possible values for the annual amounts of oil consumed worldwide during that time in millions of tons, ranging from 0 to 4,000 million (or 4 billion) tons. Usually, the y-axis appears on the left end of the x-axis, although y-axes can appear on the right end, in the middle, or on both ends of the x-axis.

The curving line on a line graph represents the measurements taken at certain time or distance intervals. In Figure 1, the curve represents changes in oil consumption between 1950 and 2009. To find the oil consumption for any year, find that year on the x-axis (a point called the *abscissa*) and run a vertical line from the axis to the curve. At the point where your line intersects the curve, run a horizontal line to the y-axis. The value at that point on the y-axis, called the *ordinate,* is the amount you are seeking. You can go through the same process in reverse to find a year in which oil consumption was at a certain point.

Questions

1. What was the total amount of oil consumed in the world in 1990? In about what year between 1950 and 2000 did oil consumption first start declining?
2. About how much oil was consumed in 2009? Roughly how many times more oil was consumed in 2009 than in 1970? How many times more oil was consumed in 2009 than in 1950?

Line graphs have several important uses. One of the most common applications is to compare two or more variables. Figure 2 compares two variables: monthly temperature and precipitation (rain and snowfall) during a typical year in a temperate deciduous forest. However, in this case the variables are measured on two different scales, so there are two y-axes. The y-axis on the left end of the graph shows a Centigrade temperature scale, and the y-axis on the right shows the range of precipitation measurements in millimeters. The x-axis displays the first letters of each of the 12 months' names.

Questions

1. In which month does most precipitation fall? Which is the driest month of the year? Which is the hottest month?

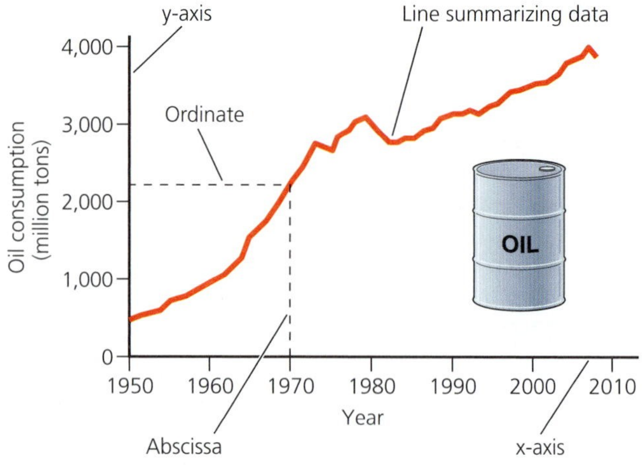

Figure 1 World oil consumption, 1950–2009.

(Compiled by the authors using data from U.S. Energy Information Administration, International Energy Agency, and United Nations.)

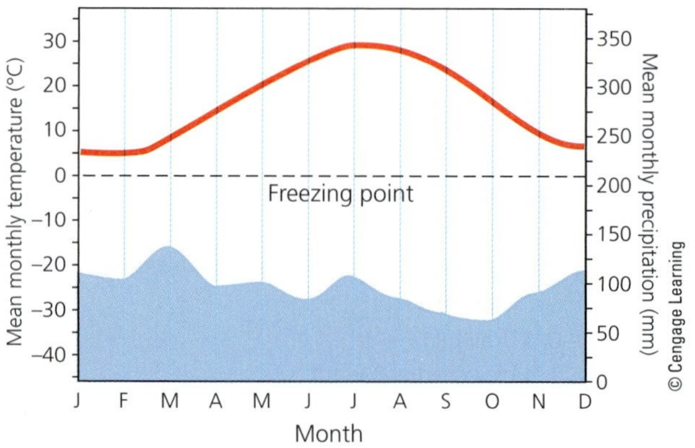

Figure 2 Climate graph showing typical variations in annual temperature (red) and precipitation (blue) in a temperate deciduous forest.

2. If the temperature curve were almost flat, running throughout the year at roughly its highest point of about 30°C, how do you think this forest would change from what it is now (see Figure 7-13, center, p. 156)? If the annual precipitation suddenly dropped and remained under 25 centimeters all year, what do you think would eventually happen to this forest?

It is also important to consider what aspect of a set of data is being displayed on a graph. The creator of a graph can take two different aspects of one data set and create two very different-looking graphs that would give two different interpretations of the same phenomenon. For example, when talking about any type of growth we must be careful to distinguish the question of whether something is growing from the question of how fast it is growing. While a quantity can keep growing continuously, its rate of growth can go up and down.

One of many important examples of growth used in this book is human population growth. For example, the graph in Figure 1-16 (p. 17) gives you the impression that human population growth has, for the most part, been continuous and uninterrupted. However, consider Figure 3, which plots the rate of growth of the human population since 1950. Note that all of the numbers on the y-axis, even the smallest ones, represent growth. The lower end of the scale represents slower growth and the higher end faster growth. Thus, while one graph tracks population growth in terms of numbers of people, the other tracks the rate of growth.

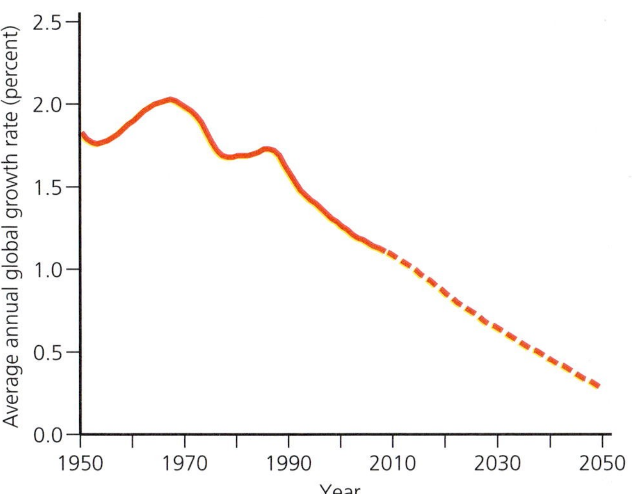

Figure 3 Annual growth rate in world population, 1950–2010, with projections to 2050.

(Compiled by the authors using data from UN Population Division, U.S. Census Bureau, and Population Reference Bureau.)

Questions

1. If this graph were presented to you as a picture of human population growth, what would be your first impression?
2. Do you think that reaching a growth rate of 0.5% would relieve those who are concerned about overpopulation? Why or why not?

Bar Graphs

The *bar graph* is used to compare measurements for one or more variables across categories. Unlike the line graph, a bar graph typically does not involve a sequence of measurements over time or distance. The measurements compared on a bar graph usually represent data collected at some point in time or during a well-defined period. For instance, we can compare the *net primary productivity (NPP),* a measure of chemical energy produced by plants in an ecosystem, for different ecosystems, as represented in Figure 4.

In most bar graphs, the categories to be compared are laid out on the x-axis, and the range of measurements for the variable under consideration lies along the y-axis. In our example in Figure 4, the categories (ecosystems) are on the y-axis, and the variable range (NPP) lies on the x-axis. In either case, reading the graph is straightforward. Simply run a line perpendicular to the bar you are reading from the top of that bar (or the right or left end, if it lies horizontally) to the variable value axis. In Figure 4, you can see that the NPP for continental shelf, for example, is close to 1,600 kcal/m^2/yr.

Questions

1. What are the two terrestrial ecosystems that are closest in NPP value of all pairs of such ecosystems? About how many times greater is the NPP in a tropical rain forest than the NPP in a savanna?
2. What is the most productive of aquatic ecosystems shown here? What is the least productive?

An important application of the bar graph used in this book is the *age-structure diagram* (see Figure 6-11, p. 131), which describes a population by showing the numbers of males and females in certain age groups (see Chapter 6, pp. 131–132).

Pie Graphs

Like bar graphs, *pie graphs,* or *pie charts,* illustrate numerical values for two or more categories. But in addition to that, they can also show each category's proportion of the total of all measurements. The categories are usually ordered on the graph from largest to smallest, for ease of comparison, although this is not always the case. Also, as with bar graphs, pie graphs are generally snapshots of a

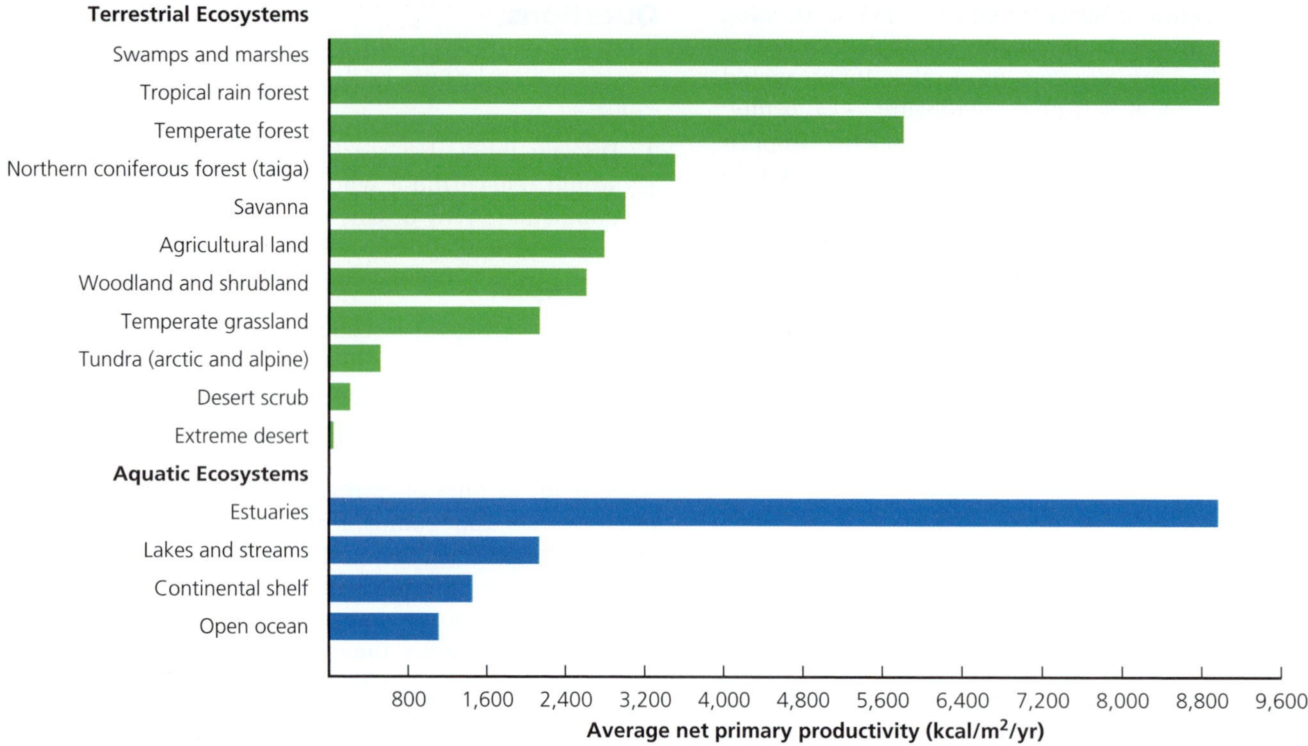

Figure 4 Estimated annual average net primary productivity (NPP) in major life zones and ecosystems, expressed as kilocalories of energy produced per square meter per year (kcal/m²/yr).

(Compiled by the authors using data from R. H. Whittaker, *Communities and Ecosystems*, 2nd ed., New York: Macmillan, 1975.)

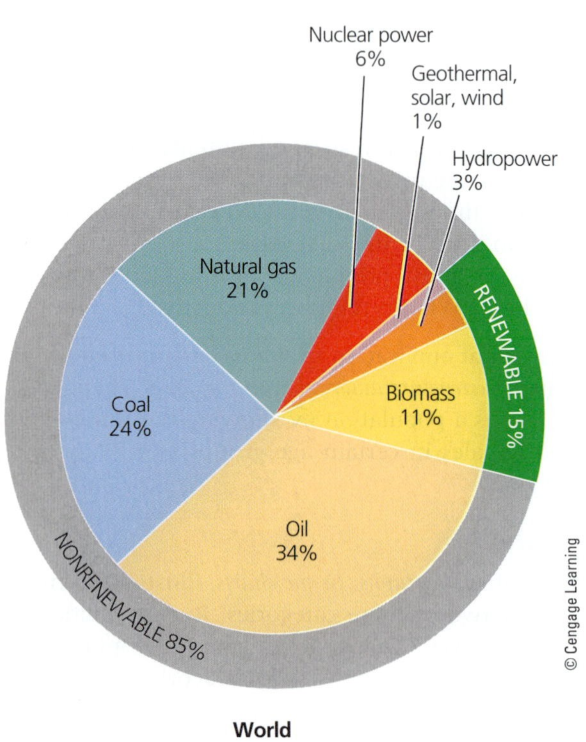

Figure 5 Pie graph showing world energy use by source in 2010.

data set at a point in time or during a defined time period. Unlike line graphs, one pie graph cannot show changes over time.

For example, Figure 5 shows how much each major energy source contributed to the world's total amount of energy used in 2010. This graph includes the numerical data used to construct it: the percentages of the total taken up by each part of the pie. But we can use pie graphs without including the numerical data and we can roughly estimate such percentages. The pie graph thereby provides a generalized picture of the composition of a data set.

Questions

1. Can you tell from this graph whether the use of renewable energy sources is growing or shrinking? Explain.
2. About how many times bigger is oil use than biomass use?

Reading Maps

We can use maps for considerably more than showing where places are relative to one another. For example, in environmental science, maps can be very helpful in comparing how people in different areas are affected by environmental problems such as air pollution and acid deposition (a form of air pollution). Figure 6 is a map of the United States showing the relative numbers of premature deaths due to air pollution in the various regions of the country.

Questions

1. Which part of the country generally has the lowest level of premature deaths due to air pollution?
2. Which part of the country has the highest level? What is the level in the area where you live or go to school?

In some of the data analysis exercises that appear at the ends of the chapters in this book, you will have opportunities to apply much of the information from this supplement.

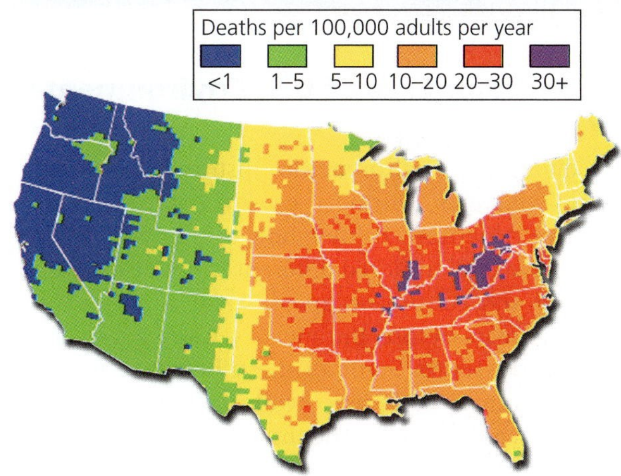

Figure 6 Map showing comparative numbers of premature deaths from air pollution in the United States.

(Compiled by the authors using data from U.S. Environmental Protection Agency.)

SUPPLEMENT 3 ENVIRONMENTAL HISTORY OF THE UNITED STATES

The Four Major Eras of U.S. Environmental History

We can divide the environmental history of the United States into four eras. During the *tribal era,* people (now called Native Americans or members of the First Nations) belonging to several hundred different tribes distinguished by language and culture occupied North America for at least 13,000 years before European settlers began arriving in the early 1600s. Some of the tribes were made up of nomadic hunter–gatherers, while others settled into communities centered on agriculture. Some of the more populous Native American societies were more sophisticated than their counterparts in Europe in terms of agricultural practices and other cultural aspects. Yet these North American tribes generally had more sustainable, low-impact ways of life because of their relatively limited numbers and modest rates of resource use per person.

Next was the *frontier era* (1607–1890) when European colonists began settling North America. Faced with a continent offering seemingly inexhaustible resources, the early colonists developed a *frontier environmental worldview.* That is, they saw a wilderness to be conquered and managed for human use.

Next came the *early conservation era* (1832–1870), which overlapped the end of the frontier era. During this period, some people became alarmed at the scope of resource depletion and degradation in the United States. They argued that part of the unspoiled wilderness on public lands should be protected as a legacy for future generations. Most of these warnings and ideas were not taken seriously.

This period was followed by an era—lasting from 1870 to the present—featuring an increased role by the federal government and private citizens in resource conservation, public health, and environmental protection.

The Frontier Era (1607–1890)

During the frontier era, European settlers spread across the land by clearing forests for cropland and settlements. In the process, they displaced the Native Americans who, for the most part, had lived on the land sustainably for thousands of years.

The U.S. government accelerated this settling of the continent and the use of its resources by transferring vast areas of public land to private interests. To encourage settlement, between 1850 and 1890, more than half of the country's public land was given away or sold cheaply by the government to railroad, timber, and mining companies, land developers, states, schools, universities, and homesteaders. This era came to an end when the government declared the frontier officially closed in 1890.

Early Conservationists (1832–1870)

Between 1832 and 1870, some citizens became alarmed at the scope of resource depletion and degradation in the United States. They urged the government to preserve part of the unspoiled wilderness on public lands owned jointly by all people (but managed by the government), and to protect it as a legacy for future generations.

Two of these early conservationists were Henry David Thoreau (1817–1862) and George Perkins Marsh (1801–1882). Thoreau (Figure 1) was alarmed at the loss of numerous wild species from his native eastern Massachusetts. To gain a better understanding of nature, he built a cabin in the woods on Walden Pond near Concord, Massachusetts, lived there alone for 2 years, and wrote *Life in the Woods,* an environmental classic.*

In 1864, Marsh, a scientist and member of Congress from Vermont, published *Man and Nature,* an extensive study of how human activities were altering the environment, which helped legislators and citizens see the need for resource conservation. Marsh questioned the idea that the country's resources were inexhaustible. He also used scientific studies and case studies to show how the rise and fall of past civilizations were linked to the use and

*I (Miller) can identify with Thoreau. I spent 10 years living in the deep woods studying and thinking about how nature works and writing early editions of the book you are reading. See the whole story of this adventure on p. 1.

Figure 1 Henry David Thoreau (1817–1862) was an American writer and naturalist who kept journals about his excursions into wild areas in parts of the northeastern United States and Canada, and at Walden Pond in Concord, Massachusetts. He sought self-sufficiency, a simple lifestyle, and a harmonious coexistence with nature.

misuse of their soils, water supplies, and other resources. Some of his resource conservation principles are still used today.

What Happened between 1870 and 1930?

Between 1870 and 1930, a number of events increased the role of the federal government and private citizens in resource conservation and public health. The *Forest Reserve Act of 1891* was a turning point in establishing the responsibility of the federal government for protecting public lands from resource exploitation.

In 1892, nature preservationist and activist John Muir (1838–1914) (Figure 2) founded the Sierra Club. He had been deeply troubled by the rapid deforestation that he had witnessed in large areas of North America. By 1870, the vast forests of the Northeast had been cut down, and between then and 1920, loggers essentially clear-cut the great pine forests of the upper Midwest—an area the size of Europe.

Muir became the leader of the *preservationist movement,* which called for protecting large areas of wilderness on public lands from human exploitation, except for low-impact recreational activities such as hiking and camping. However, this idea was not enacted into law until 1964. Muir also proposed and lobbied for creation of a national park system on public lands. In addition, he founded the Sierra Club, which is to this day a political force working on behalf of the environment.

Primarily because of political opposition, effective protection of forests and wildlife on federal lands did not begin until Theodore Roosevelt (1858–1919) (Figure 3), an ardent conservationist, became president. His term of office, 1901–1909, has been called the country's *Golden Age of Conservation.*

While in office, Roosevelt persuaded Congress to give the president power to designate public land as federal wildlife refuges. He was motivated partly by the fates of prominent wildlife species such as the passenger pigeon (extinct by 1900) and the North American bison, or American buffalo, which was nearly wiped out. In 1903, Roosevelt established the first federal refuge at Pelican Island (see Figure 9-20, p. 210) off the east coast of Florida for preservation of the endangered brown pelican, and he added 35 more wildlife reserves by 1904. He also more than tripled the size of the national forest reserves.

In 1905, Congress created the U.S. Forest Service to manage and protect the forest reserves. Roosevelt appointed Gifford Pinchot (1865–1946) as its first chief. Pinchot

Figure 2 John Muir (1838–1914) was a Scottish geologist, explorer, and naturalist. He spent 6 years studying, writing journals, and making sketches in the wilderness of California's Yosemite Valley and then went on to explore wilderness areas in Utah, Nevada, the Northwest, and Alaska. He was largely responsible for establishing Yosemite National Park in 1890. He also founded the Sierra Club and spent 22 years lobbying actively for conservation laws.

Figure 3 Theodore (Teddy) Roosevelt (1858–1919) was a writer, explorer, naturalist, avid birdwatcher, and twenty-sixth president of the United States. He was the first national political figure to bring conservation issues to the attention of the American public. According to many historians, Theodore Roosevelt contributed more than any other U.S. president to natural resource conservation in the United States.

pioneered scientific management of forest resources on public lands. In 1906, Congress passed the *Antiquities Act,* which allows the president to protect areas of scientific or historical interest on federal lands as national monuments. Roosevelt used this act to protect the Grand Canyon and other wilderness areas that would later become national parks.

Congress became upset with Roosevelt in 1907, because by then, he had added vast tracts to the national forest reserves. Congress passed a law banning further executive withdrawals of public forests. However, on the day before the bill became law, Roosevelt defiantly reserved another large block of land. Most environmental historians view Roosevelt as the country's best environmental president.

Early in the 20th century, the U.S. conservation movement split into two factions over how public lands should be used. The *wise-use,* or *conservationist,* school, led by Roosevelt and Pinchot, believed all public lands should be managed wisely and scientifically to provide needed resources. The *preservationist* school, led by Muir, wanted wilderness areas on public lands to be left untouched. This controversy over use of public lands continues today.

In 1916, Congress passed the *National Park Service Act,* which declared that parks are to be maintained in a manner that leaves them unimpaired for future generations. The act also established the National Park Service (within the Department of the Interior) to manage the park system. Under its first head, Stephen T. Mather (1867–1930), the dominant park policy was to encourage tourist visits by allowing private concessionaires to operate facilities within the parks.

After World War I, the country entered a new era of economic growth and expansion. To stimulate economic growth during the administrations of Presidents Harding, Coolidge, and Hoover, the federal government promoted the increased sales, at low prices, of timber, energy, mineral, and other resources found on public lands.

President Herbert Hoover went even further and proposed that the federal government return all remaining federal lands to the states or sell them to private interests for economic development. But the Great Depression (1929–1941) made owning such lands unattractive to state governments and private investors. The Depression was bad news for the country. But some say that without it, we might have little if any of the public lands that now make up about one-third of the total land area of the United States (see Figure 25-4, p. 687).

What Happened between 1930 and 1960?

Along with a second wave of national resource conservation, improvements in public health also began in the early 1930s as President Franklin D. Roosevelt (1882–1945) strove to bring the country out of the Great Depression. He persuaded Congress to enact federal programs to provide jobs and to help restore the country's degraded environment.

During this period, the government purchased large tracts of land from cash-poor landowners, and established the *Civilian Conservation Corps* (CCC) in 1933. According to the U.S. Forest Service, it eventually put more than 3 million unemployed people to work planting trees and developing and maintaining parks and recreation areas. During its 10 years of existence, the CCC planted an estimated 3 billion trees. It also restored silted waterways and built levees and dams for flood control.

The government also built and operated many large dams in the Tennessee River Valley, as well as in the arid western states, including Hoover Dam on the Colorado River (see Figure 13-1, p. 318). The goals were to provide jobs, flood control, cheap irrigation water, and cheap electricity for industry.

In 1935, Congress passed the Soil Conservation Act. During the Great Depression, erosion problems had ruined many farms in the Great Plains states and created a large area of degraded land known as the *Dust Bowl.* To correct these devastating erosion problems, the new law established the *Soil Erosion Service* as part of the Department of Agriculture. Its name was later changed to the *Soil Conservation Service,* and now it is called the *Natural Resources Conservation Service.* Many environmental historians praise Franklin D. Roosevelt for his efforts to get the country out of a major economic depression, partly by helping to restore environmentally degraded areas.

Also, in 1935, Aldo Leopold (1887–1948) (Figure 4), wildlife manager, professor, writer, and conservationist, helped to found the U.S. Wilderness Society. Largely through his writings, especially his 1949 essay "The Land Ethic" and his 1949 book *A Sand County Almanac,* he became one of the foremost leaders of the *conservation* and *environmental movements.* His energy and foresight helped to lay the critical groundwork for the field of environmental ethics. Leopold contended that the role of the human species should be to protect nature, not to conquer it.

Federal resource conservation policy changed little during the 1940s and 1950s, mostly because of preoccupation with World War II (1941–1945) and economic recovery after the war.

Between 1930 and 1960, improvements in public health included establishment of public health boards and agencies at the municipal, state, and federal levels; increased public education about health issues; introduction of vaccination programs; and a sharp reduction in the incidence of waterborne infectious diseases, mostly because of improved sanitation and garbage collection.

Tom Coleman/Aldo Leopold Foundation

Figure 4 Aldo Leopold (1887–1948) worked in game management early in his career, but then grew interested in the emerging scientific field of ecology. He became a leading conservationist and his book, *A Sand County Almanac* (published after his death), is considered an environmental classic that helped to inspire the modern environmental and conservation movements.

What Happened during the 1960s?

A number of milestones in American environmental history occurred during the 1960s. In 1962, biologist Rachel Carson (1907–1964) wrote *Silent Spring,* which documented the pollution of air, water, and wildlife from the use of pesticides such as DDT (see Chapter 12, pp. 296–300). This influential book helped to broaden the concept of resource conservation to include preservation of the *quality* of the planet's air, water, soil, and wildlife.

Many environmental historians mark Carson's wake-up call as the beginning of the modern *environmental movement* in the United States. It consists of citizens organized to demand that political leaders enact laws and develop policies to curtail pollution, clean up polluted environments, and protect unspoiled areas from environmental degradation.

In 1964, Congress passed the *Wilderness Act,* inspired by the vision of John Muir more than 80 years earlier. It authorized the government to protect undeveloped tracts of public land as part of the National Wilderness System. Any land in this system is to be used only for nondestructive forms of recreation such as hiking and camping, and it remains in the system unless and until Congress decides that it is needed for the national good.

Between 1965 and 1970, the emerging science of *ecology* received widespread media attention. At the same time, the popular writings of biologists such as Leopold, Carson, Paul Ehrlich, Barry Commoner, and Garrett Hardin awakened Americans to the interlocking relationships among population growth, resource use, and pollution.

During that period, a number of events increased public awareness of pollution, habitat loss, and other forms of environmental degradation. For example, many people learned about these problems when well-known wildlife species such as the American bald eagle, the grizzly bear, the whooping crane, and the peregrine falcon became endangered. Also, the stretch of the Cuyahoga River running near Cleveland, Ohio, was so polluted with oil and other flammable pollutants that it caught fire several times. Another major event was a devastating oil spill off the California coast in 1969.

In 1968, when U.S. astronauts aboard Apollo 8 became the first humans to fly to the moon, they photographed the entire earth for the first time from lunar orbit. This allowed people to see the earth as a tiny blue and white planet in the black void of space, and it led to the development of the *spaceship–earth environmental worldview.* According to this view, we live on a planetary spaceship that we should not harm because it is the only home we have.

What Happened during the 1970s? The Environmental Decade

During the 1970s, media attention, public concern about environmental problems, scientific research, and action to address environmental concerns grew rapidly. This period is sometimes called the *environmental decade,* or the *first decade of the environment.*

During the 1970s, several major U.S. environmental laws were passed. The first and possibly most important of these laws, signed by President Richard M. Nixon on January 1, 1970, was the *National Environmental Policy Act* (NEPA). It established the *environmental impact statement* (EIS) process. An EIS is now required for any action taken by the federal government that could significantly affect the environment. The EIS process forces the government and businesses that receive government contracts to carefully evaluate environmental impacts and avoid those with the potential for significant environmental damage. Many activities, including road building, dam construction, and logging on federal lands, are affected. NEPA set in motion a legislative process that produced several major environmental laws (see Figure 24-8, p. 668).

The first annual *Earth Day* was held on April 20, 1970. This event was proposed by Senator Gaylord Nelson (1916–2005) and organized by then Harvard graduate student Denis Hayes (see Individuals Matter 24.1, p. 666). Some 20 million people in more than

2,000 U.S. communities took to the streets to heighten the nation's environmental awareness and to demand improvements in environmental quality. There were also many environmental efforts that took place on a smaller scale that were greatly instructive for the environmental movement.

The *Environmental Protection Agency* (EPA) was established in 1970. In addition, the *Endangered Species Act of 1973* greatly strengthened the role of the federal government in protecting endangered species and their habitats. In addition, Congress created the Department of Energy in 1977, charging this new agency with developing a long-range energy strategy to help reduce the country's heavy dependence on imported oil.

During the 1970s, the area of land in the National Wilderness System tripled and the area in the National Park System doubled (primarily because vast tracts of land in the state of Alaska were added to these systems). In 1978, the *Federal Land Policy and Management Act* gave the *Bureau of Land Management* (BLM) its first real authority to manage the public land under its control, 85% of which is in 12 western states. This law angered a number of western interests whose use of these public lands was restricted for the first time.

In response, a coalition of ranchers, miners, loggers, developers, farmers, some elected officials, and other citizens in the affected states launched a political campaign known as the *sagebrush rebellion.* It had two major goals. *First,* sharply reduce government regulation of the use of public lands. *Second,* remove most public lands in the western United States from federal ownership and management and turn them over to the states. After that, the plan was to persuade state legislatures to sell or lease the resource-rich lands at low prices to ranching, mining, timber, land development, and other private interests. This represented a return to President Herbert Hoover's plan to get rid of all public land, which had been thwarted by the Great Depression. This political movement continues to exist.

What Happened during the 1980s? Environmental Backlash

In 1980, Congress created the *Superfund* as part of the *Comprehensive Environmental Response, Compensation, and Liability Act* (CERCLA; see Chapter 21, p. 595). Its goal was to clean up abandoned hazardous waste sites, including the Love Canal housing development in Niagara Falls, New York, which had to be abandoned when hazardous wastes from the site of a former chemical company began leaking into school grounds, yards, and basements.

During the decade of the 1980s, farmers, ranchers, and leaders of the oil, coal, automobile, mining, and timber industries strongly opposed many of the environmental laws and regulations developed in the 1960s and 1970s. They organized and funded multiple efforts to defeat environmental laws and regulations. In 1988, these industries backed a new coalition called the *wise-use movement.* Its major goals were to weaken or repeal most of the country's environmental laws and regulations, and destroy the effectiveness of the environmental movement in the United States. Backers of this movement argued that environmental laws had gone too far and were hindering economic growth.

Congress, influenced by this growing backlash, allowed for much more private energy and mineral development and timber cutting on public lands. Congress also lowered automobile gas mileage standards and relaxed federal air and water quality pollution standards.

In 1980, the United States had led the world in research and development of wind and solar energy technologies. Between 1981 and 1983, however, Congress slashed by 90% government subsidies for renewable energy research and for energy efficiency research, and eliminated tax incentives for the residential solar energy and energy conservation programs enacted in the late 1970s. As a result, the United States lost its lead in developing and selling the wind turbines and solar cells to Denmark, Germany, Japan, and China. These businesses are rapidly becoming two of the biggest and most profitable industries in the world.

At the same time, the 1980s saw rising public interest in some environmental issues, largely due to the influence of some prominent environmental scientists. For example, oceanographer Jacques Cousteau continued to tell the story of his four decades of undersea explorations and his concerns for the ocean environment through dozens of TV programs, books, and films. In Africa, zoologist Dian Fossey studied mountain gorillas and raised public awareness about these endangered animals, their shrinking habitats, and the threat of poachers. Her murder in 1985, possibly at the hands of poachers, made her story even more compelling.

What Happened from 1990 to 2013?

Between 1990 and 2013, opposition to environmental laws and regulations gained strength. This occurred because of continuing political and economic support from corporate backers, who not only argued that environmental laws were hindering economic growth, but also helped elect many members of Congress who were generally unsympathetic to environmental concerns. Since 1990, leaders and supporters of the environmental movement have had to spend much of their time and funds fighting efforts to discredit the movement and efforts to weaken or eliminate most environmental laws passed during the 1960s and 1970s.

During the 1990s, many small and mostly local grassroots environmental organizations sprang up to help deal with environmental threats in their local communities. Interest in environmental issues increased on many college and university campuses, resulting in the expansion of environmental studies courses and programs at these institutions. In addition, there was growing awareness of critical and complex environmental issues, such as sustainability, population growth, biodiversity protection, and threats from atmospheric warming and projected climate change. Whereas 20 million Americans took part in the first Earth Day in 1970, an estimated 200 million people from more than 140 countries participated in Earth Day 1990.

During the first decade of the 21st century, some environmental progress was made as a number of business leaders began to argue for a gradual shift from using fossil fuels to using a mix of renewable energy resources, including solar energy, wind power, and geothermal energy. Many of these leaders also put a high priority on saving their companies money by improving their energy efficiency and reducing their unnecessary energy waste.

In addition, during this decade, many analysts as well as business and government leaders have called for developing an effective U.S. policy for slowing projected climate change caused mostly by the burning of fossil fuels. However, the politically and economically powerful fossil fuel industries have been able to block efforts to develop a comprehensive energy policy shift toward renewables, as well as any major effort to address the projected threat of climate disruption.

Chemists Use the Periodic Table to Classify Elements on the Basis of Their Chemical Properties

Matter consists of elements and compounds (see Chapter 2, pp. 35–38). Chemists have developed a way to classify the elements according to their chemical behavior in what is called the *Periodic Table of Elements* (Figure 1). Each horizontal row in the table is called a *period*. Each vertical column lists elements with similar chemical properties and is called a *group*.

The Periodic Table in Figure 1 shows how the elements can be classified as *metals, nonmetals,* and *metalloids*. Examples of metals are sodium (Na), calcium (Ca), aluminum (Al), iron (Fe), lead (Pb), silver (Ag), and mercury (Hg).

Atoms of metals tend to lose one or more of their electrons to form positively charged ions such as Na^+, Ca^{2+}, and Al^{3+}. For example, an atom of the metallic element sodium (Na, atomic number 11) with 11 positively charged protons and 11 negatively charged electrons can lose one of its electrons. It then becomes a sodium ion with a positive charge of 1 (Na^+) because it now has 11 positive charges (protons) but only 10 negative charges (electrons).

Examples of *nonmetals* are hydrogen (H), carbon (C), nitrogen (N), oxygen (O), phosphorus (P), sulfur (S),

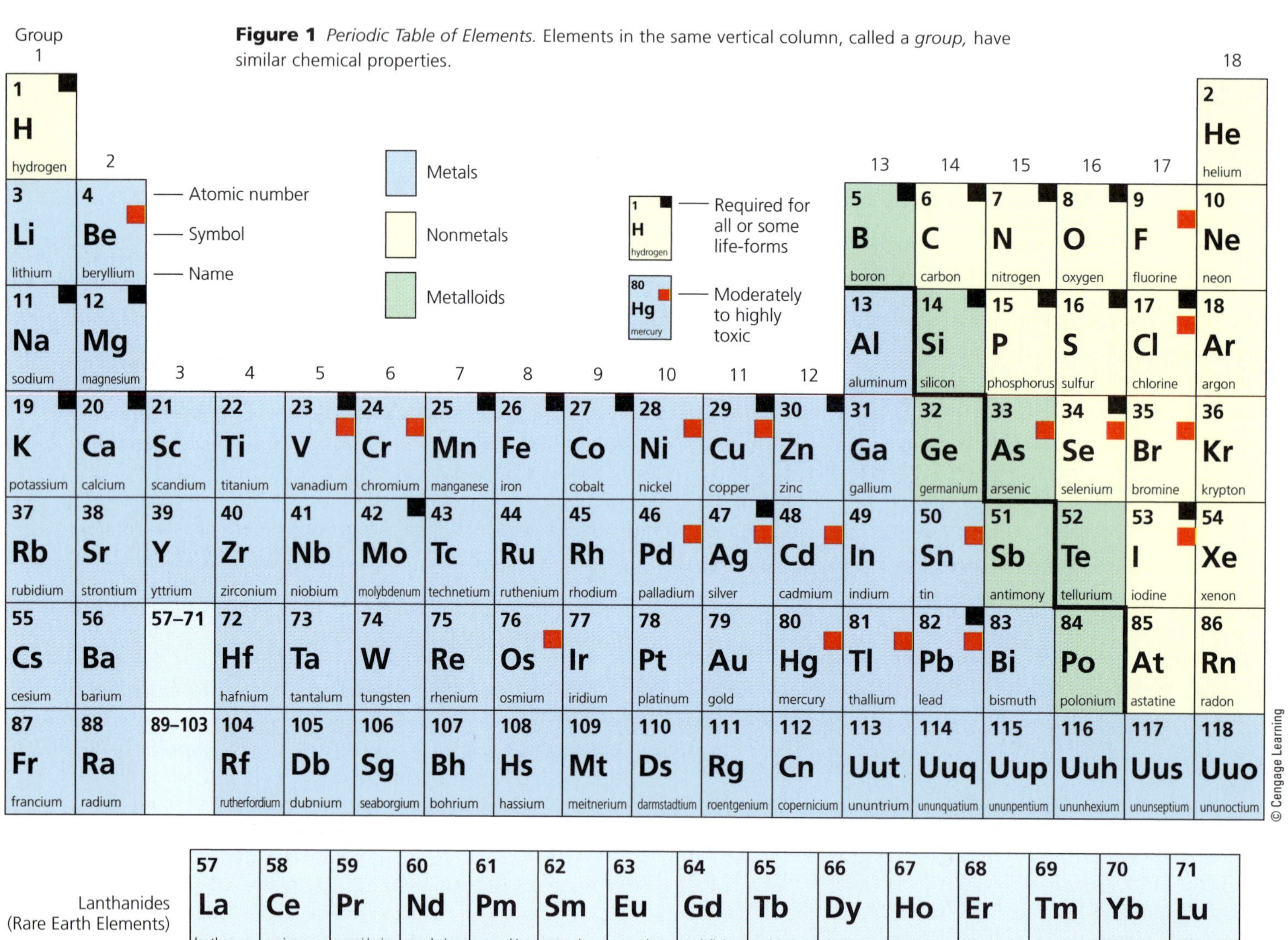

Figure 1 *Periodic Table of Elements.* Elements in the same vertical column, called a *group*, have similar chemical properties.

chlorine (Cl), and fluorine (F). Atoms of some nonmetals such as chlorine, oxygen, and sulfur tend to gain one or more electrons lost by metallic atoms to form negatively charged ions such as O^{2-}, S^{2-}, and Cl^-. For example, an atom of the nonmetallic element chlorine (Cl, atomic number 17) can gain an electron and become a chlorine ion. The ion has a negative charge of 1 (Cl^-) because it has 17 positively charged protons and 18 negatively charged electrons. Atoms of nonmetals can also combine with one another to form molecules in which they share one or more pairs of their electrons. Hydrogen, a nonmetal, is placed by itself above the center of the table because it does not fit very well into any of the groups.

The elements arranged in a diagonal staircase pattern between the metals and nonmetals have a mixture of metallic and nonmetallic properties and are called *metalloids*. Examples are germanium (Ge) and arsenic (As).

Figure 1 also identifies the elements required as *nutrients* (marked by small black squares) for all or some forms of life, and elements that are moderately or highly toxic (marked by small red squares) to all or most forms of life. Note that some elements such as copper (Cu) serve as nutrients, but can also be toxic at high enough doses. Six nonmetallic elements—carbon (C), oxygen (O), hydrogen (H), nitrogen (N), sulfur (S), and phosphorus (P)—make up about 99% of the atoms of all living things.

THINKING ABOUT The Periodic Table

Use the Periodic Table to identify by name and symbol two elements that should have chemical properties similar to those of **(a)** Ca, **(b)** potassium, **(c)** S, **(d)** lead.

Ionic and Covalent Bonds Hold Compounds Together

Sodium chloride (NaCl) consists of a three-dimensional network of oppositely charged *ions* (Na^+ and Cl^-) held together by the forces of attraction between opposite charges (see Figure 2-6, p. 38). The strong forces of attraction between such oppositely charged ions are called *ionic bonds*. They are formed when an electron is transferred from a metallic atom such as sodium (Na) to a nonmetallic element such as chlorine (Cl). Because ionic compounds consist of ions formed from atoms of metallic elements (positive ions) and nonmetallic elements (negative ions), they can be described as *metal–nonmetal compounds*.

Sodium chloride and many other ionic compounds tend to dissolve in water and break apart into their individual ions (Figure 2).

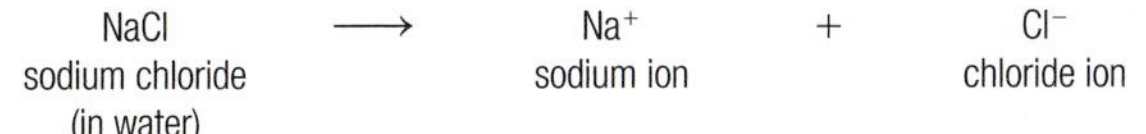

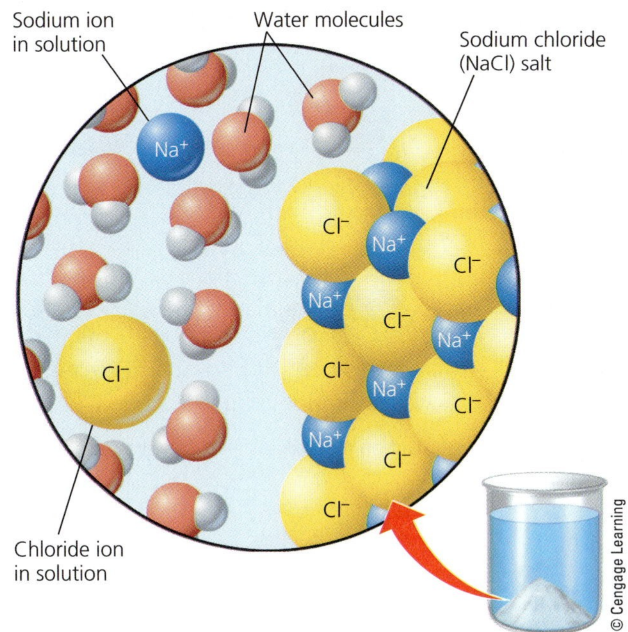

Figure 2 This model illustrates how a salt dissolves in water.

Water, a *covalent compound*, consists of molecules made up of uncharged atoms of hydrogen (H) and oxygen (O). Each water molecule consists of two hydrogen atoms chemically bonded to an oxygen atom, yielding H_2O molecules. The bonds between the atoms in such molecules are called *covalent bonds* and form when the atoms in the molecule share one or more pairs of their electrons. Because they are formed from atoms of nonmetallic elements (see Figure 1), covalent compounds can be described as *nonmetal–nonmetal compounds*. Figure 3 shows the chemical formulas and shapes of the molecules that are the building blocks for several common *covalent compounds*.

What Makes Solutions Acidic? Hydrogen Ions and pH

The *concentration*, or number of hydrogen ions (H^+) in a specified volume of a solution (typically a liter), is a measure of its acidity. Pure water (not tap water or rainwater) has an equal number of hydrogen (H^+) and hydroxide (OH^-) ions. It is called a **neutral solution**. An **acidic solution** has more hydrogen ions than hydroxide ions per liter. A **basic solution** has more hydroxide ions than hydrogen ions per liter.

Scientists use **pH** as a measure of the acidity of a solution based on its concentration of hydrogen ions (H^+). By definition, a neutral solution has a pH of 7, an acidic solution has a pH of less than 7, and a basic solution has a pH greater than 7.

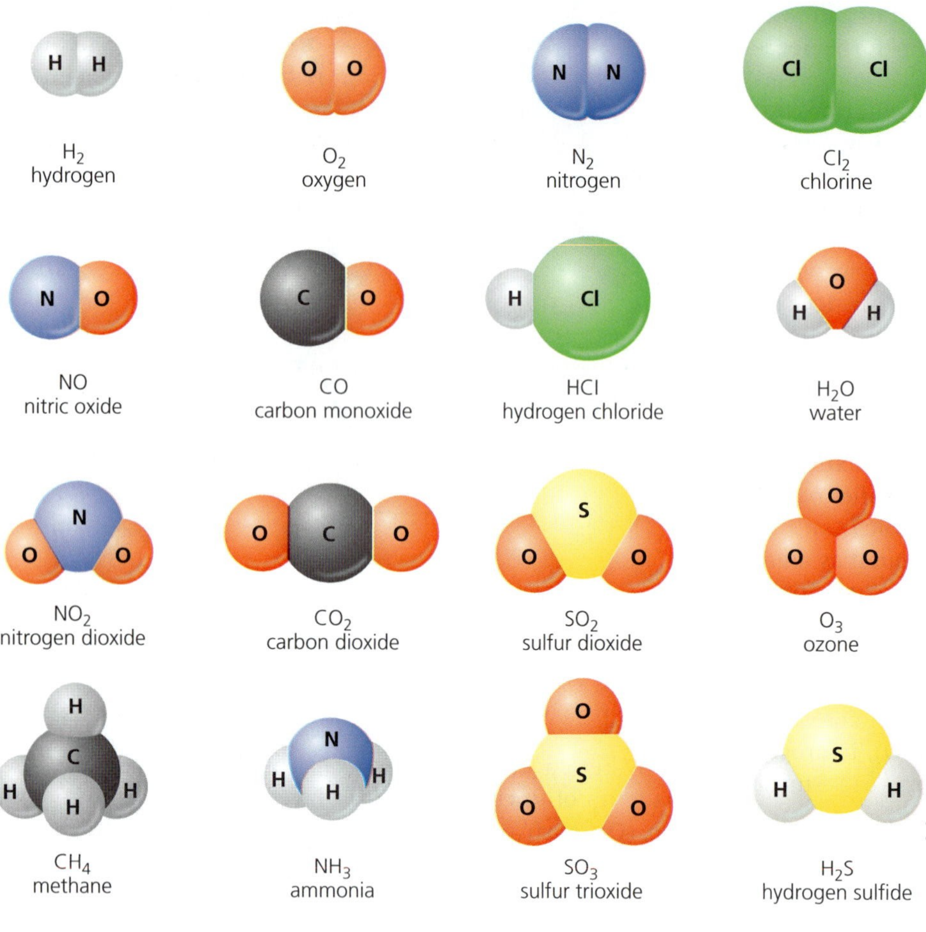

Figure 3 Chemical formulas and shapes for some *covalent compounds* formed when atoms of one or more nonmetallic elements combine with one another. The bonds between the atoms in such molecules are called *covalent bonds.*

Each single unit change in pH represents a tenfold increase or decrease in the concentration of hydrogen ions per liter. For example, an acidic solution with a pH of 3 is 10 times more acidic than a solution with a pH of 4. Figure 4 shows the approximate pH and hydrogen ion concentration per liter of solutions for various common substances.

CONSIDER THIS. . .

THINKING ABOUT pH

A solution has a pH of 2. How many times more acidic is this solution than one with a pH of 6?

The measurement of acidity is important in the study of environmental science, because environmental changes involving acidity can have serious environmental impacts. For example, when coal and oil are burned, they give off acidic compounds that can return to the earth as *acid deposition* (see Figure 18-12, p. 485, and Figure 18-13, p. 486), which has become a major environmental problem.

There Are Weak Forces of Attraction between Some Molecules

Ionic and covalent bonds form between the ions or atoms *within* a compound. There are also weaker forces of attraction *between* the molecules of covalent compounds (such as water) resulting from an unequal sharing of electrons by two atoms.

For example, an oxygen atom has a much greater attraction for electrons than does a hydrogen atom. Thus, the electrons shared between the oxygen atom and its two hydrogen atoms in a water molecule are pulled closer to the oxygen atom, but not actually transferred to the oxygen atom. As a result, the oxygen atom in a water molecule has a slightly negative partial charge and its two hydrogen atoms have a slightly positive partial charge.

The slightly positive hydrogen atoms in one water molecule are then attracted to the slightly negative oxygen atoms in another water molecule. These forces of attraction *between* water molecules are called *hydrogen bonds* (see Figure 3-A, p. 64). They account for many of water's unique properties (see Chapter 3, Science Focus 3.2, p. 64). Hydrogen bonds also form between other covalent molecules or between portions of such molecules containing hydrogen and nonmetallic atoms with a strong ability to attract electrons.

Four Types of Large Organic Compounds Are the Molecular Building Blocks of Life

Larger and more complex organic compounds, called *polymers,* consist of a number of basic structural or molecular units *(monomers)* linked by chemical bonds, somewhat like rail cars linked in a freight train. Four types of macromolecules—complex carbohydrates, proteins, nucleic acids, and lipids—are the molecular building blocks of life.

Complex carbohydrates consist of two or more monomers of *simple sugars* (such as glucose, Figure 5) linked together. One example is the starches that plants use to store energy and also to provide energy for animals that feed on plants. Another is cellulose, the earth's most abundant organic compound, which is found in the cell walls of bark, leaves, stems, and roots.

Proteins are large polymer molecules formed by linking together long chains of monomers called *amino acids* (Figure 6). Living organisms use about 20 different amino acid molecules to build a variety of proteins, which play different roles. Some help to store energy. Some are components of the *immune system* that protects the body against diseases and harmful substances by forming antibodies that make invading agents harmless. Others are *hormones* that are used as chemical messengers in the bloodstreams of animals to turn various bodily functions on or off. In animals, proteins are also components of hair, skin, muscle, and tendons. In addition, some proteins act as *enzymes* that catalyze or speed up certain chemical reactions.

Nucleic acids are large polymer molecules made by linking hundreds to thousands of four types of monomers called *nucleotides.* Two nucleic acids—DNA (**d**eoxyribo**n**ucleic **a**cid) and RNA (**r**ibo**n**ucleic **a**cid)—participate in the building of proteins and carry hereditary information used to pass traits from parent to offspring. Each nucleotide consists of a *phosphate group,* a *sugar molecule* containing five carbon atoms (deoxyribose in DNA molecules and ribose in RNA molecules), and one of four different *nucleotide bases* (represented by A, G, C, and T, the first letter in each of their names, or A, G, C, and U in RNA) (Figure 7). The four basic nucleotides used to make various forms of DNA molecules differ in the types of nucleotide bases they contain—adenine (A), guanine (G), cytosine (C), and thymine (T). (Uracil, labeled U, occurs instead of thymine in RNA.) In the cells of living organisms, these nucleotide units combine in different numbers and sequences to form *nucleic acids* such as various types of DNA and RNA (Figure 8, p. S17).

Hydrogen bonds formed between parts of the four nucleotides in DNA hold two DNA strands together like a spiral staircase, forming a double helix (see Figure 8). DNA molecules can unwind and replicate themselves.

The total weight of the DNA needed to reproduce all of the world's people is only about 50 milligrams—the weight of a small match. If the DNA coiled in your body were unwound, it would stretch about 960 million kilometers (600 million miles)—more than 6 times the distance between the sun and the earth.

The different molecules of DNA that make up the millions of species found on the earth are like a vast and diverse

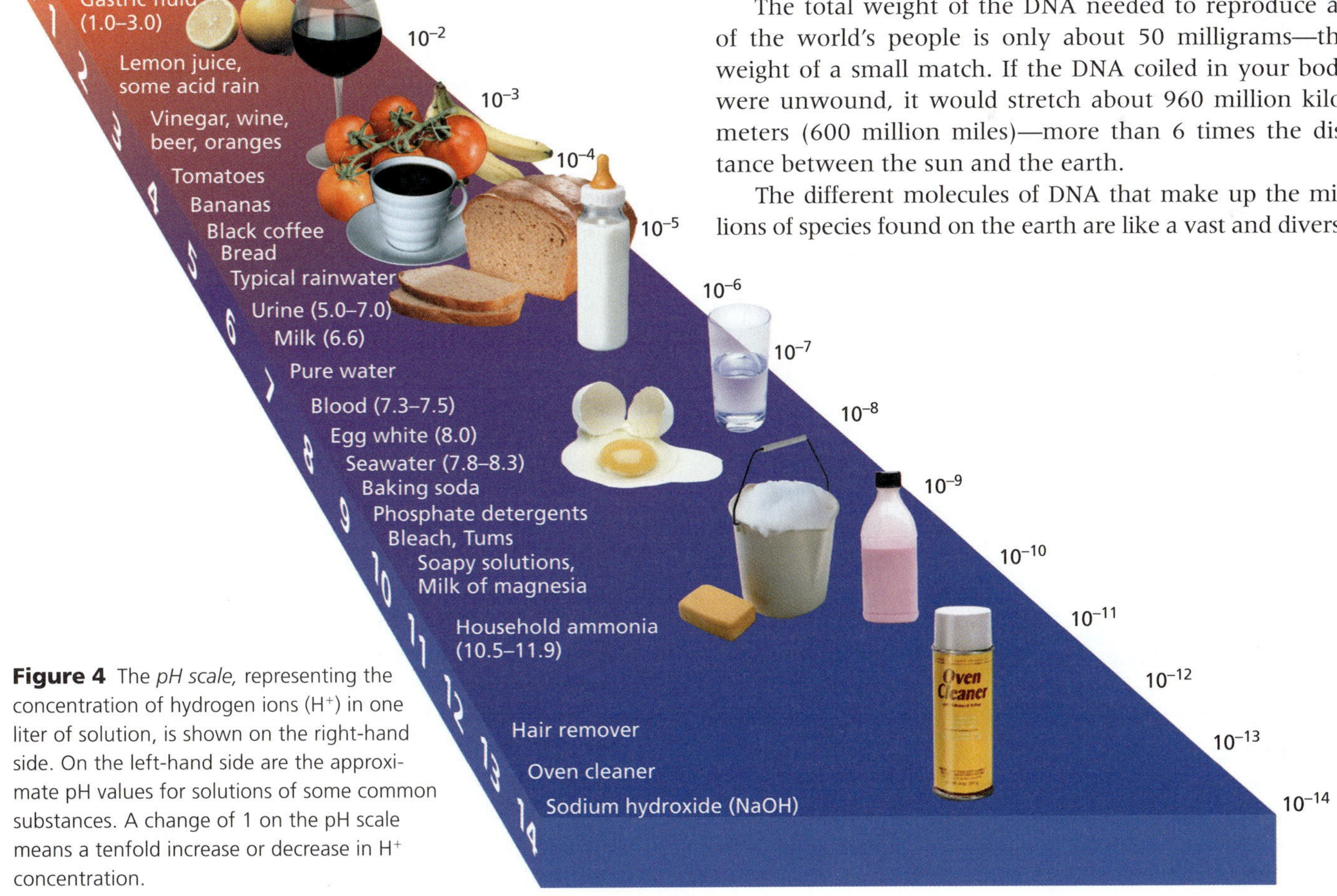

Figure 4 The *pH scale*, representing the concentration of hydrogen ions (H^+) in one liter of solution, is shown on the right-hand side. On the left-hand side are the approximate pH values for solutions of some common substances. A change of 1 on the pH scale means a tenfold increase or decrease in H^+ concentration.

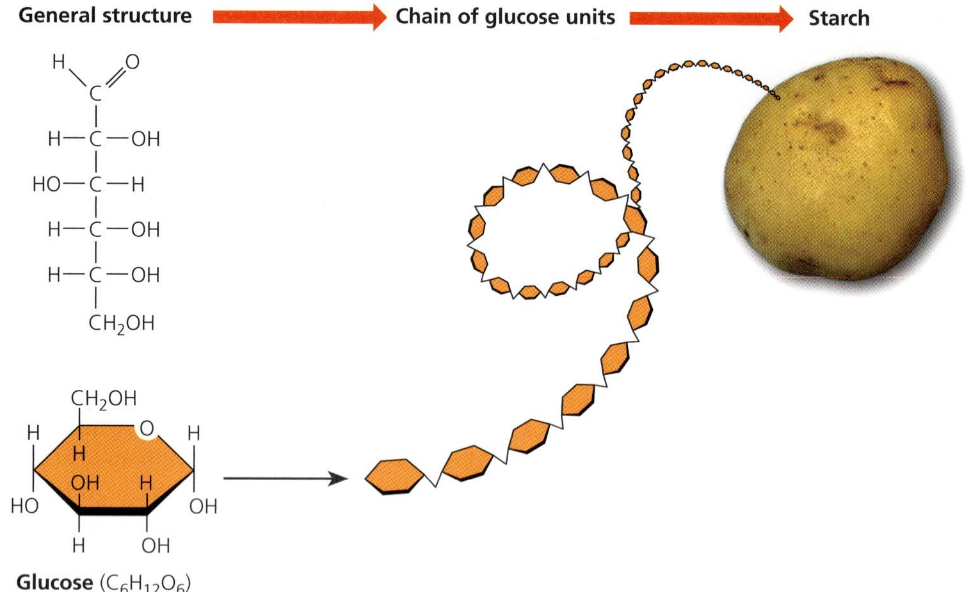

Figure 5 Straight-chain and ring structural formulas of glucose, a simple sugar that can be used to build long chains of complex carbohydrates such as starch and cellulose.

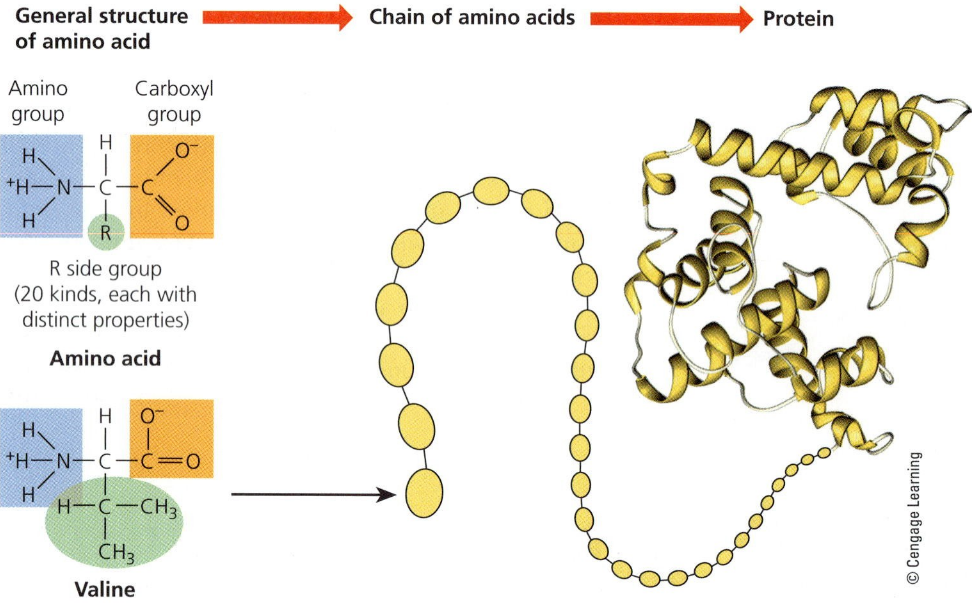

Figure 6 General structural formula of amino acids and a specific structural formula of one of the 20 different amino acid molecules that can be linked together in chains to form proteins that fold up into more complex shapes.

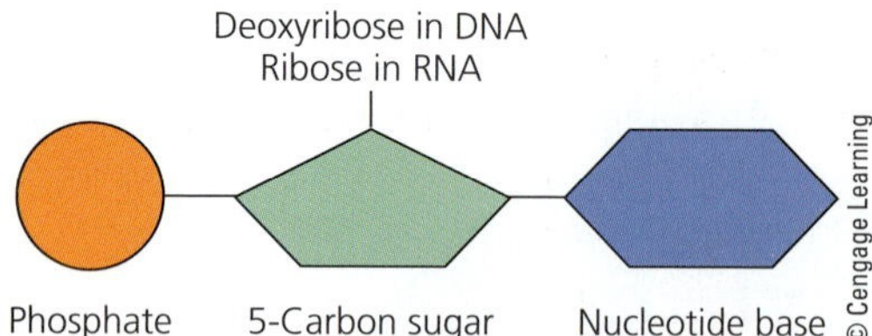

Figure 7 Generalized structures of the nucleotide molecules linked in various numbers and sequences to form large nucleic acid molecules such as various types of deoxyribonucleic acid (DNA) and ribonucleic acid (RNA). In DNA, the five-carbon sugar in each nucleotide is deoxyribose; in RNA it is ribose.

genetic library. Each species is a unique book in that library. The *genome* of a species is made up of the entire sequence of DNA "letters" or base pairs that combine to "spell out" the chromosomes in typical members of each species. In 2002, scientists were able to map out the genome for the human species by using powerful computers to help them in analyzing the 3.1 billion base sequences in human DNA.

Lipids, a fourth building block of life, are a chemically diverse group of large organic compounds that do not dissolve in water. Examples are *fats and oils* for storing energy (Figure 9), *waxes* for structure, and *steroids* for producing hormones.

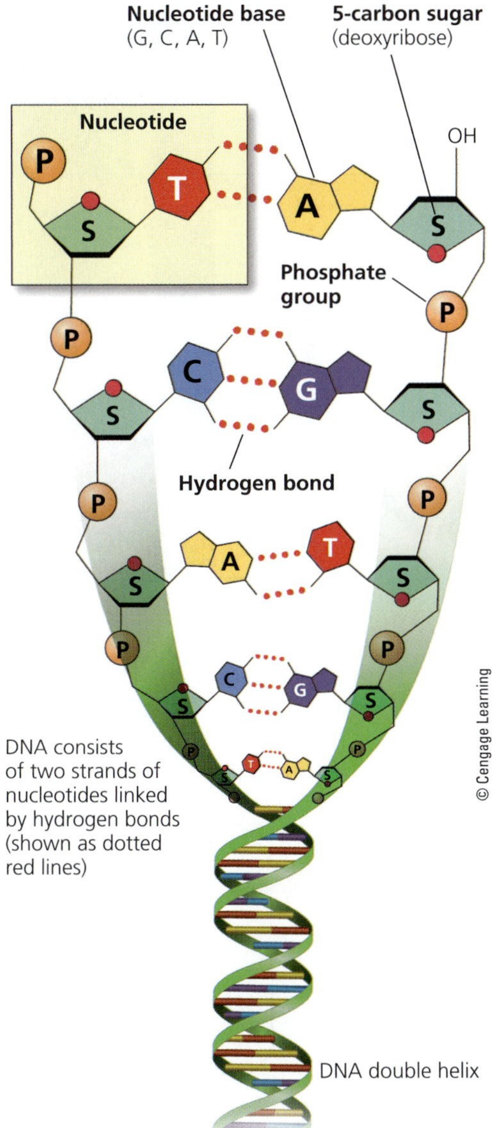

Figure 8 A portion of the double helix of a DNA molecule, which is composed of two spiral (helical) strands of nucleotides. Each nucleotide contains a unit of phosphate (P), deoxyribose (S), and one of four nucleotide bases: adenine (A), guanine (G), cytosine (C), and thymine (T). The two strands are held together by hydrogen bonds formed between various pairs of the nucleotide bases. Guanine (G) bonds with cytosine (C), and adenine (A) with thymine (T).

Figure 10 shows the relative sizes of simple and complex molecules, cells, and multi-celled organisms.

Certain Molecules Store and Release Energy in Cells

Chemical reactions occurring in plant cells during photosynthesis (see Chapter 3, p. 55) release energy that is absorbed by adenosine diphosphate (ADP) molecules and stored as chemical energy in adenosine triphosphate (ATP) molecules (Figure 11, left). When cellular processes require energy, ATP molecules release it to form ADP molecules (Figure 11, right).

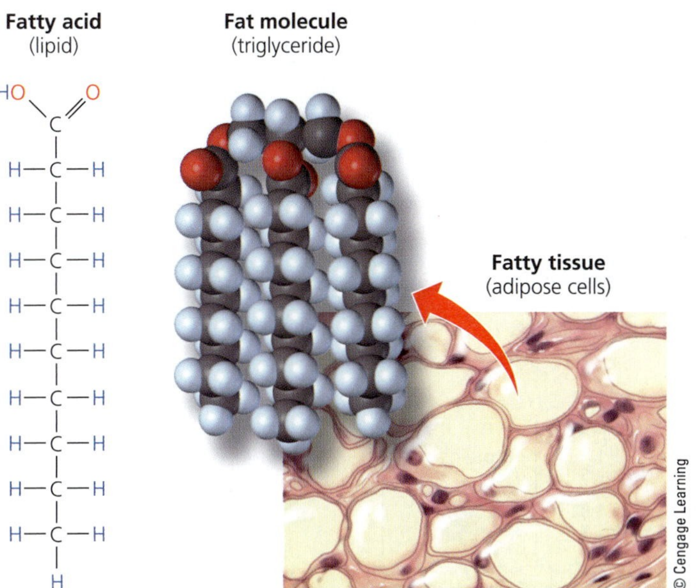

Figure 9 Structural formula of fatty acid, which is one form of lipid (left). Fatty acids are converted into more complex fat molecules (center) that are stored in adipose cells (right).

Chemists Balance Chemical Equations to Keep Track of Atoms

Chemists use a shorthand system, or equation, to represent chemical reactions. These chemical equations are also used as an accounting system to verify that no atoms are created or destroyed in a chemical reaction as required by the law of conservation of matter (see Chapter 2, p.40). As a consequence, each side of a chemical equation must have the same number of atoms or ions of each element involved. Ensuring that this condition is met leads to what chemists call a *balanced chemical equation*. The equation for the burning of carbon ($C + O_2 \rightarrow CO_2$) is balanced because one atom of carbon and two atoms of oxygen are on both sides of the equation.

Consider the following chemical reaction. When electricity passes through water (H_2O), the latter can be broken down into hydrogen (H_2) and oxygen (O_2), as represented by the following equation:

H_2O	$\longrightarrow$	H_2	+	O_2
2 H atoms		2 H atoms		2 O atoms
1 O atom				

This equation is unbalanced because one atom of oxygen is on the left side of the equation but two oxygen atoms are on the right side. We cannot change the subscripts of any of the formulas to balance this equa-

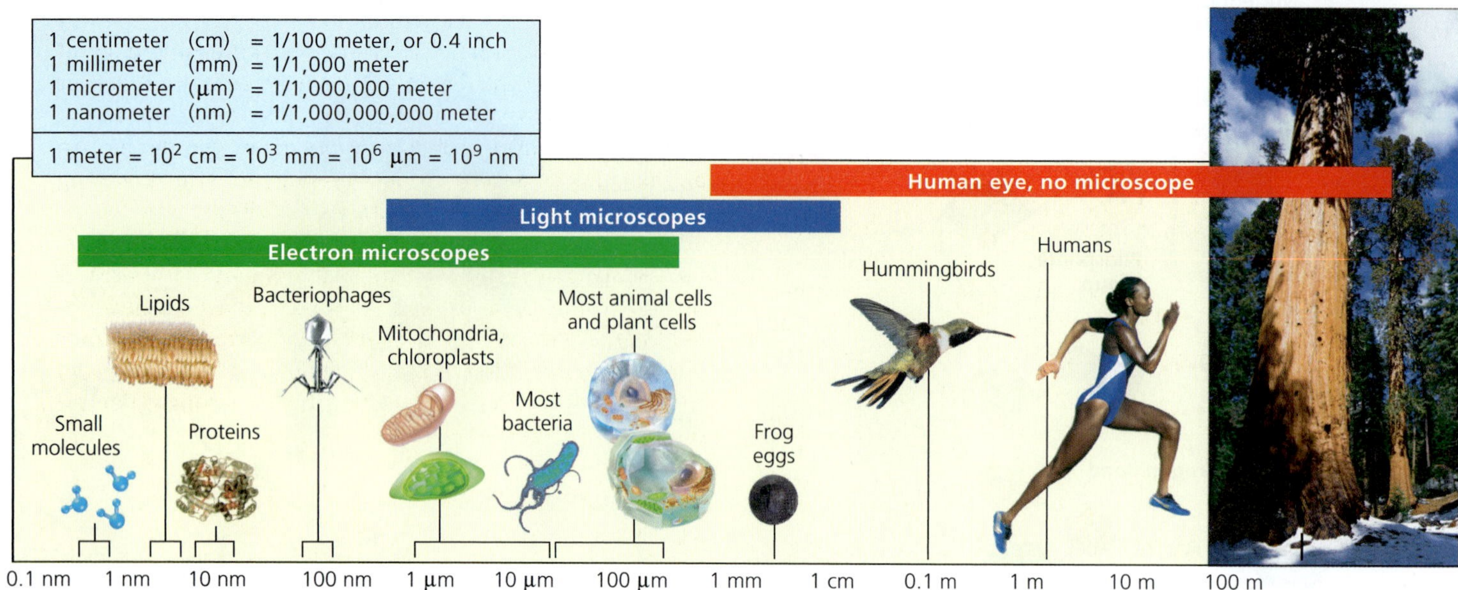

Figure 10 Comparison of the relative sizes of simple molecules, complex molecules, cells, and multicellular organisms. This scale is exponential, not linear. Each unit of measure is 10 times larger than the unit preceding it.

(Used by permission from STARR/TAGGART, Biology, 11E. © 2006 Cengage Learning.)

ATP synthesis:
Energy is stored in ATP

A P P + P Energy

ADP Phosphate

A P P P

ATP

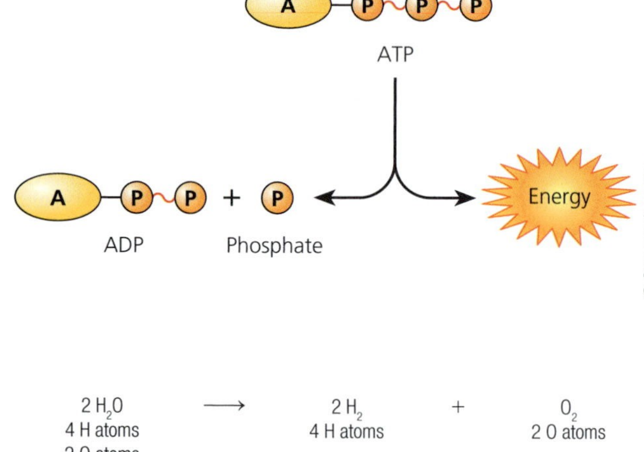

Figure 11 Models representing energy storage and release in cells.

tion because that would change the arrangements of the atoms, leading to different substances. Instead, we must use different numbers of the molecules involved to balance the equation. For example, we could use two water molecules:

$2\,H_2O$	$\longrightarrow$	H_2	$+$	O_2
4 H atoms		2 H atoms		2 O atoms
2 O atoms				

This equation is still unbalanced. Although the numbers of oxygen atoms on both sides of the equation are now equal, the numbers of hydrogen atoms are not. We can correct this problem by having the reaction produce two hydrogen molecules:

$2\,H_2O$	$\longrightarrow$	$2\,H_2$	$+$	O_2
4 H atoms		4 H atoms		2 O atoms
2 O atoms				

Now the equation is balanced, and the law of conservation of matter has been observed. For every two molecules of water through which we pass electricity, two hydrogen molecules and one oxygen molecule are produced.

CONSIDER THIS. . .

THINKING ABOUT Chemical Equations

Try to balance the chemical equation for the reaction that combines nitrogen gas (N_2) and hydrogen gas (H_2) to form ammonia gas (NH_3).

SUPPLEMENT 5 WEATHER BASICS: EL NIÑO, TORNADOES, AND TROPICAL CYCLONES

Weather Is Affected by Moving Masses of Warm and Cold Air

Weather is the set of short-term atmospheric conditions—typically those occurring over hours or days—for a particular area. Examples of atmospheric conditions include temperature, pressure, moisture content, precipitation, sunshine, cloud cover, and wind direction and speed.

Meteorologists use equipment mounted on weather balloons, aircraft, ships, and satellites, as well as radar and stationary sensors, to obtain data on weather variables. They then feed these data into computer models to draw weather maps. Other computer models project the weather for a period of several days by calculating the probabilities that air masses, winds, and other factors will change in certain ways.

Much of the weather we experience results from interactions between the leading edges of moving masses of warm and cold air driven largely by uneven heating of the earth's surface by the sun (see Figure 7-3, p. 146). Weather changes as one air mass replaces or meets another. The most dramatic changes in weather occur along a **front**, the boundary between two air masses with different temperatures and densities.

A **warm front** is the boundary between an advancing warm air mass and the cooler one it is replacing (Figure 1, left). Because warm air is less dense (weighs less per unit of volume) than cool air, an advancing warm front rises up over a mass of cool air. As the warm front rises, its moisture begins condensing into droplets, forming layers of clouds at different altitudes. Gradually, the clouds thicken, descend to a lower altitude, and often release their moisture as rainfall. A moist warm front can bring days of cloudy skies and drizzle.

A **cold front** (Figure 1, right) is the leading edge of an advancing mass of cold air. Because cold air is denser than warm air, an advancing cold front stays close to the ground and wedges underneath less dense warmer air. An approaching cold front produces rapidly moving, towering clouds called *thunderheads,* with flat, anvil-like tops.

As a cold front passes through, it may cause high surface winds and thunderstorms. After it leaves the area, it usually results in cooler temperatures and a clear sky.

Near the top of the troposphere, hurricane-force winds circle the earth. These powerful winds, called *jet streams,* follow rising and falling paths that have a strong influence on weather patterns (Figure 2).

Weather Is Affected by Changes in Atmospheric Pressure

Changes in atmospheric pressure also affect weather. *Atmospheric pressure* results from molecules of gases (mostly nitrogen and oxygen) in the atmosphere zipping around at very high speeds and hitting and bouncing off everything they encounter.

Atmospheric pressure is greater near the earth's surface because the molecules in the atmosphere are squeezed together under the weight of the air above them. An air mass with high pressure, called a **high**, contains cool, dense air that descends slowly toward the earth's surface and becomes warmer. Because of this warming, condensation of moisture usually does not take place and clouds usually do not form. Fair weather with clear skies follows as long as this high-pressure air mass remains over the area.

In contrast, a low-pressure air mass, called a **low**, produces cloudy and sometimes stormy weather. Because of

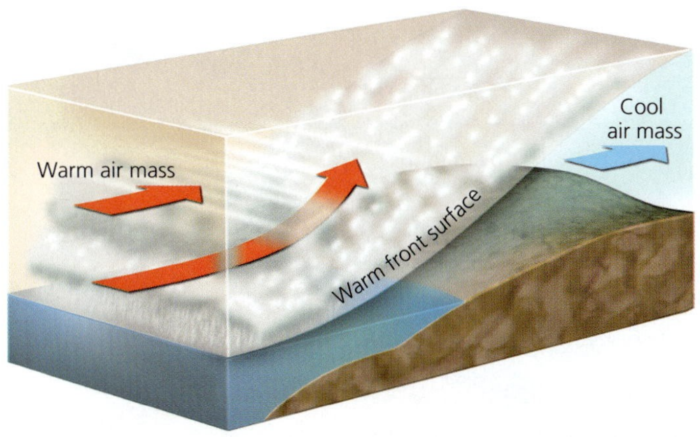

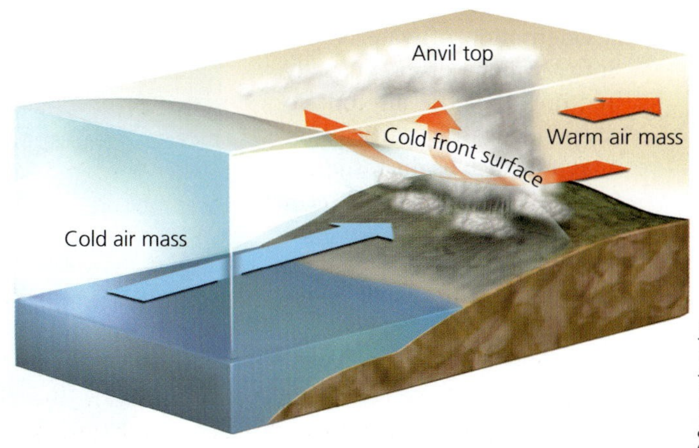

Figure 1 *Weather fronts:* A *warm front* (left) occurs when an advancing mass of warm air meets and rises up over a mass of denser cool air. A *cold front* (right) forms when a moving mass of cold air wedges beneath a mass of less dense warm air.

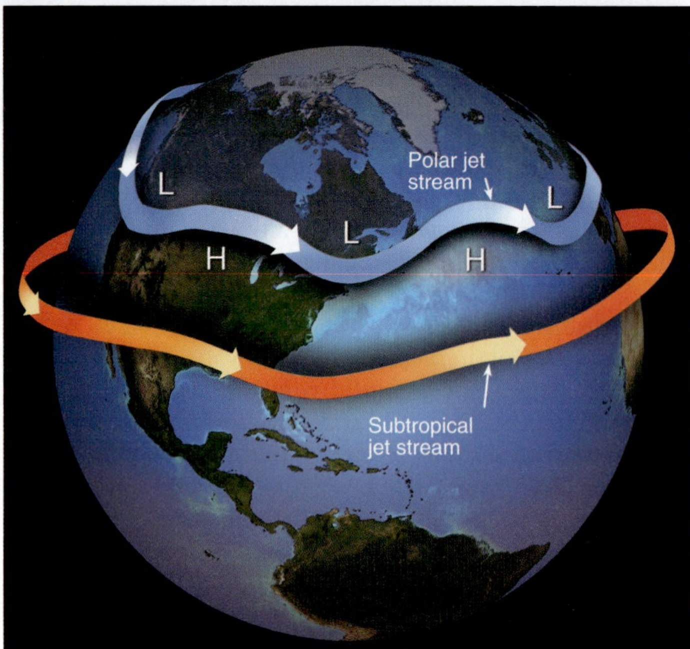

Figure 2 A *jet stream* is a rapidly flowing air current that moves west to east in a wavy pattern. This figure shows a polar jet stream and a subtropical jet stream in winter. In reality, jet streams are discontinuous and their positions vary from day to day.

its low pressure and low density, the center of a low rises, and its warm air expands and cools. When the temperature drops below a certain level where condensation takes place, called the *dew point,* moisture in the air condenses and forms clouds.

If the droplets in the clouds coalesce into larger drops or snowflakes heavy enough to fall from the sky, precipitation occurs. The condensation of water vapor into water drops usually requires that the air contain suspended tiny particles of material such as dust, smoke, sea salts, or volcanic ash. These so-called *condensation nuclei* provide surfaces on which the droplets of water can form and coalesce.

Every Few Years Major Wind Shifts in the Pacific Ocean Affect Global Weather Patterns

Ocean currents can have a strong effect on the weather. Water moves vertically, as well as horizontally from one area of the ocean to another. An **upwelling**, or upward movement of ocean water, can mix upper levels of seawater with lower levels, bringing cool and nutrient-rich water from the bottom of the ocean to the warmer surface where it supports large populations of phytoplankton, zooplankton, fish, and fish-eating seabirds.

Figure 7-2, p. 145, shows the oceans' major upwelling zones. Upwellings far from shore occur when surface currents move apart and draw water up from deeper layers. Strong upwellings are also found along the steep western coasts of some continents when winds blowing along the coasts push surface water away from the land and draw water up from the ocean bottom (Figure 3).

Every few years, normal shore upwellings in the Pacific Ocean (Figure 4, left) are affected by changes in weather patterns called the *El Niño–Southern Oscillation,* or *ENSO* (Figure 4, right). In an ENSO, often called simply *El Niño,* prevailing winds called tropical trade winds blowing east to west weaken or reverse direction. This allows the warmer waters of the western Pacific to move toward the coast of South America, which suppresses the normal upwellings of cold, nutrient-rich water (Figure 4, right). The decrease in nutrients reduces primary productivity and causes a sharp decline in the populations of some fish species.

When an ENSO lasts 12 months or longer, it can severely disrupt populations of plankton, fish, and sea-

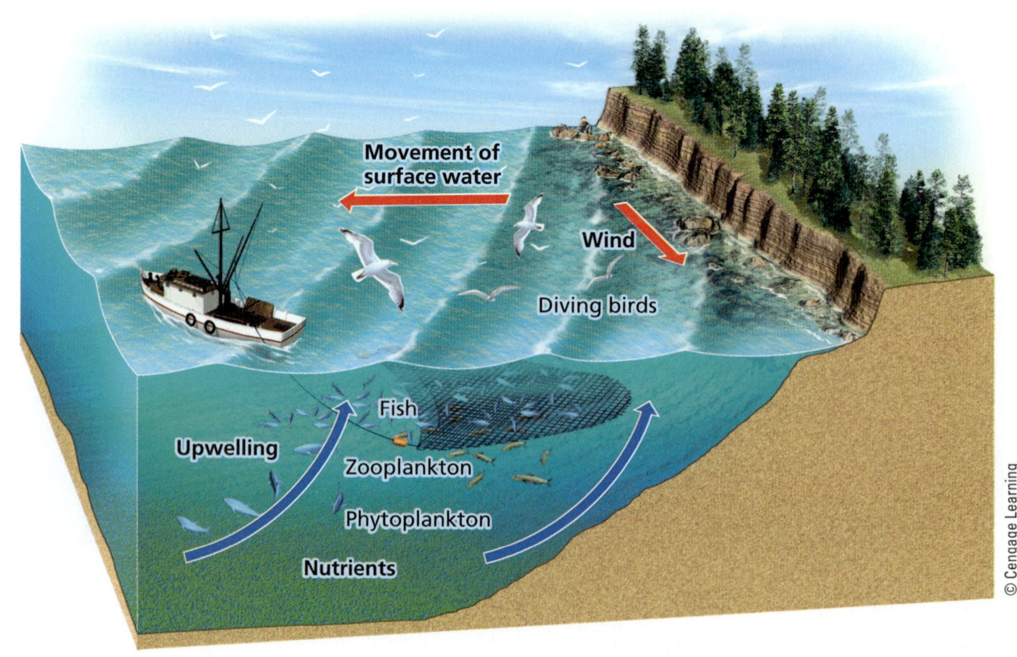

Figure 3 A shore upwelling occurs when deep, cool, nutrient-rich waters are drawn up to replace surface water moved away from a steep coast by wind flowing along the coast toward the equator.

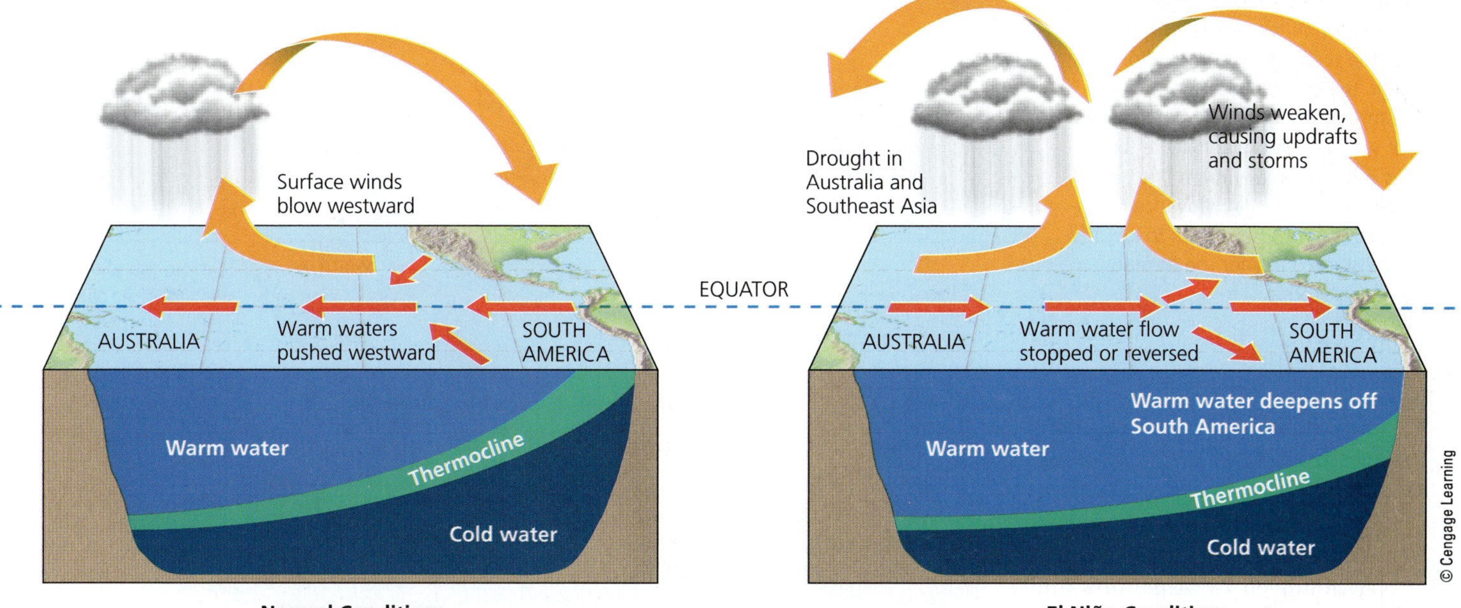

Figure 4 Normal trade winds blowing east to west cause shore upwellings of cold, nutrient-rich bottom water in the tropical Pacific Ocean near the coast of Peru (left). A zone of gradual temperature change called the *thermocline* separates the warm and cold water. Every few years, a shift in trade winds known as the *El Niño–Southern Oscillation (ENSO)* disrupts this pattern.

birds in upwelling areas. A strong ENSO can also alter weather conditions over at least two-thirds of the globe (Figure 5)—especially in lands along the Pacific and Indian Oceans. Scientists do not know exactly what causes an ENSO, but they do know how to detect its formation and track its progress.

La Niña, the reverse of El Niño, cools some coastal surface waters and brings back upwellings. Typically, La Niña brings more Atlantic Ocean hurricanes, colder winters in Canada and the northeastern United States, and warmer and drier winters in the southeastern and southwestern United States. It also usually leads to wetter winters in the Pacific Northwest, torrential rains in Southeast Asia, lower wheat yields in Argentina, and more wildfires in Florida.

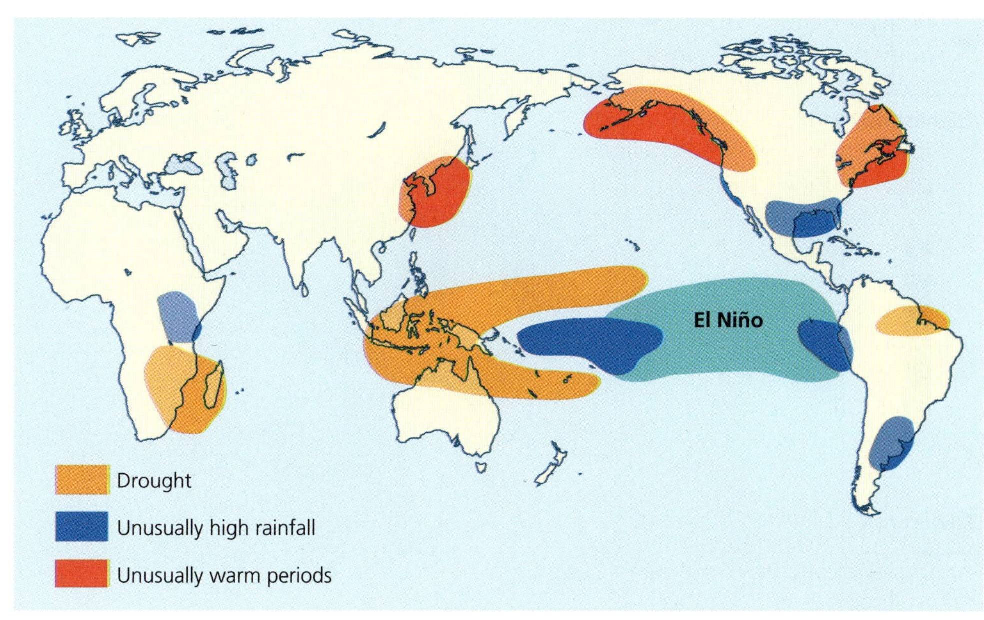

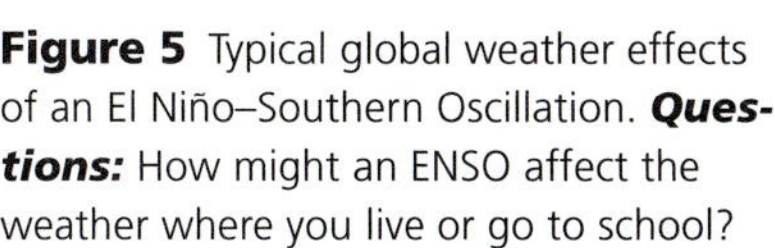

Figure 5 Typical global weather effects of an El Niño–Southern Oscillation. ***Questions:*** How might an ENSO affect the weather where you live or go to school?

(Compiled by the authors using data from United Nations Food and Agriculture Organization.)

Tornadoes and Tropical Cyclones Are Violent Weather Extremes

Sometimes we experience *weather extremes.* Two examples are violent storms called *tornadoes* (which form over land) and *tropical cyclones* (which form over warm ocean waters and sometimes pass over coastal areas).

Tornadoes, or *twisters,* are swirling, funnel-shaped clouds that form over land. They can destroy houses

Figure 6 Formation of a *tornado,* or *twister.* Although twisters can form at any time of the year, the most active tornado season in the United States is usually March through August. Meteorologists cannot yet forecast exactly where tornadoes will form at any given time, but research on tornados and advanced computer modeling can help them to identify areas at risk each day for the formation of these deadly storms.

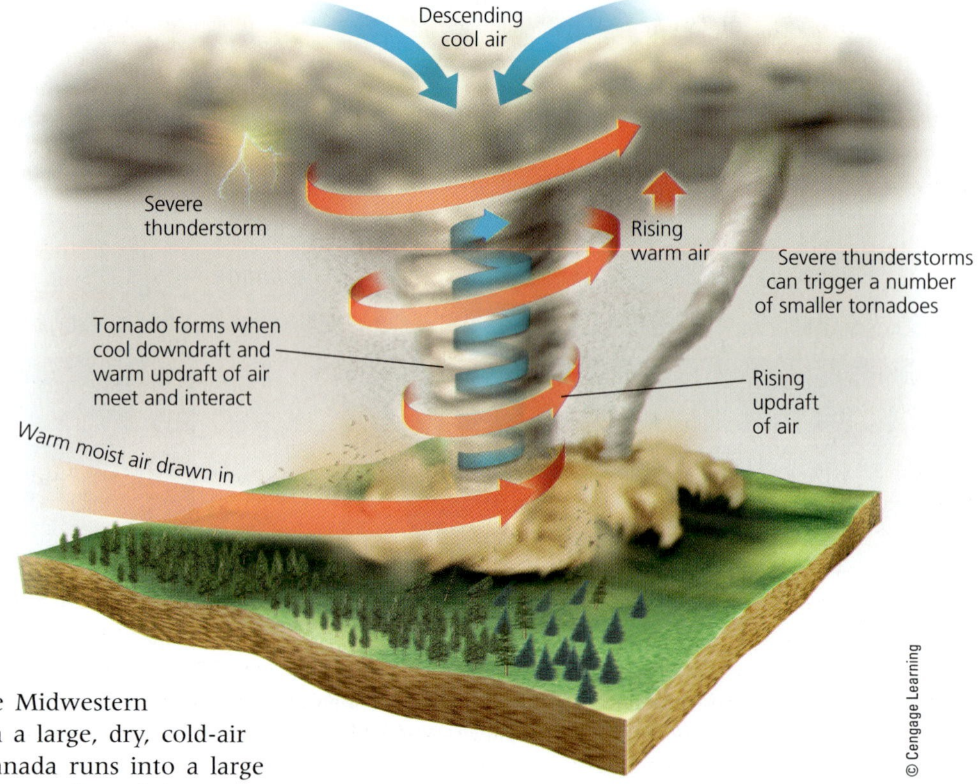

Descending cool air
Severe thunderstorm
Rising warm air
Severe thunderstorms can trigger a number of smaller tornadoes
Tornado forms when cool downdraft and warm updraft of air meet and interact
Rising updraft of air
Warm moist air drawn in
© Cengage Learning

and cause other serious damage in areas where they touch down on the earth's surface. The United States is the world's most tornado-prone country, followed by Argentina and Bangladesh. Tornadoes occur on every continent in the world except for Antarctica.

Tornadoes in the plains of the Midwestern United States usually occur when a large, dry, cold-air front moving southward from Canada runs into a large mass of warm humid air moving northward from the Gulf of Mexico. Most tornadoes occur in the spring and summer when fronts of cold air from the north penetrate deeply into the Great Plains and the Midwest.

As the large warm-air mass moves rapidly over the more dense cold-air mass, it rises swiftly and forms strong vertical convection currents that suck air upward, as shown in Figure 6. Scientists hypothesize that the interaction of the cooler air nearer the ground and the rapidly rising warmer air above causes a spinning, vertically rising air mass, or vortex.

Figure 7 shows the areas of greatest risk from tornadoes in the continental United States.

Tropical cyclones are spawned by the formation of low-pressure cells of air over warm tropical seas. Figure 8

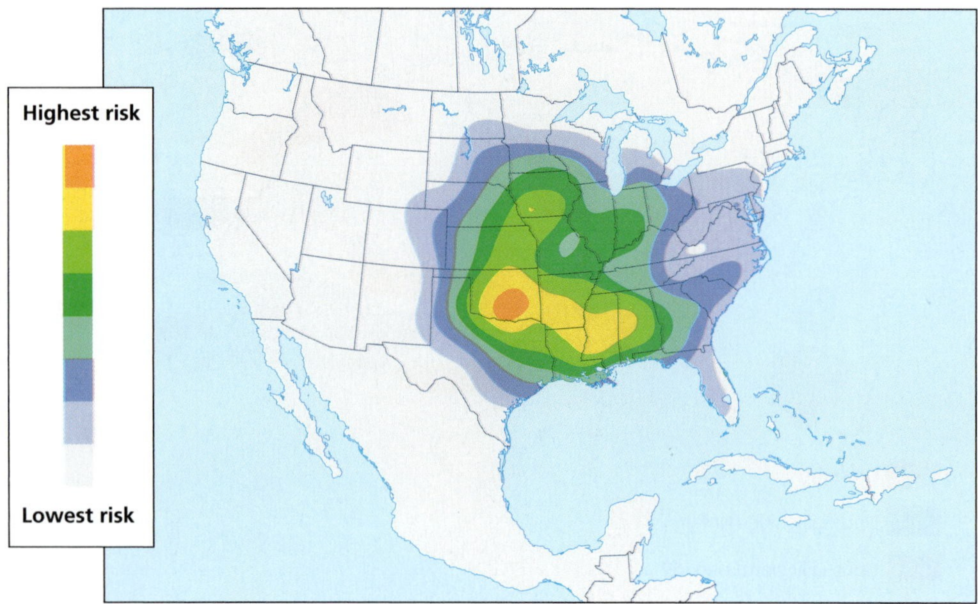

Figure 7 Comparison of the relative risk of tornados across the continental United States.

(Compiled by the authors using data from NOAA.)

shows the formation and structure of a tropical cyclone. *Hurricanes* are tropical cyclones that form in the Atlantic Ocean; those forming in the Pacific Ocean usually are called *typhoons.* Tropical cyclones take a long time to form and gain strength. As a result, meteorologists can track their paths and wind speeds, and warn people in areas likely to be hit by these violent storms.

For a tropical cyclone to form, the temperature of ocean water has to be at least 27°C (80°F) to a depth of 46 meters (150 feet). A tropical cyclone forms when areas of low pressure over the warm ocean draw in air from surrounding higher-pressure areas. The earth's rotation makes these winds spiral counterclockwise in the northern hemisphere and clockwise in the southern hemisphere (see Figure 7-3, p. 146). Moist air, warmed by the heat of the ocean, rises in a vortex through the center of the storm until it becomes a tropical cyclone (Figure 8).

The intensities of tropical cyclones are rated in different categories, based on their sustained wind speeds: *Category 1,* 119–153 kilometers per hour, or kph (74–95 miles per hour, or mph); *Category 2,* 154–177 kph (96–110 mph); *Category 3,* 178–209 kph (111–130 mph); *Category 4,* 210–249 kph (131–155 mph); and *Category 5,* greater than 249 kph (155 mph). The longer a tropical cyclone stays over warm waters, the stronger it gets. Significant hurricane-force winds can extend 64–161 kilometers (40–100 miles) from the center, or eye, of a tropical cyclone.

Hurricanes and typhoons kill and injure people and damage property and agricultural production. Sometimes, however, the long-term ecological and economic benefits of a tropical cyclone exceed its short-term harmful effects.

For example, in parts of the U.S. state of Texas along the Gulf of Mexico, coastal bays and marshes, because of their unique natural formations and the barrier islands that protect them, normally receive very limited freshwater and saltwater inflows. In August 1999, Hurricane Brett struck this coastal area. According to marine biologists, the storm flushed out excess nutrients from land runoff and swept dead sea grasses and rotting vegetation from the coastal bays and marshes. It also carved out 12 channels through the barrier islands along the coast, allowing huge quantities of fresh seawater to flood the bays and marshes.

This flushing of the bays and marshes reduced brown tides consisting of explosive growths of algae feeding on excess nutrients. It also increased growth of sea grasses, which serve as nurseries for shrimp, crabs, and fish, and provide food for millions of ducks wintering in Texas bays. Production of commercially important species of shellfish and fish also increased.

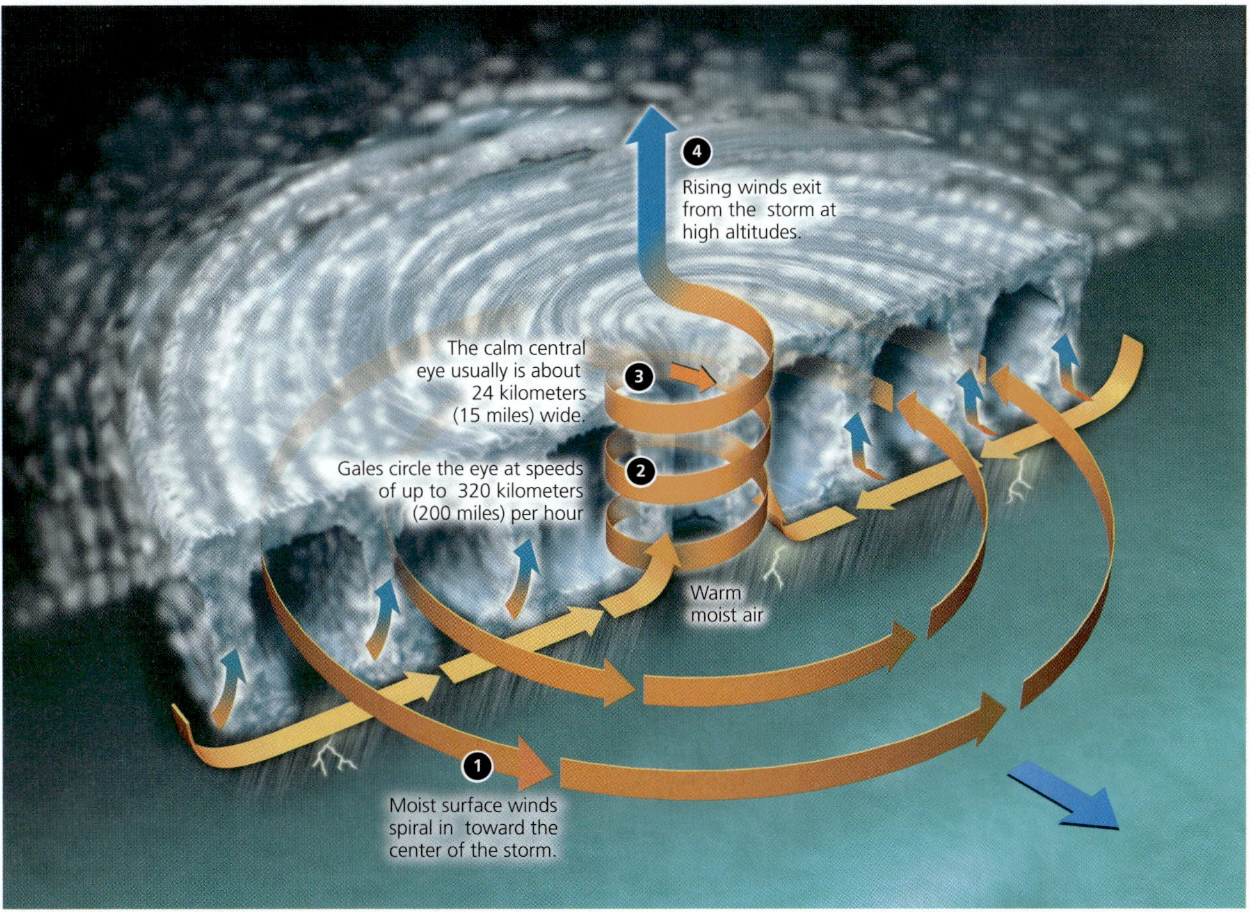

Figure 8 The formation of a *tropical cyclone.* Those forming in the Atlantic Ocean are called *hurricanes;* those forming in the Pacific Ocean are called *typhoons.*

Figure 1 Countries of the world.

Map Analysis

1. Name three countries that border the Arctic Ocean.
2. Which countries surround **(a)** China, **(b)** Mexico, **(c)** Germany, and **(d)** Sudan?

WINKEL TRIPEL PROJECTION, CENTRAL MERIDIAN 0°
KILOMETERS
STATUTE MILES
Longitude East of Greenwich
Meridian of Greenwich (London)
ARCTIC CIRCLE
TROPIC OF CANCER
EQUATOR
TROPIC OF CAPRICORN
ANTARCTIC CIRCLE
Date Line
Sunday
Monday
O C E A N
MAKAROV BASIN
FRANZ JOSEF LAND
Russia
SVALBARD
Norway
BARENTS SEA
KARA SEA
LAPTEV SEA
NEW SIBERIAN ISLANDS
EAST SIBERIAN SEA
NORWEGIAN SEA
NORTH SEA
R U S S I A
K A Z A K H S T A N
M O N G O L I A
C H I N A
I N D I A
BERING SEA
SEA OF OKHOTSK
NORTH PACIFIC OCEAN
NORTHWEST PACIFIC BASIN
PHILIPPINE SEA
SOUTH CHINA SEA
PHILIPPINES
I N D O N E S I A
PAPUA NEW GUINEA
FEDERATED STATES OF MICRONESIA
MARSHALL ISLANDS
KIRIBATI
NAURU
SOLOMON ISLANDS
TUVALU
VANUATU
FIJI
NEW CALEDONIA
A U S T R A L I A
NEW ZEALAND
TASMAN SEA
SOUTH PACIFIC OCEAN
ARABIAN SEA
BAY OF BENGAL
I N D I A N O C E A N
WHARTON BASIN
SOUTH AUSTRALIAN BASIN
SOUTH INDIAN BASIN
ATLANTIC-INDIAN BASIN
ATLANTIC - INDIAN RIDGE
SOUTHWEST INDIAN RIDGE
SOUTHEAST INDIAN RIDGE
NINETYEAST RIDGE
MID-INDIAN RIDGE
KERGUELEN PLATEAU
MADAGASCAR
TANZANIA
KENYA
ETHIOPIA
SUDAN
EGYPT
LIBYA
ALGERIA
NIGER
CHAD
NIGERIA
DEMOCRATIC REPUBLIC OF THE CONGO
ANGOLA
NAMIBIA
BOTSWANA
ZIMBABWE
MOZAMBIQUE
SOUTH AFRICA
SAUDI ARABIA
IRAN
IRAQ
TURKEY
UKRAINE
JAPAN
QUEEN MAUD LAND
ENDERBY LAND
WILKES LAND
VICTORIA LAND
TRANSANTARCTIC MOUNTAINS
South Magnetic Pole
A R C T I C A
ANTARCTIC REGION
600 km
600 mi
Azimuthal Equidistant Projection
Research station
ANTARCTICA
WEST ANTARCTICA
EAST ANTARCTICA
South Pole
Polar Plateau
Ellsworth Mts.
Vinson Massif
Queen Maud Land
Marie Byrd Land
Weddell Sea
Ross Sea
Ross Ice Shelf
Amundsen-Scott U.S.
McMurdo U.S.
Scott N.Z.
ATLANTIC OCEAN
INDIAN OCEAN
PACIFIC OCEAN

Figure 2 Composite satellite view of the earth showing its major terrestrial and aquatic features.

OCEAN
Greenwich
Longitude East of Greenwich
Franz Josef Land
North Land
Svalbard
BARENTS SEA
Novaya Zemlya
ARCTIC CIRCLE
SIBERIA
ASIA
SEA OF OKHOTSK
Kamchatka Peninsula
URAL MOUNTAINS
Scandinavia
Lake Baikal
British Isles
North Sea
EUROPE
ALPS
Aral Sea
GOBI
Black Sea
El'brus 5642
Caspian Sea
Tien Shan
NORTH PACIFIC OCEAN
JAPAN
MEDITERRANEAN SEA
Plateau of Tibet
HIMALAYA
Mt. Everest 8850
TROPIC OF CANCER
ARABIAN PENINSULA
INDIA
PHILIPPINE SEA
SAHARA
Red Sea
ARABIAN SEA
BAY OF BENGAL
SOUTH CHINA SEA
Philippine Islands
Challenger Deep -10994
MICRONESIA
AFRICA
Gulf of Guinea
Lake Victoria
Congo Basin
Kilimanjaro 5895
INDONESIA
EQUATOR
New Guinea
MELANESIA
Lake Tanganyika
Lake Malawi
NINETYEAST RIDGE
Madagascar
CORAL SEA
Fiji Islands
Kalahari Desert
INDIAN OCEAN
AUSTRALIA
New Caledonia
TROPIC OF CAPRICORN
Great Dividing Range
SOUTH PACIFIC OCEAN
Cape of Good Hope
SOUTHWEST INDIAN RIDGE
Mt. Kosciuszko 2228
Bass Strait
TASMAN SEA
North Island
Tasmania
SOUTHEAST INDIAN RIDGE
South Island
NEW ZEALAND
ANTARCTIC CIRCLE
Wilkes Land
Winkel Tripel Projection, Central Meridian 0°
KILOMETERS
STATUTE MILES
Elevations in meters
ANTARCTICA

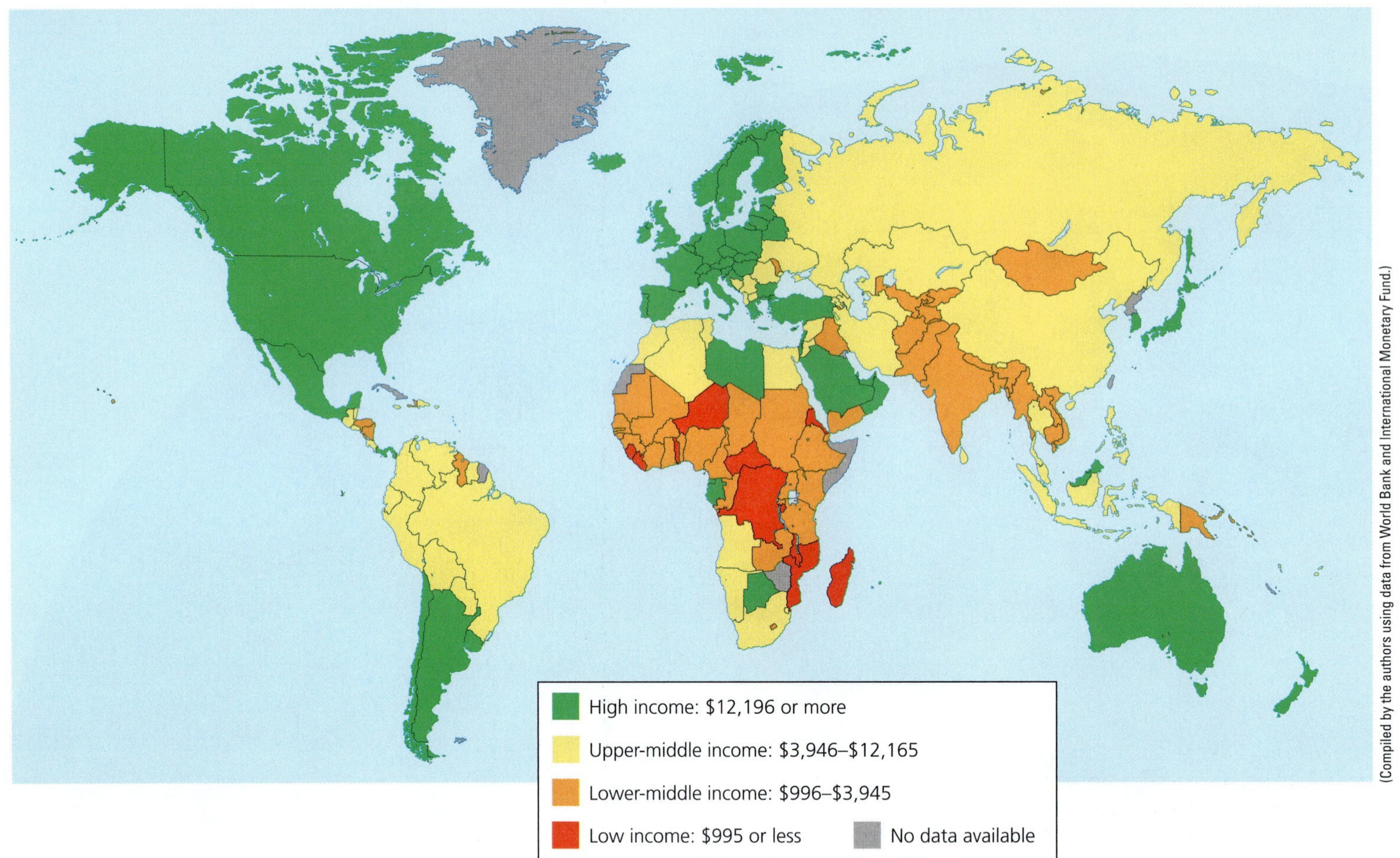

(Compiled by the authors using data from World Bank and International Monetary Fund.)

Figure 3 High-income, upper-middle-income, lower-middle-income, and low-income countries in terms of gross national income (GNI) purchasing power parity (PPP) per capita (U.S. dollars) in 2010.

Data and Map Analysis

1. In how many countries is the per capita average income $995 or less? Look at Figure 1 and find the names of three of these countries.
2. In how many instances does a lower-middle- or low-income country share a border with a high-income country? Look at Figure 1 and find the names of the countries that reflect three examples of this situation.

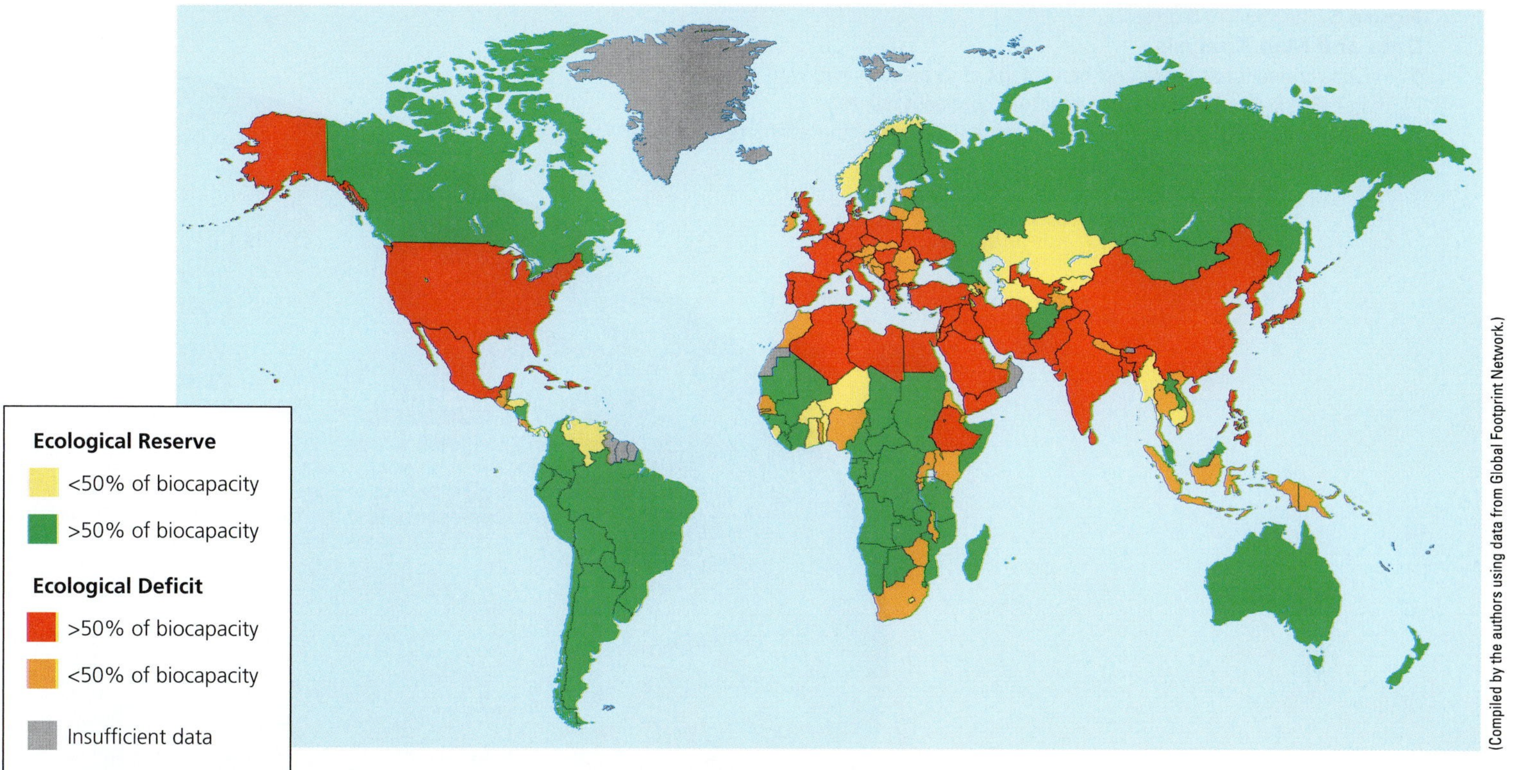

(Compiled by the authors using data from Global Footprint Network.)

Figure 4 *Ecological debtors and creditors:* The ecological footprints of some countries exceed their biocapacity, while other countries still have ecological reserves.

Data and Map Analysis

1. List five countries, including the three largest, in which the ecological deficit is greater than 50% of biocapacity. (See Figure 1 of this supplement for country names.)
2. On which two continents does land with ecological reserves of more than 50% of biocapacity occupy the largest percentage of total land area? (See Figure 2 of this supplement for continent names.)

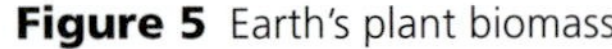

Figure 5 Earth's plant biomass.

Data and Map Analysis

1. Which continent has the largest percentage of its area covered with minimal biomass? (See Figure 2 of this supplement for continent names.)
2. Is ocean chlorophyll concentration, in general, higher near the earth's poles or near its equator?

National Geographic

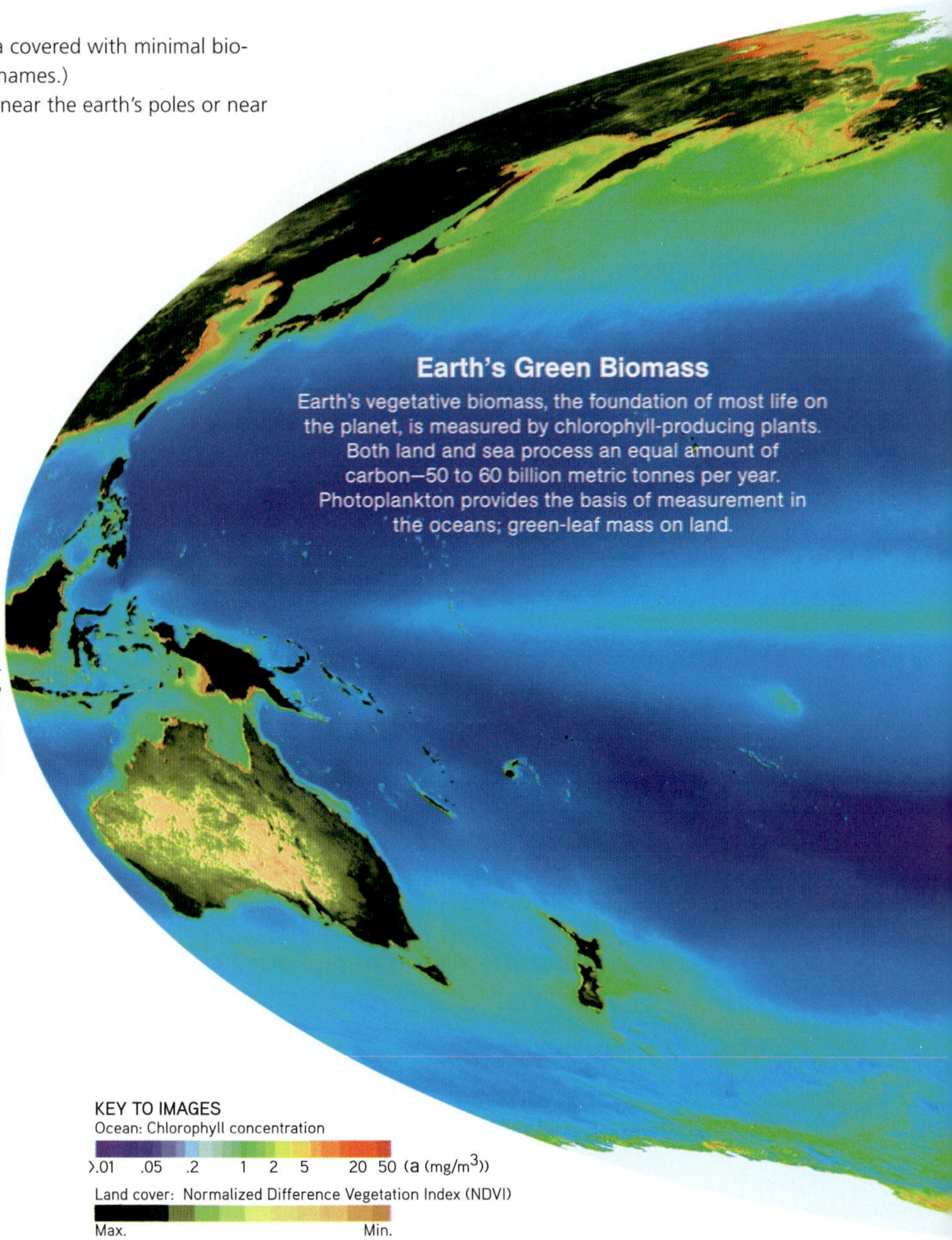

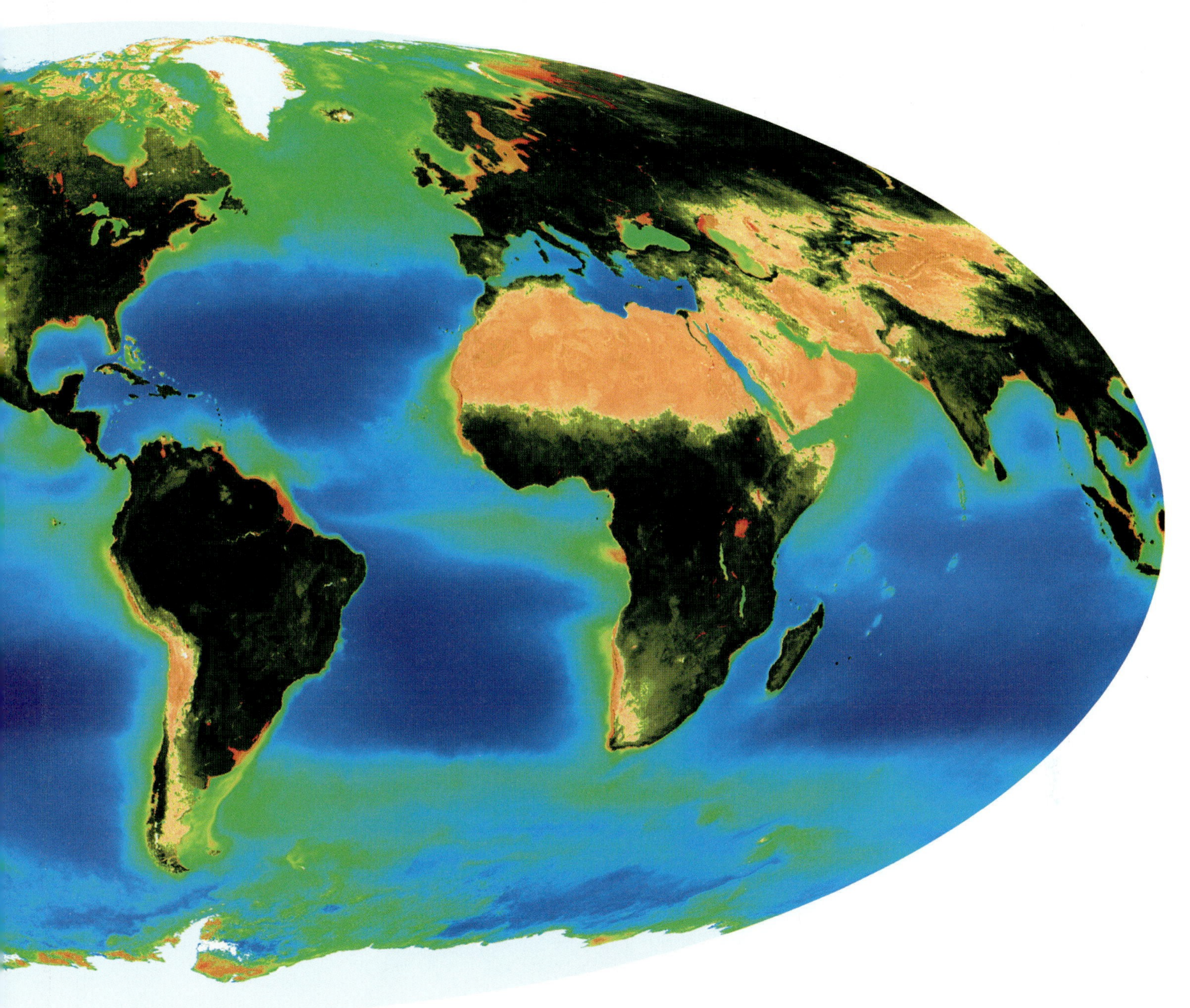

Figure 6 Land cover in North America.

Data and Map Analysis

1. Approximately what percentage of North America is covered by savanna?
2. Approximately what percentage of the continental United States is used for cropland?

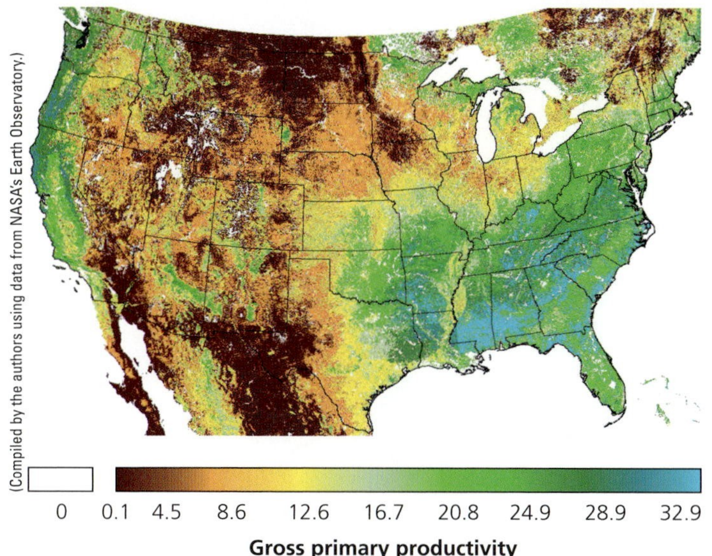

(Compiled by the authors using data from NASA's Earth Observatory.)

Figure 7 **Natural capital:** Gross primary productivity across the continental United States, based on remote satellite data. The differences roughly correlate with variations in moisture and soil types.

Data and Map Analysis

1. Comparing the five northwestern-most states with the five southeastern-most states, which of these regions has the greater variety in levels of gross primary productivity? Which of the regions has the highest levels overall?
2. Compare this map with that of Figure 6. Which biome in the United States is associated with the highest level of gross primary productivity?

Figure 8 **Natural capital degradation:** The human ecological footprint in North America. Colors represent the percentage of each area influenced by human activities.

(Compiled by the authors using data from Wildlife Conservation Society and the Center for International Earth Science Information Network at Columbia University.)

Data and Map Analysis

1. Which general area of the United States has the highest human footprint values?
2. What is the relative value of the human ecological footprint in the area where you live or go to school?

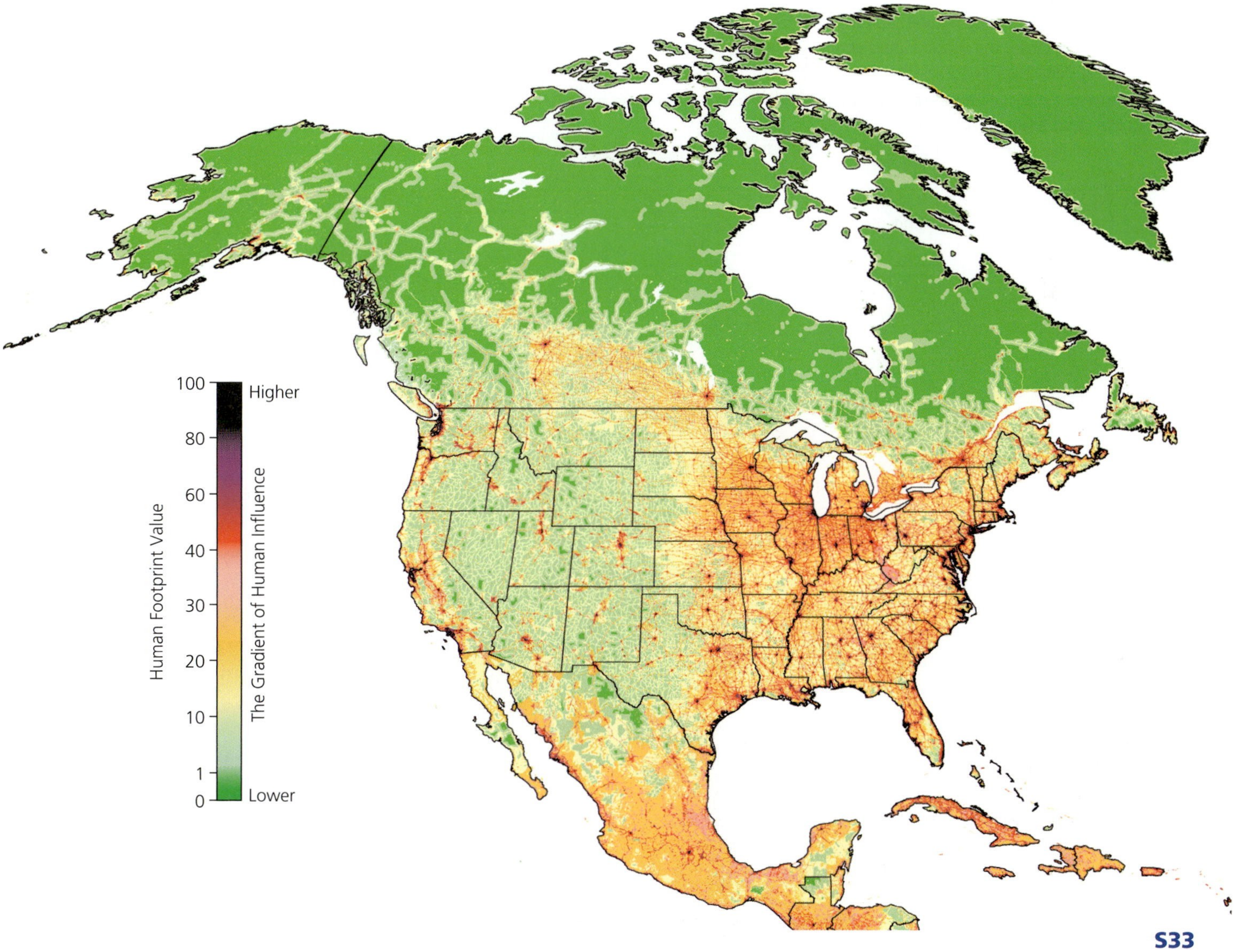

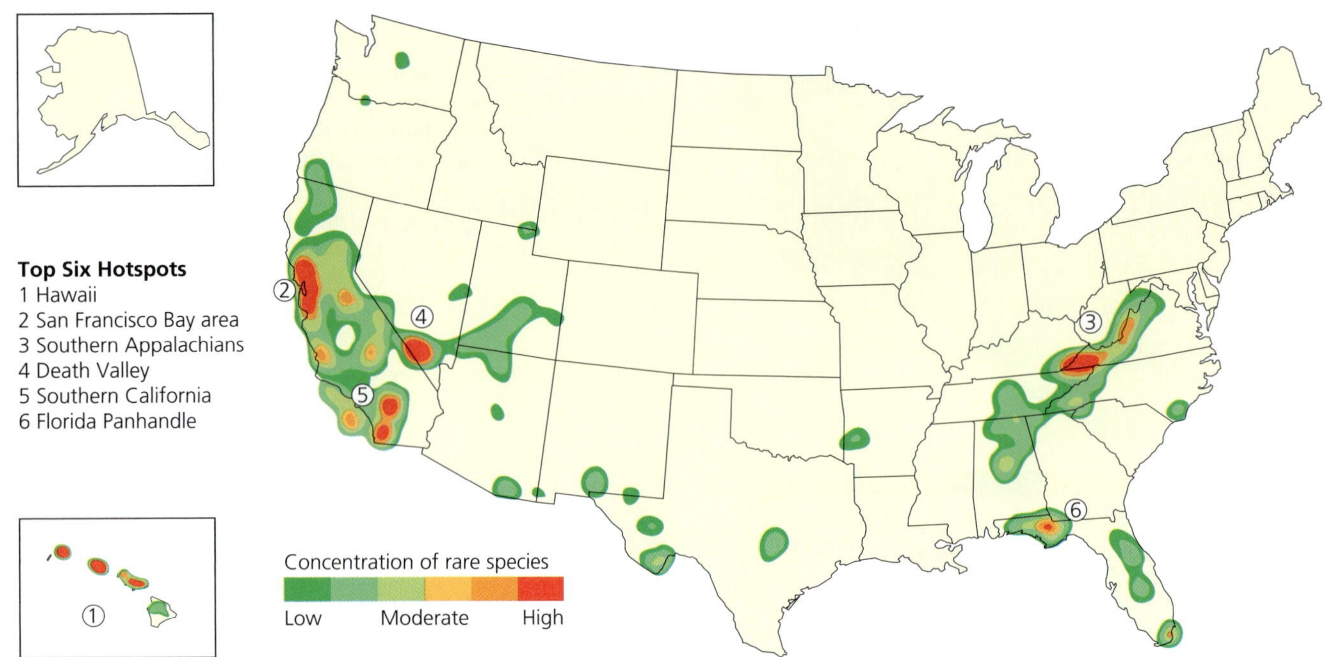

Figure 9 **Endangered natural capital:** Major biodiversity hotspots in the United States that need emergency protection. The shaded areas contain the largest concentrations of rare and potentially endangered species.

(Compiled by the authors using data from State Natural Heritage Programs, the Nature Conservancy, and Association for Biodiversity Information.)

Data and Map Analysis

1. If you live in the United States, which of the top six hotspots is closest to where you live or go to school?
2. Which general part of the country has the highest overall concentration of rare species? Which part has the second-highest concentration?

Figure 10 Land cover in South America.

Data and Map Analysis

1. Which type of land cover would you say occupies the largest area in South America?
2. Brazil was once almost completely covered by rain forest (evergreen broadleaf), much of which has been converted to savanna as it has been cleared by humans and has not grown back as forest. About what percentage of Brazil would you say is occupied by savanna?

Figure 11 Land cover in Europe.

Data and Map Analysis

1. What are the two most common types of land cover in Europe?
2. What percentage of European land would you estimate to be used for growing crops?

Figure 12 Land cover in Asia.

Data and Map Analysis

1. About what percentage of China's land area would you estimate to be classified as barren or sparsely vegetated?
2. About what percentage of India's land would you say is used for growing crops?

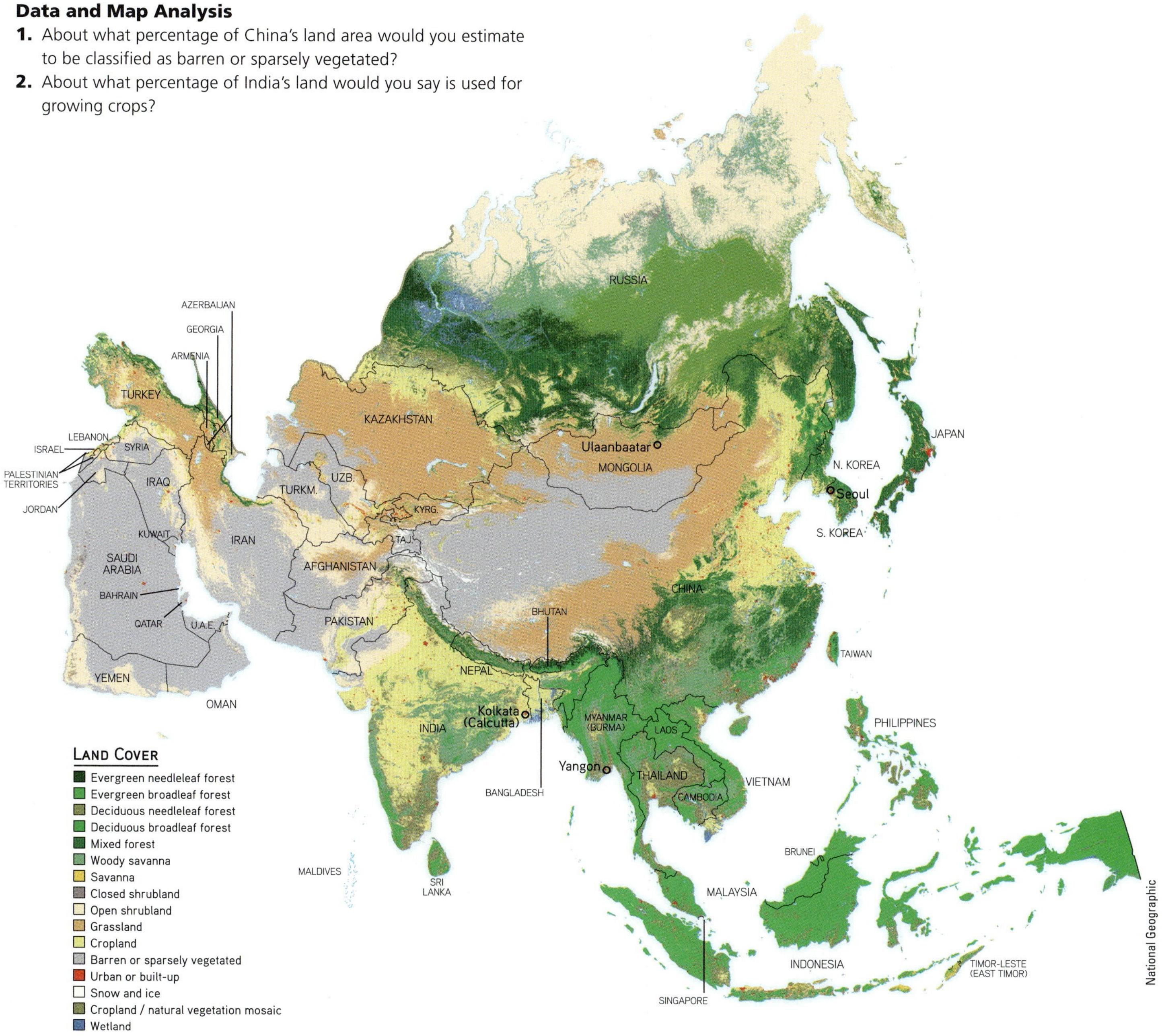

Land Cover

- Evergreen needleleaf forest
- Evergreen broadleaf forest
- Deciduous needleleaf forest
- Deciduous broadleaf forest
- Mixed forest
- Woody savanna
- Savanna
- Closed shrubland
- Open shrubland
- Grassland
- Cropland
- Barren or sparsely vegetated
- Urban or built-up
- Snow and ice
- Cropland / natural vegetation mosaic
- Wetland

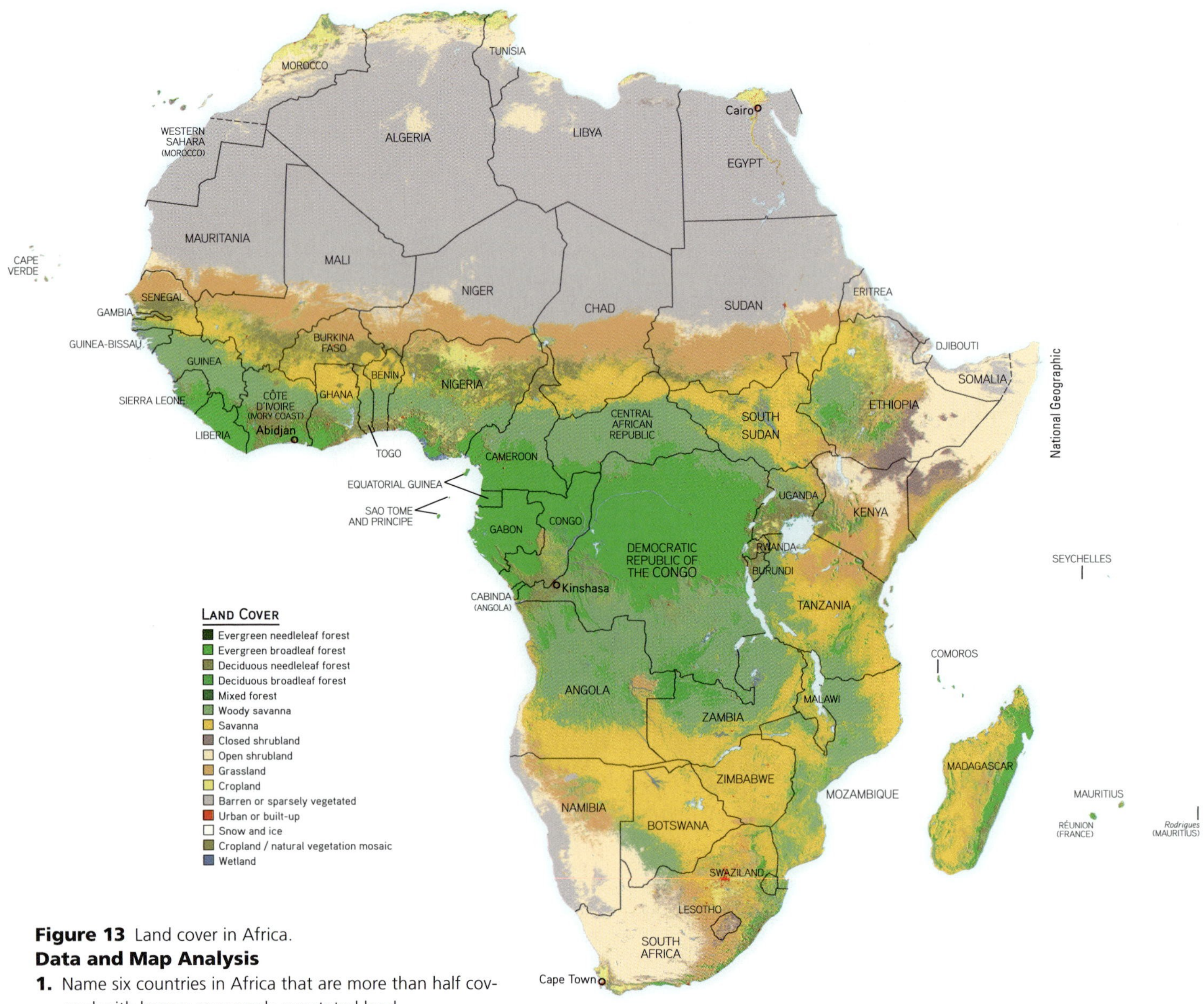

Figure 13 Land cover in Africa.

Data and Map Analysis

1. Name six countries in Africa that are more than half covered with barren or sparsely vegetated land.
2. Which African country has the largest area of forest land?

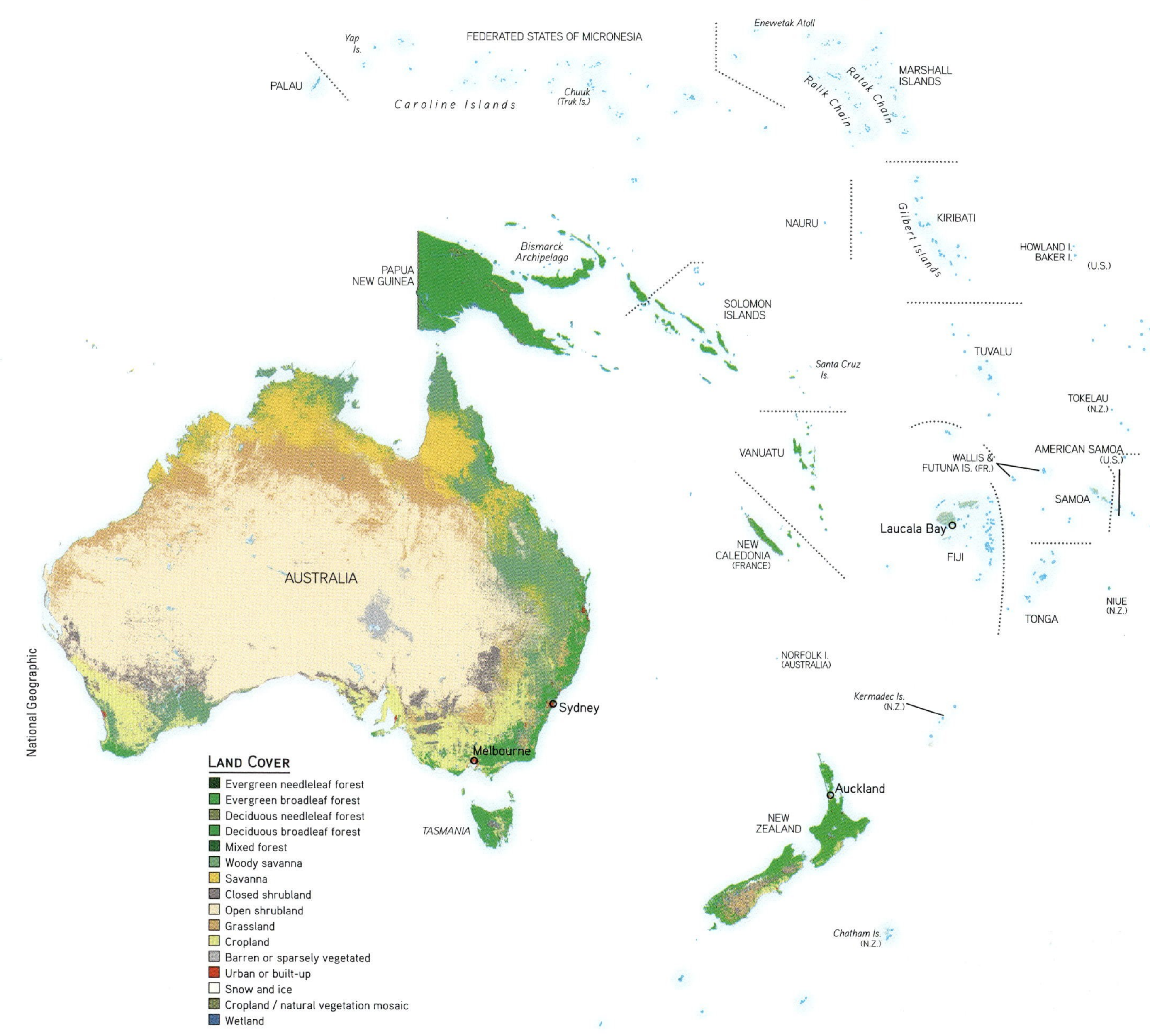

Figure 14 Land cover in Oceania and Australia.

Data and Map Analysis

1. What type of land cover occupies the most land in Australia?
2. What do all urban areas shown in red on this map have in common?

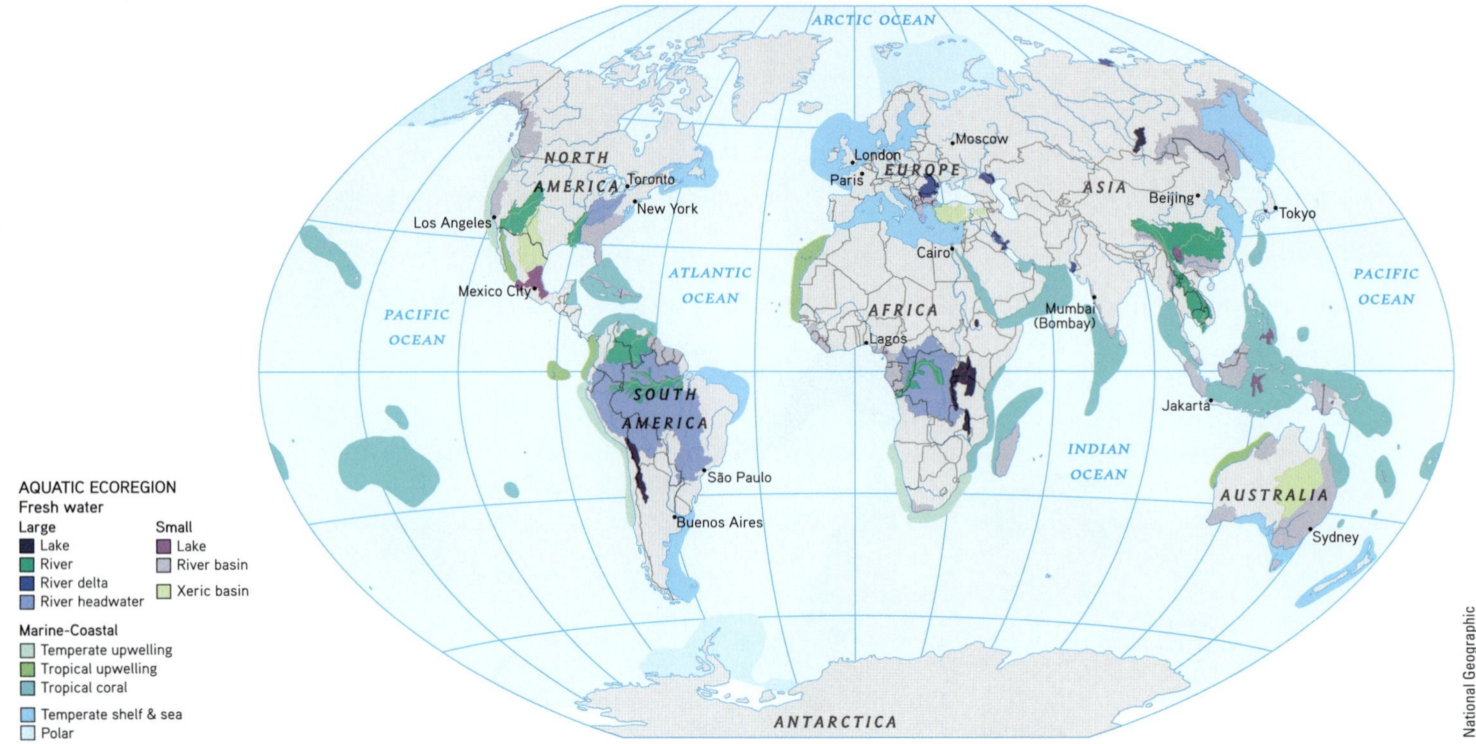

Figure 15 Major types of marine and freshwater habitats.

Data and Map Analysis

1. On which continent is the largest area of river habitat?
2. List the locations of three tropical upwelling areas.

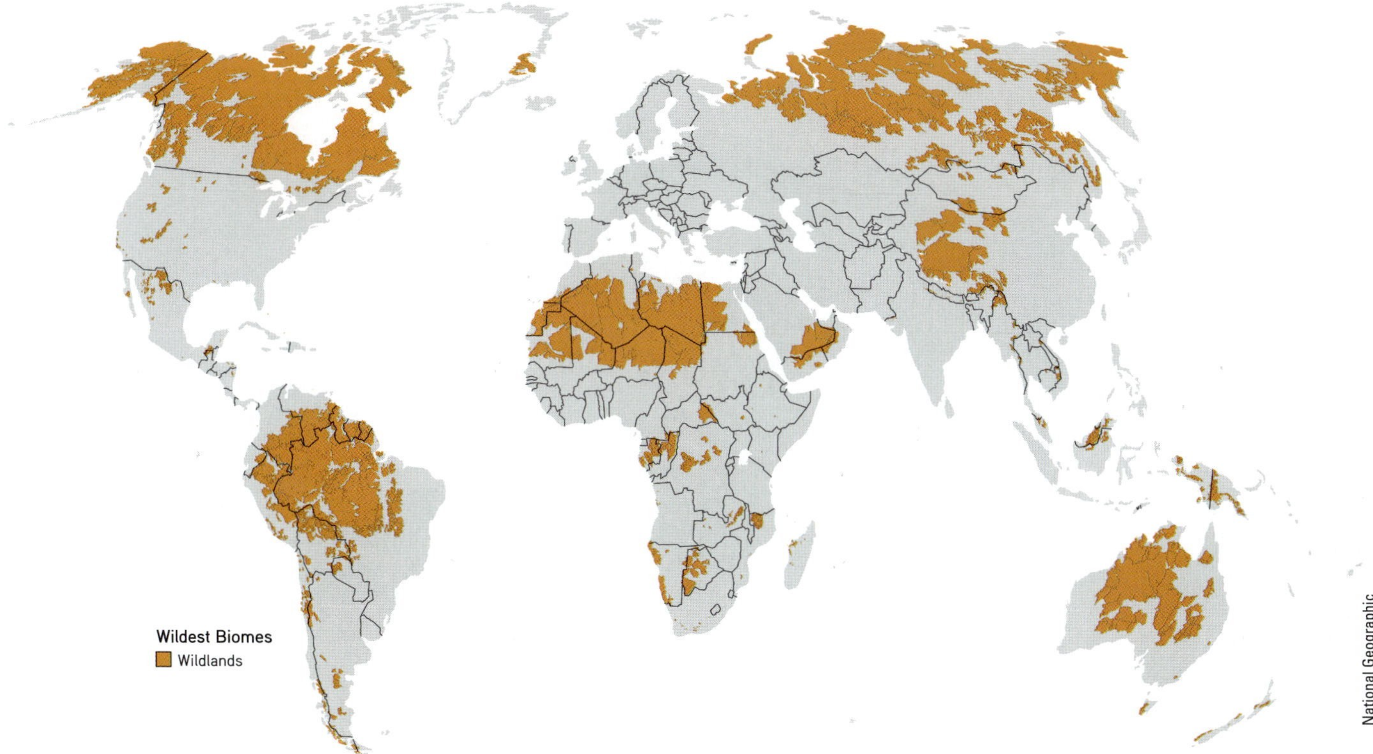

Figure 16 World's remaining wildlands.

Data and Map Analysis

1. Comparing this map with those in Figures 12–14, list the locations of three wild areas that are also classified as barren or sparsely vegetated. (See Figures 1 and 2 of this supplement for country and continent names.)
2. What continent appears to have the largest percentage of wildland?

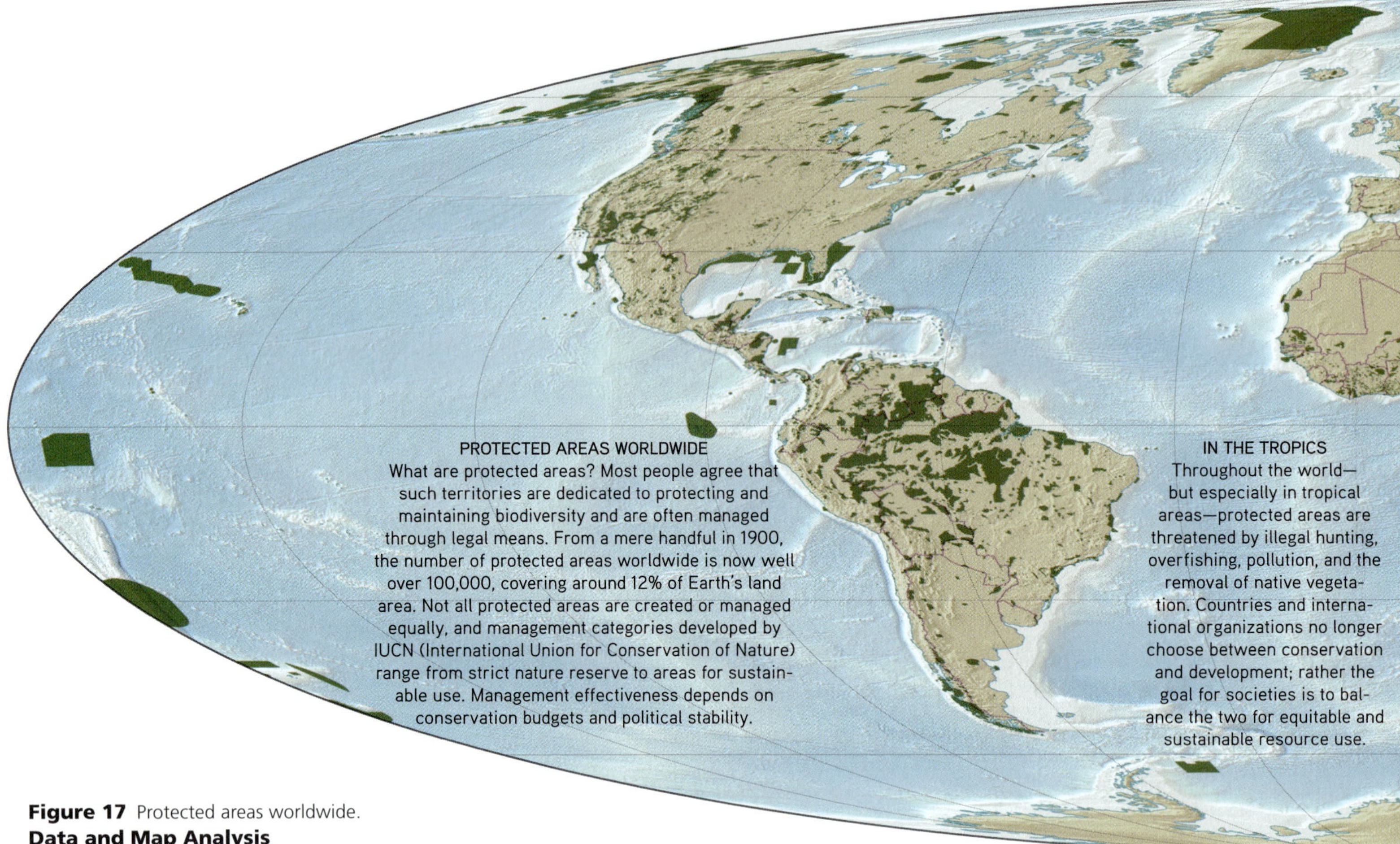

Figure 17 Protected areas worldwide.

Data and Map Analysis

1. List the locations of three relatively large protected areas. (See Figures 1 and 2 of this supplement for country and continent names.)
2. What continent appears to have the largest percentage of protected land? What continent appears to have the smallest percentage of protected land?

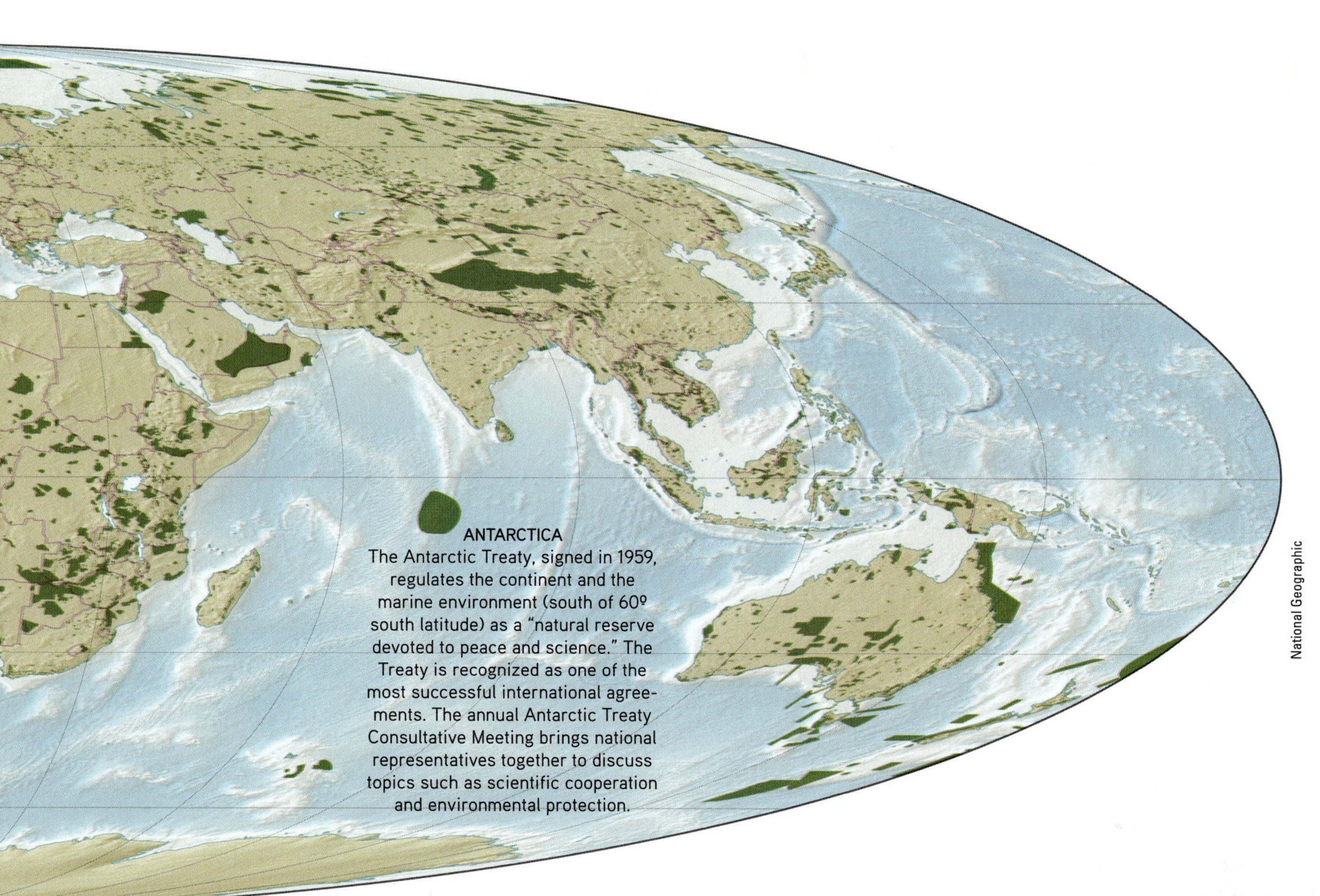

National Geographic

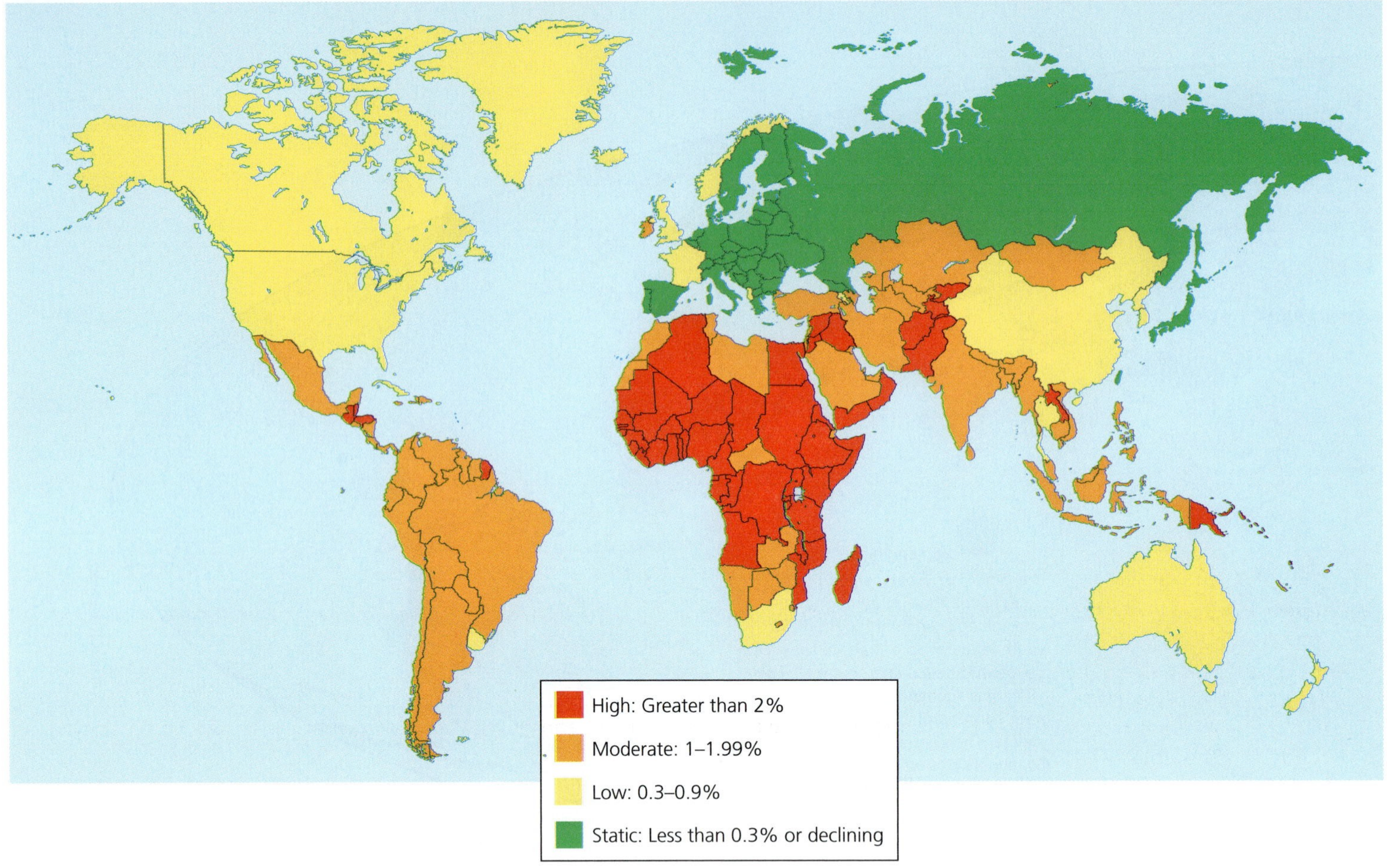

Figure 18 Comparative rates of population growth (%) throughout the world in 2012.

(Compiled by the authors using data from Population Reference Bureau and United Nations Population Division.)

Data and Map Analysis

1. Which continent has the greatest number of countries with high rates of population increase? Which continent has the greatest number of countries with static rates? (See Figure 2 of this supplement for continent names.)
2. For each category on this map, name the two countries that you think are largest in terms of total area (see Figure 1 for country names).

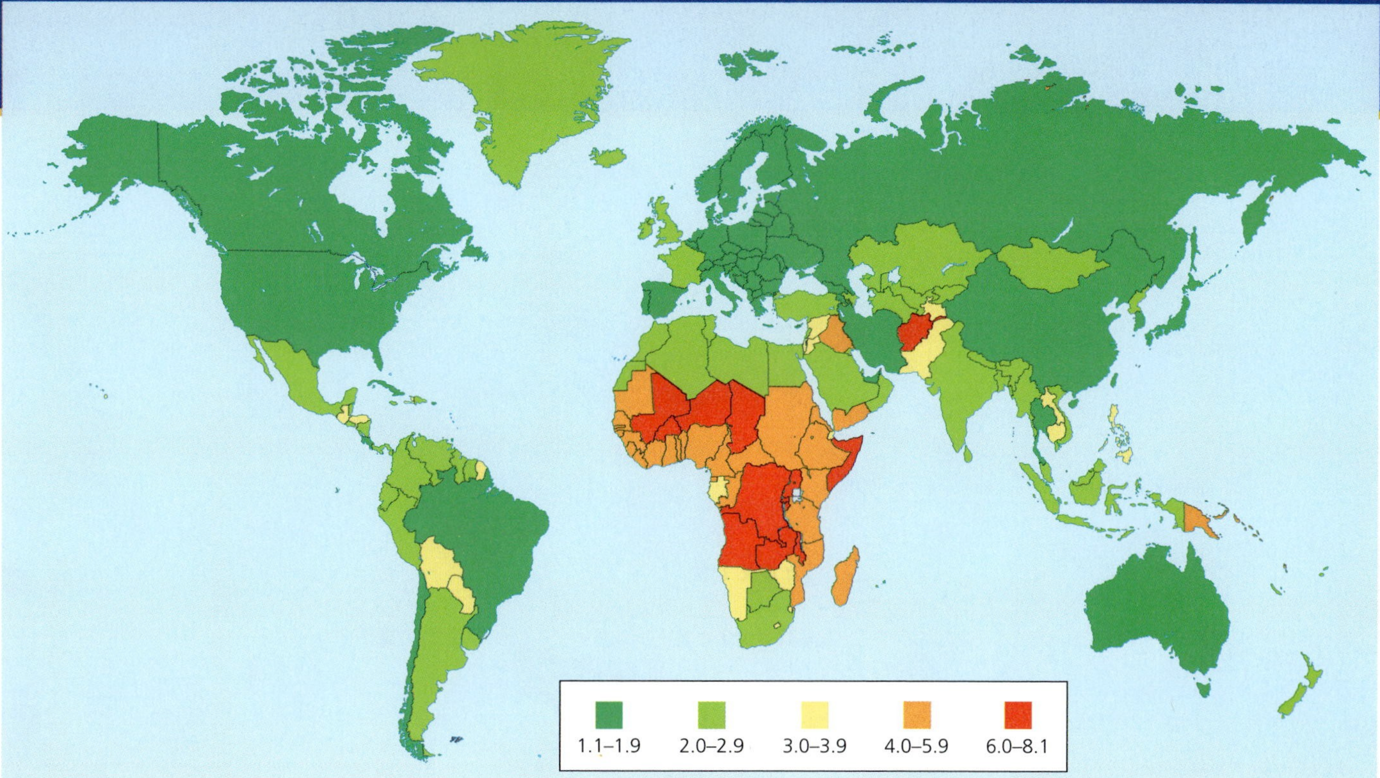

Figure 19 Comparison of total fertility rate (TFR), or average number of children born to the world's women throughout their lifetimes, around the world as measured in 2012.

(Compiled by the authors using data from Population Reference Bureau and United Nations Population Division.)

Data and Map Analysis

1. Which country in the highest TFR category borders two countries in the lowest TFR category? What are those two countries? (See Figure 1 of this supplement for country names.)
2. Do you see any geographic patterns on this map? Explain.

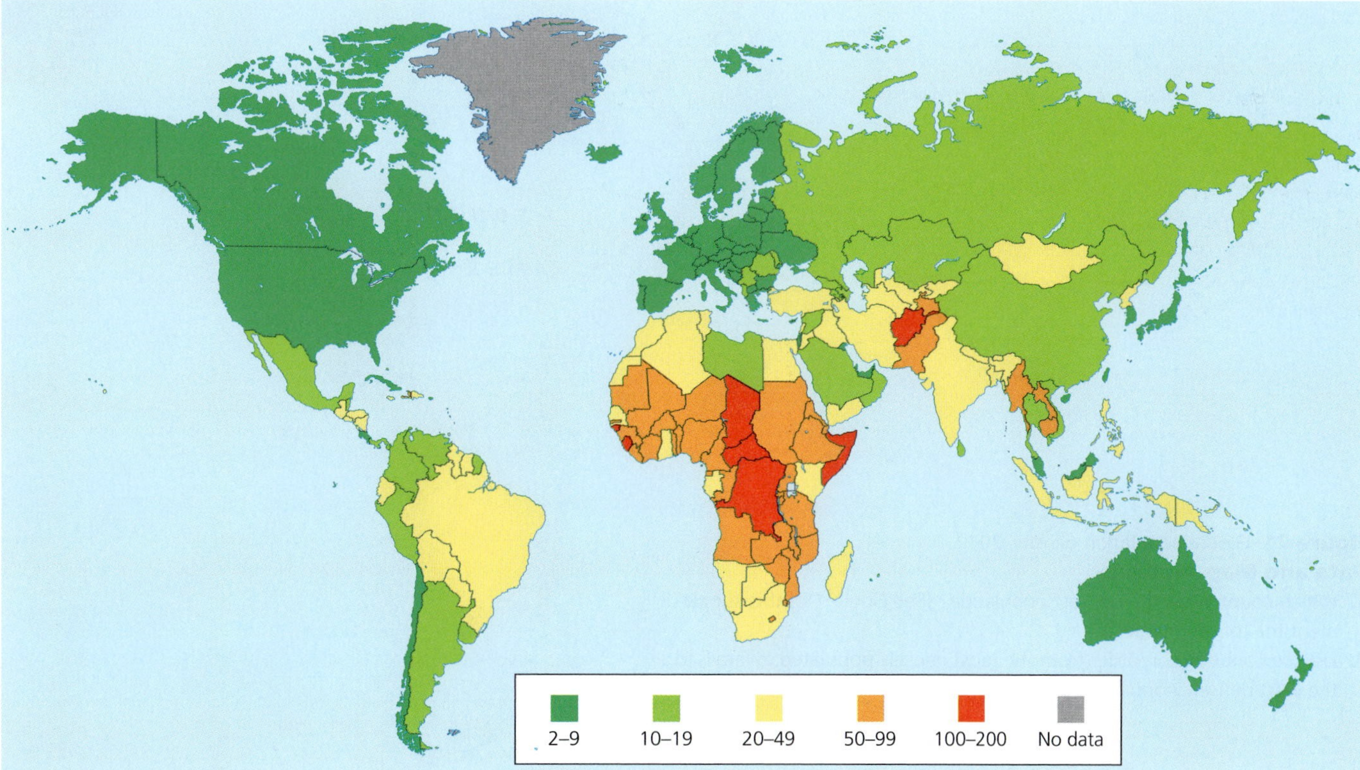

Figure 20 Comparison of global infant mortality rates around the world in 2012.

(Compiled by the authors using data from Population Reference Bureau and United Nations Population Division.)

Data and Map Analysis

1. Do you see a geographic pattern related to infant mortality rates as reflected on this map? Explain.
2. List any similarities that you see in geographic patterns between this map and the one in Figure 19.

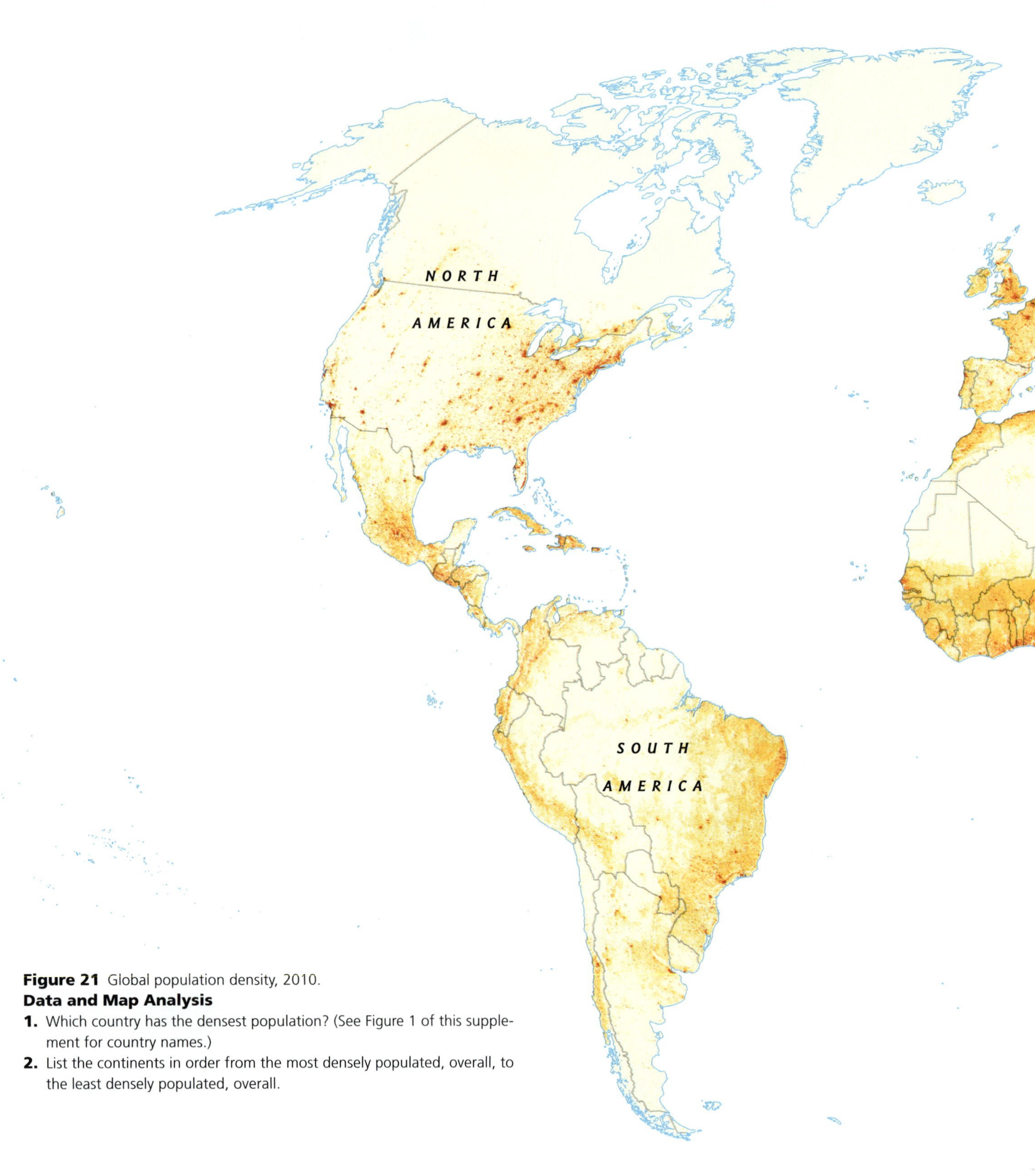

Figure 21 Global population density, 2010.

Data and Map Analysis

1. Which country has the densest population? (See Figure 1 of this supplement for country names.)
2. List the continents in order from the most densely populated, overall, to the least densely populated, overall.

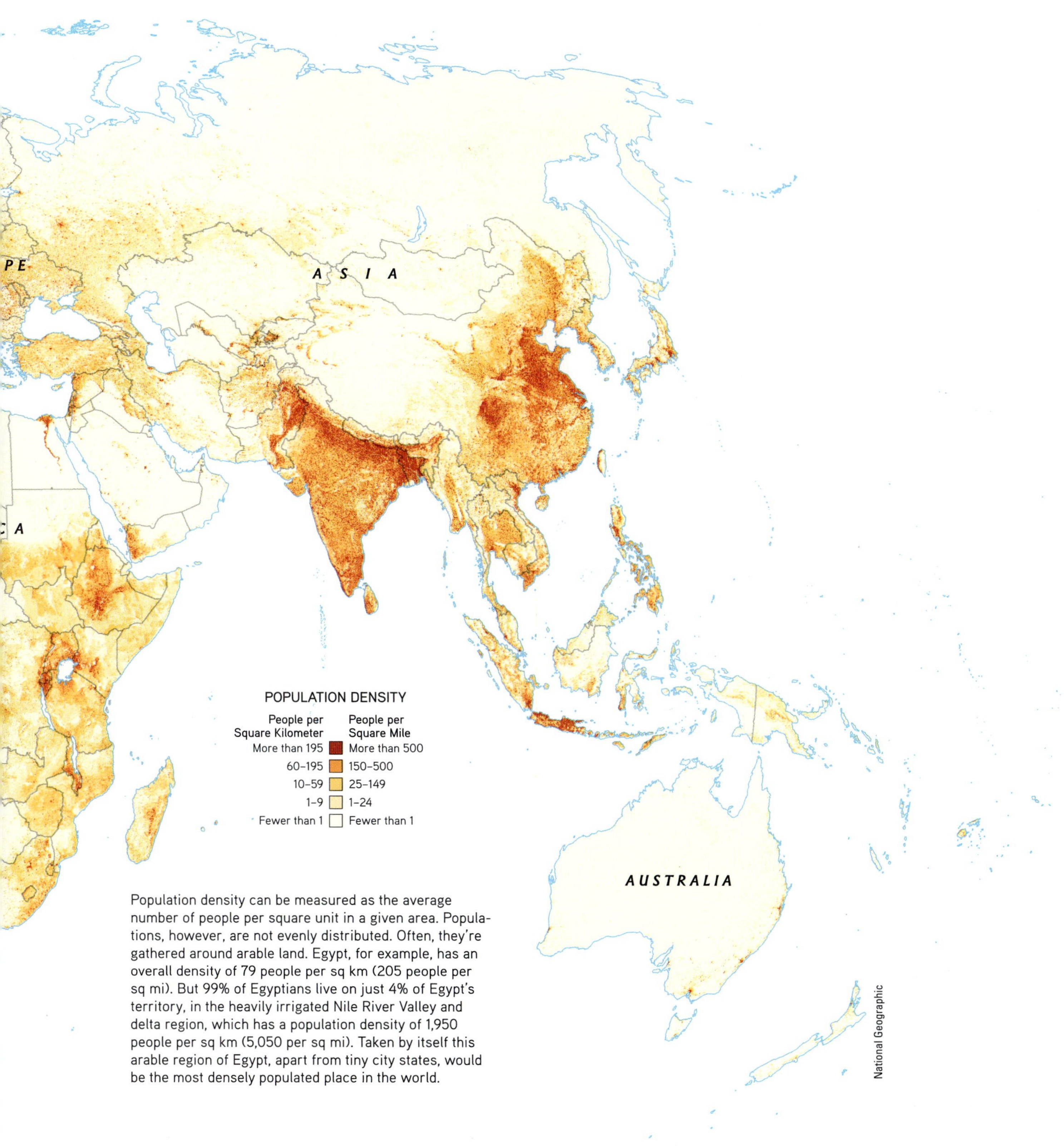

Population density can be measured as the average number of people per square unit in a given area. Populations, however, are not evenly distributed. Often, they're gathered around arable land. Egypt, for example, has an overall density of 79 people per sq km (205 people per sq mi). But 99% of Egyptians live on just 4% of Egypt's territory, in the heavily irrigated Nile River Valley and delta region, which has a population density of 1,950 people per sq km (5,050 per sq mi). Taken by itself this arable region of Egypt, apart from tiny city states, would be the most densely populated place in the world.

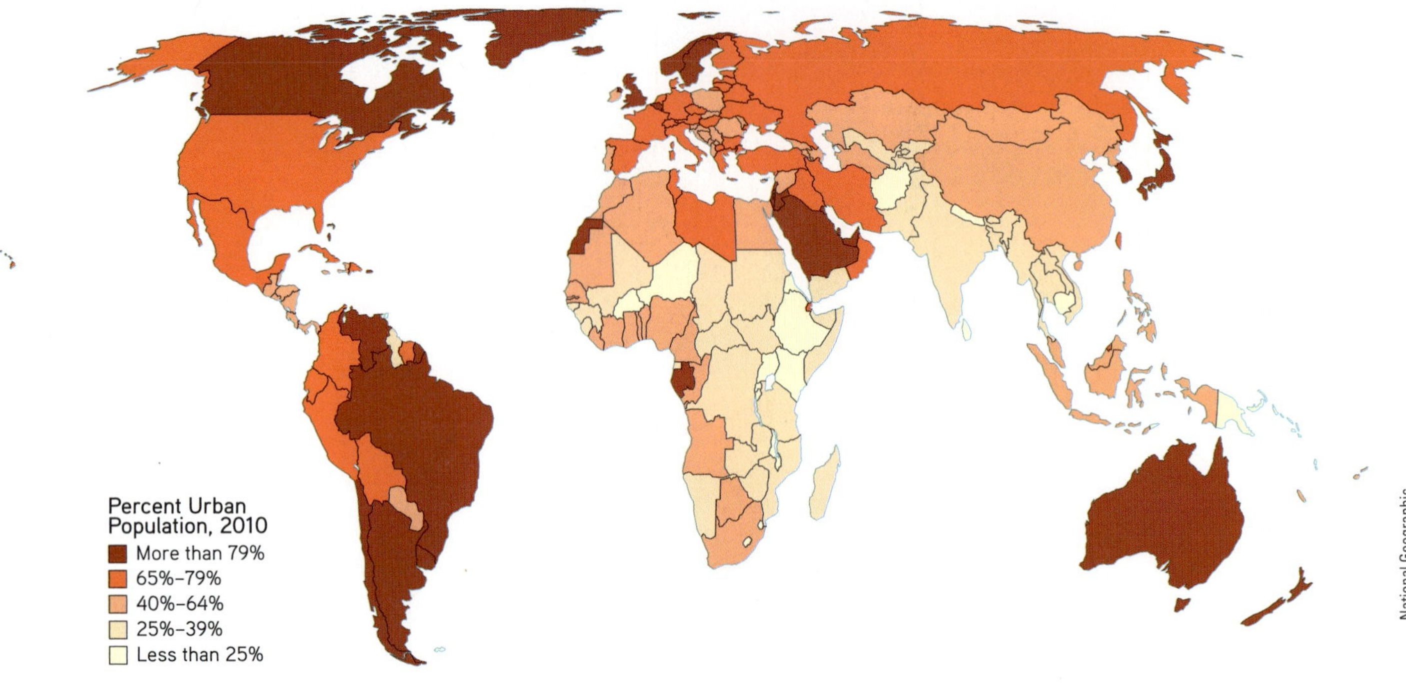

Figure 22 Urban population, 2010.

Data and Map Analysis

1. Name six countries whose populations are more than 79% urban. (See Figure 1 of this supplement for country names.)
2. List three pairs of bordering countries in which the population of one country is more than 79% urban and the population of the other is 39% or less urban.

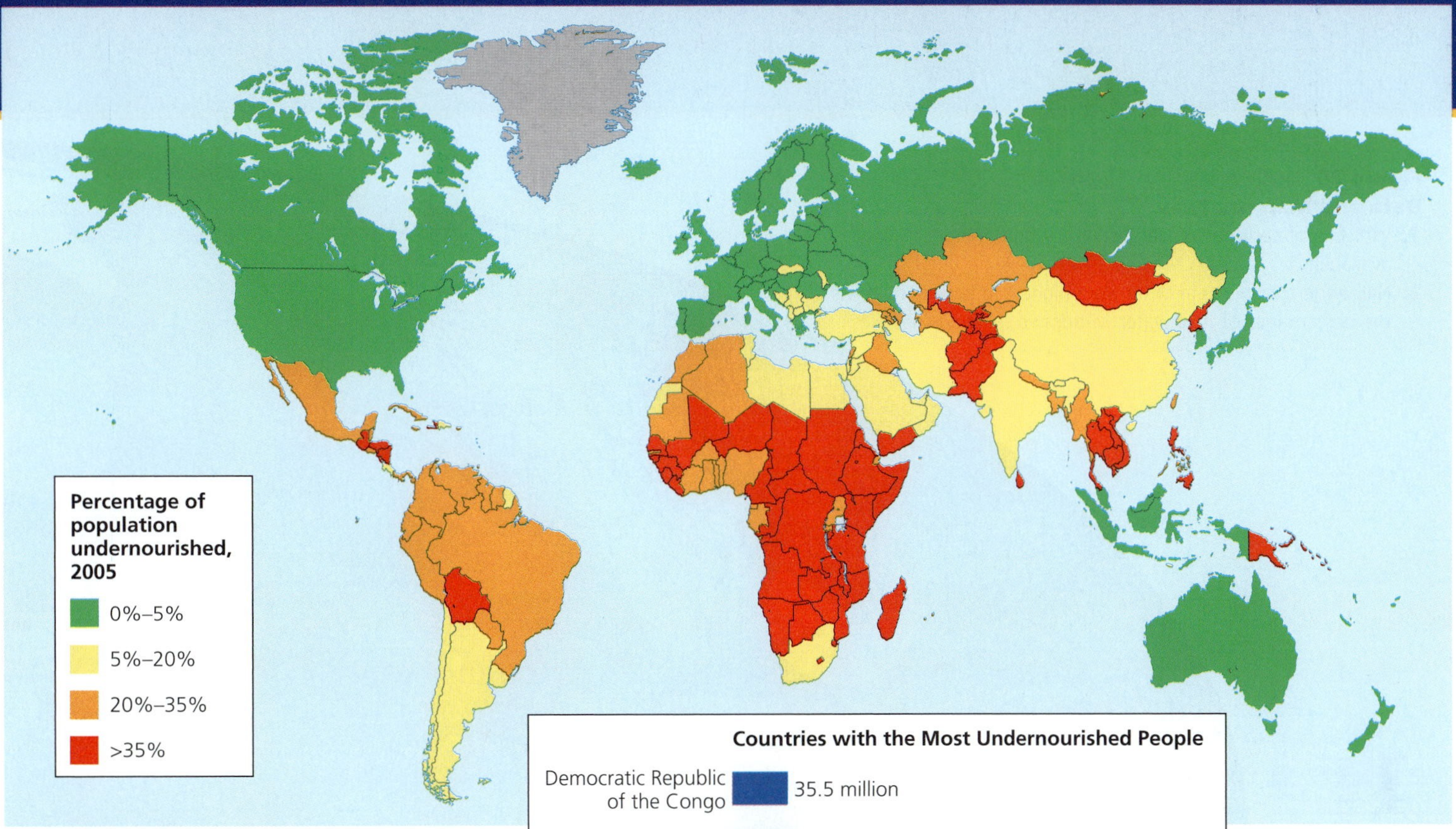

Percentage of population undernourished, 2005

- 0%–5%
- 5%–20%
- 20%–35%
- >35%

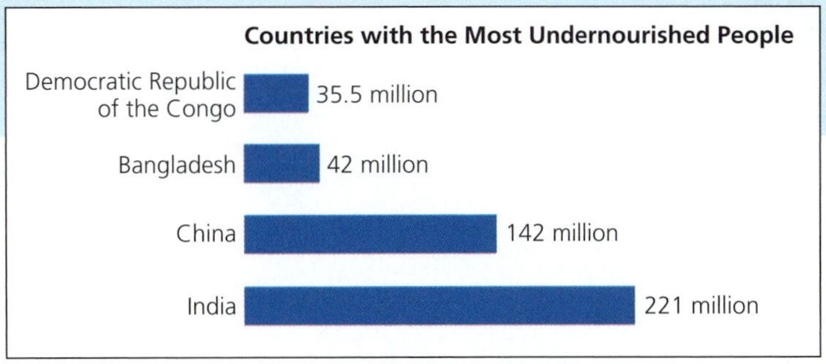

Countries with the Most Undernourished People

Country	Undernourished people
Democratic Republic of the Congo	35.5 million
Bangladesh	42 million
China	142 million
India	221 million

Figure 23 World hunger expressed as percentages of populations suffering from chronic hunger and malnutrition.

(Compiled by the authors using data from the UN Food and Agriculture Organization, World Health Organization, and U.S. Department of Agriculture.)

Data and Map Analysis

1. List the continents in order, starting with the one that has the highest percentage of undernourished people and ending with the one that has the lowest such percentage. (See Figure 2 of this supplement for continent names.)
2. On which continent is the largest block of countries that suffer the highest levels of undernourishment? List five of these countries.

Figure 24 U.S. farm area, 2007.

Data and Map Analysis

1. Name six states that are primarily more than 70% covered by farmland.
2. Name six states where there is a high variation in farmland cover, ranging from less than 10% to more than 70%.

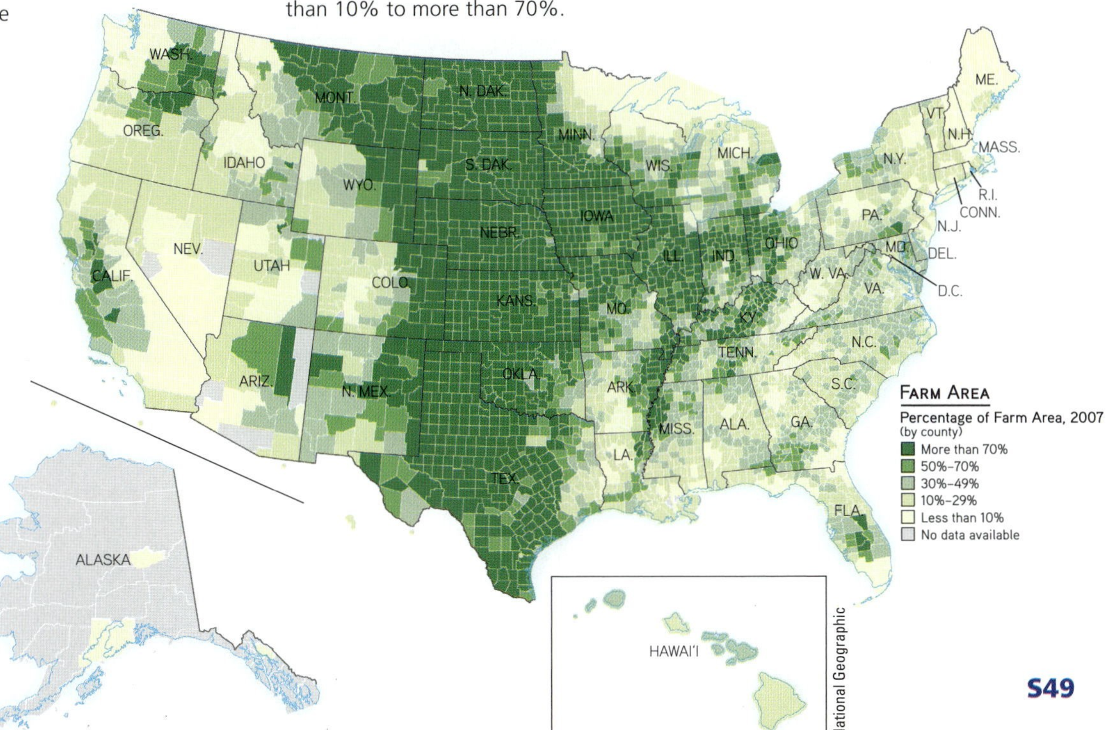

Figure 25 Global availability of freshwater.

Data and Map Analysis

1. Which continent has the largest area with the highest water availability?
2. Name the continents in order from highest to lowest, in terms of the percentages of freshwater withdrawn for industrial uses.

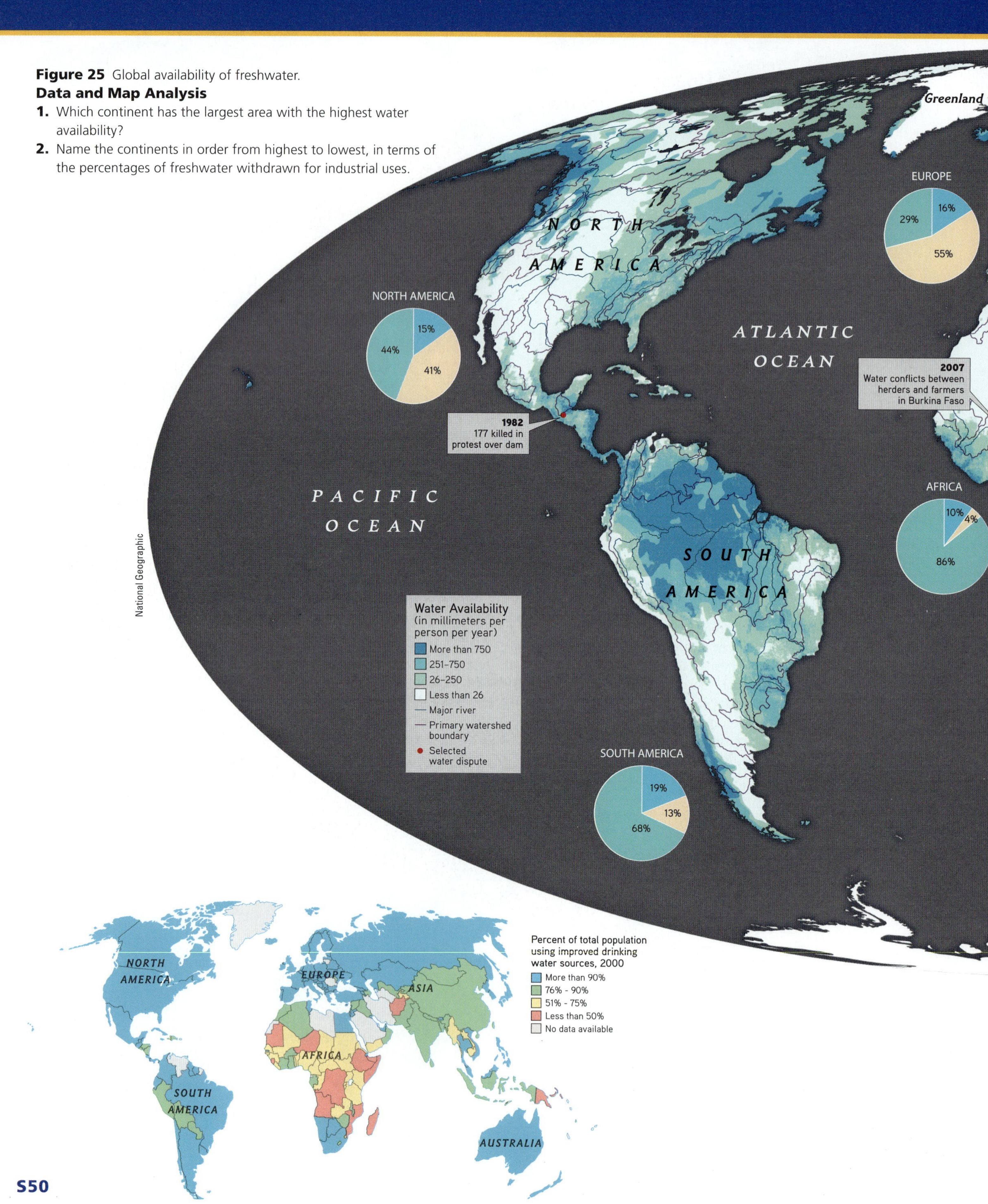

National Geographic

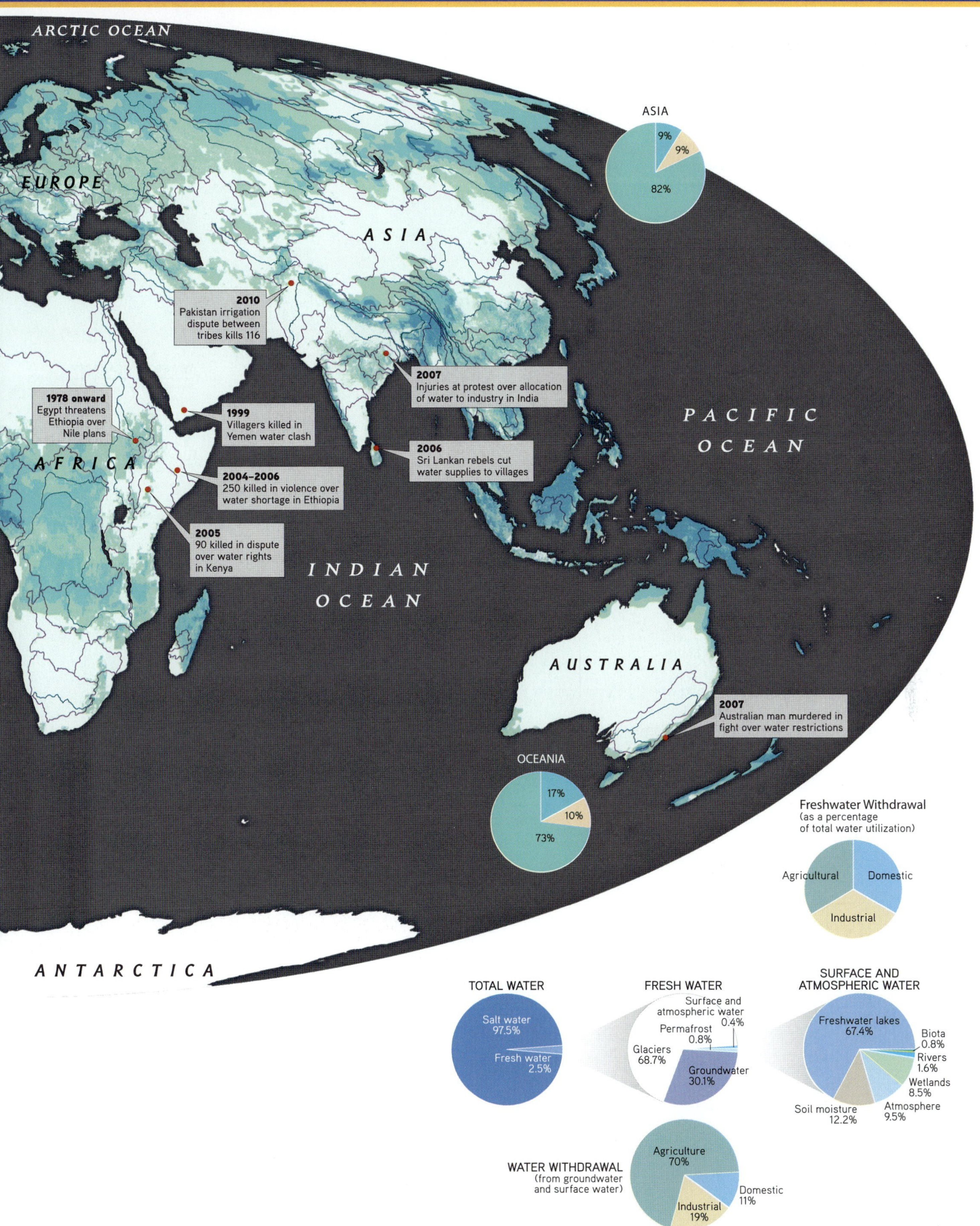
ARCTIC OCEAN
EUROPE
ASIA
AFRICA
PACIFIC OCEAN
INDIAN OCEAN
AUSTRALIA
ANTARCTICA
ASIA
9%
9%
82%
2010
Pakistan irrigation dispute between tribes kills 116
2007
Injuries at protest over allocation of water to industry in India
1978 onward
Egypt threatens Ethiopia over Nile plans
1999
Villagers killed in Yemen water clash
2006
Sri Lankan rebels cut water supplies to villages
2004–2006
250 killed in violence over water shortage in Ethiopia
2005
90 killed in dispute over water rights in Kenya
2007
Australian man murdered in fight over water restrictions
OCEANIA
17%
10%
73%
Freshwater Withdrawal
(as a percentage of total water utilization)
Agricultural
Domestic
Industrial
TOTAL WATER
Salt water 97.5%
Fresh water 2.5%
FRESH WATER
Surface and atmospheric water 0.4%
Permafrost 0.8%
Glaciers 68.7%
Groundwater 30.1%
SURFACE AND ATMOSPHERIC WATER
Freshwater lakes 67.4%
Biota 0.8%
Rivers 1.6%
Wetlands 8.5%
Atmosphere 9.5%
Soil moisture 12.2%
WATER WITHDRAWAL
(from groundwater and surface water)
Agriculture 70%
Domestic 11%
Industrial 19%

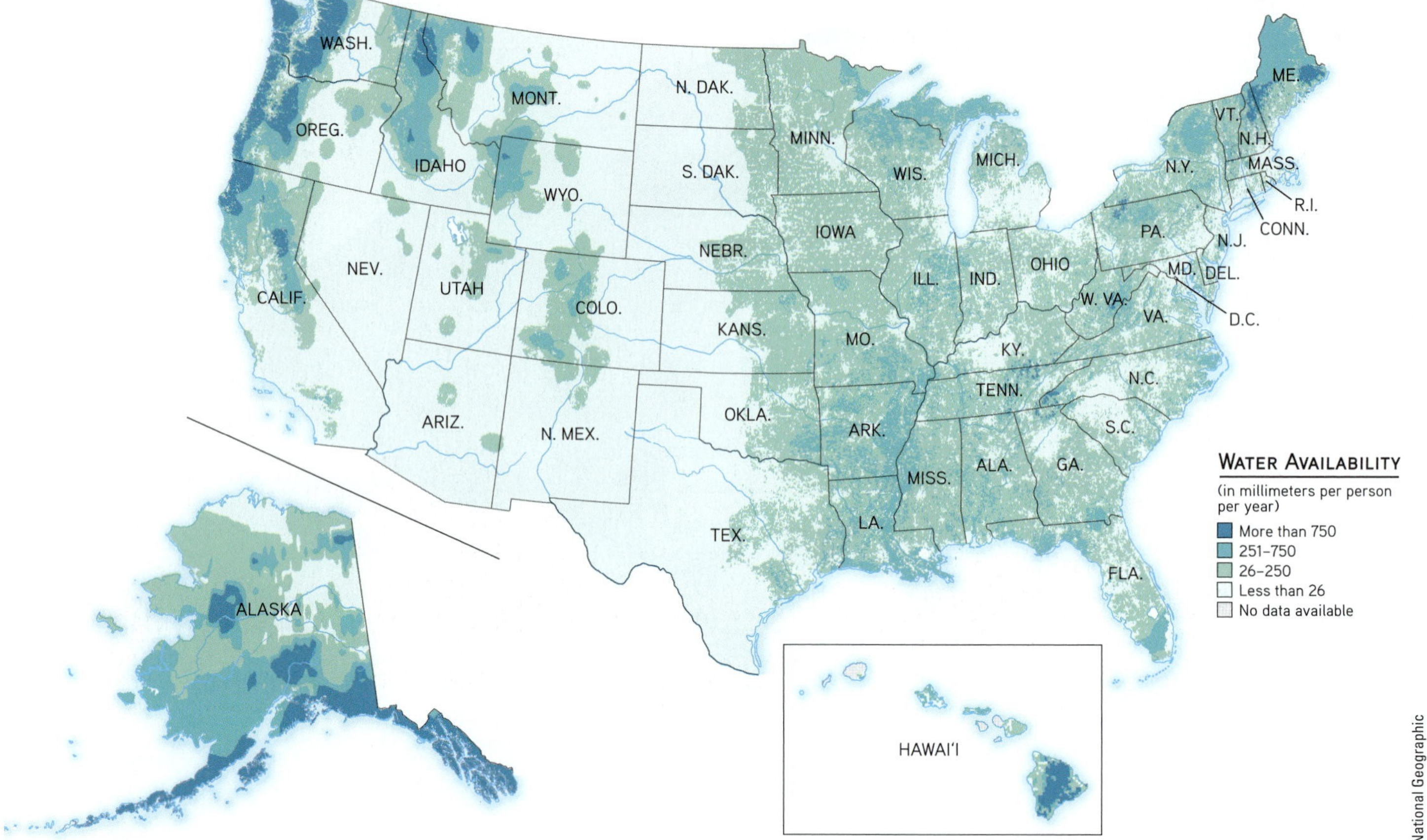

Figure 26 Availability of freshwater in the United States.

Data and Map Analysis

1. Name the three states with the largest areas of land where water availability is more than 750 milliliters per person per year.
2. Name six states where there are no areas of water availability of more than 250 milliliters per person per year.

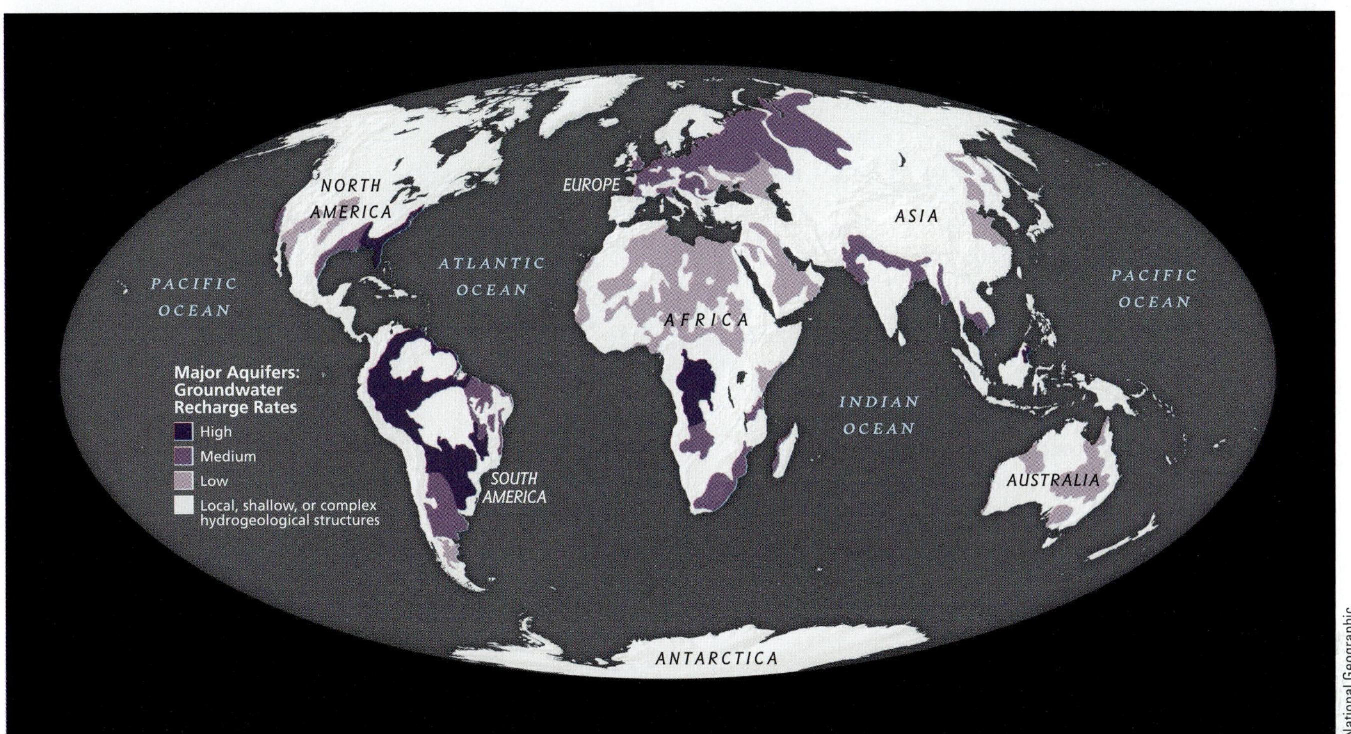

National Geographic

Figure 27 Recharge rates of the world's major aquifers.

Data and Map Analysis

1. Which continent has the largest area of aquifers with high recharge rates?
2. Which continents have no aquifers with high recharge rates?

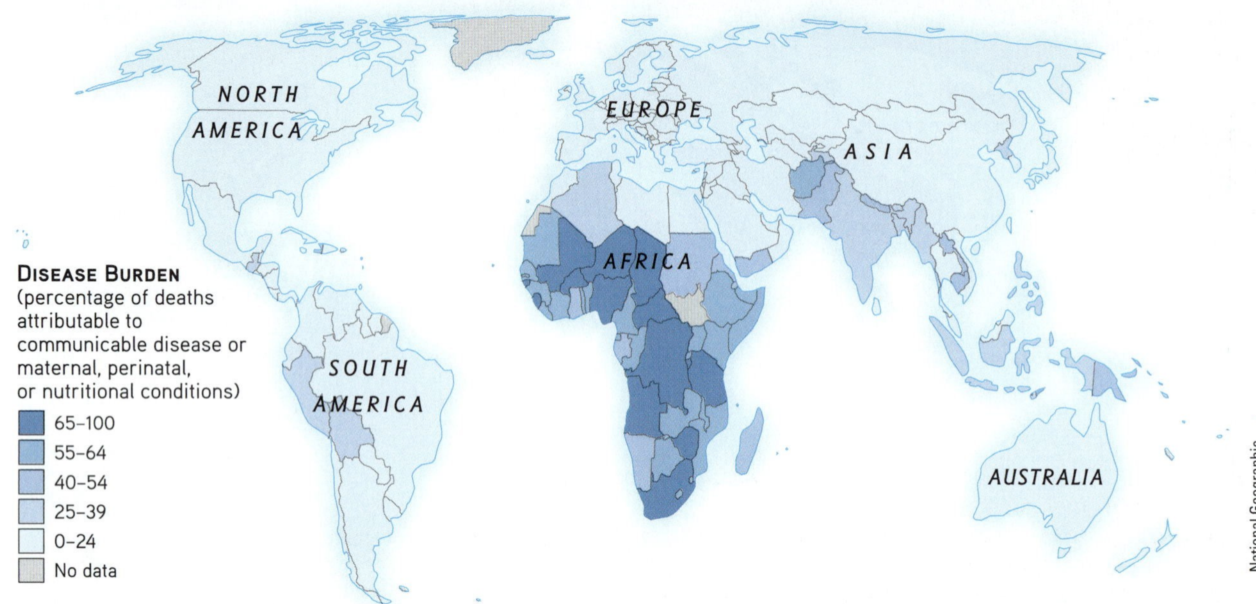

Figure 28 Global disease burden.

Data and Map Analysis

1. Name five countries that have the highest rates of communicable diseases. (See Figure 1 to find the names of countries.)
2. Name five countries that have the highest rates of noncommunicable diseases.

Highest risk
Lowest risk
(Compiled by the authors using data from U.S. Geological Survey.)

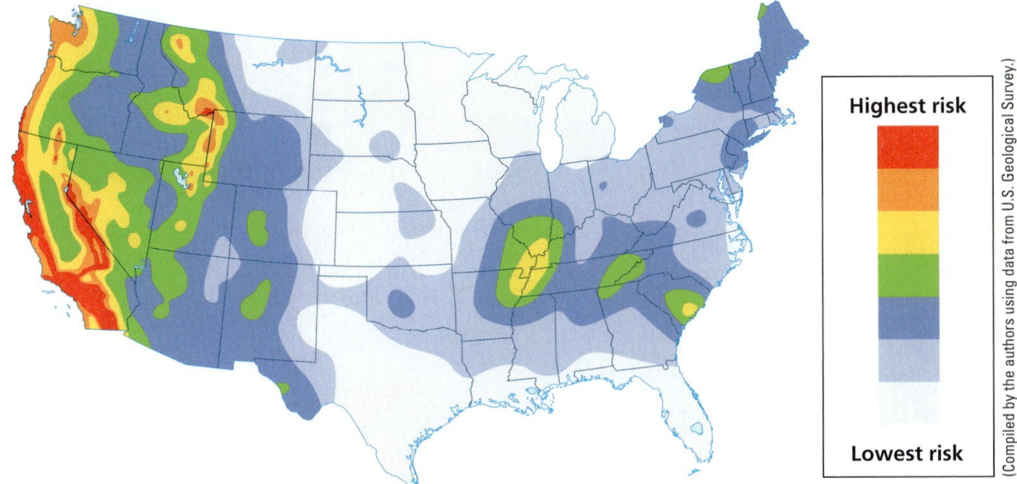

Figure 29 Earthquake (seismic) risk in various areas of the continental United States.

Data and Map Analysis

1. Speaking in general terms (northeast, southeast, central, west coast, etc.) which area has the highest earthquake risk and which area has the lowest risk?
2. For each of the categories of risk, list the number of states that fall into the category.

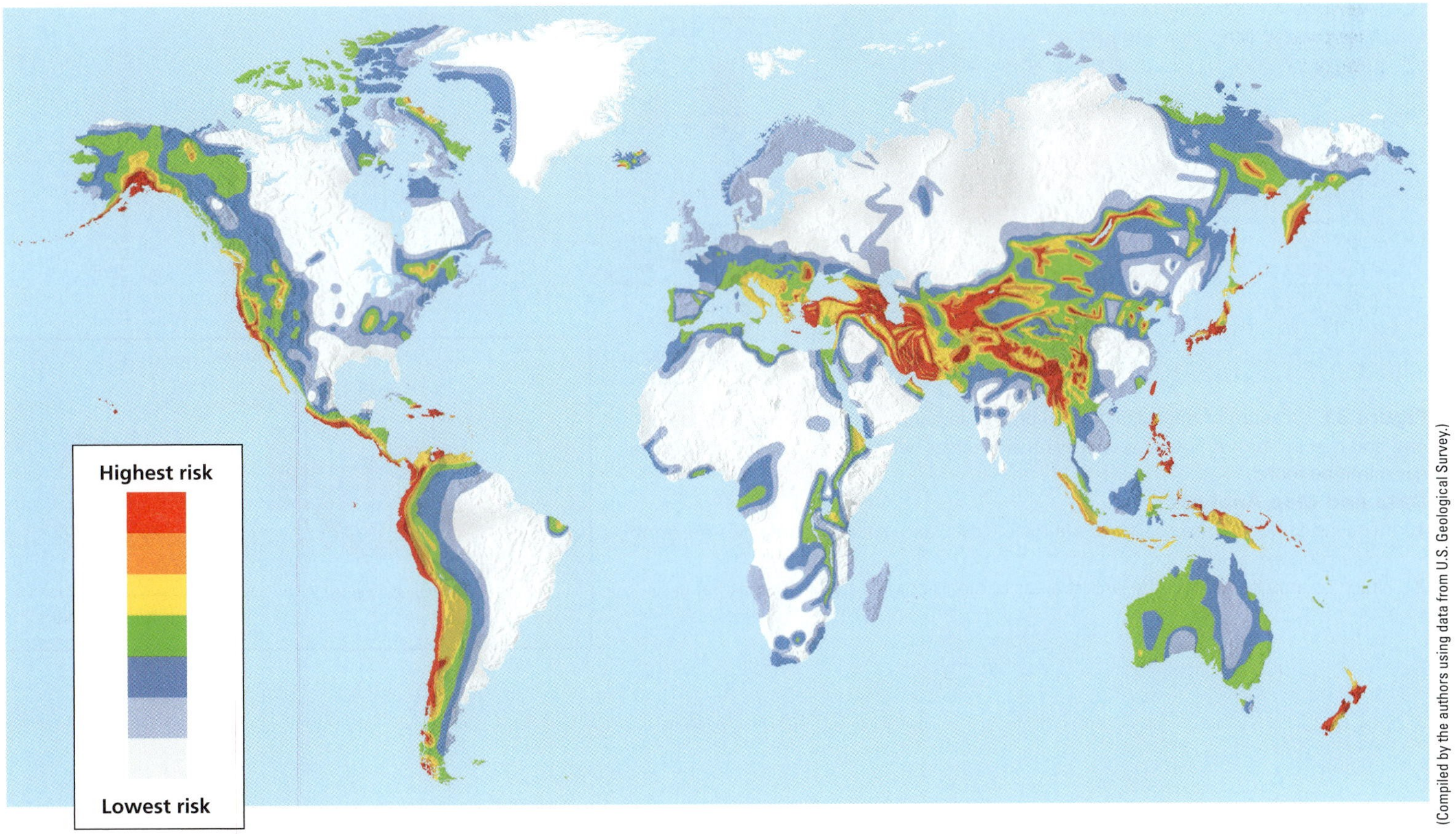

Highest risk
Lowest risk
(Compiled by the authors using data from U.S. Geological Survey.)

Figure 30 Earthquake (seismic) risk in the world.

Data and Map Analysis

1. How are these areas related to the boundaries of the earth's major tectonic plates as shown in Figure 14-18, p. 366?
2. Which continent has the longest coastal area subject to the highest possible risk? Which continent has the second longest such coastal area? (See Figure 2 of this supplement for continent names.)

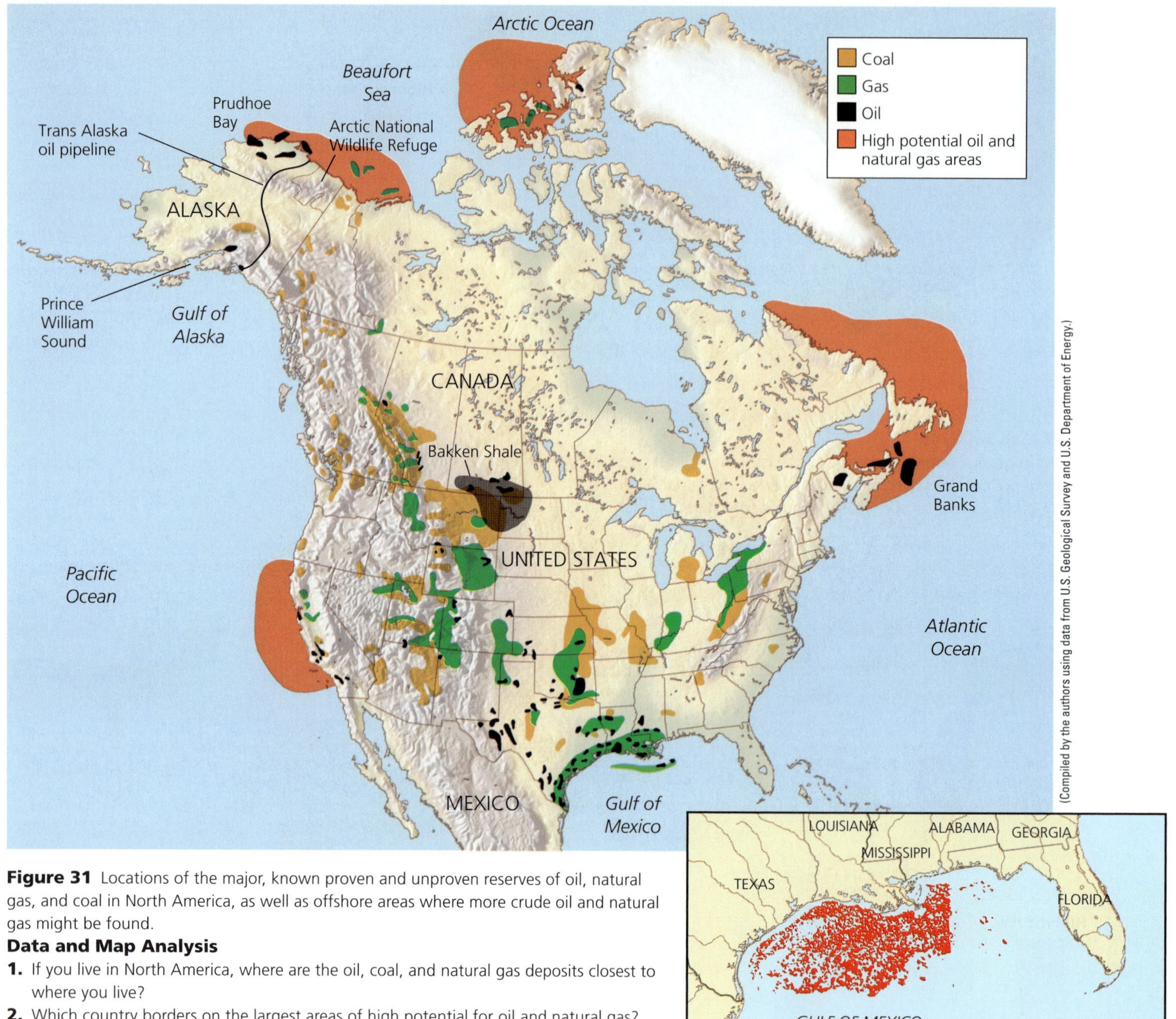

Figure 31 Locations of the major, known proven and unproven reserves of oil, natural gas, and coal in North America, as well as offshore areas where more crude oil and natural gas might be found.

Data and Map Analysis

1. If you live in North America, where are the oil, coal, and natural gas deposits closest to where you live?
2. Which country borders on the largest areas of high potential for oil and natural gas?

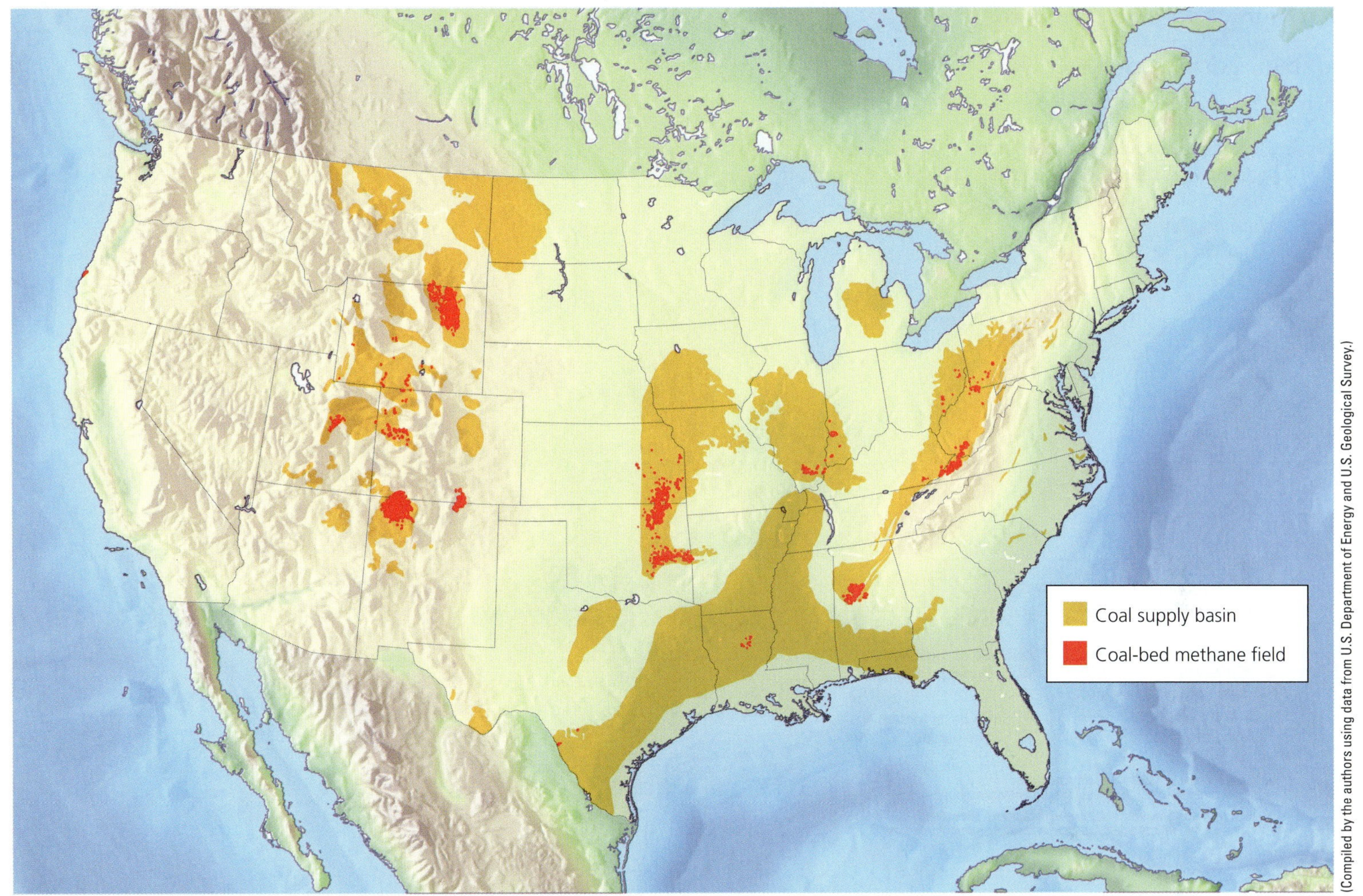

Figure 32 Major coal supply basins and coal-bed methane fields in the lower 48 states of the United States.

Data and Map Analysis

1. If you live in the United States, where are the coal-bed methane deposits closest to where you live?
2. Removing these deposits requires lots of water. Compare the locations of the major deposits of coal-bed methane with water-deficit areas shown in Figure 13-7, p. 322, and Figure 13-8, p. 323.

Figure 33 Major natural gas shale deposits in North America.

Data and Map Analysis

1. What state has the largest area of natural gas shale deposits?
2. Describe the locations of three areas where two states or two countries share a border over a natural gas shale deposit.

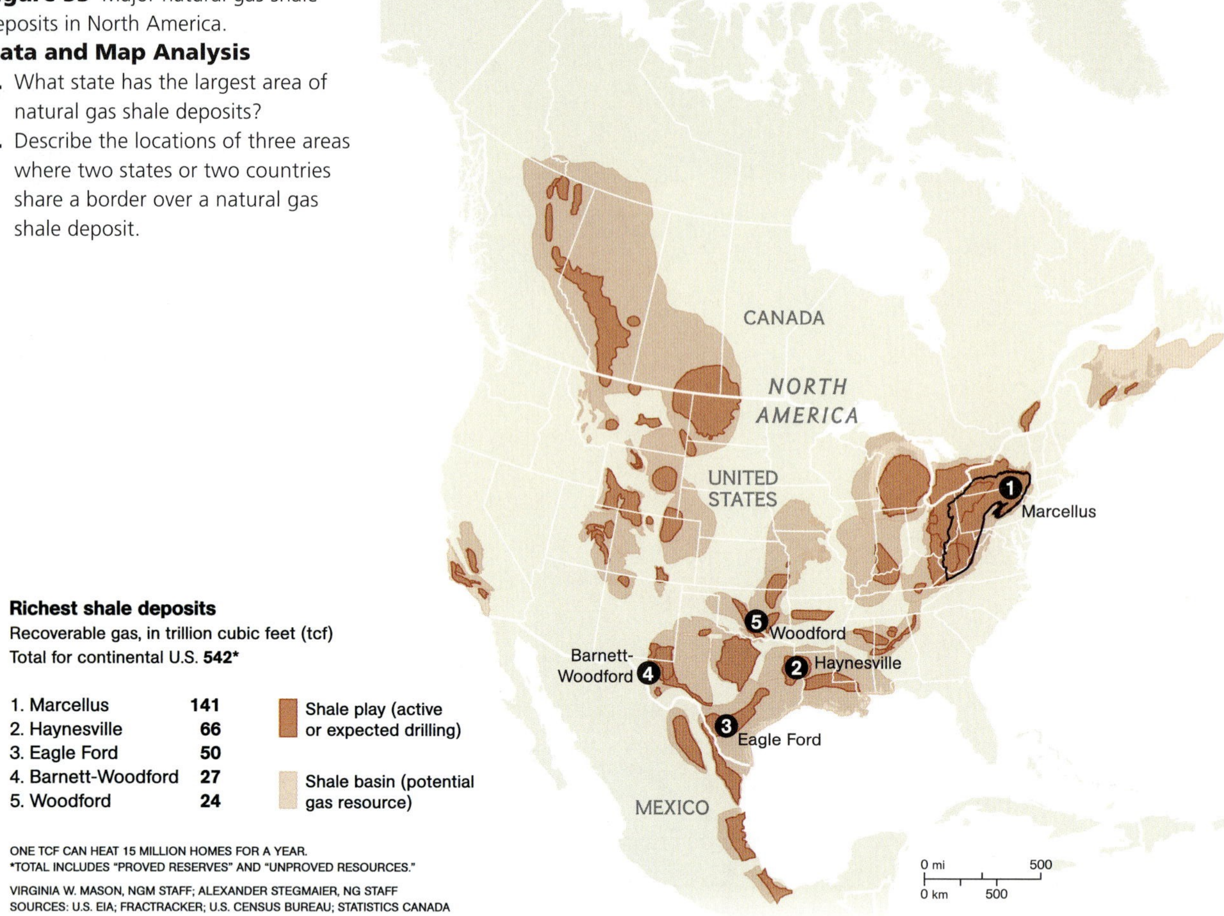

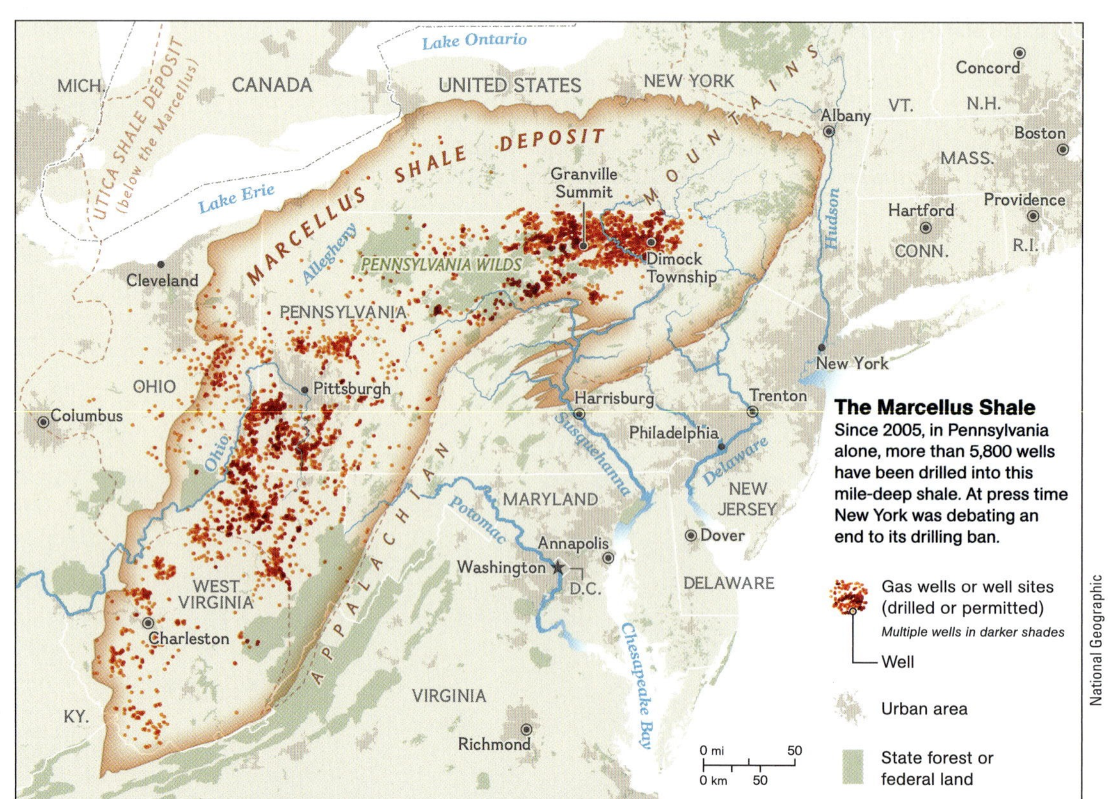

National Geographic

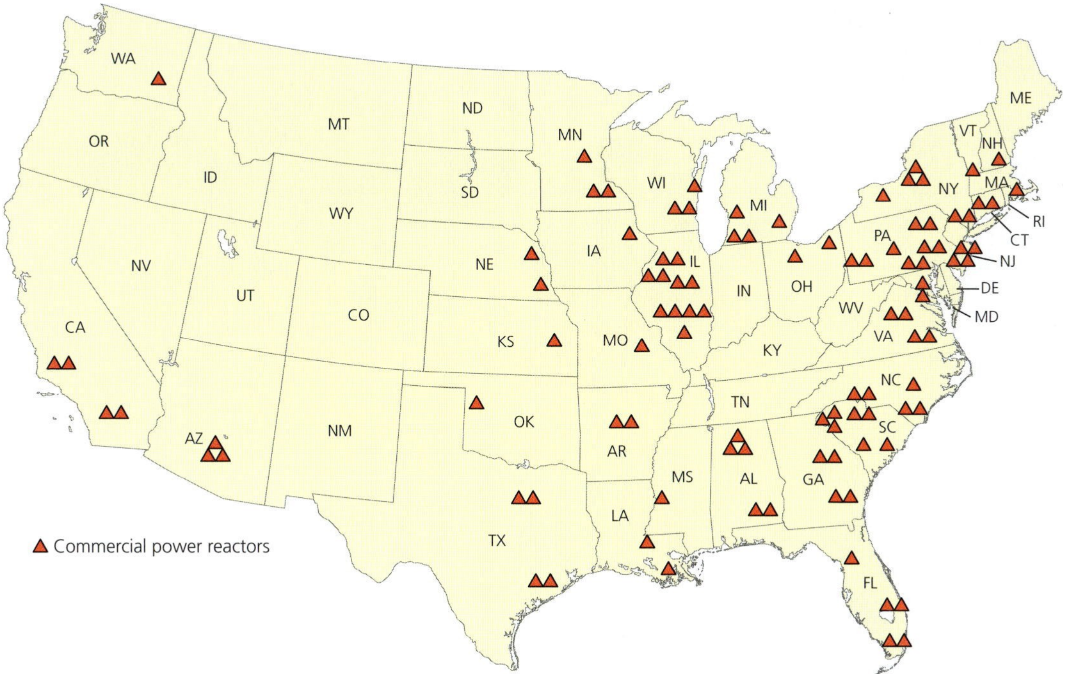

Figure 34 Locations of the 100 commercial nuclear power reactors in the United States.

(Compiled by the authors using data from U.S. Nuclear Regulatory Commission and U.S. Department of Energy.)

Data and Map Analysis

1. If you live in the United States, do you live or go to school within about 97 kilometers (60 miles) of a commercial nuclear power reactor?
2. Which state has the largest number of commercial nuclear power reactors?

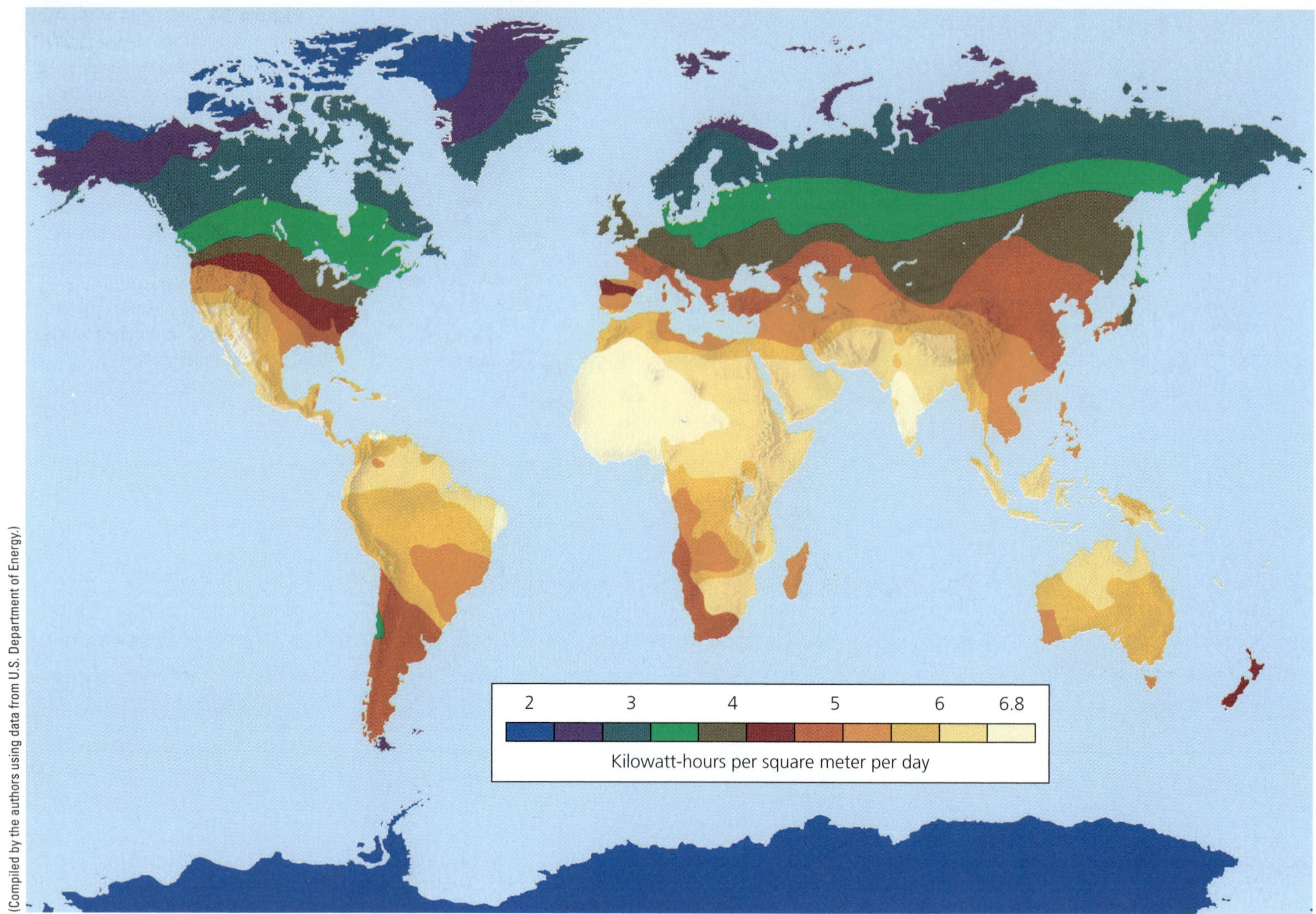

Figure 35 Global availability of direct solar energy. Areas with more than 3.5 kilowatt-hours per square meter per day (see scale) are good candidates for passive and active solar heating systems and use of solar cells to produce electricity.

Data and Map Analysis

1. What is the potential for making greater use of solar energy to provide heat and produce electricity (with solar cells) where you live or go to school?
2. List the continents in order of overall availability of direct solar energy, from those with the highest to those with the lowest. (See Figure 2 of this supplement for continent names.)

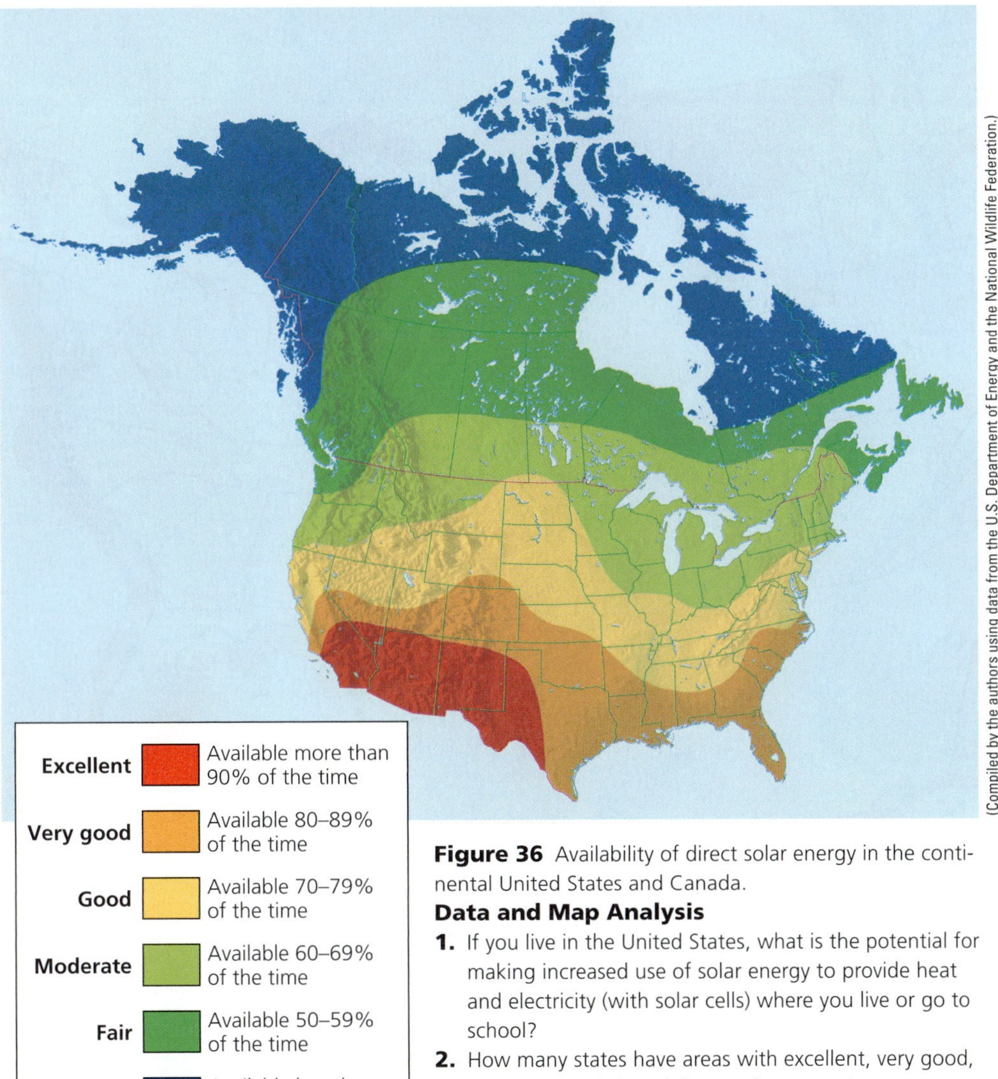

Figure 36 Availability of direct solar energy in the continental United States and Canada.

Data and Map Analysis

1. If you live in the United States, what is the potential for making increased use of solar energy to provide heat and electricity (with solar cells) where you live or go to school?
2. How many states have areas with excellent, very good, or good availability of direct solar energy?

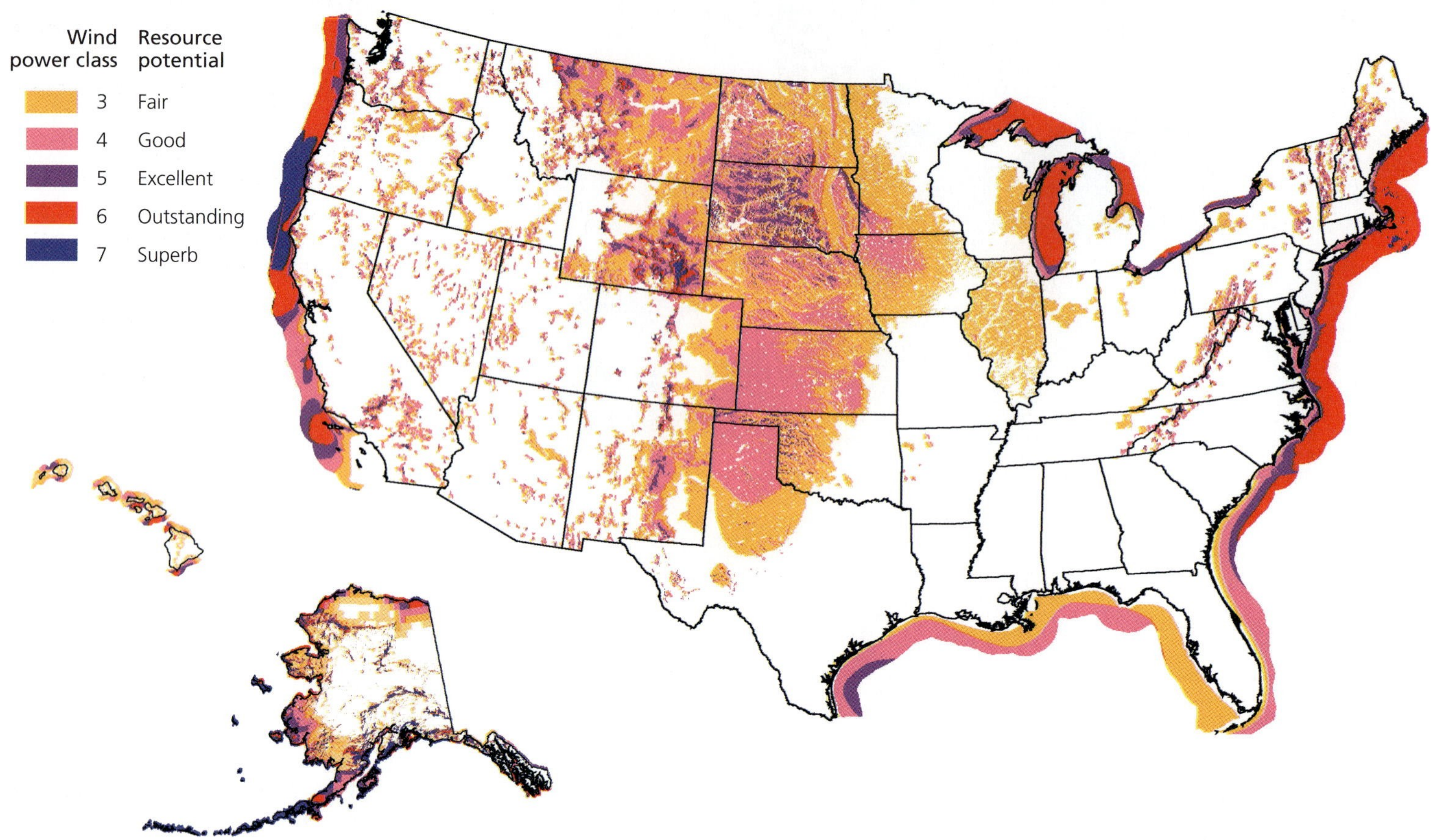

Figure 37 Potential supply of land- and ocean-based wind energy in the United States.

(Compiled by the authors using data from U.S. Geological Survey and U.S. Department of Energy.)

Data and Map Analysis

1. If you live in the United States, what is the general wind energy potential where you live or go to school?
2. How many states have areas with good or excellent potential for wind energy?

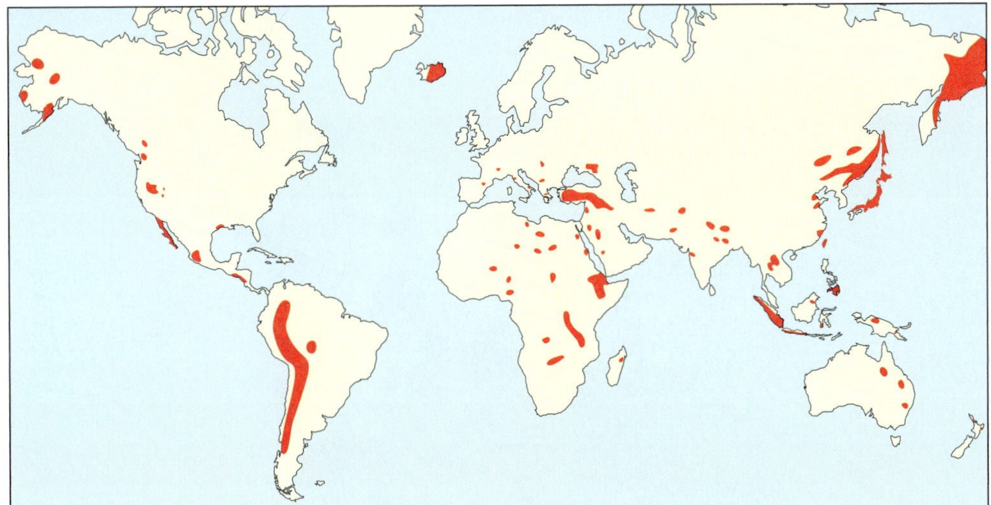

Figure 38 Known global reserves of moderate- to high-temperature geothermal energy.

(Compiled by the authors using data from Canadian Geothermal Resources Council, U.S. Geological Survey, and U.S. Department of Energy.)

Data and Map Analysis

1. Between North and South America, which continent appears to have the greater total potential for geothermal energy? (See Figure 2 of this supplement for continent names.)
2. Which country in Asia has the greatest known reserves of geothermal energy? (See Figure 1 of this supplement for country names.)

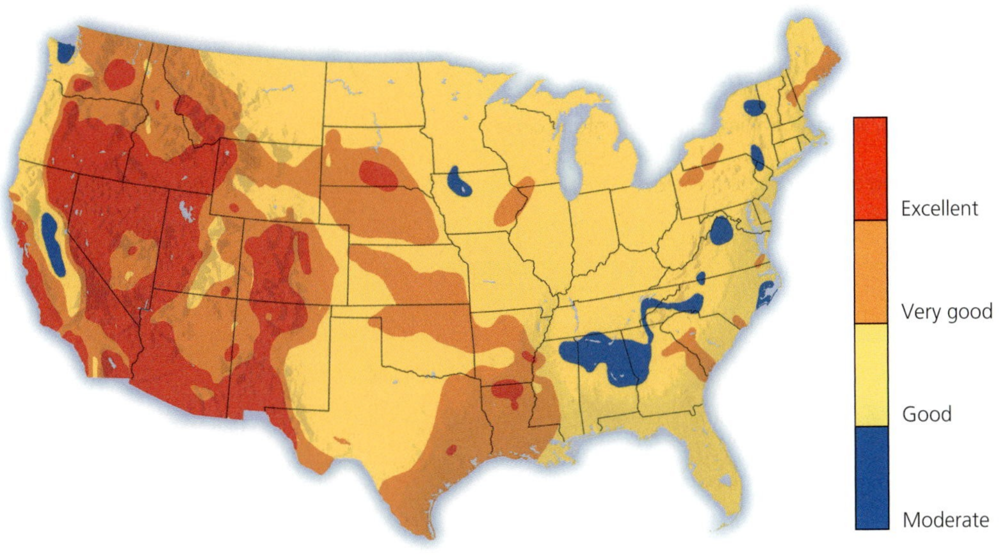

Figure 39 Potential geothermal energy resources in the continental United States.

(Compiled by the authors using data from U.S. Department of Energy and U.S. Geological Survey.)

Data and Map Analysis

1. If you live in the United States, what is the potential for using geothermal energy to provide heat or to produce electricity where you live or go to school?
2. How many states have areas with very good or excellent potential for using geothermal energy?

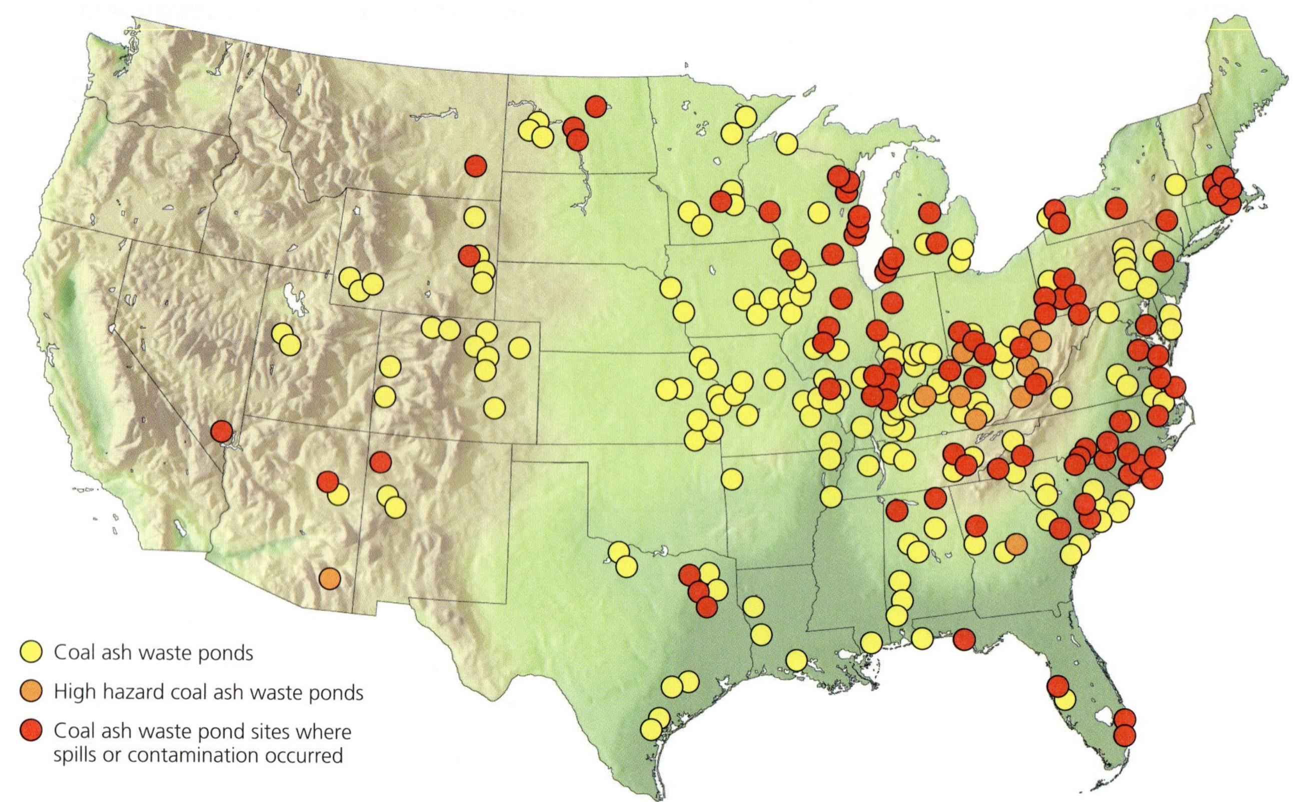

Figure 40 Coal ash waste pond sites in the United States. ***Question:*** Do you live or go to school near any of these ponds?

(Compiled by the authors using data from U.S. Environmental Protection Agency and the Sierra Club.)

Data and Map Analysis

1. In what part of the country (eastern third, central third, or western third) have most of the cases of contamination and spills occurred?
2. In what part of the country are most of the high-hazard coal ash ponds located?

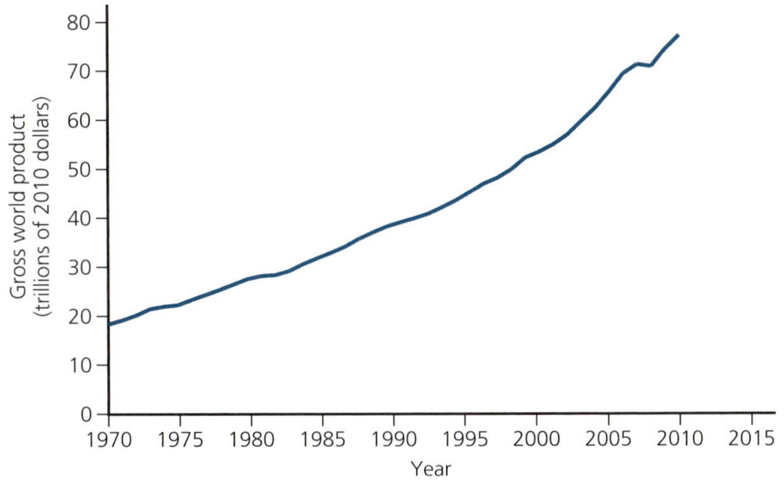

Figure 1 Gross world product (GWP), 1970–2011, in 2010 dollars.

(Compiled by the authors using data from the International Monetary Fund, World Bank, World Economic Fund, UN Population Division, and Earth Policy Institute, using purchasing power parity terms.)

Data and Graph Analysis

1. Roughly how many times bigger than the gross world product of 1985 was the gross world product of 2011?
2. If the current trend continues, about what do you think the gross world product will be in 2015, in trillions of dollars?

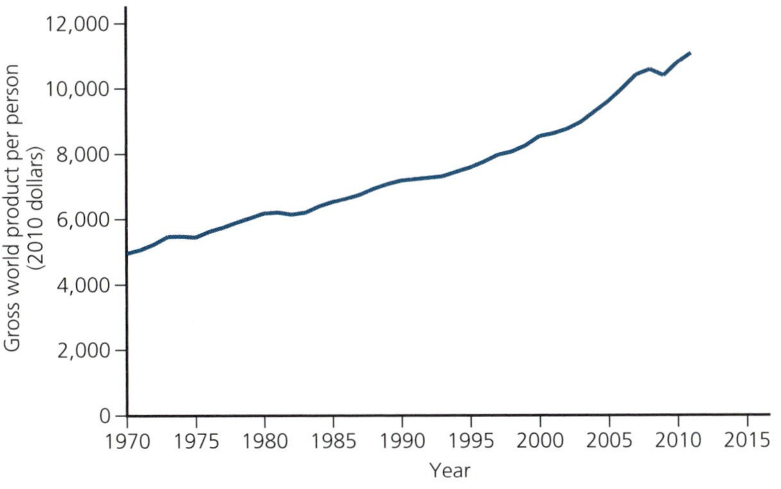

Figure 2 Gross world product (GWP) per person, 1970–2011, in 2010 dollars, using purchasing power parity terms.

(Compiled by the authors using data from the International Monetary Fund, World Bank, World Economic Fund, UN Population Division, and Earth Policy Institute.)

Data and Graph Analysis

1. Roughly how many years did it take for the GWP per person value in 1970 to double?
2. If the current trend continues, about what do you think the GWP per person will be in 2015?

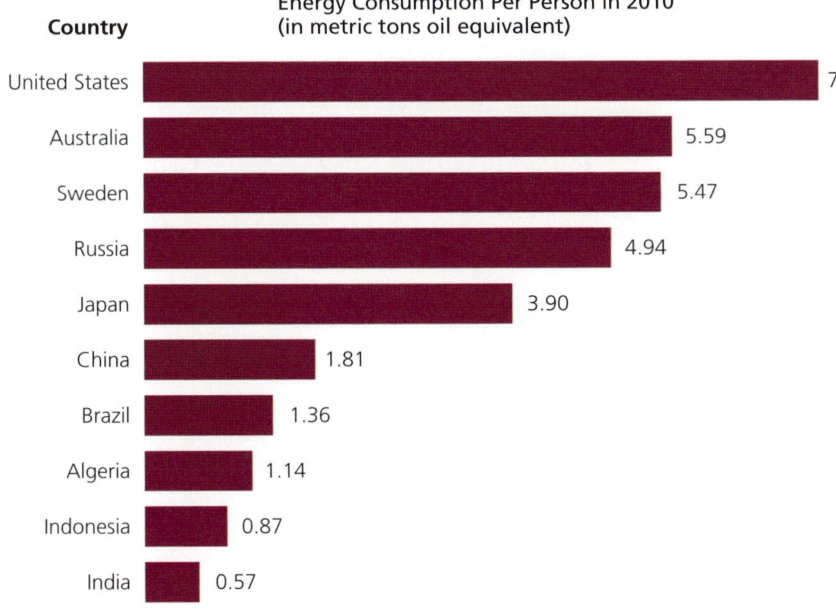

Figure 3 Energy consumption per person in 2010 for selected countries.

(Compiled by the authors using data from the World Bank.)

Data and Graph Analysis

1. On average, how many times more energy does an American consume per year than does a person in **(a)** China, **(b)** India, **(c)** Japan, and **(d)** Brazil?
2. On average, how many times more energy does a Chinese citizen consume per year than does a person in **(a)** India and **(b)** Algeria?

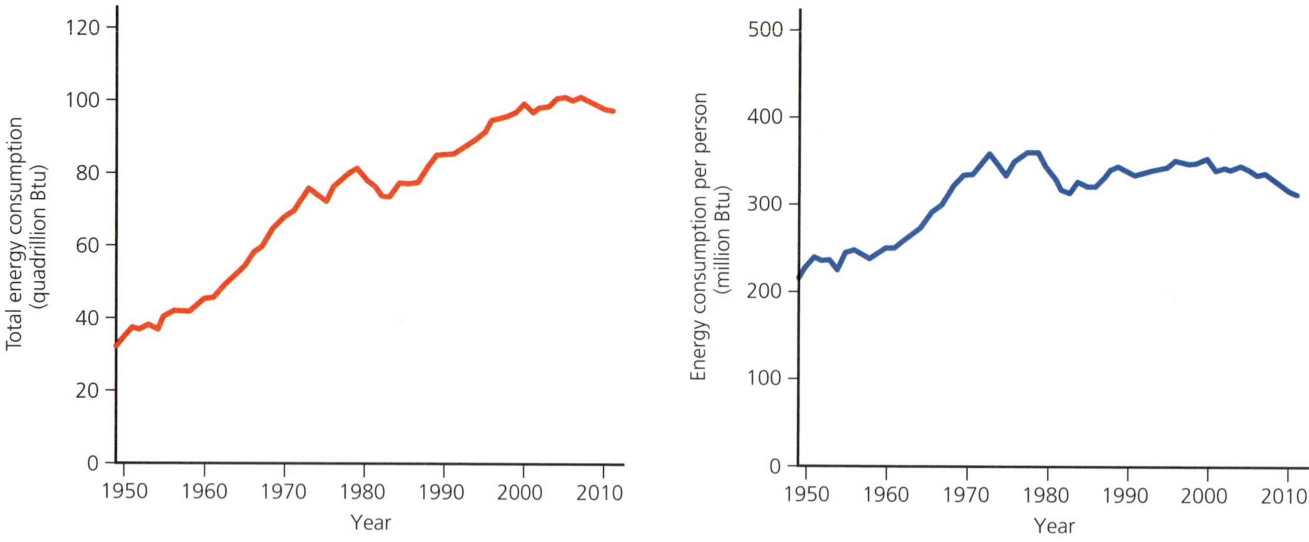

Figure 4 Total (left) and per capita (right) energy consumption in the United States, 1950–2011.

(Compiled by the authors using data from U.S. Energy Information Administration and the U.S. Census Bureau.)

Data and Graph Analysis

1. In what year or years did total U.S. energy consumption reach 80 quadrillion Btus?
2. In what year did energy consumption per person reach its highest level shown on this graph, and about what was that level of consumption?

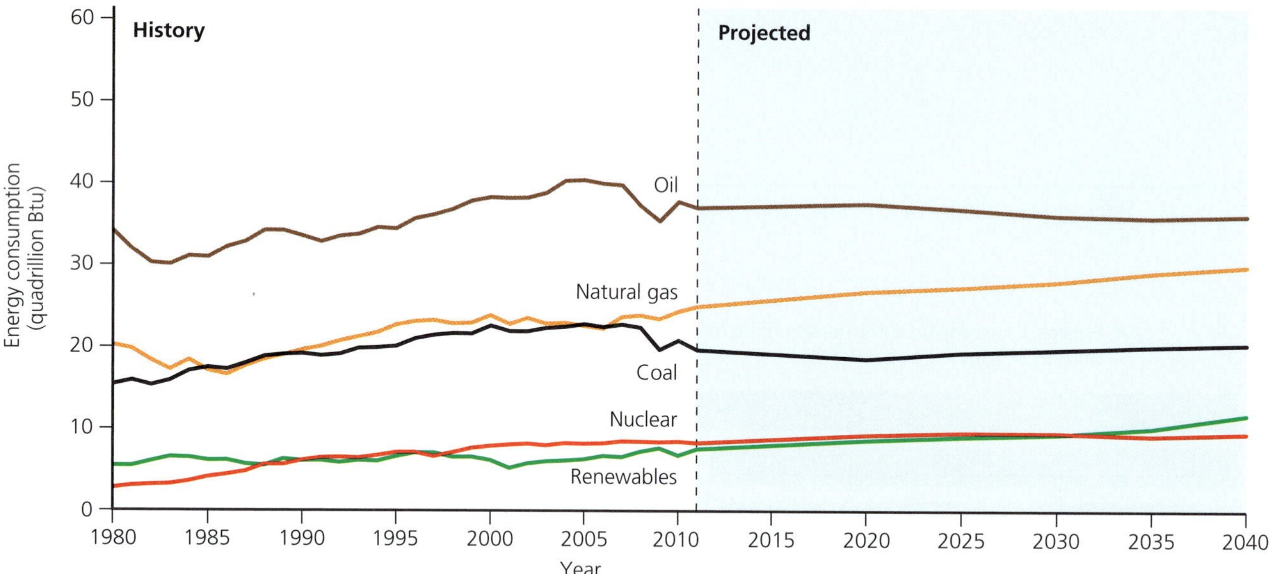

Figure 5 Energy consumption by fuel in the United States, 1980–2011, with projections to 2040.

(Compiled by the authors using data from U.S. Energy Information Administration Primary Energy Consumption Estimates by Source, 1949–2011, and Annual Energy Outlook 2013.)

Data and Graph Analysis

1. Usage of which energy source grew the most between 1990 and 2011? How much did it grow (in quadrillion Btus)?
2. For which two energy sources is usage expected to grow the fastest between 2011 and 2040?

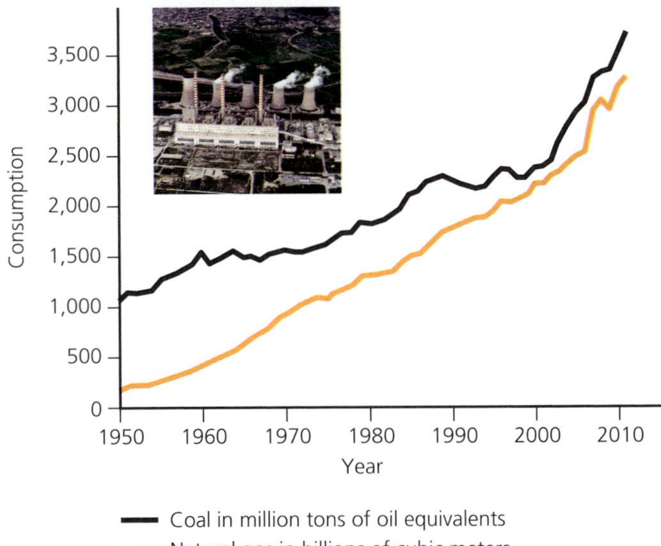

Figure 6 World coal and natural gas consumption for the period 1950–2011.

(Compiled by the authors using data from British Petroleum and International Energy Agency.)

Photo: airphoto.gr/Shutterstock.com

Data and Graph Analysis

1. Which energy source has grown more steadily—coal or natural gas? In what years has coal use grown most sharply?
2. In what year did coal use reach a level twice as high as it was in 1960?

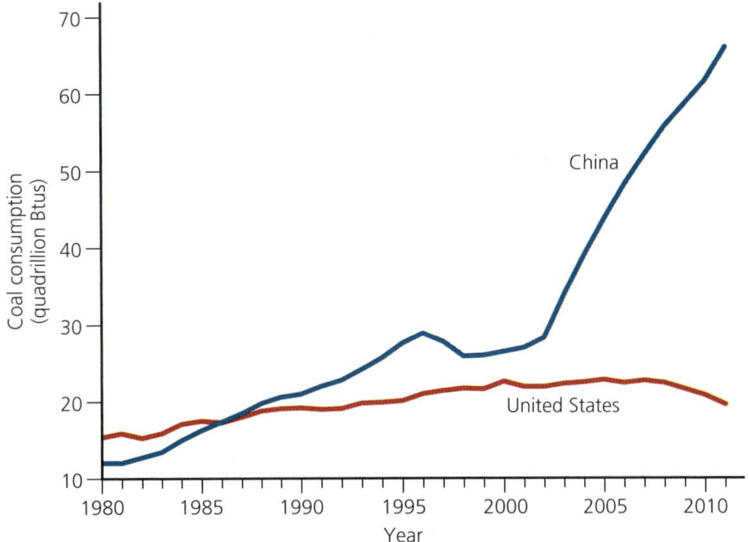

Figure 7 Coal consumption in China and the United States, 1980–2011.

(Compiled by the authors using data from Earth Policy Institute and British Petroleum.)

Data and Graph Analysis

1. By what percentage did coal consumption increase in China between 1980 and 2011?
2. By what percentage did coal consumption increase in the United States between 1980 and 2011?

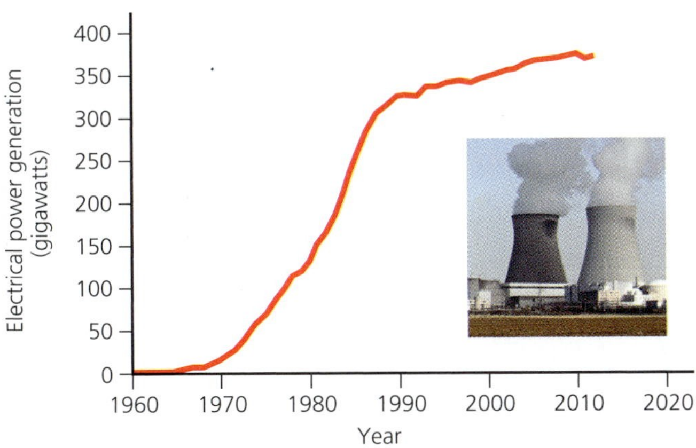

Figure 8 Global electrical generating capacity of nuclear power plants, 1960–2012.

(Compiled by the authors using data from International Energy Agency, Worldwatch Institute, and Earth Policy Institute.)

Photo: SpaceKris/Shutterstock.com

Data and Graph Analysis

1. After 1980, how long did it take to double that year's generating capacity?
2. Considering the decades of the 1970s, 1980s, 1990s, and 2000s, which decade saw the sharpest growth in generating capacity? During which decade did this growth level off?

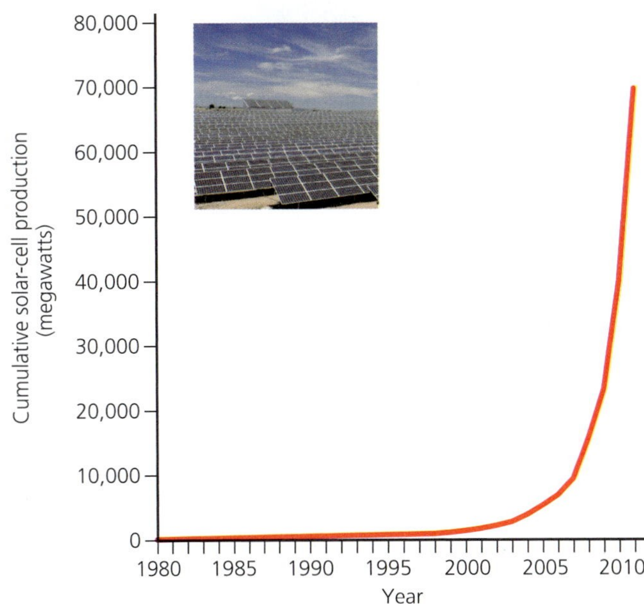

Figure 9 Global cumulative solar (photovoltaic) installations, 1980–2011.

(Compiled by the authors using data from U.S. Energy Information Administration, International Energy Agency, Worldwatch Institute, and Earth Policy Institute.)

Photo: pedrosala/Shutterstock.com

Data and Graph Analysis

1. About how many times more photovoltaic capacity was there in 2011, compared to the capacity in 2000?
2. How long did it take the world to go **(a)** from 0 to 5,000 megawatts in its cumulative installation of photovoltaic capacity and **(b)** from 5,000 to 40,000 megawatts?

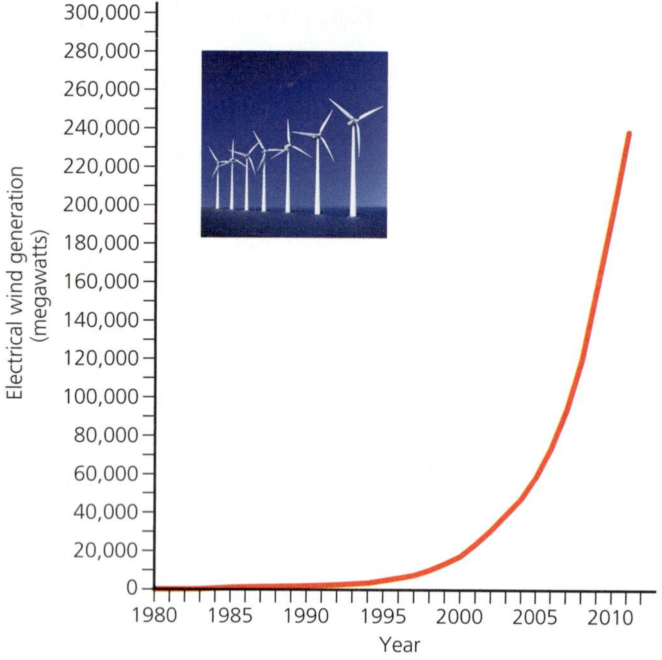

Figure 10 Global installed capacity for generation of electricity by wind energy, 1980–2011.

(Compiled by the authors using data from Global Wind Energy Council, European Wind Energy Association, American Wind Energy Association, Worldwatch Institute, World Wind Energy Association, and Earth Policy Institute.)

Photo: TebNad/Shutterstock.com

Data and Graph Analysis

1. How long did it take for the world to go from 0 to 100,000 megawatts of installed capacity for wind-generated electricity?
2. In 2011, the world's installed capacity for generating electricity by wind power was about how many times more than it was in 1995?

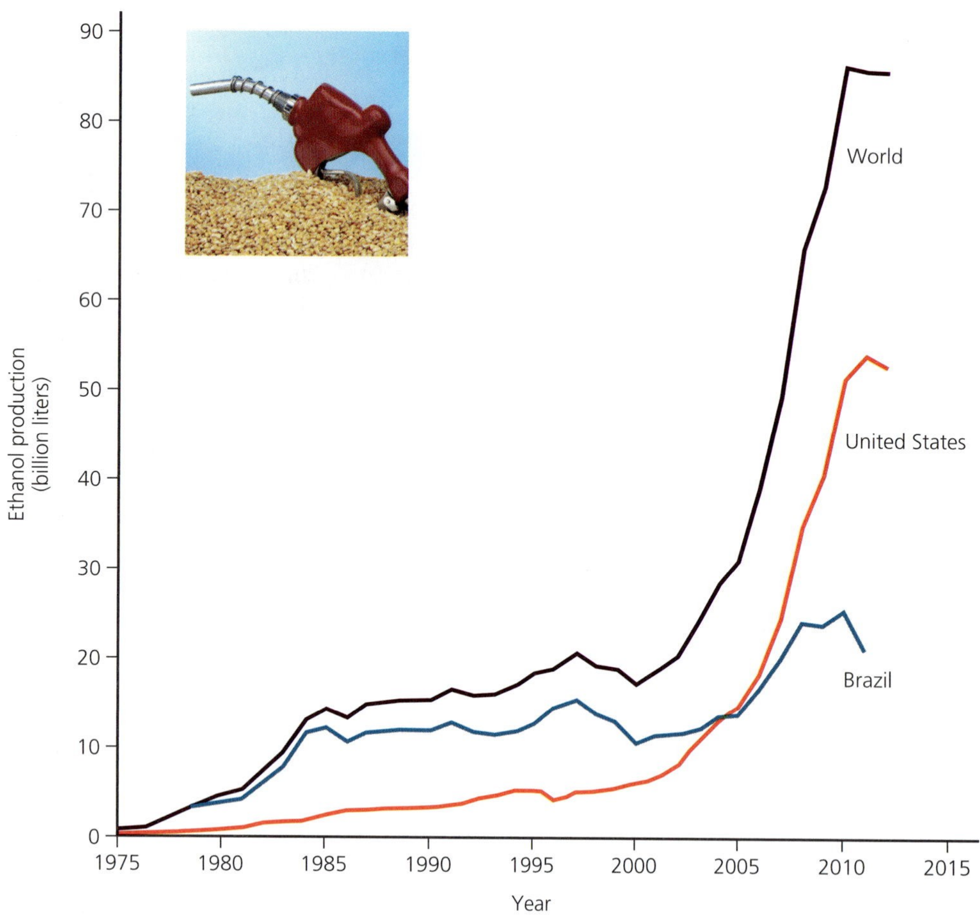

Figure 11 Production of ethanol motor fuel in the world, in Brazil, and in the United States, 1975–2012.

(Compiled by the authors using data from USDA, U.S. Energy Information Administration, Worldwatch Institute, and Earth Policy Institute.)

Photo: Jim Barber/Shutterstock.com

Data and Graph Analysis

1. By roughly what percentage did global ethanol production increase between 2000 and 2010?
2. In what years did ethanol production peak in **(a)** the United States, **(b)** Brazil, and **(c)** the world?

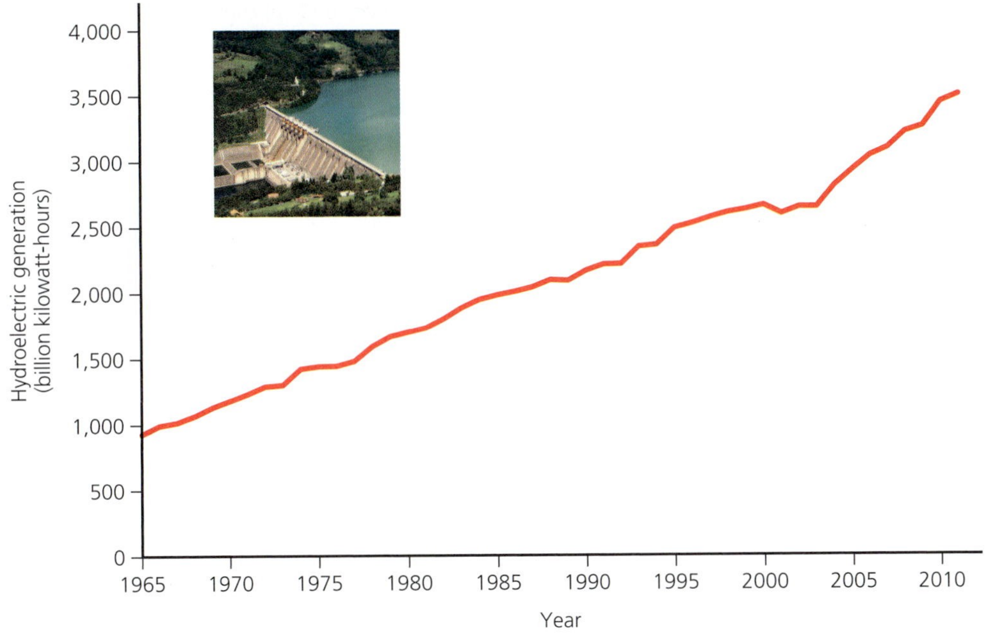

Figure 12 World hydroelectric generation, 1965–2011.

(Compiled by the authors using data from International Energy Agency, British Petroleum, Worldwatch Institute, and Earth Policy Institute.)

Photo: Zeljko Radokjo/Shutterstock.com

Data and Graph Analysis

1. After 1965, how many years did it take for the world to double its hydroelectric capacity?
2. Between what years was growth in hydroelectric capacity the sharpest?

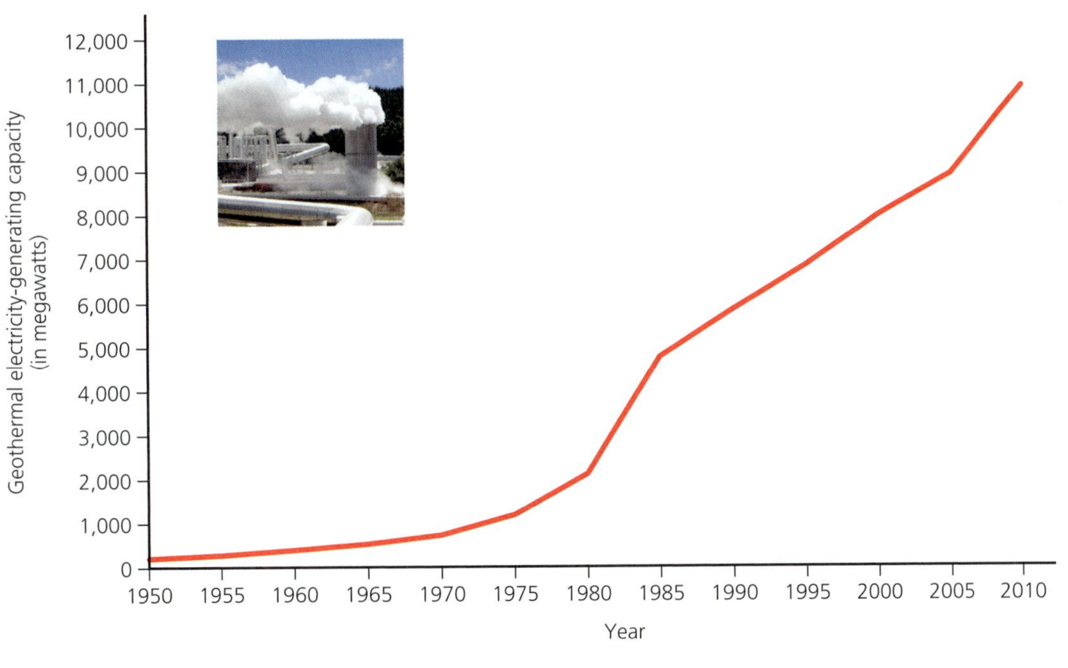

Figure 13 Global cumulative installed geothermal electricity–generating capacity, 1950–2010.

(Compiled by the authors using data from International Energy Agency, Worldwatch Institute, Earth Policy Institute, and Ruggero Bertani.)

Photo: N.Minton/Shutterstock.com

Data and Graph Analysis

1. About how many times more geothermal electricity–generating capacity was available in 2010 as there was in 1965?
2. Between what years was growth in this power source the sharpest?

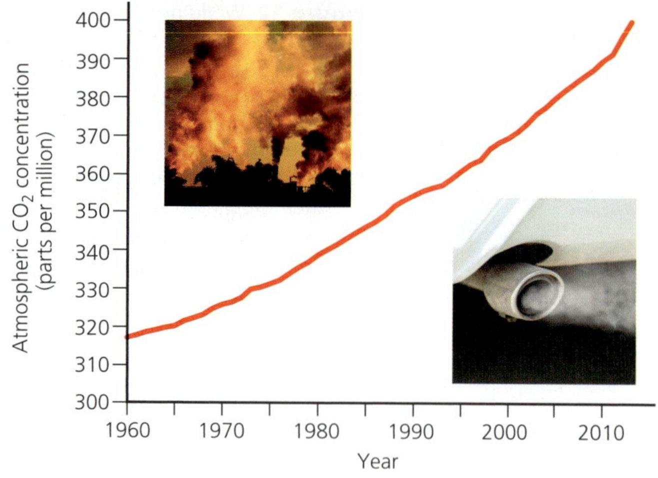

Figure 14 Atmospheric concentration of carbon dioxide (CO_2) measured at a major atmospheric research center in Mauna Loa, Hawaii, 1960–2013.

(Compiled by the authors using data from Scripps Institute of Oceanography, U.S. Energy Information Agency, and Earth Policy Institute.)

Top photo: Peter Weber/Shutterstock.com

Bottom photo: INSAGO/Shutterstock.com

Data and Graph Analysis

1. By how much did atmospheric CO_2 concentrations grow between 1960 and 2011 (in parts per million)?
2. Assuming that atmospheric CO_2 concentrations continue growing as is reflected on this graph, estimate the year in which such concentrations will reach 450 parts per million.

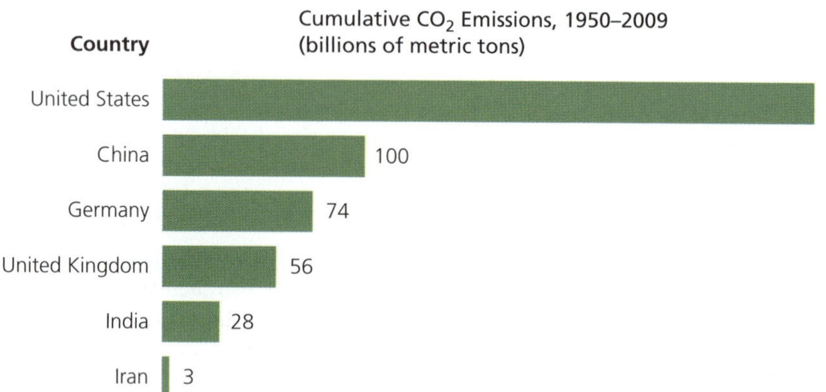

Figure 15 Cumulative carbon dioxide (CO_2) emissions from use of fossil fuels for selected countries, 1950–2009.

(Compiled by the authors using data from International Panel on Climate Change, World Resources Institute, Earth Policy Institute, and British Petroleum.)

Data and Graph Analysis

1. How many times higher are the cumulative CO_2 emissions of the United States than those of **(a)** China and **(b)** India?
2. How many times higher are the cumulative CO_2 emissions of China than those of **(a)** the United Kingdom and **(b)** India?

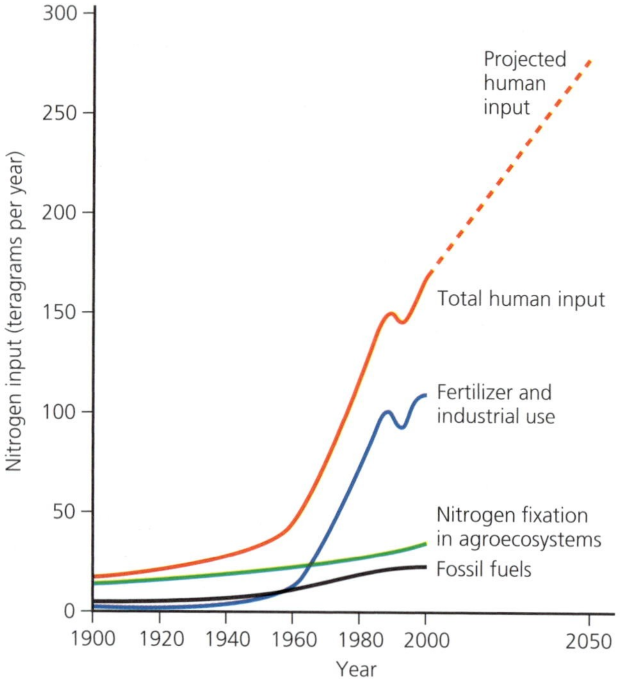

Figure 16 Global trends in the annual inputs of nitrogen into the environment from human activities, with projections to 2050.

(Compiled by the authors using data from Millennium Ecosystem Assessment and International Fertilizer Industry Association.)

Data and Graph Analysis

1. About how many times higher than the total human input of nitrogen to the environment in 1900 was the human input in 2000?
2. If nitrogen inputs continue as projected, about how many times higher will the total inputs be in 2050 than they were in 1980?

GLOSSARY

abiotic Nonliving. Compare *biotic.*

acid See *acid solution.*

acidity Chemical characteristic that helps determine how a substance dissolved in water (a solution) will interact with and affect its environment; based on the comparative amounts of hydrogen ions (H^+) and hydroxide ions (OH^-) contained in a particular volume of the solution. See *pH.*

acid solution Any water solution that has more hydrogen ions (H^+) than hydroxide ions (OH^-); any water solution with a pH less than 7. Compare *basic solution, neutral solution.*

adaptation Any genetically controlled structural, physiological, or behavioral characteristic that helps an organism survive and reproduce under a given set of environmental conditions. It usually results from a beneficial mutation. See *biological evolution, differential reproduction, mutation, natural selection.*

adaptive trait See *adaptation.*

aerobic respiration Complex process that occurs in the cells of most living organisms, in which nutrient organic molecules such as glucose ($C_6H_{12}O_6$) combine with oxygen (O_2) to produce carbon dioxide (CO_2), water (H_2O), and energy. Compare *photosynthesis.*

affluence wealth that results in high levels of consumption and unnecessary waste of resources, based mostly on the assumption that buying more and more material goods will bring fulfillment and happiness.

age structure Percentage of the population (or number of people of each sex) at each age level in a population.

air pollution One or more chemicals in high enough concentrations in the air to harm humans, other animals, vegetation, or materials. Excess heat is also considered a form of air pollution. Such chemicals or physical conditions are called air pollutants. See *primary pollutant, secondary pollutant.*

alien species See *nonnative species.*

altitude Height above sea level. Compare *latitude.*

anaerobic respiration Form of cellular respiration in which some decomposers get the energy they need through the breakdown of glucose (or other nutrients) in the absence of oxygen. Compare *aerobic respiration.*

annual plant Plant that grows, sets seed, and dies in one growing season. Compare *perennial.*

Anthropocene A new era in which humans have become major agents of change in the functioning of the earth's life-support systems as their ecological footprints have spread over the earth. See *ecological footprint.* Compare *Holopocene.*

anthropocentric Human-centered.

aquaculture Growing and harvesting of fish and shellfish for human use in freshwater ponds, irrigation ditches, and lakes, or in cages or fenced-in areas of coastal lagoons and estuaries or in the open ocean.

aquatic Pertaining to water. Compare *terrestrial.*

aquatic life zone Marine and freshwater portions of the biosphere. Examples include freshwater life zones (such as lakes and streams) and ocean or marine life zones (such as estuaries, coastlines, coral reefs, and the open ocean).

aquifer Porous, water-saturated layers of sand, gravel, or bedrock that can yield an economically significant amount of water.

arable land Land that can be cultivated to grow crops.

arid Dry. A desert or other area with an arid climate has little precipitation.

artificial selection Process by which humans select one or more desirable genetic traits in the population of a plant or animal species and then use *selective breeding* to produce populations containing many individuals with the desired traits. Compare *genetic engineering, natural selection.*

atmosphere Whole mass of air surrounding the earth. See *stratosphere, troposphere.* Compare *biosphere, geosphere, hydrosphere.*

atmospheric pressure Force or mass per unit area of air, caused by the bombardment of a surface by the molecules in air.

atom Minute unit made of subatomic particles that is the basic building block of all chemical elements and thus all matter; the smallest unit of an element that can exist and still have the unique characteristics of that element. Compare *ion, molecule.*

atomic number Number of protons in the nucleus of an atom. Compare *mass number.*

atomic theory Idea that all elements are made up of atoms; the most widely accepted scientific theory in chemistry.

autotroph See *producer.*

background extinction rate Normal extinction of various species as a result of changes in local environmental conditions. Compare *mass extinction.*

bacteria Prokaryotic, one-celled organisms. Some transmit diseases. Most act as decomposers and get the nutrients they need by breaking down complex organic compounds in the tissues of living or dead organisms into simpler inorganic nutrient compounds.

basic solution Water solution with more hydroxide ions (OH^-) than hydrogen ions (H^+); water solution with a pH greater than 7. Compare *acid solution, neutral solution.*

benthos Bottom-dwelling organisms. Compare *decomposer, nekton, plankton.*

bioaccumulation An increase in the concentration of a chemical in specific organs or tissues at a level higher than would normally be expected. Compare *biomagnification.*

biocentric Life-centered. Compare *anthropocentric.*

biodegradable Capable of being broken down by decomposers.

biodiversity Variety of different species (*species diversity*), genetic variability among individuals within each species (*genetic diversity*), variety of ecosystems (*ecological diversity*), and functions such as energy flow and matter cycling needed for the survival of species and biological communities (*functional diversity*).

biodiversity hot spot An area especially rich in plant species that are found nowhere else and are in great danger of extinction. Such areas suffer serious ecological disruption, mostly because of rapid human population growth and the resulting pressure on natural resources.

biogeochemical cycle Natural processes that recycle nutrients in various chemical forms from the nonliving environment to living organisms and then back to the nonliving environment. Examples include the carbon, oxygen, nitrogen, phosphorus, sulfur, and hydrologic cycles.

biological amplification See *biomagnification.*

biological community See *community.*

biological diversity See *biodiversity.*

biological evolution Change in the genetic makeup of a population of a species in successive generations. If continued long enough, it can lead to the formation of a new species. Note that populations, not individuals, evolve. See also *adaptation, differential reproduction, natural selection, theory of evolution.*

biological extinction Complete disappearance of a species from the earth. It happens when a species cannot adapt and successfully reproduce under new environmental conditions or when a species evolves into one or more new species. Compare *speciation.* See also *endangered species, mass extinction, threatened species.*

biomagnification Increase in concentration of DDT, PCBs, and other slowly degradable, fat-soluble chemicals in organisms at successively higher trophic levels of a food chain or web. Compare *bioaccumulation.*

biomass Organic matter produced by plants and other photosynthetic producers; total dry weight of all living organisms that can be supported at each trophic level in a food chain or web; dry weight of all organic matter in plants and animals in an ecosystem; plant materials and animal wastes used as fuel.

biome Terrestrial regions inhabited by certain types of life, especially vegetation. Examples include various types of deserts, grasslands, and forests.

biomimicry Process of observing certain changes in nature, studying how natural systems have responded to such changing conditions over many millions of years, and applying what is learned to dealing with some environmental challenge.

biosphere Zone of the earth where life is found. It consists of parts of the atmosphere (the troposphere), hydrosphere (mostly surface water and groundwater), and lithosphere (mostly soil and surface rocks and sediments on the bottoms of oceans and other bodies of water) where life is found. Compare *atmosphere, geosphere, hydrosphere.*

biotic Living organisms. Compare *abiotic.*

biotic pollution The effect of invasive species that can reduce or wipe out populations of many native species and trigger ecological disruptions.

biotic potential Maximum rate at which the population of a given species can increase when there are no limits on its rate of growth. See *environmental resistance.*

birth rate See *crude birth rate.*

broadleaf deciduous plants Plants such as oak and maple trees that survive drought and cold by shedding their leaves and becoming dormant. Compare *broadleaf evergreen plants, coniferous evergreen plants.*

broadleaf evergreen plants Plants that keep most of their broad leaves year-round. An example is the trees found in the canopies of tropical rain forests. Compare *broadleaf deciduous plants, coniferous evergreen plants.*

calorie Unit of energy; amount of energy needed to raise the temperature of 1 gram of water by 1 C° (unit on Celsius temperature scale). See also *kilocalorie.*

cancer Group of more than 120 different diseases, one for each type of cell in the human body. Each type of cancer produces a tumor in which cells multiply uncontrollably and invade surrounding tissue.

carbon cycle Cyclic movement of carbon in different chemical forms from the environment to organisms and then back to the environment.

carnivore Animal that feeds on other animals. Compare *herbivore, omnivore.*

carrying capacity Maximum population of a particular species that a given habitat can support over a given period. Compare *cultural carrying capacity.*

cell Smallest living unit of an organism. Each cell is encased in an outer membrane or wall and contains genetic material (DNA) and other parts to perform its life function. Organisms such as bacteria consist of only one cell, but most organisms contain many cells.

cell theory The idea that all living things are composed of cells; the most widely accepted scientific theory in biology.

chemical change Interaction between chemicals in which the chemical composition of the elements or compounds involved changes. Compare *nuclear change, physical change.*

chemical cycling The continual cycling of chemicals necessary for life through natural processes such as the water cycle and feeding interactions; processes that evolved due to the fact that the earth gets essentially no new inputs of these chemicals.

chemical formula Shorthand way to show the number of atoms (or ions) in the basic structural unit of a compound. Examples include H_2O, NaCl, and $C_6H_{12}O_6$.

chemical reaction See *chemical change.*

chemosynthesis Process in which certain organisms (mostly specialized bacteria) extract inorganic compounds from their environment and convert them into organic nutrient compounds without the presence of sunlight. Compare *photosynthesis.*

chromosome A grouping of genes and associated proteins in plant and animal cells that carry certain types of genetic information. See *genes.*

chronic malnutrition Faulty nutrition, caused by a diet that does not supply an individual with enough protein, essential fats, vitamins, minerals, and other nutrients needed for good health. Compare *overnutrition, chronic undernutrition.*

chronic undernutrition Condition suffered by people who cannot grow or buy enough food to meet their basic energy needs. Most chronically undernourished children live in developing countries and are likely to suffer from mental retardation and stunted growth and to die from infectious diseases. Compare *chronic malnutrition, overnutrition.*

clear-cutting Method of timber harvesting in which all trees in a forested area are removed in a single cutting. Compare *selective cutting, strip cutting.*

climate Physical properties of the troposphere of an area based on analysis of its weather records over a long period (at least 30 years). The two main factors determining an area's climate are its average *temperature,* with its seasonal variations, and the average amount and distribution of *precipitation.* Compare *weather.*

climax community See *mature community.*

coastal wetland Land along a coastline, extending inland from an estuary that is covered with saltwater all or part of the year. Examples include marshes, bays, lagoons, tidal flats, and mangrove swamps. Compare *inland wetland.*

coastal zone Warm, nutrient-rich, shallow part of the ocean that extends from the high-tide mark on land to the edge of a shelflike extension of continental land masses known as the continental shelf. Compare *open sea.*

coevolution Evolution in which two or more species interact and exert selective pressures on each other that can lead each species to undergo adaptations. See *evolution, natural selection.*

cold front Leading edge of an advancing mass of cold air. Compare *warm front.*

colony collapse disorder (CCD) Loss through death or disappearance of all or most of the European honey bees in a particular colony due to unknown causes; a phenomenon that has resulted in large losses of European honey bees in the United States and in parts of Europe.

commensalism An interaction between organisms of different species in which one type of organism benefits and the other type is neither helped nor harmed to any great degree. Compare *mutualism.*

commercial extinction Depletion of the population of a wild species used as a resource to a level at which it is no longer profitable to harvest the species.

commercial forest See *tree plantation.*

common-property resource Resource that is owned jointly by a large group of individuals. One example is the roughly one-third of the land in the United States that is owned jointly by all U.S. citizens and held and managed for them by the government. Another example is an area of land that belongs to a whole village and that can be used by anyone for grazing cows or sheep. Compare *open-access renewable resource.* See *tragedy of the commons.*

community Populations of all species living and interacting in an area at a particular time.

competition Two or more individual organisms of a single species (*intraspecific competition*) or two or more individuals of different species (*interspecific competition*) attempting to use the same scarce resources in the same ecosystem.

compound Combination of atoms, or oppositely charged ions, of two or more elements held together by attractive forces called chemical bonds. Examples are NaCl, CO_2, and $C_6H_{12}O_6$. Compare *element.*

concentration Amount of a chemical in a particular volume or weight of air, water, soil, or other medium.

coniferous evergreen plants Cone-bearing plants (such as spruces, pines, and firs) that keep some of their narrow, pointed leaves (needles) all year. Compare *broadleaf deciduous plants, broad-leaf evergreen plants.*

coniferous trees Cone-bearing trees, mostly evergreens, that have needle-shaped or scale-like leaves. They produce wood known commercially as softwood. Compare *deciduous plants.*

conservation Sensible and careful use of natural resources by humans. People with this view are called *conservationists.*

conservation biology Multidisciplinary science created to deal with the crisis of maintaining the genes, species, communities, and ecosystems that make up earth's biological diversity. Its goals are to investigate human impacts on biodiversity and to develop practical approaches to preserving biodiversity.

conservationist Person concerned with using natural areas and wildlife in ways that sustain them for current and future generations of humans and other forms of life.

consumer Organism that cannot synthesize the organic nutrients it needs and gets its organic nutrients by feeding on the tissues of producers or of other consumers; generally divided into *primary consumers* (herbivores), *secondary consumers* (carnivores), *tertiary* (*higher-level*) *consumers, omnivores,* and *detritivores* (decomposers and detritus feeders). In economics, one who uses economic goods. Compare *producer.*

controlled burning Deliberately set, carefully controlled surface fires that reduce flammable litter and decrease the chances of damaging *crown fires.* See *ground fire, surface fire.*

coral reef Formation produced by massive colonies containing billions of tiny coral animals, called polyps, that secrete a stony substance (calcium carbonate) around themselves for protection. When the corals die, their empty outer skeletons form layers and cause the reef to grow. Coral reefs are found in the coastal zones of warm tropical and subtropical oceans.

core Inner zone of the earth. It consists of a solid inner core and a liquid outer core. Compare *crust, mantle.*

corrective feedback loop See *negative feedback loop.*

crown fire Extremely hot forest fire that burns ground vegetation and treetops. Compare *controlled burning, ground fire, surface fire.*

crude birth rate Annual number of live births per 1,000 people in the population of a geographic area at the midpoint of a given year. Compare *crude death rate.*

crude death rate Annual number of deaths per 1,000 people in the population of a geographic area at the midpoint of a given year. Compare *crude birth rate.*

crust Solid outer zone of the earth. It consists of oceanic crust and continental crust. Compare *core, mantle.*

cultural carrying capacity The limit on population growth that would allow most people in an area or the world to live in reasonable comfort and freedom without impairing the ability of the planet to sustain future generations. Compare *carrying capacity.*

cultural eutrophication Overnourishment of aquatic ecosystems with plant nutrients (mostly nitrates and phosphates) because of human activities such as agriculture, urbanization, and discharges from industrial plants and sewage treatment plants. See *eutrophication.*

culture Whole of a society's knowledge, beliefs, technology, and practices.

dam A structure built across a river to control the river's flow or to create a reservoir. See *reservoir.*

data Factual information collected by scientists.

death rate See *crude death rate.*

deciduous plants Trees, such as oaks and maples, and other plants that survive during dry or cold seasons by shedding their leaves. Compare *coniferous trees, succulent plants.*

decomposer Organism that digests parts of dead organisms, and cast-off fragments and wastes of living organisms by breaking down the complex organic molecules in those materials into simpler inorganic compounds and then absorbing the soluble nutrients. Producers return most of these chemicals to the soil and water for reuse. Decomposers consist of various bacteria and fungi. Compare *consumer, detritivore, producer.*

deforestation Removal of trees from a forested area.

demographic transition Hypothesis that countries, as they become industrialized, have declines in death rates followed by declines in birth rates.

density Mass per unit volume.

desert Biome in which evaporation exceeds precipitation and the average amount of precipitation is less than 25 centimeters (10 inches) per year. Such areas have little vegetation or have widely spaced, mostly low vegetation. Compare *forest, grassland.*

detritivore Consumer organism that feeds on detritus, parts of dead organisms, and cast-off fragments and wastes of living organisms. Examples include earthworms, termites, and crabs. Compare *decomposer.*

detritus Parts of dead organisms and cast-off fragments and wastes of living organisms.

detritus feeder See *detritivore.*

developed country See *more-developed country.*

developing country See *less-developed country.*

dieback Sharp reduction in the population of a species when its numbers exceed the carrying capacity of its habitat. See *carrying capacity.*

differential reproduction Phenomenon in which individuals with adaptive genetic traits produce more living offspring than do individuals without such traits. See *natural selection.*

disturbance An event that disrupts an ecosystem or community. Examples of *natural disturbances* include fires, hurricanes, tornadoes, droughts, and floods. Examples of *human-caused disturbances* include deforestation, overgrazing, and plowing.

DNA (deoxyribonucleic acid) Large molecules in the cells of living organisms that carry genetic information.

domesticated species Wild species tamed or genetically altered by crossbreeding for use by humans for food (cattle, sheep, and food crops), as pets (dogs and cats), or for enjoyment (animals in zoos and plants in botanical gardens). Compare *wild species.*

doubling time Time it takes (usually in years) for the quantity of something growing exponentially to double. It can be calculated by dividing the annual percentage growth rate into 70.

drainage basin See *watershed.*

drought Condition in which an area does not get enough water because of lower-than-normal precipitation or higher-than-normal temperatures that increase evaporation.

earthquake Shaking of the ground resulting from the fracturing and displacement of subsurface rock, which produces a fault, or from subsequent movement along the fault.

ecological diversity The variety of forests, deserts, grasslands, oceans, streams, lakes, and other biological communities interacting with one another and with their nonliving environment. See *biodiversity.* Compare *functional diversity, genetic diversity, species diversity.*

ecological footprint Amount of biologically productive land and water needed to supply a population with the renewable resources it uses and to absorb or dispose of the wastes from such resource use. It is a measure of the average environmental impact of populations in different countries and areas. See *per capita ecological footprint.*

ecological niche Total way of life or role of a species in an ecosystem. It includes all physical, chemical, and biological conditions that a species needs to live and reproduce in an ecosystem.

ecological restoration Deliberate alteration of a degraded habitat or ecosystem to restore as much of its ecological structure and function as possible.

ecological succession Process in which communities of plant and animal species in a particular area are replaced over time by a series of different and often more complex communities. See *primary ecological succession, secondary ecological succession.*

ecological tipping point Point at which an environmental problem reaches a threshold level, which causes an often irreversible shift in the behavior of a natural system.

ecologist Biological scientist who studies relationships between living organisms and their environment.

ecology Biological science that studies the relationships between living organisms and their environment; study of the structure and functions of nature.

economic development Improvement of human living standards by economic growth. Compare *economic growth, environmentally sustainable economic development.*

economic growth Increase in the capacity to provide people with goods and services; an increase in gross domestic product (GDP). Compare *economic development, environmentally sustainable economic development.* See *gross domestic product.*

ecosphere See *biosphere.*

ecosystem One or more communities of different species interacting with one another and with the chemical and physical factors making up their nonliving environment.

ecosystem services Natural services or natural capital that support life on the earth and are essential to the quality of human life and the functioning of the world's economies. Examples are the chemical cycles, natural pest control, and natural purification of air and water. See *natural resources.*

electromagnetic radiation Forms of kinetic energy traveling as electromagnetic waves. Examples include radio waves, TV waves, microwaves, infrared radiation, visible light, ultraviolet radiation, X-rays, and gamma rays. Compare *ionizing radiation, nonionizing radiation.*

electron (e) Tiny particle moving around outside the nucleus of an atom. Each electron has one unit of negative charge and almost no mass. Compare *neutron, proton.*

element Chemical, such as hydrogen (H), iron (Fe), sodium (Na), carbon (C), nitrogen (N), or oxygen (O), whose distinctly different atoms serve as the basic building blocks of all matter. Two or more elements combine to form the compounds that make up most of the world's matter. Compare *compound.*

elevation Distance above sea level.

emigration movement of people out of a specific geographic area. Compare *immigration, migration.*

endangered species Wild species with so few individual survivors that the species could soon become extinct in all or most of its natural range. Compare *threatened species.*

endemic species Species that is found in only one area. Such species are especially vulnerable to extinction.

energy Capacity to do work by performing mechanical, physical, chemical, or electrical tasks or to cause a heat transfer between two objects at different temperatures.

energy conservation Reducing or eliminating the unnecessary waste of energy.

energy efficiency Percentage of the total energy input that does useful work and is not converted into low-quality, generally useless heat in an energy conversion system or process. See *energy quality, net energy.* Compare *material efficiency.*

energy productivity See *energy efficiency.*

energy quality Ability of a form of energy to do useful work. High-temperature heat and the chemical energy in fossil fuels and nuclear fuels are concentrated high-quality energy. Low-quality energy such as low-temperature heat is dispersed or diluted and cannot do much useful work. See *high-quality energy, low-quality energy.*

environment All external conditions, factors, matter, and energy, living and nonliving, that affect any living organism or other specified system.

environmental degradation Depletion or destruction of a potentially renewable resource such as soil, grassland, forest, or wildlife that is used faster than it is naturally replenished. If such use continues, the resource becomes nonrenewable (on a human time scale) or nonexistent (extinct). See also *sustainable yield.*

environmental ethics Human beliefs about what is right or wrong with how we treat the environment.

environmentalism Social movement dedicated to protecting the earth's life-support systems for us and other species.

environmentalist Person who is concerned about the impacts of human activities on the environment.

environmental justice Fair treatment and meaningful involvement of all people, regardless of race, color, sex, national origin, or income, with respect to the development, implementation, and enforcement of environmental laws, regulations, and policies.

environmentally sustainable economic development Development that meets the basic needs of the current generations of humans and other species without preventing future generations of humans and other species from meeting their basic needs. It is the economic component of an *environmentally sustainable society.* Compare *economic development, economic growth.*

environmentally sustainable society Society that meets the current and future needs of its people for basic resources in a just and equitable manner without compromising the ability of future generations of humans and other species from meeting their basic needs.

environmental resistance All of the limiting factors that act together to limit the growth of a population. See *biotic potential, limiting factor.*

environmental science Interdisciplinary study that uses information and ideas from the physical sciences (such as biology, chemistry, and geology) with those from the social sciences and humanities (such as economics, politics, and ethics) to learn how nature works, how we interact with the environment, and how we can to help deal with environmental problems.

environmental scientist Scientist who uses information from the physical sciences and social sciences to understand how the earth works, learn how humans interact with the earth, and develop solutions to environmental problems. See *environmental science.*

environmental wisdom worldview Worldview holding that humans are part of and totally dependent on nature and that nature exists for all species, not just for us. Our success depends on learning how the earth sustains itself and integrating such environmental wisdom into the ways we think and act. Compare *frontier worldview, planetary management worldview, stewardship worldview.*

environmental worldview Set of assumptions and beliefs about how people think the world works, what they think their role in the world should be, and what they believe is right and wrong environmental behavior (environmental ethics). See *environmental wisdom worldview, frontier worldview, planetary management worldview, stewardship worldview.*

EPA U.S. Environmental Protection Agency; responsible for managing federal efforts to control air and water pollution, radiation and pesticide hazards, environmental research, hazardous waste, and solid waste disposal.

epiphyte Plant that uses its roots to attach itself to branches high in trees, especially in tropical forests.

erosion Process or group of processes by which loose or consolidated earth materials, especially topsoil, are dissolved, loosened, or worn away and removed from one place and deposited in another.

estuary Partially enclosed coastal area at the mouth of a river where its freshwater, carrying fertile silt and runoff from the land, mixes with salty seawater.

eukaryotic cell Cell that is surrounded by a membrane and has a distinct nucleus. Compare *prokaryotic cell.*

euphotic zone Upper layer of a body of water through which sunlight can penetrate and support photosynthesis.

eutrophication Physical, chemical, and biological changes that take place after a lake, estuary, or slow-flowing stream receives inputs of plant nutrients—mostly nitrates and phosphates—from natural erosion and runoff from the surrounding land basin. See *cultural eutrophication.*

eutrophic lake Lake with a large or excessive supply of plant nutrients, mostly nitrates and phosphates. Compare *mesotrophic lake, oligotrophic lake.*

evaporation Conversion of a liquid into a gas.

evergreen plants Plants that keep some of their leaves or needles throughout the year. Examples include cone-bearing trees (conifers) such as firs, spruces, pines, redwoods, and sequoias. Compare *deciduous plants, succulent plants.*

evolution See *biological evolution.*

exhaustible resource See *nonrenewable resource.*

exotic species See *nonnative species.*

experiment Procedure a scientist uses to study some phenomenon under known conditions. Scientists conduct some experiments in the laboratory and others in nature. The resulting scientific data or facts must be verified or confirmed by repeated observations and measurements, ideally by several different investigators.

exponential growth Growth in which some quantity, such as population size or economic output, increases at a constant rate per unit of time. An example is the growth sequence 2, 4, 8, 16, 32, 64, and so on, which increases by 100% at each interval. When the increase in quantity over time is plotted, this type of growth yields a curve shaped like the letter J. Compare *linear growth.*

extinction See *biological extinction.*

extinction rate Percentage or number of species that go extinct within a certain period of time such as a year.

family planning Providing information, clinical services, and contraceptives to help people choose the number and spacing of children they want to have.

feedback Any process that increases (positive feedback) or decreases (negative feedback) a change to a system.

feedback loop Occurs when an output of matter, energy, or information is fed back into the system as an input and leads to changes in that system. See *positive feedback loop* and *negative feedback loop.*

fermentation See *anaerobic respiration.*

fertility rate Number of children born to an average woman in a population during her lifetime. Compare *replacement-level fertility.*

first law of thermodynamics In any physical or chemical change, no detectable amount of energy is created or destroyed, but energy can be changed from one form to another; you cannot get more energy out of something than you put in; in terms of energy quantity, you cannot get something for nothing. This law does not apply to nuclear changes, in which large amounts of energycan be produced from small amounts of matter. See *second law of thermodynamics.*

fishery Concentration of particular aquatic species suitable for commercial harvesting in a given ocean area or inland body of water.

fish farming See *aquaculture.*

fishprint Area of ocean needed to sustain the consumption of an average person, a nation, or the world. Compare *ecological footprint.*

floodplain Flat valley floor next to a stream channel. For legal purposes, the term often applies to any low area that has the potential for flooding, including certain coastal areas.

flows See *throughputs.*

food chain Series of organisms in which each eats or decomposes the preceding one. Compare *food web.*

food web Complex network of many interconnected food chains and feeding relationships. Compare *food chain.*

forest Biome with enough average annual precipitation to support the growth of tree species and smaller forms of vegetation. Compare *desert, grassland.*

fossil fuel Products of partial or complete decomposition of plants and animals; occurs as crude oil, coal, natural gas, or heavy oils as a result of exposure to heat and pressure in the earth's crust over millions of years. See *coal, crude oil, natural gas.*

fossils Skeletons, bones, shells, body parts, leaves, seeds, or impressions of such items that provide recognizable evidence of organisms that lived long ago.

foundation species Species that plays a major role in shaping a community by creating and enhancing a habitat that benefits other species. Compare *indicator species, keystone species, native species, nonnative species.*

free-access resource See *open-access renewable resource.*

freshwater Water that contains very low levels of dissolved salts.

freshwater life zones Aquatic systems where water with a dissolved salt concentration of less than 1% by volume accumulates on or flows through the surfaces of terrestrial biomes. Examples include *standing* (lentic) bodies of freshwater such as lakes, ponds, and inland wetlands and *flowing* (lotic) systems such as streams and rivers. Compare *biome.*

front The boundary between two air masses with different temperatures and densities. See *cold front, warm front.*

frontier science See *tentative science.*

full-cost pricing Finding ways to include the harmful environmental and health costs of producing and using goods in their market prices. See *external cost, internal cost.*

functional diversity Biological and chemical processes or functions such as energy flow and matter cycling needed for the survival of species and biological communities. See *bio-diversity, ecological diversity, genetic diversity, species diversity.*

game species Type of wild animal that people hunt or fish as a food source or for sport or recreation.

GDP See *gross domestic product.*

gene mutation See *mutation.*

gene pool Sum total of all genes found in the individuals of the population of a particular species.

generalist species Species with a broad ecological niche. They can live in many different places, eat a variety of foods, and tolerate a wide range of environmental conditions. Examples include flies, cockroaches, mice, rats, and humans. Compare *specialist species.*

genes Coded units of information about specific traits that are passed from parents to offspring during reproduction. They consist of segments of DNA molecules found in chromosomes.

gene splicing See *genetic engineering.*

genetic adaptation Changes in the genetic makeup of organisms of a species that allow the species to reproduce and gain a competitive advantage under changed environmental conditions. See *differential reproduction, evolution, mutation, natural selection.*

genetically modified organism (GMO) Organism whose genetic makeup has been altered by genetic engineering.

genetic diversity Variability in the genetic makeup among individuals within a single species. See *biodiversity.* Compare *ecological diversity, functional diversity, species diversity.*

genetic engineering Insertion of an alien gene into an organism to give it a beneficial genetic trait. Compare *artificial selection, natural selection.*

geographic isolation Separation of populations of a species into different areas for long periods of time.

geosphere Earth's intensely hot core, thick mantle composed mostly of rock, and thin outer crust that contains most of the earth's rock, soil, and sediment. Compare *atmosphere, biosphere, hydrosphere.*

global climate change Broad term referring to long-term changes in any aspects of the earth's climate, especially temperature and precipitation. Compare *weather.*

global warming Warming of the earth's lower atmosphere (troposphere) because of increases in the concentrations of one or more greenhouse gases. It can result in climate change that can last for decades to thousands of years. See *greenhouse effect,* *greenhouse gases, natural greenhouse effect.*

GMO See *genetically modified organism.*

GPP See *gross primary productivity.*

grassland Biome found in regions where there is enough annual average precipitation to support the growth of grass and small plants but not enough to support large stands of trees. Compare *desert, forest.*

greenhouse effect Natural effect that releases heat in the atmosphere (troposphere) near the earth's surface. Water vapor, carbon dioxide, ozone, and other gases in the lower atmosphere (troposphere) absorb some of the infrared radiation (heat) radiated by the earth's surface. Their molecules vibrate and transform the absorbed energy into longer-wavelength infrared radiation (heat) in the troposphere. If the atmospheric concentrations of these greenhouse gases increase and other natural processes do not remove them, the average temperature of the lower atmosphere will increase. Compare *global warming.* See also *natural greenhouse effect.*

greenhouse gases Gases in the earth's lower atmosphere (troposphere) that cause the greenhouse effect. Examples include carbon dioxide, chlorofluorocarbons, ozone, methane, water vapor, and nitrous oxide.

gross domestic product (GDP) Annual market value of all goods and services produced by all firms and organizations, foreign and domestic, operating within a country. See *per capita GDP.* Compare *genuine progress indicator (GPI).*

gross primary productivity (GPP) Rate at which an ecosystem's producers capture and store a given amount of chemical energy as biomass in a given length of time. Compare *net primary productivity.*

ground fire Fire that burns decayed leaves or peat deep below the ground's surface. Compare *crown fire, surface fire.*

groundwater Water that sinks into the soil and is stored in slowly flowing and slowly renewed underground reservoirs called *aquifers;* underground water in the zone of saturation, below the water table. Compare *runoff, surface water.*

habitat Place or type of place where an organism or population of organisms lives. Compare *ecological niche.*

habitat fragmentation Breakup of a habitat into smaller pieces, usually as a result of human activities.

hazard Something that can cause injury, disease, economic loss, or environmental damage. See also *risk.*

hazardous chemical Chemical that can cause harm because it is flammable or explosive, can irritate or damage the skin or lungs (such as strong acidic or alkaline substances), or can cause allergic reactions of the immune system (allergens). See also *toxic chemical.*

heat Total kinetic energy of all randomly moving atoms, ions, or molecules within a given substance, excluding the overall motion of the whole object. Heat always flows spontaneously from a warmer sample of matter to a colder sample of matter. This is one way to state the *second law of thermodynamics.* Compare *temperature.*

herbicide Chemical that kills a plant or inhibits its growth.

herbivore Plant-eating organism. Examples include deer, sheep, grasshoppers, and zooplankton. Compare *carnivore, omnivore.*

heterotroph See *consumer.*

high Air mass with a high pressure. Compare *low.*

high-quality energy Energy that is concentrated and has great ability to perform useful work. Examples include high-temperature heat and the energy in electricity, coal, oil, gasoline, sunlight, and nuclei of uranium-235. Compare *low-quality energy.*

high-quality matter Matter that is concentrated and contains a high concentration of a useful resource. Compare *low-quality matter.*

HIPPCO Acronym used by conservation biologists for the six most important secondary causes of premature extinction: **H**abitat destruction, degradation, and fragmentation; **I**nvasive (nonnative) species; **P**opulation growth (too many people consuming too many resources); **P**ollution; **C**limate change; and **O**verexploitation.

Holocene A geological period of relatively stable climate and other environmental conditions following the last glacial period. It began about 12,000 years ago and continues today. Compare *Anthropocene.*

host Plant or animal on which a parasite feeds.

hunger See chronic undernutrition.

hydrocarbon Organic compound made of hydrogen and carbon atoms. The simplest hydrocarbon is methane (CH_4), the major component of natural gas.

hydrologic cycle Biogeochemical cycle that collects, purifies, and distributes the earth's fixed supply of water from the environment to living organisms and then back to the environment.

hydrosphere Earth's *liquid water* (oceans, lakes, other bodies of surface water, and underground water), *frozen water* (polar ice caps, floating ice caps, and ice in soil, known as permafrost), and *water vapor* in the atmosphere. See also *hydrologic cycle.* Compare *atmosphere, biosphere, geosphere.*

immature community Community at an early stage of ecological succession. It usually has a low number of species and ecological niches and cannot capture and use energy and cycle critical nutrients as efficiently as more complex, mature communities. Compare *mature community.*

immigrant species See *nonnative species.*

immigration Migration of people into a country or area to take up permanent residence. Compare *emigration.*

indicator species Species whose decline serves as early warnings that a community or ecosystem is being degraded. Compare *foundation species, keystone species, native species, nonnative species.*

inertia See *persistence.*

inexhaustible resource See *perpetual resource.* Compare *nonrenewable resource, renewable resource.*

infant mortality rate Number of babies out of every 1,000 born each year who die before their first birthday.

inland wetland Land away from the coast, such as a swamp, marsh, or bog, that is covered all or part of the time with freshwater. Compare *coastal wetland.*

inorganic compounds All compounds not classified as organic compounds. See *organic compounds.*

input Matter, energy, or information entering a system. Compare *output, throughput.*

input pollution control See *pollution prevention.*

insecticide Chemical that kills insects.

interspecific competition Attempts by members of two or more species to use the same limited resources in an ecosystem. See *competition, intraspecific competition.*

intertidal zone The area of shoreline between low and high tides.

intraspecific competition Attempts by two or more organisms of a single species to use the same limited resources in an ecosystem. See *competition, interspecific competition.*

intrinsic rate of increase Rate at which a population could grow if it had unlimited resources. Compare *environmental resistance.*

invasive species See *nonnative species.*

invertebrates Animals that have no backbones. Compare *vertebrates.*

ion Atom or group of atoms with one or more positive (+) or negative (−) electrical charges. Examples are Na^+ and Cl^-. Compare *atom, molecule.*

isotopes Two or more forms of a chemical element that have the same number of protons but different mass numbers because they have different numbers of neutrons in their nuclei.

J-shaped curve Curve with a shape similar to that of the letter J; can represent prolonged exponential growth. See *exponential growth.*

junk science See *unreliable science.*

keystone species Species that play roles affecting many other organisms in an ecosystem. Compare *foundation species, indicator species, native species, nonnative species.*

kilocalorie (kcal) Unit of energy equal to 1,000 calories. See *calorie.*

kinetic energy Energy that matter has because of its mass and speed, or velocity. Compare *potential energy.*

lake Large natural body of standing freshwater formed when water from precipitation, land runoff, or groundwater flow fills a depression in the earth created by glaciation, earth movement, volcanic activity, or a giant meteorite. See *eutrophic lake, mesotrophic lake, oligotrophic lake.*

land degradation Decrease in the ability of land to support crops, livestock, or wild species in the future as a result of natural or human-induced processes.

latitude Distance from the equator. Compare *altitude.*

law of conservation of energy See *first law of thermodynamics.*

law of conservation of matter In any physical or chemical change, matter is neither created nor destroyed but merely changed from one form to another; in physical and chemical changes, existing atoms are rearranged into different spatial patterns (physical changes) or different combinations (chemical changes).

law of nature See *scientific law.*

law of tolerance Existence, abundance, and distribution of a species in an ecosystem are determined by whether the levels of one or more physical or chemical factors fall within the range tolerated by the species. See *threshold effect.*

LDC See *less-developed country.* Compare *more-developed country.*

less-developed country Country that has low to moderate industrialization and low to moderate per capita GDP. Most are located in Africa, Asia, and Latin America. Compare *more-developed country.*

life expectancy Average number of years a newborn infant can be expected to live.

limiting factor Single factor that limits the growth, abundance, or distribution of the population of a species in an ecosystem. See *limiting factor principle.*

limiting factor principle Too much or too little of any abiotic factor can limit or prevent growth of a population of a species in an ecosystem, even if all other factors are at or near the optimal range of tolerance for the species.

linear growth Growth in which a quantity increases by some fixed amount during each unit of time. An example is growth that increases by 2 units in the sequence 2, 4, 6, 8, 10, and so on. Compare *exponential growth.*

lithosphere Outer shell of the earth, composed of the crust and the rigid, outermost part of the mantle outside the asthenosphere; material found in the earth's plates. See *crust, geosphere, mantle.*

logistic growth Pattern in which exponential population growth occurs when the population is small, and population growth decreases steadily with time as the population approaches the carrying capacity. See *S-shaped curve.*

low Air mass with a low pressure. Compare *high.*

low-quality energy Energy that is dispersed and has little ability to do useful work. An example is low-temperature heat. Compare *high-quality energy.*

low-quality matter Matter that is dilute or dispersed or contains a low concentration of a useful resource. Compare *high-quality matter.*

malnutrition See *chronic malnutrition.*

mangrove swamps Swamps found on the coastlines in warm tropical climates. They are dominated by mangrove trees, any of about 55 species of trees and shrubs that can live partly submerged in the salty environment of coastal swamps.

mantle Zone of the earth's interior between its core and its crust. Compare *core, crust.* See *geosphere, lithosphere.*

marine life zone See *saltwater life zone.*

mass Amount of material in an object.

mass extinction Catastrophic, widespread, often global event in which major groups of species are wiped out over a short time compared with normal (background) extinctions. Compare *background extinction.*

mass number Sum of the number of neutrons (n) and the number of protons (p) in the nucleus of an atom. It gives the approximate mass of that atom. Compare *atomic number.*

matter Anything that has mass (the amount of material in an object) and takes up space. On the earth, where gravity is present, we weigh an object to determine its mass.

matter quality Measure of how useful a matter resource is, based on its availability and concentration. See *high-quality matter, low-quality matter.*

mature community Fairly stable, self-sustaining community in an advanced stage of ecological succession; usually has a diverse array of species and ecological niches; captures and uses energy and cycles critical chemicals more efficiently than simpler, immature communities. Compare *immature community.*

maximum sustainable yield See *sustainable yield.*

MDC See *more developed country.*

mesotrophic lake Lake with a moderate supply of plant nutrients. Compare *eutrophic lake, oligotrophic lake.*

metabolism Ability of a living cell or organism to capture and transform matter and energy from its environment to supply its needs for survival, growth, and reproduction.

microorganisms Organisms such as bacteria that are so small that it takes a microscope to see them.

migration Movement of people into and out of specific geographic areas. Compare *emigration and immigration.*

mineral resource Concentration of naturally occurring solid, liquid, or gaseous material in or on the earth's crust in a form and amount such that extracting and converting it into useful materials or items is currently or potentially profitable. Mineral resources are classified as *metallic* (such as iron and tin ores) or *nonmetallic* (such as fossil fuels, sand, and salt).

model Approximate representation or simulation of a system being studied.

molecule Combination of two or more atoms of the same chemical element (such as O_2) or different chemical elements (such as H_2O) held together by chemical bonds. Compare *atom, ion.*

more-developed country Country that is highly industrialized and has a high per capita GDP. Compare *less-developed country.*

mutation Random change in DNA molecules making up genes that can alter anatomy, physiology, or behavior in offspring.

mutualism Type of species interaction in which both participating species generally benefit. Compare *commensalism.*

native species Species that normally live and thrive in a particular ecosystem. Compare *foundation species, indicator species, keystone species, nonnative species.*

natural capital Natural resources and natural services that keep us and other species alive and support our economies. See *natural resources, natural services.*

natural capital degradation See *environmental degradation.*

natural greenhouse effect Heat buildup in the troposphere caused by the presence of certain gases, called greenhouse gases. Without this effect, the earth would be nearly as cold as Mars, and life as we know it could not exist. See *global warming.*

natural income Renewable resources such as plants, animals, and soil provided by natural capital.

natural law See *scientific law.*

natural rate of extinction See *background extinction.*

natural resources Materials such as air, water, and soil and energy in nature that are essential or useful to humans. See *natural capital.*

natural selection Process by which a particular beneficial gene (or set of genes) is reproduced in succeeding generations more than other genes. The result of natural selection is a population that contains a greater proportion of organisms better adapted to certain environmental conditions. See *adaptation, biological evolution, differential reproduction, mutation.*

natural services Processes of nature, such as purification of air and water and pest control, which support life and human economies. See *natural capital.*

negative feedback loop Feedback loop that causes a system to change in the opposite direction from which is it moving. Compare *positive feedback loop.*

nekton Strongly swimming organisms found in aquatic systems. Compare *benthos, plankton.*

net primary productivity (NPP) Rate at which all the plants in an ecosystem produce net useful chemical energy; equal to the difference between the rate at which the plants in an ecosystem produce useful chemical energy (gross primary productivity) and the rate at which they use some of that energy through cellular respiration. Compare *gross primary productivity.*

neutral solution Water solution containing an equal number of hydrogen ions (H^+) and hydroxide ions (OH^-); water solution with a pH of 7. Compare *acid solution, basic solution.*

neutron (n) Elementary particle in the nuclei of all atoms (except hydrogen-1). It has a relative mass of 1 and no electric charge. Compare *electron, proton.*

niche See *ecological niche.*

nitrogen cycle Cyclic movement of nitrogen in different chemical forms from the environment to organisms and then back to the environment.

nitrogen fixation Conversion of atmospheric nitrogen gas, by lightning, bacteria, and cyanobacteria, into forms useful to plants; it is part of the nitrogen cycle.

nonnative species Species that migrate into an ecosystem or are deliberately or accidentally introduced into an ecosystem by humans. Compare *native species.*

nonpoint sources Broad and diffuse areas, rather than points, from which pollutants enter bodies of surface water or air. Examples include runoff of chemicals and sediments from cropland, livestock feedlots, logged forests, urban streets, parking lots, lawns, and golf courses. Compare *point source.*

nonrenewable energy Energy from resources that can be depleted and are not replenished by natural processes within a human time scale. Examples are energy produced by the burning of oil, coal, and natural gas, and nuclear energy released when the nuclei of heavy elements such as uranium are split apart (nuclear fission) or when the nuclei of light atoms such as hydrogen are forced together (nuclear fusion). Compare *renewable energy.*

nonrenewable resource Resource that exists in a fixed amount (stock) in the earth's crust and has the potential for renewal by geological, physical, and chemical processes taking place over hundreds of millions to billions of years. Examples include copper, aluminum, coal, and oil. We classify these resources as exhaustible because we are extracting and using them at a much faster rate than they are formed. Compare *renewable resource.*

NPP See *net primary productivity.*

nuclear change Process in which nuclei of certain isotopes spontaneously change, or are forced to change, into one or more different isotopes. The three principal types of nuclear change are natural radioactivity, nuclear fission, and nuclear fusion. Compare *chemical change, physical change.*

nuclear fission Nuclear change in which the nuclei of certain isotopes with large mass numbers (such as uranium-235 and plutonium-239) are split apart into lighter nuclei when struck by a neutron. This process releases more neutrons and a large amount of energy. Compare *nuclear fusion.*

nuclear fusion Nuclear change in which two nuclei of isotopes of elements with a low mass number (such as hydrogen-2 and hydrogen-3) are forced together at extremely high temperatures until they fuse to form a heavier nucleus (such as helium-4). This process releases a large amount of energy. Compare *nuclear fission.*

nucleus Extremely tiny center of an atom, making up most of the atom's mass. It contains one or more positively charged protons and one or more neutrons with no electrical charge (except for a hydrogen-1 atom, which has one proton and no neutrons in its nucleus).

nutrient Any chemical an organism must take in to live, grow, or reproduce.

nutrient cycle See *biogeochemical cycle.*

nutrient cycling The circulation of chemicals necessary for life, from the environment (mostly from soil and water) through organisms and back to the environment.

ocean acidification Increasing levels of acid in world's oceans due to their absorption of much of the CO_2 emitted into the atmosphere by human activities, especially the burning of carbon-containing fossil fuels. The CO_2 reacts with ocean water to form a weak acid and decreases the levels of carbonate ions (CO_3^{2-}) needed to form coral and the shells and skeletons of organisms such as crabs, oysters, and some phytoplankton.

ocean currents Mass movements of surface water produced by prevailing winds blowing over the oceans.

old-growth forest Virgin and old, second-growth forests containing trees that are often hundreds—sometimes thousands—of years old. Examples include forests of Douglas fir, western hemlock, giant sequoia, and coastal redwoods in the western United States. Compare *second-growth forest, tree plantation.*

oligotrophic lake Lake with a low supply of plant nutrients. Compare *eutrophic lake, mesotrophic lake.*

omnivore Animal that can use both plants and other animals as food sources. Examples include pigs, rats, cockroaches, and humans. Compare *carnivore, herbivore.*

open-access renewable resource Renewable resource owned by no one and available for use by anyone at little or no charge. Examples include clean air, underground water supplies, the open ocean and its fish, and the ozone layer. Compare *common-property resource.*

open sea Part of an ocean that lies beyond the continental shelf. Compare *coastal zone.*

organic compounds Compounds containing carbon atoms combined with each other and with atoms of one or more other elements such as hydrogen, oxygen, nitrogen, sulfur, phosphorus, chlorine, and fluorine. All other compounds are called *inorganic compounds.*

organism Any form of life.

output Matter, energy, or information leaving a system. Compare *input, throughput.*

output pollution control See *pollution cleanup.*

overfishing Harvesting so many fish of a species, especially immature individuals, that not enough breeding stock is left to replenish the species and it becomes unprofitable to harvest them.

overgrazing Destruction of vegetation when too many grazing animals feed too long on a specific area of pasture or rangeland and exceed the carrying capacity of a rangeland or pasture area.

overnutrition Diet so high in calories, saturated (animal) fats, salt, sugar, and processed foods, and so low in vegetables and fruits that the consumer runs a high risk of developing diabetes, hypertension, heart disease, and other health hazards. Compare *chronic malnutrition, chronic undernutrition.*

ozone (O_3) Colorless and highly reactive gas and a major component of photochemical smog. Also found in the ozone layer in the stratosphere. See *photochemical smog.*

ozone depletion Decrease in concentration of ozone (O_3) in the stratosphere. See *ozone layer.*

ozone layer Layer of gaseous ozone (O_3) in the stratosphere that protects life on earth by filtering out most harmful ultraviolet radiation from the sun.

parasite Consumer organism that lives on or in, and feeds on, a living plant or animal, known as the host, over an extended period. The parasite draws nourishment from and gradually weakens its host; it may or may not kill the host. See *parasitism.*

parasitism Interaction between species in which one organism, called the parasite, preys on another organism, called the host, by living on or in the host. See *host, parasite.*

pasture Managed grassland or enclosed meadow that usually is planted with domesticated grasses or other forage to be grazed by livestock. Compare *feedlot.*

peer review Process of scientists reporting details of the methods and models they used, the results of their experiments, and the reasoning behind their hypotheses for other scientists working in the same field (their peers) to examine and criticize.

per capita ecological footprint Amount of biologically productive land and water needed to supply each person or population with the renewable resources they use and to absorb or dispose of the wastes from such resource use. It measures the average environmental impact of individuals or populations in different countries and areas. Compare *ecological footprint.*

per capita GDP Annual gross domestic product (GDP) of a country divided by its total population at midyear. It gives the average slice of the economic pie per person. Used to be called per capita gross national product (GNP). See *gross domestic product.* Compare *genuine progress indicator (GPI).*

perennial Plant that can live for more than two years. Compare *annual.*

permafrost Perennially frozen layer of the soil that forms when the water there freezes. It is found in arctic tundra.

perpetual resource Essentially inexhaustible resource on a human time scale because it is renewed continuously. Solar energy is an example. Compare *nonrenewable resource, renewable resource.*

persistence Ability of a living system such as a grassland or forest to survive moderate disturbances. Compare *resilience.*

pH Numeric value that indicates the relative acidity or alkalinity of a substance on a scale of 0 to 14, with the neutral point at 7. Acid solutions have pH values lower than 7; basic or alkaline solutions have pH values greater than 7.

phosphorus cycle Cyclic movement of phosphorus in different chemical forms from the environment to organisms and then back to the environment.

photosynthesis Complex process that takes place in cells of green plants. Radiant energy from the sun is used to combine carbon dioxide (CO_2) and water (H_2O) to produce oxygen (O_2), carbohydrates (such as glucose, $C_6H_{12}O_6$), and other nutrient molecules. Compare *aerobic respiration, chemosynthesis.*

physical change Process that alters one or more physical properties of an element or a compound without changing its chemical composition. Examples include changing the size and shape of a sample of matter (crushing ice and cutting aluminum foil) and changing a sample of matter from one physical state to another (boiling and freezing water). Compare *chemical change, nuclear change.*

phytoplankton Small, drifting plants, mostly algae and bacteria, found in aquatic ecosystems. Compare *plankton, zooplankton.*

pioneer community First integrated set of plants, animals, and decomposers found in an area undergoing primary ecological succession. See *immature community, mature community.*

pioneer species First hardy species—often microbes, mosses, and lichens—that begin colonizing a site as the first stage of ecological succession. See *ecological succession, pioneer community.*

planetary management worldview Worldview holding that humans are separate from nature, that nature exists mainly to meet our needs and increasing wants, and that we can use our ingenuity and technology to manage the earth's life-support systems, mostly for our benefit. It assumes that economic growth is unlimited. Compare *environmental wisdom worldview, stewardship worldview.*

plankton Small plant organisms (phytoplankton) and animal organisms (zooplankton) that float in aquatic ecosystems.

point source Single identifiable source that discharges pollutants into the environment. Examples include the smokestack of a power plant or an industrial plant, drainpipe of a meatpacking plant, chimney of a house, or exhaust pipe of an automobile. Compare *nonpoint source.*

poison Chemical that adversely affects the health of a living human or animal by causing injury, illness, or death.

pollutant Particular chemical or form of energy that can adversely affect the health, survival, or activities of humans or other living organisms. See *pollution.*

pollution Undesirable change in the physical, chemical, or biological characteristics of air, water, soil, or food that can adversely affect the health, survival, or activities of humans or other living organisms.

pollution cleanup Device or process that removes or reduces the level of a pollutant after it has been produced or has entered the environment. Examples include automobile emission control devices and sewage treatment plants. Compare *pollution prevention.*

pollution prevention Device, process, or strategy used to prevent a potential pollutant from forming or entering the environment or to sharply reduce the amount entering the environment. Compare *pollution cleanup.*

population Group of individual organisms of the same species living in a particular area.

population change Increase or decrease in the size of a population. It is equal to (Births + Immigration) − (Deaths + Emigration).

population crash Dieback of a population that has used up its supply of resources, exceeding the carrying capacity of its environment. See *carrying capacity.*

population density Number of organisms in a particular population found in a specified area or volume.

population dispersion General pattern in which the members of a population are arranged throughout its habitat.

population distribution Variation of population density over a particular geographic area or volume. For example, a country has a high population density in its urban areas and a much lower population density in its rural areas.

population dynamics Major abiotic and biotic factors that tend to increase or decrease the population size and affect the age and sex composition of a species.

population size Number of individuals making up a population's gene pool.

positive feedback loop Feedback loop that causes a system to change further in the same direction. Compare *negative feedback loop.*

potential energy Energy stored in an object because of its position or the position of its parts. Compare *kinetic energy.*

poverty Inability of people to meet their basic needs for food, clothing, and shelter.

precautionary principle When there is significant scientific uncertainty about potentially serious harm from chemicals or technologies, decision makers should act to prevent harm to humans and the environment. See *pollution prevention.*

precipitation Water in the form of rain, sleet, hail, and snow that falls from the atmosphere onto land and bodies of water.

predation Interaction in which an organism of one species (the predator) captures and feeds on some or all parts of an organism of another species (the prey).

predator Organism that captures and feeds on some or all parts of an organism of another species (the prey).

predator–prey relationship Relationship that has evolved between two organisms, in which one organism has become the prey for the other, the latter called the predator. See *predator, prey.*

prey Organism that is killed by an organism of another species (the predator) and serves as its source of food.

primary consumer Organism that feeds on some or all parts of plants (herbivore) or on other producers. Compare *detritivore, omnivore, secondary consumer.*

primary ecological succession Ecological succession in an area without soil or bottom sediments See *ecological succession.* Compare *secondary ecological succession.*

primary forest See *old-growth forest.*

primary productivity See *gross primary productivity, net primary productivity.*

principles of sustainability See *scientific principles of sustainability, social science principles of sustainability.*

probability Mathematical statement about how likely it is that something will happen.

producer Organism that uses solar energy (green plants) or chemical energy (some bacteria) to manufacture the organic compounds it needs as nutrients from simple inorganic compounds obtained from its environment. Compare *consumer, decomposer.*

prokaryotic cell Cell containing no distinct nucleus or organelles. Compare *eukaryotic cell.*

proton (p) Positively charged particle in the nuclei of all atoms. Each proton has a relative mass of 1 and a single positive charge. Compare *electron, neutron.*

pyramid of energy flow Diagram representing the flow of energy through each trophic level in a food chain or food web. With each energy transfer, only a small part (typically 10%) of the usable energy entering one trophic level is transferred to the organisms at the next trophic level.

radioactive decay Change of a radioisotope to a different isotope by the emission of radioactivity.

rain shadow effect Low precipitation on the leeward side of a mountain when prevailing winds flow up and over a high mountain or range of high mountains, creating semiarid and arid conditions on the leeward side of a high mountain range.

rangeland Land that supplies forage or vegetation (grasses, grasslike plants, and shrubs) for grazing and browsing animals and is not intensively managed. Compare *feedlot, pasture.*

range of tolerance Range of chemical and physical conditions that must be maintained for populations of a particular species to stay alive and grow, develop, and function normally. See *law of tolerance.*

rare species Species that has naturally small numbers of individuals (often because of limited geographic ranges or low population densities) or that has been locally depleted by human activities.

recharge area Any area of land allowing water to percolate down through it and into an aquifer. See *aquifer, natural recharge.*

reconciliation ecology Science of inventing, establishing, and maintaining habitats to conserve species diversity in places where people live, work, or play.

recycle To collect and reprocess a resource so that it can be made into new products; one of the four R's of resource use. An example is collecting aluminum cans, melting them down, and using the aluminum to make new cans or other aluminum products. See *primary recycling, secondary recycling.* Compare *reduce* and *reuse.*

reduce To consume less of a good or service in order to reduce one's environmental impact and to save money. Compare *recycle, refuse, reuse.*

reforestation Renewal of trees and other types of vegetation on land where trees have been removed; can be done naturally by seeds from nearby trees or artificially by planting seeds or seedlings.

refuse To refrain from buying or using a good or service in order to reduce one's ecological impact and to save money. Compare *recycle, reduce, reuse.*

reliable runoff Surface runoff of water that generally can be counted on as a stable source of water from year to year. See *runoff.*

reliable science Concepts and ideas that are widely accepted by experts in a particular field of the natural or social sciences. Compare *tentative science, unreliable science.*

renewable energy Energy that comes from resources that are replenished by natural processes continually or in a relatively short time. Examples are solar energy (sunlight), wind, moving water, heat from the earth's interior (geothermal energy), firewood from trees, tides, and waves. Compare *nonrenewable energy,*

renewable resource Resource that can be replenished rapidly (hours to several decades) through natural processes as long as it is not used up faster than it is replaced. Examples include trees in forests, grasses in grasslands, wild animals, fresh surface water in lakes and streams, most groundwater, fresh air, and fertile soil. If such a resource is used faster than it is replenished, it can be depleted and converted into a nonrenewable resource. Compare *nonrenewable resource* and *perpetual resource.* See also *environmental degradation.*

replacement-level fertility rate Average number of children a couple must bear to replace themselves. The average for a country or the world usually is slightly higher than two children per couple (2.1 in the United States and 2.5 in some developing countries) mostly because some children die before reaching their reproductive years. See also *total fertility rate.*

reproduction Production of offspring by one or more parents.

reproductive isolation Long-term geographic separation of members of a particular sexually reproducing species.

reproductive potential See *biotic potential.*

reservoir Artificial lake created when a stream is dammed. See *dam.*

resilience Ability of a living system such as a forest or pond to be restored through secondary ecological succession after a severe disturbance. See *secondary ecological succession.* Compare *persistence.*

resource Anything obtained from the environment to meet human needs and wants. It can also be applied to other species.

resource partitioning Process of dividing up resources in an ecosystem so that species with similar needs (overlapping ecological niches) use the same scarce resources at different times, in different ways, or in different places. See *ecological niche.*

respiration See *aerobic respiration.*

response Amount of health damage caused by exposure to a certain dose of a harmful substance or form of radiation. See *dose, dose-response curve, median lethal dose.*

restoration ecology Research and scientific study devoted to restoring, repairing, and reconstructing damaged ecosystems.

reuse To use a product over and over again in the same form. An example is collecting, washing, and refilling glass beverage bottles. One of the 4 Rs. Compare *recycle, reduce, and refuse.*

riparian zone A thin strip or patch of vegetation that surrounds a stream. These zones are very important habitats and resources for wildlife.

rule of 70 Doubling time (in years) = 70/(percentage growth rate). See *doubling time, exponential growth.*

runoff Freshwater from precipitation and melting ice that flows on the earth's surface into nearby streams, lakes, wetlands, and reservoirs. See *reliable runoff, surface runoff, surface water.* Compare *groundwater.*

salinity Amount of various salts dissolved in a given volume of water.

saltwater life zones Aquatic life zones associated with oceans: oceans and their accompanying bays, estuaries, coastal wetlands, shorelines, coral reefs, and mangrove forests.

scavenger Organism that feeds on dead organisms that were killed by other organisms or died naturally. Examples include vultures, flies, and crows. Compare *detritivore.*

science Attempts to discover order in nature and use that knowledge to make predictions about what is likely to happen in nature. See *reliable science, scientific data, scientific hypothesis, scientific law, scientific methods, scientific model, scientific theory, tentative science, unreliable science.*

scientific data Facts obtained by making observations and measurements. Compare *scientific hypothesis, scientific law, scientific methods, scientific model, scientific theory.*

scientific hypothesis An educated guess that attempts to explain a scientific law or certain scientific observations. Compare *scientific data, scientific* *law, scientific methods, scientific model, scientific theory.*

scientific law Description of what scientists find happening in nature repeatedly in the same way, without known exception. See *first law of thermodynamics, law of conservation of matter, second law of thermodynamics.* Compare *scientific data, scientific hypothesis, scientific methods, scientific model, scientific theory.*

scientific methods The ways scientists gather data and formulate and test scientific hypotheses, models, theories, and laws. See *scientific data, scientific hypothesis, scientific law, scientific model, scientific theory.*

scientific model A simulation of complex processes and systems. Many are mathematical models that are run and tested using computers.

scientific principles of sustainability To live more sustainably we need to rely on solar energy, preserve biodiversity, and recycle the chemicals that we use. These three principles of sustainability are scientific lessons from nature based on observing how life on the earth has survived and thrived for 3.5 billion years. See *biodiversity, chemical cycling, and solar energy.* Compare *social science principles of sustainability.*

scientific theory A well-tested and widely accepted scientific hypothesis. Compare *scientific data, scientific hypothesis, scientific law, scientific methods, scientific model.*

secondary consumer Organism that feeds only on primary consumers. Compare *detritivore, omnivore, primary consumer.*

secondary ecological succession Ecological succession in an area in which natural vegetation has been removed or destroyed but the soil or bottom sediment has not been destroyed. See *ecological succession.* Compare *primary ecological succession.*

second-growth forest Stands of trees resulting from secondary ecological succession. Compare *old-growth forest, tree farm.*

second law of thermodynamics In any conversion of heat energy to useful work, some of the initial energy input is always degraded to lower-quality, more dispersed, less useful energy—usually low-temperature heat that flows into the environment; you cannot break even in terms of energy quality. See *first law of thermodynamics.*

selective cutting Cutting of intermediate-aged, mature, or diseased trees in an uneven-aged forest stand, either singly or in small groups. This encourages the growth of younger trees and maintains an uneven-aged stand. Compare *clear-cutting, strip cutting.*

social science principles of sustainability To live more sustainably we **(1)** need to include the harmful health and environmental costs of producing the goods and services in their market prices (*full-cost pricing*), **(2)** learn to work together to focus on solutions to environmental problems that will benefit the largest number of people and the environment now and in the future (*win-win solutions*), and **(3)** accept our responsibility to future generations to leave the planet's life-support systems in at least as good a shape as what we now enjoy (*responsibility to future generations*).

soil Complex mixture of inorganic minerals (clay, silt, pebbles, and sand), decaying organic matter, water, air, and living organisms.

soil erosion Movement of soil components, especially topsoil, from one place to another, usually by wind, flowing water, or both. This natural process can be greatly accelerated by human activities that remove vegetation from soil. Compare *soil conservation.*

solar energy Direct radiant energy from the sun and a number of indirect forms of energy produced by the direct input of such radiant energy. Principal indirect forms of solar energy include wind, falling and flowing water (hydropower), and biomass (solar energy converted into chemical energy stored in the chemical bonds of organic compounds in trees and other plants)—none of which would exist without direct solar energy.

sound science See *reliable science.*

specialist species Species with a narrow ecological niche. They may be able to live in only one type of habitat, tolerate only a narrow range of climatic and other environmental conditions, or use only one type or a few types of food. Compare *generalist species.*

speciation Formation of two species from one species because of divergent natural selection in response to changes in environmental conditions; usually takes thousands of years. Compare *extinction.*

species Group of similar organisms, and for sexually reproducing organisms, they are a set of individuals that can mate and produce fertile offspring. Every organism is a member of a certain species.

species diversity Number of different species (species richness) combined with the relative abundance of individuals within each of those species (species evenness) in a given area. See *biodiversity, species evenness, species richness.* Compare *ecological diversity, genetic diversity.*

species equilibrium model See *theory of island biogeography.*

species evenness Comparative numbers of individuals of each of the species present in a community. See *species diversity.* Compare *species richness.*

species richness Number of different species contained in a community. See *species diversity.* Compare *species evenness.*

S-shaped curve Leveling off of an exponential, J-shaped curve when a rapidly growing population reaches or exceeds the carrying capacity of its environment and ceases to grow.

stewardship worldview Worldview holding that we can manage the earth for our benefit but that we have an ethical responsibility to be caring and responsible managers, or *stewards,* of the earth. It calls for encouraging environmentally beneficial forms of economic growth and discouraging environmentally harmful forms. Compare *worldview, environmental wisdom worldview, planetary management worldview.*

stratosphere Second layer of the atmosphere, extending about 17–48 kilometers (11–30 miles) above the earth's surface. It contains small amounts of gaseous ozone (O_3), which filters out about 95% of the incoming harmful ultraviolet radiation emitted by the sun. Compare *troposphere.*

stream Flowing body of surface water. Examples are creeks and rivers.

strip cutting Variation of clear-cutting in which a strip of trees is clear-cut along the contour of the land, with the corridor being narrow enough to allow natural regeneration within a few years. After regeneration, another strip is cut above the first, and so on. Compare *clear-cutting, selective cutting.*

subatomic particles Extremely small particles—electrons, protons, and neutrons—that make up the internal structure of atoms.

succession See *ecological succession, primary ecological succession, secondary ecological succession.*

succulent plants Plants, such as desert cacti, that survive in dry climates by having no leaves, thus reducing the loss of scarce water through *transpiration.* They store water and use sunlight to produce the food they need in the thick, fleshy tissue of their green stems and branches. Compare *deciduous plants, evergreen plants.*

sulfur cycle Cyclic movement of sulfur in various chemical forms from the environment to organisms and then back to the environment.

surface fire Forest fire that burns only undergrowth and leaf litter on the forest floor. Compare *crown fire, ground fire.* See *controlled burning.*

surface water Precipitation that does not infiltrate the ground or return to the atmosphere by evaporation or transpiration. See *runoff.* Compare *groundwater.*

surface runoff Water flowing off the land into bodies of surface water. See *reliable runoff.*

sustainability Ability of earth's various systems, including human cultural systems and economies, to survive and adapt to changing environmental conditions indefinitely.

sustainability revolution Major cultural change in which people learn how to reduce their ecological footprints and live more sustainably, largely by copying nature and using the six principles of sustainability to guide their lifestyles and economies. See *principles of sustainability.*

sustainable development See *environmentally sustainable economic development.*

sustainable living Taking no more potentially renewable resources from the natural world than can be replenished naturally and not overloading the capacity of the environment to cleanse and renew itself by natural processes.

sustainable society Society that manages its economy and population size without doing irreparable environmental harm by overloading the planet's ability to absorb environmental insults, replenish its resources, and sustain human and other forms of life over a specified period, indefinitely. During this period, the society satisfies the needs of its people without depleting natural resources and thereby jeopardizing the prospects of current and future generations of humans and other species.

sustainable yield (sustained yield) Highest rate at which a potentially renewable resource can be used indefinitely without reducing its available supply. See also *environmental degradation.*

synergistic interaction Interaction of two or more factors or processes so that the combined effect is greater than the sum of their separate effects.

synergy See *synergistic interaction.*

system Set of components that function and interact in some regular and theoretically predictable manner.

temperature Measure of the average speed of motion of the atoms, ions, or molecules in a substance or combination of substances at a given moment. Compare *heat.*

tentative science Preliminary scientific data, hypotheses, and models that have not been widely tested and accepted. Compare *reliable science, unreliable science.*

terrestrial Pertaining to land. Compare *aquatic.*

tertiary (higher-level) consumers Animals that feed on animal-eating animals. They feed at high trophic levels in food chains and webs. Examples include hawks, lions, bass, and sharks. Compare *detritivore, primary consumer, secondary consumer.*

theory of evolution Widely accepted scientific idea that all life-forms developed from earlier life-forms. It is the way most biologists explain how life has changed over the past 3.6–3.8 billion years and why it is so diverse today.

theory of island biogeography Widely accepted scientific theory holding that the number of different species (species richness) found on an island is determined by the interactions of two factors: the rate at which new species immigrate to the island and the rate at which species become *extinct,* or cease to exist, on the island. See *species richness.*

thermal energy The energy generated and measured by heat. See *heat.*

threatened species Wild species that is still abundant in its natural range but is likely to become endangered because of a decline in numbers. Compare *endangered species.*

threshold effect Harmful or fatal effect of a small change in environmental conditions that exceeds the limit of tolerance of an organism or population of a species. See *law of tolerance.*

throughput Rate of flow of matter, energy, or information through a system. Compare *input, output.*

time delay In a complex system, the period of time between the input of a feedback stimulus and the system's response to it. See *tipping point.*

tipping point Threshold level at which an environmental problem causes a fundamental and irreversible shift in the behavior of a system. See *ecological tipping point.*

tolerance limits Minimum and maximum limits for physical conditions (such as temperature) and concentrations of chemical substances beyond which no members of a particular species can survive. See *law of tolerance.*

total fertility rate (TFR) Estimate of the average number of children who will be born alive to a woman during her lifetime if she passes through all her childbearing years (ages 15–44) conforming to age-specific fertility rates of a given year. More simply, it is an estimate of the average number of children that women in a given population will have during their childbearing years.

tragedy of the commons Depletion or degradation of a potentially renewable resource to which people have free and unmanaged access. An example is the depletion of commercially desirable fish species in the open ocean beyond areas controlled by coastal countries. See *common-property resource, open-access renewable resource.*

trait Characteristic passed on from parents to offspring during reproduction in an animal or plant.

transgenic organisms See *genetically modified organisms.*

transpiration Process in which water is absorbed by the root systems of plants, moves up through the plants, passes through pores (stomata) in their leaves or other parts, and evaporates into the atmosphere as water vapor.

tree farm See *tree plantation.*

tree plantation Site planted with one or only a few tree species in an even-aged stand. When the stand matures it is usually harvested by clear-cutting and then replanted. These farms normally raise rapidly growing tree species for fuelwood, timber, or pulpwood. Compare *old-growth forest, second-growth forest.*

trophic level All organisms that are the same number of energy transfers away from the original source of energy (for example, sunlight) that enters an ecosystem. For example, all producers belong to the first trophic level and all herbivores belong to the second trophic level in a food chain or a food web.

troposphere Innermost layer of the atmosphere. It contains about 75% of the mass of earth's air and extends about 17 kilometers (11 miles) above sea level. Compare *stratosphere.*

tsunami Series of large waves generated when part of the ocean floor suddenly rises or drops.

turbidity Cloudiness in a volume of water; a measure of water clarity in lakes, streams, and other bodies of water.

undernutrition See *chronic undernutrition.*

unreliable science Scientific results or hypotheses presented as reliable science without having undergone the rigors of the peer review process. Compare *reliable science, tentative science.*

warm front Boundary between an advancing warm air mass and the cooler one it is replacing. Because warm air is less dense than cool air, an advancing warm front rises over a mass of cool air. Compare *cold front.*

water cycle See *hydrologic cycle.*

watershed Land area that delivers water, sediment, and dissolved substances via small streams to a major stream (river).

weather Short-term changes in the temperature, barometric pressure, humidity, precipitation, sunshine, cloud cover, wind direction and speed, and other conditions in the troposphere at a given place and time. Compare *climate.*

wetland Land that is covered all or part of the time with saltwater or freshwater, excluding streams, lakes, and the open ocean. See *coastal wetland, inland wetland.*

wilderness Area where the earth and its ecosystems have not been seriously disturbed by humans and where humans are only temporary visitors.

wildlife All free, undomesticated species. Sometimes the term is used to describe animals only.

wildlife resources Wildlife species that have actual or potential economic value to people.

wild species Species found in the natural environment. Compare *domesticated species.*

worldview How people think the world works and what they think their role in the world should be. See *environmental wisdom worldview, planetary management worldview, stewardship worldview.*

zooplankton Animal plankton; small floating herbivores that feed on plant plankton (phytoplankton). Compare *phytoplankton.*

INDEX

Note: Page numbers in **boldface** refer to boldface terms in the text. Page numbers followed by italicized *f, t, or b* indicate figures, tables, and boxes.

A

D

T